Astronomy: a Handbook

ASTRONOMY
A HANDBOOK

Edited by G. D. Roth

with the collaboration of

A. Güttler
W. D. Heintz
W. Jahn
R. Kühn
F. Link
R. Müller
W. Petri
N. B. Richter
W. Sandner
F. Schmeidler
H. A. Schmid
K. Schütte
P. Wellmann

translated and revised by

Arthur Beer

Springer-Verlag
New York Heidelberg Berlin
1975

Dipl.-Kfm. Günter Dietmar Roth
Portiastrasse 10
D–8000 München 90
BRD

Library of Congress Cataloging in Publication Data

Roth, Günter Dietmar.
  Astronomy: a handbook.

  Based on the 2nd German ed. of Handbuch für Stern-
freunde, published in 1967.
  Bibliography: p.
  1. Astronomy—Handbooks, manuals, etc. I. Beer,
Arthur, 1900–   ed. II. Title.
  QB64.R59   520   74–11408
  ISBN 0–387–06503–2

ISBN 0–387–06503–2 Springer-Verlag New York Heidelberg Berlin

ISBN 3–540–06503–2 Springer-Verlag Berlin Heidelberg New York

Preface to the First Edition

Several decades have elapsed since the publication of any similar book in the German language. The lack of such a book has been felt keenly by all friends of astronomy. In our space age, astronomical knowledge arouses public interest more and more. Practical observation at the telescope depends more than anything else on such knowledge. The educational value of such a training is undisputed. On the other hand, the work of the amateur astronomer can also contribute essentially to the work of the professionals.

It is from these points of view that this handbook aims to help with versatile advice. At the same time, the book intends to show the wide range of applied astronomy, as it presents itself to the friend of the stars; in mathematical-physical fields, in precision mechanics and optics, and last but not least in the area of social relations. Beyond the circle of amateur astronomers the book is addressed to lecturers, teachers, students and pupils. It wishes to serve them as a guide to "astronomical experiments", which we suggest should be performed in primary and secondary schools, specialist colleges, and extramural courses.

The wide range of astronomy compels us to restrict the ground covered. The presentation of general astronomical facts has been restricted to make more room for instructions required by independent astronomical workers. But even here the richness of the available material forces us to make selections. Elementary astronomical, mathematical, and physical knowledge, as taught in the higher schools, was assumed as practical basis. The bibliography at the end of our book and references in the text provide sufficient indications to direct the reader to further sources of factual knowledge.

The use of the book should be facilitated by the structure of the table of contents and by the index. The division of the contents into theory and practice must not be taken very literally. Essentially this division is between general and special sections. The individual flavour of the different chapters has been maintained; numerous cross-references in the various sections provide the necessary unification.

The selection of the material has been adapted to modern astronomical requirements. New instruments, e.g. the Maksutov-telescope and the radio telescopes, are dealt with, as is also the whole new topical field of artificial satellites. Possibilities of celestial photography are treated in detail. A special chapter on applied mathematics for amateur astronomers is intended to give the fundamentals for an independent reduction of the observations.

Scientifically productive fields for these observers, e.g. solar work, star occultations, planets, stellar photometry, have received specially detailed consideration. The use of an astronomical yearbook or almanac to supplement our handbook must become to every reader just as natural as the use of astronomical charts and catalogs. The reader is also advised to consult astronomical journals to combine the more or less "timeless" help of the handbook with new developments and results.

As the editor of our handbook, I wish at this point to thank sincerely all my collaborators for their understanding and friendly cooperation during the planning stages and the execution of the work involved. With these thanks to my collaborators, I combine a word of grateful memory to Professor WILHELM RABE of Munich, who from the very beginning warmly supported the project, but whose untimely death prevented his collaboration.

Munich   GÜNTER D. ROTH

Preface to the Second Edition

The new edition has been revised and amplified. The additions concern particularly radio astronomy, which now occupies an independent chapter, as well as a new chapter on sundials and an appendix of tables. The section on lunar eclipses has been completely rewritten.

The new edition has benefited from numerous suggestions received from our readers. There is now a much amplified chapter on the observing instruments of the amateur astronomer, and the new material on solar, lunar, and planetary observations will also be welcomed by the reader.

The interest in the exploration of the Universe has grown greatly during the past few years and with it the wish of scientifically inclined laymen to be able to make their own astronomical observations. Each year, new observatories come into being to cater to schools and other groups. Here, too, our handbook provides essential information by its up-to-date bibliography and by references to astronomical organizations and instruments and educational tools.

Again I must express my sincere thanks for the fine cooperation of all the collaborators. I also welcome as new collaborators Prof. P. WELLMANN, Director of the University Observatory at Munich, Dr. F. LINK of the Astronomical Institute of the Academy of Sciences in Prague, and H. A. SCHMID of the University Observatory at Munich. Furthermore, my thanks are due to Prof. N. B. RICHTER, the Director of the Schwarzschild Observatory at Tautenburg-Jena, DDR, and Dr. W. KREIDLER, Munich, for their help with the reading of the manuscript and suggestions for improvement.

We all recall the sad loss of two collaborators of the first edition, Dr. A. GÜTTLER of the University Observatory at Munich and Dr. R. KÜHN; to both of them, amateurs and popular astronomy will always remain grateful.

We have had the good fortune that Dr. W. D. HEINTZ, Prof. N. B. RICHTER and Prof. F. SCHMEIDLER have kindly looked through GÜTTLER's and KÜHN's contributions and provided revisions where necessary.

Munich   GÜNTER D. ROTH

Translator's Preface

It has been a particular pleasure to work on the "Handbuch für Sternfreunde" and produce this revised English edition to make the volume available to a wider readership.

The production of this work fills a very real gap in astronomical literature. The transformation of the German original into a "Handbook" available to all English-speaking readers, resulted from the widespread interest shown by astronomers and nonastronomers alike in many lands; its success will be ample reward for the effort involved.

I owe debts of gratitude to many people. "He who befriends a traveller is not easily forgotten, and I am very grateful to everyone who helped me on a long journey": these words of PETER FLEMING, in his "Travels in Tartary", very well suit the present case. Here I only wish to say how extremely grateful I am especially to the following friends and colleagues who have given me their help in the critical reading and improvement of various chapters:

NOEL ARGUE (The Observatories, University of Cambridge), PETER BEER (External Services, British Broadcasting Corporation, London), DAVID DEWHIRST (The Observatories, University of Cambridge), WULFF HEINTZ (Sproul Observatory, Swarthmore College, Pennsylvania), JUDITH FIELD (The Observatories, University of Cambridge), JANET HERRING (The Observatories, University of Cambridge), MICHAEL INGHAM (lately of the University Observatory, Oxford), EDWARD KIBBLEWHITE (The Observatories, University of Cambridge), ZDENĚK KOPAL (Department of Astronomy, University of Manchester), BRIAN MARSDEN (Smithsonian Astrophysical Observatory, Harvard College), JOHN PORTER (lately of H.M. Nautical Almanac Office, Royal Greenwich Observatory), PETER RADO (Royal Holloway College, University of London), RICHARD RADO (University of Reading), ANTHONY RUSSELL-CLARK (Computer Laboratory, University of Cambridge), PATRICK WAYMAN (Dunsink Observatory, Dublin), and GORDON WORRALL (The Observatories, University of Cambridge).

In particular, I most gratefully acknowledge the unfailing helpfulness of GÜNTER ROTH, the editor of the German "Handbuch", and his collaborating authors, and last but not least that of Springer-Verlag Heidelberg, where Mr. W. BERGSTEDT always gave every possible consideration to my suggestions.

University of Cambridge   ARTHUR BEER

Contributors

Arthur Beer University of Cambridge, 188 Huntingdon Road, Cambridge, England

A. Güttler (deceased) University Observatory, München, BRD

Wulff D. Heintz Sproul Observatory, Swarthmore College, Swarthmore, Pennsylvania 19081, U.S.A.

Wilhelm Jahn Hackländerstrasse 7, D–8 München 80, BRD

R. Kühn (deceased) University Observatory, München, BRD

F. Link Institut d'Astrophysique, Paris, France

Rolf Müller Neubeuererstrasse 1, D–8201 Nussdorf/Inn BRD

Winfried Petri Institut für Geschichte der Naturwissenschaften, Universität München, Unterleiten 2, D–8162 Schliersee/Obb., BRD

N. B. Richter Karl-Schwarzschild Observatory, DDR–6901, Tautenburg-Jena, DDR

Günter D. Roth Portiastrasse 10, D–8000 München 90, BRD

Werner Sandner D–8018 Grafing/Bahnhof/Obb., BRD

Felix Schmeidler Universitäts-Sternwarte, Scheinerstrasse 1, D–8 München 27, BRD

Hans A. Schmid Universitäts-Sternwarte, Scheinerstrasse 1, D–8 München 27, BRD

Karl Schütte (deceased) Leuchtfeuersteig 4, D–2 Hamburg 56, BRD

Peter Wellmann Universitäts-Sternwarte, Scheinerstrasse 1, D–8 München 27, BRD

Contents

5 / **The Terrestrial Atmosphere and Its Effects**

F. Schmeidler

6 / **Fundamentals of Spherical Astronomy**

K. Schütte

7 / **Modern Sundials**

K. Schütte

8 / **Applied Mathematics for Amateur Astronomers**

F. Schmeidler

14 / **Artificial Earth Satellites**

W. Petri

15 / **The Observation of the Planets**

W. Sandner

16 / **The Observation of Comets**

A. Güttler, revised by *N. B. Richter*

17 / **Meteors and Fireballs**

F. Schmeidler

18 / Noctilucent Clouds, Aurorae, Zodiacal Light

W. Sandner

19 / The Photometry of Stars and Planets

W. Jahn

20 / Double Stars

W. D. Heintz

21 / The Milky Way and the Galaxies

R. Kühn

22 / Appendix 500

23 / Bibliography 535

Index 557

1 | Introduction to Astronomical Literature and Nomenclature

W. D. Heintz

1.1. The Universe and the Amateur Astronomer

The objective of astronomy is the exploration of all phenomena outside the Earth. Its realm is the whole of space, out to the remotest distances. Thus the astronomer is spatially separated from his research objects (except those few celestial bodies now within reach of spacecraft exploration); he cannot approach them at will in order to experiment with them. He must make his observations often at those times and under those conditions that Nature prescribes. Furthermore, the objects and the movements he has to measure in astrometric work, and the amount of light available for astrophysical analysis are usually so small that observations are subject to considerable natural uncertainty. If, in spite of all this, in speaking of the numerical conquest of celestial motions, the term astronomical accuracy has become a household word, the reason lies in the fact that these calculations can be based on a large amount of observed data and on a careful consideration of all possible sources of error.

Astronomy is a mathematical science, an "exact" science: its methods and results are usually quantitative, expressed in numbers. In some fields the reduction of the observations (the derivation of a result) requires lengthy computations and the use of extensive tables; the astronomical theories that depend on the observations frequently use the most advanced mathematical methods and high-speed computers.

Of course, as higher mathematics comes into play, the difficulties of the amateur astronomer grow. Also, in the use of complicated and expensive instruments (spectrographs and electronic techniques), his possibilities remain limited. Even so, the field for the amateur astronomer at his telescope is wide enough, provided he uses his observing time carefully and systematically and does not waste it on problems which have already been settled, or which can only be advanced with extensive instrumental means. Nevertheless, without some numerical work and simple calculation nothing can be done. The present book is not intended for those who would like to discuss relativistic models of the Universe, but consider elementary astronomical, physical, and mathematical concepts inessential.

1

Even the purely contemplative view through a pair of binoculars, just for the pleasure of looking at the starry sky, would be only half the fun without a knowledge of the astronomical fundamentals. One wants to explain what is seen. At this stage it may be useful to summarize a few points, to provide a guide for the amateur astronomer using this book, and in this way to further independent observations which can contribute to our knowledge.

(a) Astronomical fundamentals: some knowledge of the introductory literature but also of important specialized textbooks and useful tools, such as maps, catalogs, and almanacs; we shall talk about these in this first chapter.

(b) The instrumental means, their performance and suitable application (pages 7–105).

(c) Motions on the celestial sphere and their basic mathematical treatment, as well as the influence of the Earth's atmosphere on the observations (pages 149–220).

(d) The special features of the observations of different kinds of celestial objects—dealt with in the later part of the book.

(e) It is, of course, routine and patience which make a good observer, but nevertheless he should face his observations always most critically. A simple mathematical help can here prove very advantageous: the basic theory of errors (see page 204). For special projects, cooperation with other experienced observers (professional or amateur) with knowledge of the particular field can prove very useful.

Our wish for the amateur astronomer is that his knowledge of the facts may be accompanied by a steadily increasing joy in his celestial observations.

1.2. Astronomical Literature

There exist numerous popular treatments of the whole field, many of which are listed in the Bibliography.

The various branches of research are already subdivided and have grown so much that their treatment in all the detail necessary for professional purposes requires specialized books. Some years ago it was still possible to summarize astrophysics in multivolume handbooks (*Handbuch der Astrophysik*, 1928; *Handbuch der Physik, Gruppe VII: Astrophysik*), and large parts of mathematical astronomy, for example, in the *Enzyklopädie der Mathematischen Wissenschaften*. Beyond these, more specialized works have to be consulted.

There are certain types of work where professional and popular astronomical journals are indispensable; the most important of these are listed in the Appendix, together with their usual abbreviations. In addition, the various observatories exchange series of their own publications. Amateur-operated observatories cooperating in research projects sometimes also share in the exchange of publications.

A broad survey of the astronomical literature is provided by annual review journals, in particular by the *Astronomy and Astrophysics Abstracts* (semiannual volumes, since 1969). Their predecessor, the *Astronomische Jahresbericht* (annual publication in German) covered the literature from 1899 to 1968.

1.3. Charts and Catalogs

Indispensable for the observer is a good star atlas, containing all naked-eye stars. There are, of course, extensive maps for the fainter stars. However, observers of such objects should normally use setting circles and thus refer to catalog positions. Large special star catalogs are generally only to be found in observatory libraries.

A few comments on celestial mapping may be given at this point. A grid for this purpose is illustrated in Fig. 1–1.

The division of the sky into constellations began in the distant past, but not until the Eighteenth and Nineteenth Centuries was it essentially completed and unified, particularly for the southern sky. We distinguish today 88 constellations, the boundaries of which are exactly defined by international agreement, and follow essentially the meridians and the parallel circles. They are usually quoted by their Latin names or their standard abbreviations to three or four letters (see Table 18 of the Appendix). The designation of the brighter stars uses the small letters of the Greek alphabet, following BAYER; he attached Greek letters in the order of the apparent magnitude of the stars, and then within a given magnitude class according to the position within the constellation. For example: α Leonis = the brightest star in Leo (Regulus). FLAMSTEED numbered the stars down to about fifth magnitude within each constellation in the order of right ascension (for instance, 61 Cygni). Latin letters, which were given by BAYER in several cases where the Greek ones did not suffice, have dropped out of use, as they lead to confusion. They are replaced by FLAMSTEED numbers.

Popular names (mostly in garbled Arabic versions) are known for some 130 of the brightest stars of the northern sky. Sometimes we even find the same name given to different stars, or different names or spellings for the same star. It is therefore best to use only those names (like Sirius, Antares, Altair) which are familiar and unambiguous.

GOULD extended the numbering of stars to the southern sky. His numbers are followed by the letter "G." (for instance, 38G. Puppis). Sometimes we also find numbers followed by the letter "H.", according to a corresponding catalog of the northern sky due to HEVELIUS (for instance, 19 H.Cam). There are other designations, which however are hardly used nowadays (for instance, H^1 = HEIS, B. = BODE). More convenient, for the fainter stars, is the designation by the current number in one of the large catalogs, for instance, *Bonner Durchmusterung (BD)*, *Henry Draper Catalogue (HD)*, *General Catalogue (GC)*, where the stars are numbered (without regard to the constellation in which they are located) in order of right ascension, sometimes

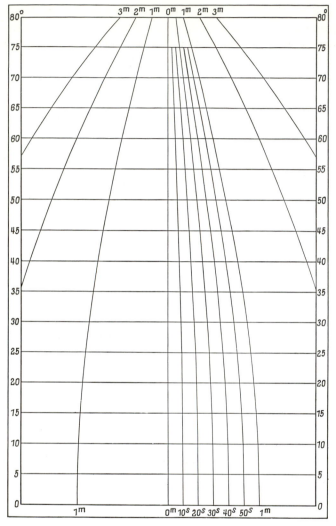

Fig. 1–1 Grid for the drawing of star maps. From Hilfstafeln der Hamburger Sternwarte (1916).

including an additional subdivision into zones of declination, for instance, star HD 111123, or BD + 53°779 (which is the star 779 in the zone +53° of the BD). Star clusters and nebulae are numbered according to a list given by MESSIER (M) or in the *New General Catalogue* by DREYER (NGC), for instance, M 31 = NGC 224 = the Great Andromeda Nebula. Variables have designations of their own with one or two capital letters (for instance, U Gem, RR Lyr). Occasionally, BAYER letters have been assigned to variables (δ Cep, β Lyr), and even to clusters (χ Per, ω Cen), and are retained.

Next to the name, a star is characterized by its position in the sky (right ascension and declination). Since the position changes (mainly due to precession, see page 176), the time to which this position refers is stated. All positions in a catalog are referred to the same time, the so-called *equinox* or *epoch* of the catalog. As discussed on page 176 it is possible to take account of the precession (and for higher accuracy also of the proper motion, aberration, etc.) and convert any positions into those of other epochs.

Other characteristics of the star are its visual or photographic brightness and its spectrum; some catalogs provide data on motions, variability, duplicity, etc.

To avoid misidentification caused by printing or reading errors in publications, references, or enquiries, one should always try to give *two independent* designations, for instance, name and coordinates, or numbers in two different catalogs, or, in the case of faint variable stars, brightness and small identification charts. To distinguish the component of a double star or of close neighbors the following notations can be used:

For—	*Use—*
North	N or Bo (Borealis)
South	S or Au (Australis)
West	pr or p (praecedens, preceding)
East	sq (sequens) or f (following)
Brighter star	major
Fainter star	minor
Center of light (in the case of close double stars)	m. or med.

The list of objects in the Appendix of this book will suffice for many purposes; specialized recent catalogs, which also may be used occasionally, are listed on page 547.

1.4. Almanacs

All the changing phenomena are to be found in the astronomical almanacs: the courses of the Sun, Moon, and planets, Sidereal Time, eclipses, tables for the exact conversion of star positions, data for planetary observations (central meridians, position of the axes, illumination, positions of the Moons), etc. The most widely used almanacs, the *Astronomical Ephemeris* (U.K., former title Nautical Almanac, established 200 years ago) and the *American Ephemeris* (U.S.A.) are two prints of almost identical contents. The detailed ephemerides (predictions) of the almanacs are computed to an accuracy which, in general, the amateur observer will not need. Amateur yearbooks based on the exact ephemerides as calculated for the almanacs are just as reliable and for many purposes sufficient. On the other hand, the almanacs also contain auxiliary tables which are omitted in smaller yearbooks, although they may prove useful to the amateur observer, especially as they are not restricted to the particular year: transformation of Solar Time into Sidereal Time, and vice versa, calculation of fractions of the day and the year, twilight tables, positions of the Pole Star for the determination of azimuths, auxiliary tables for interpolation, etc. (see also page 502, etc).

Each almanac contains an Appendix with explanations of the notations used. Apart from these, the contents are largely as self-explanatory as the well-known calendars for the amateur's use.

1.5. Reduction of Observations

The reduction of series of observations and important single observations (provided, of course, that they are valuable with respect to their accuracy and the optical means employed) must usually be left to the expert who has the necessary mathematical tools, the relevant literature, and will collect and coordinate the results of several observers. It depends on the individual project whether preprocessing the data is desirable (combining the single observations into means, applying systematic corrections, etc.) in order to facilitate their handling, and also whether publication will be worthwhile.

If a manuscript is intended for a central bureau or for publication, the results have to be presented very concisely, though they should show in a well-arranged manner everything that might be relevant for the reduction, and of course also any uncertainties, or numerical probable errors. The reductions should be checked most carefully for errors and omissions, to avoid later corrections and amendments, which would only give rise to confusion or might get lost altogether.

Essential for the evaluation of the observations are data concerning the instrument: aperture and optical quality, focal length, magnification. The observer should always make full notes about the atmospheric conditions. Here we distinguish between seeing (steady or diffuse, or "jumping" stars) and transparency (classified on a scale ranging from completely clear sky through haze and mist of different density up to cloud formation). Scales divided into five or ten steps characterize the quality of the image and the kind of seeing. Objective calibration, particularly that of the very important seeing scale, is hardly possible; too many factors are involved. Every experienced observer has his own reliable scale of estimates, taking telescope aperture and magnification into account (see page 547).

In the rare cases of important, unpredicted phenomena (a local meteor shower, a new comet, or the appearance of a nova), the nearest research observatory should be contacted right away. It will relay the message to the central offices in charge, and possibly confirm it. But do not get so enthusiastic as to forget to double-check! It actually has happened that Jupiter was reported as a nova, or the beams of airport lights as new comets, to mention but a few extreme cases. This sort of "discovery" is unlikely to arouse appreciation!

2 / The Observing Instruments of the Amateur Astronomer

W. Jahn

2.1. The Performance of Astronomical Observing Instruments

2.1.1. Problems and Types of Astronomical Instruments

The aim of the astronomer is to see more and to see better with his observing instruments. What exactly is meant by the phrase "to see more and to see better"? This question can be answered in various ways: to see why large appears small; to increase the record produced by faint light, directly or indirectly, so as to achieve a clear impression; to make invisible radiation visible; to see clearly what has only been received blurred; to separate double images from what appears single; to determine directions; and to measure angular distances. Indeed, every optical instrument serves several of these purposes.

However, astronomical observing instruments carry names that were given to them from a somewhat different point of view. They are named according to the kind of image—whether it is produced by refraction, by reflection, or by both phenomena (e.g., refractor, reflector)—to the purpose (comet finder, coronograph), to the kind of mounting (zenith telescope, vertical circle), to its relevant parts (sextant, meridian circle), or to the method of observation (transit circle, astrograph). These designations already indicate the variety of the properties involved.

A useful principle of classification is a purely optical one: we can distinguish:

1. Dioptric systems (lens telescopes, lens cameras) in which an image is produced by refraction.
2. Catoptric systems (mirror telescopes, mirror cameras) in which it is formed by reflection.
3. Catadioptric systems (mirror lens telescopes and cameras) in which both refraction and reflection essentially contribute to the formation of the image.

Furthermore, observing instruments are also constructed according to different mechanical principles. Here we distinguish azimuthal and equatorial mountings, as well as fixed and transportable instruments. Thus, a complete description of each instrument actually requires a consideration of very different points of view, which, of course, would exceed the framework of the present article. It is for this reason that we shall deal with the properties of the instruments rather than with the instruments themselves.

Performance is judged by the quality of the image, which itself depends on the resolving power (see page 13), and on the reproduction of contrast. The measure of the latter is the contrast transfer factor D, which is the ratio of the object contrast to the image contrast. In this definition, the contrast is given by the ratio of two light densities (see page 18). For instance, we can measure on a rectangular grating the ratio of light to dark, firstly direct and then through the optical system. Continuous variation of the grating constant and repeated measurements of the light ratio give other D-values. The behavior of these factors in their dependence on the grating constant, or the number of slits per

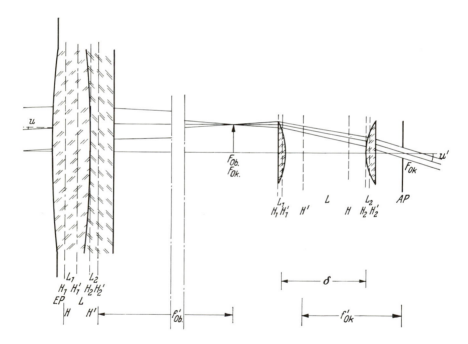

Fig. 2–1 Path of rays in KEPLER's astronomical telescope. The objective is a Littrow-lens; the eyepiece is of RAMSDEN's design. L_1, L_2 are the single lenses; H_1, H'_1, H_2, H'_2 are their principal planes. With L a composite system is denoted: objective and eyepiece; H and H' are their principal planes: δ is the separation of the principal planes of the eyepiece. EP is the entry pupil, AP the exit pupil, f' the focal lengths, and F the focal points. The angles to the axes of the light beams are called u and u'. If the objective is moved to the left at the distance $6 \times f'_{OB}$, the diameter of the incoming light beam is as large as the entry pupil EP. The lenses of the eyepiece are drawn separated. The plane of the field of view diaphragm passes through $F_{Ob} = F_{Ok}$.

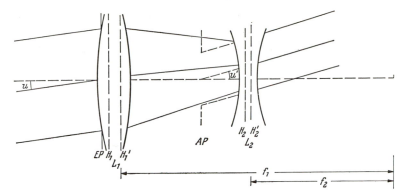

Fig. 2–2 Paths of rays in GALLILEO's telescope with a single objective lens L_1 and a single eyepiece L_2; H_1, H_1', H_2, H_2' are their principal planes. *EP* is the entrance pupil and *AP* the exit pupil, while f_1, f_2 are the focal lengths of the objective and of the eyepiece, respectively. F' is the common focus of objective and eyepiece. Magnification 2-times. The virtual diaphragm of the field of view is formed by the objective ring.

unit length, represents the transfer function of contrast. An ideal optical system would have the value 1 for these contrast factors. Good coated systems (see page 26) show a decrease of contrast of 1% to 2%, and in the case of an uncoated surface of 6% to 8%.[1]

Reference should be made here to the permanent feature dealing with amateur telescope making under the heading "ATM Gleanings" in *Sky and Telescope* (Sky Publishing Corporation, 49–51 Bay State Road, Cambridge, Massachussetts 02138).

2.1.2. *Types of Instrumental Performance*

2.1.2.1. *The Magnification of the Telescope*

The "astronomical telescope" consists of at least two "optical systems"— lenses and mirrors. Incoming parallel "rays" (parallel light beams) reappear again as parallel rays; see Figs. 2–1 and 2–2.

The astronomical telescope produces an enlarged image of the observed object. The angular magnification V can be measured or calculated by one of the following four formulae:

$$V = \frac{\text{angle of view to the very distant object as seen with the telescope}}{\text{the same angle without telescope}}, \quad (1)$$

$$V = \frac{\text{focal length of the objective}}{\text{focal length of the eyepiece}}, \quad (2)$$

[1] KÖNIG, A., and KÖHLER, H.: Die Fernrohre und Entfernungsmesser. Berlin: Springer, 1959.

$$V = \frac{\text{diameter of an entering parallel light beam}}{\text{diameter of the emerging parallel light beam}}, \tag{3}$$

$$V = \frac{\text{diameter of the entrance pupil}}{\text{diameter of the exit pupil}}. \tag{4}$$

The measurement of the telescope magnification (observers wearing spectacles should keep them on when measuring) is indicated by the above formulae. The observer looks with one eye only through the telescope and with the other alongside of it to a distant object (e.g., to a wooden fence or to a tiled roof) and counts the number of measured units (boards, bricks), which coincide exactly with one or several of the units as observed with the eye at the telescope. The first number divided by the second leads, with Eq. (1), to the magnification V. Otherwise, we may measure the focal length of the objective and the eyepiece from one of its cardical points H and H' and divide, according to Eq. (2), the first quantity by the second (see above and Figs. 2–1 and 2–2). Or we may also measure with the help of a magnifying lens attached to the eyepiece the image of the opening of a diaphragm which is placed immediately in front of the objective (and may for the purpose of easier measurement be shaped as a square); we then divide, according to formula (3), the diameter of this opening by the diameter of the image. Finally, we may take as the diameter of the entrance pupil the diameter of the objective, by using for the size of the exit pupil the diameter of the image of the objective, as produced by the eyepiece, and projected on a screen when we point the telescope to the bright sky (Ramsden disk). Thus we measure the diameter of the free aperture and divide it, according to Eq. (4), by that of the bright circle which appears well-defined on the projection screen with the telescope pointing to the sky.

2.1.2.2. Normal Magnification

Equations (3) and (4) show that the diameter of the parallel light beam which leaves the eyepiece can be increased or decreased by the choice of the magnification V. If this diameter is as large as the diameter of the pupil of the eye, we speak of "normal magnification." The knowledge of this quantity is of importance for photometric measurements, as we shall further see below.

To calculate the normal magnification we assume, if no other data are given, that the diameter of the pupil of the eye is 5 mm. At night the diameter of the pupil is larger than 5 mm, and in complete darkness it even reaches 8 mm. In the case of normal magnification the eye receives the total exit beam. In the case of *subnormal magnification* the diameter of the exit beam is too large to be received as a whole, and this causes a change in the diameter of the pupil, i.e., in the intensity of the radiation measured by our eye. It is therefore advisable, in the case of photometric measurement, to choose a not too large focal length for the eyepiece, so that the magnification keeps the diameter of the light beam that enters the eye no more than 4 mm ("*photo-metric eyepiece*").

To give an example: a small telescope of 50 mm aperture and 540 mm focal length gives, with an eyepiece of 40 mm focal length, according to Eq. (2), a magnification of 13.5 and, according to Eq. (3), a diameter of the light beam of 3.7 mm. The normal magnification would be 10-fold, reached with an eyepiece of 54 mm focal length. Under very favorable atmospheric conditions a magnification of at the most 10 times the normal magnification can still be regarded as useful.

2.1.2.3. Highest and Lowest Magnification

The image of the objective (the Ramsden disk), as produced by the eyepiece, has the following approximate diameter:

Diameter of the Ramsden disk in mm
  = focal length of eyepiece in mm × aperture ratio of the objective.   (5)

This diameter is as large as that of the parallel light beam which leaves the eyepiece during the observation of a star; it should not be larger than the diameter of the pupil of the eye. If the latter is taken to be 8 mm, our Eq. (5) gives

$$\text{Lowest applicable magnification} = \frac{\text{diameter of the objective in mm}}{8 \text{ mm}}. \quad (6)$$

If we use a still smaller magnification, then part of the light of the star is used for the illumination of the iris.

If the diameter of the parallel light beam leaving the eyepiece is about 3 mm, the eye attains its greatest resolving power. This leads to Eq. (7), giving a favorable magnification for the observation:

$$\text{Highest applicable magnification} = \frac{\text{diameter of the objective in mm}}{3 \text{ mm}}. \quad (7)$$

Experience has shown that in fact doubling the magnification deduced from Eq. (7) appears to be more suitable as the highest convenient magnification.

2.1.2.4. Field of View

The angle of view to an object in the telescope becomes the apparent *diameter of the field of view* if the object completely fills this field. The angle of view without a telescope, A, is then

$$A = \frac{\text{half height of the object (measured perpendicularly)}}{\text{distance of the object (measured horizontally)}} \times (57°2958) \times 2. \quad (8)$$

It is then possible, with the help of Eqs. (1) and (8), to proceed as follows:

Apparent diameter of the field of view in the telescope in degrees
  = magnification of the telescope
  × angle of view without telescope to an object which exactly fills
          the field of view of the telescope.   (9)

If we divide the apparent diameter of the field of view by the magnification, we obtain the true diameter of the field of view, i.e., the diameter of the surveyed field of the sky in degrees:

True diameter of the field of view in degrees

$$= \frac{\text{apparent diameter of the field of view in degrees}}{\text{magnification}}. \quad (10)$$

As a general rule of thumb for telescopes whose eyepieces do not have a very large field of view we may say

$$\text{True diameter of the field of view} = \frac{30°}{\text{magnification}}. \quad (11)$$

These formulae demonstrate that the higher the magnification, V, the smaller the surveyed field of the sky.

The measurement of the apparent diameter of the field of view is carried out according to Eq. (9); the true diameter of the field of view is derived from Eq. (10), or can be obtained as follows. We first determine the time required by a star near the equator to cross centrally the field of view of a fixed telescope. This time, in minutes or seconds, multiplied by 15, gives the diameter of the true field of view in minutes or seconds of arc, respectively.

2.1.2.5. Geometrical Light-Gathering Power

Equation (4) and Figs. 2–1 and 2–2 demonstrate that, keeping the entrance pupil unchanged, an increase of the magnification brings about a decrease in the exit pupil. Furthermore, the object viewed with high magnification appears more extended and therefore less bright than with small magnification. The square of the diameter of the exit pupil, expressed in millimeters, is the *geometric light-gathering power* of the telescope. It can be calculated according to formula (4), whereby the entrance pupil is represented by the diameter of the objective (in millimeters), which is then divided by the known magnification. The square of this quotient is the geometrical light-gathering power of the telescope. The loss of light by absorption and reflection within the telescope is not taken into account here.

2.1.2.6. The Diffraction Disk

The light of a point-like star, which may be received through a circular opening (pinhole camera, holder of the objective lens or of the main mirror), is imaged in the focal plane as a bright disk surrounded by concentric rings with colored rims (diffraction disks); see Figs. 2–3 and 2–4. If we use monochromatic light, we obtain bright rings which are interspersed by dark zones. The diameter b of the innermost dark ring also becomes the diameter of the central diffraction disk; it is

b (in mm)

$$= 2.44 \frac{\text{wavelength of the particular light (in mm)}}{\text{aperture ratio of the pinhole camera or of the objective or mirror}}. \quad (12)$$

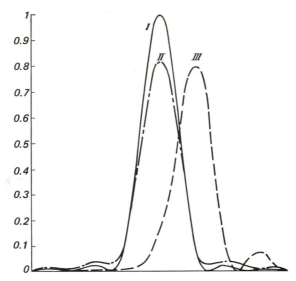

Fig. 2–3 The light distribution (light density) in the diffraction image of a star. I: In the case of a circular aperture of the objective; II: with spherical error; III: with coma. From *Handbuch der Astrophysik*, Vol. 1, p. 118. Berlin: Springer, 1931.

In the case of visual observation without a color filter the wavelength is taken to be 0.000550 mm. The aperture ratio of an objective lens or a mirror is

$$\text{Aperture ratio} = \frac{\text{diameter of the objective or mirror}}{\text{focal length of the objective or mirror}}. \qquad (13)$$

The diameter of the diffraction disk is smaller, according to Eq. (12), the smaller the focal length and the larger the diameter of the objective or of the mirror.

The shape and brightness of the diffraction disk and the diffraction rings can be considerably changed by variations of the shape and the size of the entrance opening. Sometimes the diffraction rings can only be made clearly visible by a reduction of the aperture ratio. This applies particularly to mirrors, where they are more difficult to see than in the case of objective lenses. The brightness of the rings is sometimes, at short moments of unsteady atmospheric conditions, larger than that of the diffraction disks, and more frequently so for mirrors than for lenses. Also some deviations from the theoretical shape of the diffraction disks and rings are observed, which cannot be traced back to errors in the optics.

2.1.2.7. The Resolving Power

Two equally large diffraction disks can be clearly separated from each other, i.e., resolved into two disks, if they touch on the outside; then their mutual distance is as large as their diameter. Since, however, the brightness

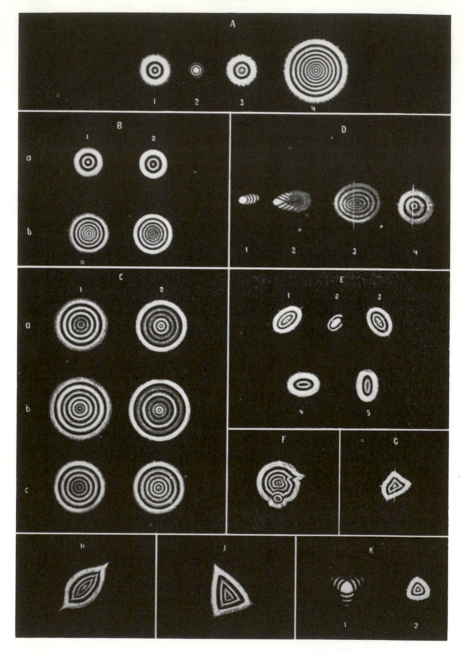

Fig. 2–4 Diffraction images of a star. A for a good objective; B in the case of spherical aberration; C in the case of zones; D in the case of coma; E if astigmatism present; F if striae are present; G, H, and J if there is tension; K flexure of the lens. After COOKE and Sons. From *Handbuch der Astrophysik*, Vol. 1, p. 186. Berlin: Springer, 1931.

of the diffraction disks decreases appreciably toward the rims it is possible to separate them at a still smaller distance. The limit of separability is called the *resolving power* of the telescope. This limit is generally reached when the distance of the centers of the disks of equally bright star images is just as large as their radius $b/2$. The angle at which this radius is viewed from the objective is as large as the angle in the sky between the two still separable stars. We have

Resolving power in seconds of arc

$$= \frac{13.8 \text{ seconds of arc}}{\text{diameter of the objective or of the mirror (in cm)}}. \quad (14)$$

This rule can replace the often used "DAWES' Rule" which, as experience shows, mainly relates to objectives and mirrors with an aperture ratio of at least 1:12 and which therefore yields a somewhat better resolving power:

Resolving power in seconds of arc

$$= \frac{4.56 \text{ seconds of arc}}{\text{diameter of the objective or mirror in inches}}$$

$$= \frac{11.6 \text{ seconds of arc}}{\text{diameter in cm}}. \quad (15)$$

Through the unsteadiness of the air, the imperfections of the optics, and the scattering of the light in the emulsion, the diffraction disks are widened to "star disks." For aperture ratios of 1:20 and smaller, the diameters of the diffraction disks exceed 0.02 mm, and the diffraction disks and rings can be recorded in all their details on the photographic plate (resolving power 0.03 mm).

Our eye is capable of separating two points if their mutual distance is not smaller than a certain minimum value. For daylight observations the limiting angles for the resolving power of the eye are

> 1 minute of arc for absolutely sharp vision
> 2 minutes of arc for clear vision
> 4 minutes of arc for comfortable vision

For night observations at the telescope we have

$$\text{Resolving power of the eye} = \frac{720 \text{ seconds of arc}}{\text{diameter of the pupil of the eye in mm}}. \quad (16)$$

If the diameter of the pupil is 6 mm the eye is able to separate 120 seconds of arc. A 2-inch telescope achieves—according to DAWES' Rule (15)—clear separation for two stars at a distance of 2.28 seconds of arc. These 2.28 seconds of arc must be magnified by the telescope to 120 seconds of arc, i.e., 53 times, if they should still be separated by the eye. This 53-fold magnification therefore exceeds the normal magnification (which in this case would be 8½) by a factor of 6.

The particular magnification which reaches the limiting angle of the resolving power of our eye may be called the *useful magnification*, and there is the following rule of thumb:

$$\text{Useful magnification by day} \quad = 2 \times \text{objective diameter in mm,} \quad (17)$$

$$\text{Useful magnification by night} = 1 \times \text{objective diameter in mm.} \quad (18)$$

Or shorter,

$$\text{Useful magnification} = 6 \times \text{normal magnification.} \quad (19)$$

Any value exceeding this useful magnification is called a *"dead"* or *"empty"* *magnification*. It furnishes no contribution to a better resolution. The diffraction disks and rings have no sharp rims. If they are viewed with an eyepiece at very short focal length, that is with a very high magnification, they appear diffuse and of a changing extent. This hinders an accurate measurement or estimate of distance. Furthermore, a large telescope magnification also magnifies the influence of unsteady air. Caution is therefore recommended in the use of high magnification.

The resolving power with which we may be provided during visual observations can be measured with the help of double stars whose components are, as nearly as possible, of the same brightness and whose angular distances are known (Table 2.3, page 13, and Table 22, page 526). The larger the brightness difference of the two components of a double star, the more difficult is their "separation." At small angular distances double stars are considered as separate when they are seen as an elongated image. The use of a diaphragm in front of the objective (i.e., a small decrease of the aperture ratio) can improve the definition of the image.

A change in the resolving power takes place in the following cases:

(1) By attaching a limiting diaphragm in a thin glass plate in front of the image plane (E. LAU, 1937); here the object reveals more detail.

(2) The addition of an absorbing layer to the surface of the objective, which diminishes the light in the rim regions stronger than at the center, leads to a magnification of the diffraction disks, and to a weakening of the light of the diffraction rings. This is important for the observation of planets passing in front of the sun's disk.

(3) The attachment of a central diaphragm, which obliterates light passing through the central regions of the objective or mirror, leads to a decrease in the size of the diffraction disks and to a brightening of the diffraction rings. This is the case when the secondary mirror is fixed in front of the main mirror. It is for this reason that reflectors (except for the largest ones) do not give as good results in planetary observations as do refractors. COUDER[2] showed in 1954 how diffraction patterns produced by secondary mirrors on photographic images of bright stars can be avoided; see Fig. 2–5.

[2] See *Amateur Telescope Making*, Book 2, p. 620. New York: Scientific American, Inc., 1954.

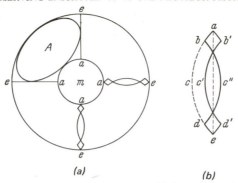

(a)   (b)

Fig. 2–5 (a) Changes in the four arms *ae* which support the secondary mirror *m*. The 4 identical openings of elliptical form (*A*) produce a star image without the spiked arms in the diffraction pattern. The loss of light is 20%. (After A. DANJON.) (b) Protective shielding of the secondary supports by little screens reduce such diffraction patterns; loss of light 4%. It is $c'c'' = 3bb'/4$. After A. COUDER. From *Amateur Telescope Making*, Book 2, p. 621. New York: Scientific American, Inc., 1954.

2.1.2.8. The Brightness of Telescopic Images

The brightness of the images produced by the telescope depends on several factors: on the apparent brightness and the size of the celestial object, on the aperture and focal length of the objective or the mirror, and on their imaging quality, as well as on the loss of light and the amount of scattered light inside the instrument. To these factors have to be further added the properties of the eyepiece and of the radiation receiver (eye, photographic emulsion, photoelectric cell). All this determines the limiting brightness, i.e., the magnitude of a point-source or extended object on the sky, which is just accessible to the particular telescope. We now turn to some theoretical calculations.

The total amount of light received by an objective or a mirror of the diameter *d* is (d^2/p^2)-times larger than the amount which enters the pupil (diameter *p*) of the eye. A two-inch telescope ($= 2 \times 2.53998$ cm) receives therefore $(2 \times 2.53998)^2/(0.5)^2 = 102.7$, i.e., 103 times as much light as our eye-pupil of 0.5 cm diameter. If the eye can still see stars of 5^m, then this telescope should still show stars which are by $2.5 \times \log 103 = 5.03$ magnitude-classes fainter, i.e., stars of $5.0 + 5.03 = 10^m.03$. This, of course, is only a rough calculation, but although it does not allow for most of the above factors, it leads to nearly correct results.

A small telescope of aperture $d = 50$ mm and focal length $f = 540$ mm produces from a point-like light source (star) a diffraction disk of the diameter *b*, which is, according to Eq. (12),

$$b \text{ (in mm)} = 2.44 \times \frac{0.000550 \times 540}{50} \text{ mm} = 0.0146 \text{ mm.}$$

Viewed with an eyepiece of 40 mm focal length (or 25, 16, 10, and 6 mm, respectively) the diffraction disk appears to subtend an angle of $(0.0146/40) \times 57°2958 = 0°02 = 1'2$ (or 2.0, 3.2, 5.0, and 8.0 minutes of arc, respectively). In the above-mentioned cases—furnishing an image on the retina of at the most 20′ diameter—the eye is not affected by the angular size of the diffraction disk as long as its neighborhood is absolutely dark. The eye only responds to the total amount of the light received. This amount is mainly determined by the diameter of the objective on which therefore the limiting magnitude also depends. If, however, the surrounding of the diffraction disk is not completely dark, the limiting magnitude also depends on this neighborhood. The magnitude is the smaller the brighter the latter. If the diameter of the diffraction disk exceeds 20′, the limiting magnitude is reduced the more—even in the case of a dark neighborhood—the larger the size of the diffraction disk. This dependence of the limiting magnitude on the size of the disk and on the brightness of its neighborhood is still more prominent if the celestial object is an extended one (nebula, cometary head).[3]

Calculation of the brightness of visual images

Diameter of an object in the sky in angular measure $\qquad w^0,\ w',\ w''$

Diameter of an object in the sky expressed in radian $\qquad w^0\dfrac{\pi}{180},\ w'\dfrac{\pi}{180\cdot 60},\ w''\dfrac{\pi}{180\cdot 60\cdot 60}$

Diameter of entrance pupil and of exit pupil of telescope $\qquad d,\ d'$

Focal length of objective and eyepiece $\qquad f,\ f'$

Transparency of objective and eyepiece $\qquad k,\ k'$

Energy density of radiation from a pointlike object (star) received per square centimeter on the Earth's surface $\qquad J$

Energy density of radiation from an extended object, or from the sky per angular unit, received per square centimeter on the Earth's surface $\qquad N,\ H$

Energy density of radiation due to scattered light in the objective or from the mirror $\qquad S$

Sensitivity factor of the eye $\qquad K$

Diameter of the pupil of the eye $\qquad p$

Linear size y' of the image as received from the objective or mirror: size of Airy disk $\qquad f\cdot w^0\dfrac{\pi}{180}$

Light intensity of a pointlike object as seen with the naked eye $\qquad \dfrac{K\cdot J\cdot p^2\cdot \pi}{4}$

Light intensity of an extended object as seen with the naked eye $\qquad K\cdot N\cdot\dfrac{(p\cdot w\cdot \pi)^2}{4^2}$

[3] Siedentopf, H.: Grundriss der Astrophysik, p. 32, Fig. 18. Stuttgart: Wissenschaftliche Verlagsgesellschaft, 1950.
Struve, O.: Elementary Astronomy, New York, Oxford University Press, 1968.

Light intensity of a pointlike image in the telescope, if $d' \leqq p$
$$\frac{K \cdot J \cdot d^2 \cdot \pi \cdot k \cdot k'}{4}$$

Light intensity of a pointlike image in the telescope, if $d' > p$
$$\frac{K \cdot J \cdot p^2 \cdot d^2 \cdot \pi \cdot k \cdot k'}{4 d'^2}$$

Light intensity of an extended circular object in the telescope, if $d' \leqq p$
$$K \cdot N \cdot \frac{p^2 \cdot d^2 \cdot w^2 \cdot \pi^3 \cdot k \cdot k'}{4^3}$$

Luminous energy of an extended circular object in the telescope, if $d' > p$
$$K \cdot N \cdot \frac{p \cdot d^2 \cdot w^2 \cdot \pi^2 \cdot k \cdot k'}{d'^2 4^2}$$

Diameter of the diffraction disk of the star
$$\frac{2.44 \cdot \lambda \cdot f}{d}$$

Luminous energy of the image of an object (ϕ of size w') in the focal plane of the objective
$$\frac{N \cdot k \cdot d^2 (180 \cdot 60)^2}{f^2 \cdot w'^2 \cdot 4 \cdot \pi}$$

Luminous energy of one square degree of the sky in the focal plane of the objective
$$\frac{H \cdot k \cdot d^2 \cdot 180^2}{f^2 \cdot \pi \cdot 4}$$

Luminous energy of the scattered light in the telescope
$$K \cdot S \cdot k' \cdot \log\left(1 + \frac{1}{4}\frac{d^2}{f^2}\right)^2$$

Contrast of a diffraction image, if the linear field of view is y'
$$\frac{J \cdot 4 \cdot y'^2 \cdot d^2}{H \cdot \pi \cdot 2.44^2 \cdot \lambda^2 \cdot f^2}$$

Contrast of the image of a surface-like object, of size w in the focal plane
$$\frac{N \cdot y'^2}{H \cdot \tan^2 w \cdot f^2}$$

Let us compare the surface brightness of the two images on the retina, which are produced if we view an extended object first with the telescope (diameter d' of the exit pupil), and then with the unaided eye (diameter p of the pupil). The first value becomes d'^2/p^2-times larger than the second one.[4] If we choose such a large magnification that the exit pupil is smaller than the pupil of the eye, then the extended object appears less bright in the telescope than for the naked eye. If, on the other hand, d' is made larger than p, then the eye does not take in the whole bundle of rays that leaves the eyepiece, and the observed object appears again darker than as seen with the naked eye. Only if the magnification is chosen in such a way that the exit pupil is equal to the pupil of the eye, i.e., taking the normal magnification, is the object (nebula, comet, etc.) at its brightest (see also page 10). Concerning the empirically determined visual, photographic, and photoelectric limiting magnitudes see page 29.

---

[4] *Handbuch der Experimentalphysik*, Vol. 26 (Astrophysik), p. 107. Leipzig: Akademische Verlagsgesellschaft, 1937.

Calculation of the Brightness of Photographic Images
(According to H. KNAPP)

Diameter of a blackening disk in linear measure $\qquad\qquad b$

Diameter of a blackening disk in angular measure $\qquad \dfrac{b}{f}\left(\dfrac{180°}{\pi}\right)$

Exposure time $\qquad\qquad t$

"SCHWARZSCHILD Factor" (see page 454) $\qquad\qquad \alpha$

Energy density of radiation received from the star, per square degree $\qquad \dfrac{J\cdot d^2\cdot k}{\left(\dfrac{b}{f}\cdot\dfrac{180}{\pi}\right)^2}$

Energy density of radiation from the sky, per square degree $\qquad \dfrac{H\cdot d^2\cdot k\cdot\pi^2}{f^2\cdot 180^2}$

Area of sky corresponding to the blackening disk $\qquad \left(\dfrac{b}{f}\cdot\dfrac{180°}{\pi}\right)^2\cdot\dfrac{\pi}{4}$

Sensitivity of plate or film, in DIN-units (see page 114) $\qquad D/10$

Magnitude difference between the radiation energies
m (star) $- m(H)$ $\qquad 2.5\cdot\log\dfrac{H}{J}+5\log\dfrac{b}{f}-8.5$

Approximate limiting magnitude m_0 of a star image $\qquad m_H+5\cdot\log\dfrac{b}{f}-8.5$

Blackening of a surface by illumination m_F, if an equally large surface illuminated by m_1 would receive unit blackening per unit of time $\qquad 2.512^{(m_1-m_F)}\cdot 10^{D/10}\cdot t^\alpha\cdot\dfrac{d^2}{f^2}$

Attained magnitude m_F of an extended object $\qquad m_1+2.5\cdot\alpha\cdot\log t+\dfrac{D}{4}+5\cdot\log\dfrac{d}{f}$

Exposure time necessary to reach magnitude m $\qquad \dfrac{2.512^{(m-m_1)}}{10^{D/10}\cdot\left(\dfrac{d}{f}\right)^2}$

Attained magnitude m_0 of a star $\qquad m_1+2.5\cdot\alpha\cdot\log t+\dfrac{D}{4}+5\cdot\log\dfrac{d}{b}-8.5$

Attained magnitude in the case of fixed camera $\qquad m_1+2.5\cdot\alpha\cdot\log\dfrac{b}{f}+\dfrac{D}{4}+5\cdot\log\dfrac{d}{b}-8.5+10.4\alpha$

Exposure time in seconds, in which the star image has reached the length b $\qquad \dfrac{b}{f}\cdot 13800$

Table 2.1 lists some b- and m_0-values of well-known objective lenses according to H. KNAPP.[5]

[5] KNAPP, H.: Uber die Reichweite von Objectiven bei Astroaufnahmen mit kleinen Montierungen. *Sterne und Weltraum* 3, 262 (1964).

Table 2.1. Diameter of the Blackening Disks b and the Attained Limiting Magnitude m_0

No.	Objective lens	$b_{\mu m}$	f/b	d/b	$m_0 - m_H$
	A. Photographic objectives				
1	Xenar 1:3.5/5 cm	20	2500	730	8.5
2	Xenar 1:2.8/5 cm	20	2500	900	8.5
3	Xenon 1:2/5 cm	20	2500	1250	8.5
4	Biotar 1:2/5.8 cm	20	2900	1450	8.8
5	Ennalyt 1:1.5/8.5 cm	30	2800	1850	8.7
6	Symmar 1:5.6/21 cm	25	8400	1500	11.1
7	Tele-Ennalyt 1:4.5/24 cm	25	9600	2100	11.5
8	Petzval 1:3.4/31 cm	60	5200	1500	10.0
9	Mirror-lens objective 1:4/50 cm	20	25,000	6250	13.5
	B. Special astronomical objectives				
10	2″-Refractor 1:16/86.5 cm	50	17,000	1100	12.5
11	SCHMIDT mirror 1:1.15/12 cm	20	6000	5200	10.5
12	SCHMIDT mirror 1:1/21 cm	30	7000	7000	10.7
13	SCHMIDT mirror at Bonn 1:4/134 cm	20	67,000	17,000	15.5
14	Four-lens SONNEFELD objective 1:5/200 cm	50	40,000	8000	14.5
15	SCHMIDT mirror at Sonneberg 1:3.5/175 cm	10	175,000	50,000	17.7

NOTE: f = focal length; d = free aperture; m_H = brightness of the sky in magnitude-classes per square degree.

2.1.2.9. The Influence of Air Disturbance on the Star Images

Air disturbance leads to variations of the brightness, the size, the color, and the position of the telescopic images. The *brightness variations* (intensity changes and scintillation) range from the smallest variations to complete short-time disappearance of the star. The *variations of size* (diameter) consist of a blowing-up of the images. *Color variations* occur only for stars near to the horizon and should not be confused with the atmospheric spectrum (see page 156). The variations in position (usually summarized as "seeing") range from 0″.5 to several seconds of arc.

In the case of visual observations, mainly the *"visual seeing"* (comprising essentially intensity variations, but also the blowing-up of the images and the changes in color) is of great importance. In photographic work, the main effects of the *"photographic seeing"* are intensity- and position-variations.

A photographic objective or a mirror of 1000 mm focal length records a change in position of 0.5 second of arc as a linear displacement by $(0″.5 \times 1000 \text{ mm})/206265″ = 0.0025$ mm. The resolving power of the photographic plate lies at 0.03 mm. Variations in position of the order of 0″.5 are therefore not noticed, and it is only 12 times this amount, i.e., 6 seconds of arc, which are recorded with this telescope of 1000 mm focal length.

Because of the complex character of air disturbances, we do not possess a general measure for them. In the case of visual observations, the variations of brightness are frequently stated on a scale of five steps 1–5 (best to worst atmospheric conditions). The changes in brightness and position of the observed star image depend widely on the aperture of the objective or the mirror. If the aperture is doubled, the variations are reduced to one half. Only for very large apertures (exceeding 10 inches) is the brightness variation smaller than 50% of the mean brightness (Fig. 2–6). Also the air motions within the tube itself must be taken into account (see page 73).

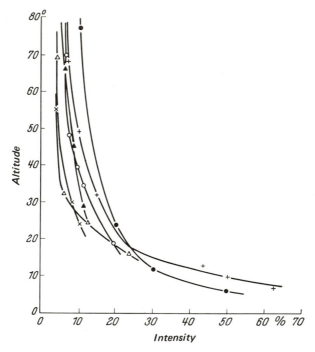

Fig. 2–6 Intensity variations (scintillation) of the total light in its dependence on the altitude of the star above the horizon, produced by an aperture of 36 inches. The various curves were obtained on different nights. The amplitude of the scintillation is never smaller than ±5%, and it is larger for smaller apertures. From M. A. ELLISON and H. SEDDON: Some experiments on the scintillation of stars and planets. *Monthly Notices* **112**, 73 (1952).

Recently, TOMBAUGH and SMITH[6] proposed a scale for an objective judgment of the quality of telescopic images, which is supposed to depend only on the *diameter* of the visually observed stars. These diameters can best be found by comparison with the mutual distances of the components of double stars. The TOMBAUGH-SMITH scale is reproduced in Table 2.2.

[6] TOMBAUGH, C. W., and SMITH, B. A.: A seeing scale for visual observers. *Sky and Telescope* **17**, 449 (1958).

Table 2.2. Scale of Image Quality According to Tombaugh-Smith (Relevant for This Scale Is the Image Diameter in Seconds of Arc)

Image quality	Image diameter in seconds of arc	Image quality	Image diameter in seconds of arc
−4	50	+3	2.0
−3	32	+4	1.3
−2	20	+5	0.79
−1	12.6	+6	0.50
0	7.9	+7	0.32
+1	5.0	+8	0.20
+2	3.2	+9	0.13

To estimate the diameter of the images we can use some of the double stars of the Northern Hemisphere, which are listed in Table 2.3 and in Fig. 2–7; they have been updated by W. D. HEINTZ.[7]

The air transparency or transmission τ must be distinguished from the transparency T. The transmission ratio of two atmospheric layers, expressed in magnitude differences, is

$$\frac{\tau_1}{\tau_0} = 2.512^{(m_0 - m_1)}$$

where one layer is at the observed zenith distance z_1, and the other at $z_0 = 0^0$. Here, m_0 denotes the limiting magnitude which can be attained by the naked eye, under the best atmospheric conditions, at the zenith of the observing site; m_1 is the limiting magnitude of the stars near to the observed object

Table 2.3. Thirty Visual Double Stars, North of Declination +77.5

No.	$\alpha_{(1950)}$	$\delta_{(1950)}$	m	ρ	No.	$\alpha_{(1950)}$	$\delta_{(1950)}$	m	ρ
1	12h09m	+82°0	6.5 − 8.5	67″	16	22h48m	+82°9	5.0 − 9.7	3″5
2	2 07	79.5	6.5 − 7.1	54	17	16 45	77.6	6.1 − 9.4	2.9
3	15 32	80.6	6.9 − 7.7	31	18	14 41	86.2	7.0 −10	2.4
4	21 17	78.4	7.2 −10	26	19	22 43	78.3	7.5 − 9.3	2.3
5	3 04	81.3	6.0 −10	24	20	21 20	80.1	7.8 − 8.5	2.0
6	12 49	83.7	5.3 − 5.8	21.6	21	8 46	88.8	7.1 −10	1.7
7	18 04	80.0	5.8 − 6.2	19.2	22	15 55	83.8	7.6 − 8.1	1.4
8	1 49	89.0	2.1 − 9.1	18.4	23	2 04	81.3	6.9 − 9.1	1.2
9	12 14	80.4	7.3 − 7.8	14.4	24	14 54	78.4	6.5 −10	1.2
10	22 00	82.6	7.0 − 7.5	13.7	25	13 12	80.7	6.3 −10	1.0
11	19 32	78.2	7.6 − 8.3	11.3	26	1 14	80.6	8.0 − 8.0	0.9
12	12 54	82.8	6.9 −11	10.5	27	4 01	80.6	5.7 − 6.4	0.9
13	20 11	77.6	4.4 − 8.4	7.4	28	15 42	80.1	7.4 − 8.2	0.7
14	2 59	79.2	5.8 − 9.0	4.6	29	0 07	79.4	6.8 − 7.1	0.5
15	20 18	80.4	6.8 −11	4.0	30	1 44	80.6	7.8 − 8.1	0.3

[7] HEINTZ, W. D.: Doppelsterne in der Polumgebung. *Sterne und Weltraum* **4**, 118 (1965).

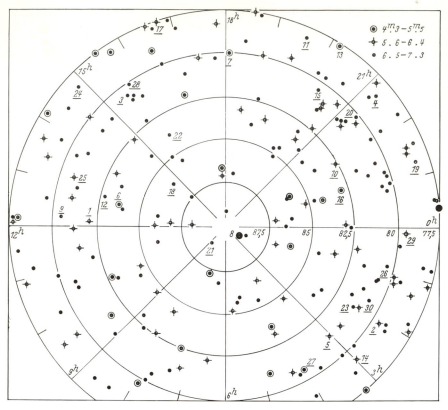

Fig. 2–7 Visual double stars in the neighborhood of the North Pole. HEINTZ, W. D.: *Sterne und Weltraum* **4**, 118 (1965).

(e.g., a planet). If we call $\tau_0 = 1$ the transparency at the zenith, we obtain

$$\tau_1 = 2.512^{m_0} \cdot 2.512^{-m_1},$$

$$\log \tau_1 = 0.4(m_0 - m_1).$$

The constant 2.512^{m_0}, or $(0.4 \times m_0)$, applies only for the particular observing site at which it had been determined. Its value represents the constant contribution to the transparency, while the other contributions are affected by zenith distance, scattered light, Moonlight, air opacity, and bright night sky; it is equal to 2.512^{-m_1}, or $-0.4\,m_1$.

If we put $m_1 = 6 - \Delta m$, we have

$$\tau_1 = 2.512^{(\Delta m - 6 + m_0)}.$$

The constant is 2.512^{m_0} and the variable contribution to the transparency is $2.512^{(\Delta m - 6)}$. The quantity Δm can be considered as a measure of the transparency T, namely $T = \Delta m$. This is a proposal made by L. J.

ROBINSON.[8] The same author has also given a scale for the unsteadiness of the image; another such scale has been published by C. H. GIFFEN.[9]

2.1.2.10. The Loss of Light by Absorption Within the Optical Equipment

The loss of light by absorption in glass of 1 cm thickness is expressed as a percentage of the light which enters the system perpendicularly, as shown in Table 2.4.[10]

Table 2.4. Loss of Light by Absorption in Optical Glass of 1 cm Thickness

Wavelengths (mm/1,000,000)	357	388	415	442	500	640
Borosilicate crown glass	4.7%	2.5%	1.2%	—	0.7%	0.5%
Calcium-silicate crown glass	3.4	2.5	1.8	1.4%	0.5	0.3
Flint glass for telescopes	49	30	12	3.6	0.7	0.7
Light flint glass	28	9.6	4.1	—	0.0	0.0
Dense flint glass	41	28	6.9	—	0.9	0.5
Uviol glass	98	98	—	—	—	—

Quartz glass is only transparent for short-wave radiation of 150 to 4200 nm wavelengths, rock-salt for light between 170 and 18000 nm, and calcite for the region of 200 to 3100 nm (1 nm $= 10^{-9}m = 10^{-6}mm$).

2.1.2.11. The Loss of Light by Reflection Within the Optical Equipment

The loss of light by reflection from a polished surface is listed in Table 2.5 as a percentage of the perpendicularly incident light. It is of no importance whether the light enters the glass from the air, or the air from the glass. If

Table 2.5. Loss of Light by Reflection from Polished Surfaces of Optical Glass

Wavelengths (mm/1,000,000)	420	450	500	550	600	650
Crown glass					4.2%	
Flint glass					5.5–6.0%	
Fresh silver	13.4%	9.5%	8.7%	7.3%	7.4%	6.5%
Old silver	27.0	18.9	16.1	15.0	13.7	11.4
Aluminium	8.6	8.5	8.0	8.9	9.7	10.4
Chrome-nickel-steel	15.0	19.0	24.0	28.5	32.0	35.0

[8] ROBINSON, L. J.: An analysis of the seeing and transparency scales as used by amateur observers. *The Strolling Astronomer* **17**, 205–212 (1961).

[9] GIFFEN, C. H.: Planetary seeing and transparency. *The Strolling Astronomer* **17**, 113–120 (1963).

[10] PFLÜGER, A.: Absorptionsvermögen einiger Gläser. *Annalen der Physik* **11**, 561 (1903).

two glass surfaces are cemented together with a liquid which has a refractive index equal to that of glass, e.g., Canada balsam, there will be no reflection by this intermediate layer.

The transparency of glass of 1 cm thickness for perpendicularly incident light can be calculated with the help of Table 2.4 as follows:

Transparency of borosilicate crown glass is, for a wavelength of 357 nm, $100 - 4.7\% = 95.3\% = 0.953$. If the length of the path is not 1 cm but s cm, the resulting total transparency is 0.953^s.

The loss by reflection in a crown-glass disk is, since it occurs on two surfaces, $0.042 + [(1 - 0.042)] \cdot 0.042$. See Table 2.5.

The amount of light which passes through the borosilicate glass at perpendicular incidence is at the wavelength 550 nm:

$$0.958 \times 0.994 \times 0.958 = 91.2\%.$$

The total loss of light amounts in the case of a *Galilean telescope* to about 20%, and for normal prism binoculars to about 40%. It consists of the losses by absorption and by reflection, and can be decreased by special precautions.

The reflection on surfaces of transparent bodies can be diminished by sputtering on spray-coating with layers of a certain thickness (optical thickness $= \frac{1}{4}$ of the wavelength λ) of a certain refractive index n (whereby n^2 of the layer $= n$ of the glass) in such a way that the losses by reflection on each glass surface are reduced from 5% to 1%. In the case of a four-fold reflection in the optical instrument, the transparency can be increased from 0.95^4 to 0.99^4, that is from 82% to 97%, if the free glass surfaces have been "bloomed" in this manner. Because of other light losses, however, the actual transparency remains a little more restricted, e.g., in binoculars to about 80%.

Metal layers are either not neutral (they show some color, and their transparency and reflection depend on the wavelengths of the light, as in the case of silver and aluminum), or absorb too much light, as, for example, rhodium (Rh) or chromium (Cr). Some of these layers change their properties in the course of time, or they are very sensitive to mechanical damage and chemical influence. It is possible to obtain a multilayer arrangement free from absorption and also color-free by sputtering several nonmetallic layers (with, for instance, TiO_2, SiO_2), in which the reflected light is obliterated (interference-layers, $\frac{1}{4}\lambda$ multilayers); see Fig. 2–8. Also metallic layers (Cr, Ni) can produce a greater light output by the insertion of a nonmetallic layer between metal and glass.[11]

2.1.2.12. The Limiting Magnitude of Stars

With the term "limiting magnitude" we denote the smallest energy density of radiation (see page 19) arriving from a pointlike (star) or extended (nebula) image, which can still be seen and recorded by the available radiation

---

[11] Pohlack, H.: Ueber neue Methoden der chromatischen und achromatischen Strahlenteilung. *Jenaer Rundschau* **3**, 82 (1958).

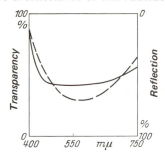

Fig. 2–8 The dependence of the degree of transparency and reflection on wavelength (broken line) is avoided within a wide range (solid line), if instead of one layer of metal oxide several such layers are used; the figure represents $\lambda/4$-layers of TiO_2-SiO_2. From *Jenaer Rundschau* **3**, 83 (1958), see Fig. 5.

receiver. It is given by the contrast between image and the surrounding sky background. The photometric brightness contrast is defined by the ratio:

$$\frac{\left(\begin{array}{c}\text{Brightness of the image}\\\text{including brightness of its neighborhood}\end{array}\right) - \left(\begin{array}{c}\text{brightness of}\\\text{the neighborhood}\end{array}\right)}{\text{Brightness of the neighborhood}}$$

This is identical with the ratio between the image brightness and the neighborhood brightness. Only if this ratio exceeds a certain size, can the image of the star (or nebula) be distinguished from its neighborhood.

A relation valid for the various radiation receivers (eye, photographic plate, photoelectric cell, image tube) which permits the calculation of the limiting brightness should take into account the following circumstances: seeing (see page 21), intensity variations (scintillation, see page 152), radiation of the bright night sky (see page 161), atmospheric extinction (see page 158), range of wavelengths (bandwidth) of the effective radiation, duration of the impact, storage capacity, degree of saturation, exclusion of the background, instrumental "noise" (plate fog, background noise), and also the quality of the image. In what follows, only a few simplified expressions for the limiting magnitude m_0 can be given.

The brightness of the image and its surroundings is produced by a large number of photons, whose effective number on the light-sensitive emulsion varies from place to place and with time. Let n be the average number of photons per surface-unit; then the smallest noticeable image brightness (threshold) is produced by a number of photons which is proportional to the square root of n.[12] The ratio (threshold brightness divided by the neighborhood brightness) is then proportional to

$$\frac{\sqrt{n}}{n} = \frac{1}{\sqrt{n}}.$$

[12] TRUMPLER, J., and WEAVER, H.: *Statistical Astronomy*, p. 167. Berkeley and Los Angeles: University of California Press, 1953.

If, for instance, the image on the photographic plate has a diameter corresponding to w'' (seconds of arc) on the sky, and if a background-surface of 1 in. diameter contains n blackened grains, then the smallest noticeable intensity ratio is $(w'')^2/\sqrt{n}$. From the definition of magnitude classes, i.e.,

$$\frac{I_0}{I_1} = 2.512^{(m_1 - m_0)},$$

it follows according to BAUM[13] for the limiting magnitude m_0 of a pointlike object (neglecting fog and assuming that the saturation limit of the light-sensitive emulsion has not been reached) that

$$m_0 = m_1 + 2.5 \log d - 2.5 \log w, \tag{20}$$

where d denotes the free aperture of the objective lens. The value of m_1 depends on the properties of the radiation receiver, on the background brightness, on the exposure time, and on the units of the quantities d and w. If the brightness of the pointlike light source m_1 increases by 1 magnitude, then m_0 increases by half a magnitude. The general relationship (20) has been confirmed by visual observation.

I. S. BOWEN (see Footnote 13) observed visually a number of stars at good seeing, using refractors and reflectors up to 150 cm aperture and with exit pupils of $0.8 - 8.0$ mm diameter, in order to find the limiting magnitude in its dependence on aperture and magnification. His results agree with the above formula (20), if we take for m_1 the value 5.5, express d and d' in cm, and replace w by the reciprocal of the magnification V, i.e.,

$$m_0 = 5.5 + 2.5 \times \log d + 2.5 \times \log V,$$

or, using Eq. (4),

$$m_0 = 5.5 + 5 \times \log d - 2.5 \times \log d'.$$

If we are using a higher magnification, the brightness of the surrounding field remains practically unchanged, and the limiting magnitude then only depends on d.

The relationship (20) changes when the sensitivity of the emulsion approaches saturation. If N is the number of blackened grains on the plate per unit area of surface, that is, according to Eq. (32), a surface of diameter $w = 1/f$, it has been shown by BAUM that the following equation holds for long-focus instruments (neglecting the fogging of the plate):

$$m_0 = m_1 - 2.5 \times \log w + 2.5 \times \log f + 1.25 \times \log N, \tag{21}$$

and for short-focus instruments

$$m_0 = m_1 + 5 \times \log f + 1.25 \times \log N. \tag{22}$$

The expressions (21) and (22) show different dependence on the seeing, which

---

[13] In KUIPER, G. P., and MIDDLEHURST, B.: *Stars and Stellar Systems*, Vol. 2, p. 6. Chicago: University of Chicago Press, 1960.

increases w. However, as we approach the saturation limit it is the focal length f and not any longer the aperture d which affects the limiting magnitude of the star.

Table 2.6 (a, b, and c) lists empirical limiting magnitudes for visual, photoelectric, and photographic observation, in accordance with data given by SIEDENTOPF (see Footnote 14).

Table 2.6. (a) Limiting Magnitude of A-type Stars for Visual Observation in a Dark Field

Entrance pupil d	Limiting magnitude
Naked eye	6^m
50mm	10.3
100	11.7
200	13.0
300	13.8
500	14.5
1000	15.0

Table 2.6. (b) Limiting Magnitude of A-type Stars for Photoelectric Photometry (Multiplier; Photometric Accuracy About 1%)

Entrance pupil d	Limiting magnitude
200 mm	$12^m\!.5$
500	14.5
1000	16.0
2500	18.0

Table 2.6. (c) Limiting Magnitude of A-type Stars on Photographs of Difference Exposure Time

Entrance pupil d	10 min	30 min	100 min
200 mm	$14^m\!.0$	$15^m\!.0$	$16^m\!.0$
400	15.5	16.5	17.5
1000	17.5	18.5	19.5
2500	19.5	20.5	21.5
5000	21.0	22.0	23.0

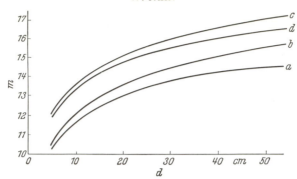

Fig. 2–9 Limiting magnitude of stars during visual observation: (a) in the dark field according to H. Siedentopf; (b) with the diameter $p = 0.76$ cm of the pupil of the eye, according to J. B. Sidgwick; (c) according to W. H. Steavenson; (d) with allowance for the loss of light in the optical system according to J. B. Sidgwick.

In Fig. 2–9 we find a plot of the limiting magnitudes against the entrance pupil for visual observation; also shown is the influence of the loss of light within the optical arrangements.

We conclude this section with a statement due to A. Güttler: The results obtained by Siedentopf[14] and his collaborators concerning their physiological threshold contrast can be used to show that (with the usual supernormal magnification $\bar{V}$) the contribution of the telescope to the increase of the limiting magnitude m (if there is not too much scattered light) amounts to

$$\Delta m_0 = 1.9 \log \kappa\kappa' + 3.8 \log \frac{d}{p_{\max}} + 1.2 \log \bar{V},$$

where $\kappa\kappa'$ denotes the degree of transparency of objective lens and eyepiece, d the diameter of the objective, $p_{\max}$ the largest diameter of the pupil of the eye. The first term, which is always negative, shows the gain achieved by the use of "bloomed" optics: each bloomed surface pair contributes a gain of $0\overset{m}{.}07$. The second term shows that a doubling of the aperture brings about a gain of $1\overset{m}{.}41$.

A noticeable contribution is also yielded by the magnification which appears as the third term. Thus we see, for instance, that the visual limiting magnitude changes by $0\overset{m}{.}36$ if we double the magnification of the eyepiece. All these figures actually apply strictly only to binocular observations, and should also, because of the extrapolation of the physiological measurements, only be regarded as approximate. It can be seen, however, that binoculars, even though they have low magnification and small aperture, can still be useful to the amateur observer.

[14] Siedentopf, H.: Neue Messungen der visuellen Kontrastschwelle. *Astronomische Nachrichten* **271**, 193 (1941); and in *Grundriss der Astrophysik*, Fig. 18, p. 32. Stuttgart: Wissenschaftliche Verlagsgesellschaft, 1950.

2.2. The Components of Astronomical Observing Instruments

2.2.1. Determination of the Focus of Lenses and Mirrors

The most important notation referring to lenses and mirrors is illustrated in Figs. 2–1 and 2–2. Frequently, several lenses are cemented together with Canada balsam of high viscosity, or combined (with very small separation) into an optical system. In this case the separate lenses must be positioned in such a way that their axes coincide in a straight line (*centered system*), i.e., when the mirror images of a small light bulb (in cemented achromats there are two such images) can be seen on a straight line which passes through the center of the aperture. Single lenses of a doublet which is not cemented are usually separated by three tiny pieces of tinfoil, whose accurate position is marked at the rim of the lenses and must be carefully taken into account when they are combined.

A discussion of the focal length of a composite optical system is given on page 69. Occasionally we also require the *lateral magnification* of a system:

$$\text{Imaging-scale} = \frac{\text{size of image}}{\text{size of object}} = \frac{\text{distance of image}}{\text{distance of object}}.$$

It is important to know the image size of a surface-like object (Sun, Moon, planet, nebula), given the focal length of the imaging optical system. We have

Size of the image, in mm = focal length, in mm

$\times$ tangent of the angle of view to the object, in degrees   (23)

For small angles up to 8° the above equation (23) can be replaced by

Size of image, in mm

$$= \text{focal length, in mm} \times \frac{\text{angle to the object, in degrees}}{57\overset{\circ}{.}2958} \quad (24)$$

EXAMPLE. The objective of a telescope of $d = 50$ mm and $f = 540$ mm gives images of the Sun or Moon (both of which have angular diameters of $0\overset{\circ}{.}52$) as disks of the following size:

$$\text{Image size, in mm} = 540 \text{ mm} \times \frac{0\overset{\circ}{.}52}{57\overset{\circ}{.}2958} = 4.9 \text{ mm}.$$

The *position of the focus* of an optical system can be found if we examine on a screen the image of a very distant object produced on the optical axis. This is best done with a magnifying lens with cross wire: The focus lies where image and cross wire appear sharp. If it is impossible to produce a real image of an object (as in dispersing optical systems and negative eyepieces), the resulting virtual image of a distant object is viewed with a small viewing telescope which can be focused for small distances. Afterwards, the examined system (lens, eyepiece) is removed, and replaced by a finely ruled screen (millimeter paper), which is displaced until it is seen sharply in focus. The required focus then lies in the surface of this screen.

The *focal length of a lens* with a real focus (*positive system*) can, following BESSEL, be determined as follows:[15] If the object is at a constant distance *l* from a screen, which exceeds four times the focal length, there exist two intermediate positions at which the lens produces on the screen a clear image of the object. If we call the distance between these two positions *e*, then the focal length of the lens is approximately

$$f = \frac{l^2 - e^2}{4l}. \tag{25}$$

The *focal length of a negative eyepiece* (e.g., a HUYGENS eyepiece) follows easily, by Eq. (2), from the magnification it leads to in the telescope and the focal length of the telescope objective or mirror.

The *focal length of a mirror* can be determined with the so-called knife-edge test which is discussed further below.

The *focal length of a negative lens* can be obtained from the size of the Sun's circular image on a screen. A description is given on page 333 of the preceding reference. The diameter *b* of the circle at a distance *l* is measured. If *d* is the diameter of the aperture we have

$$f' = \frac{l \cdot d}{d - b + 0.0094l} \tag{26}$$

Here, 0.0094 is twice the value of the tangent of the apparent radius of the sun. For not too small lenses the quantity $0.0094l$ can be neglected, and we thus obtain the simple rule: The particular distance of the screen at which the image has twice the diameter the lens is equal to the focal length.

2.2.2. Imaging-Errors and the Knife-Edge Test

It is impossible for any system of lenses or mirrors to image an area of the sky at any point onto a plane completely sharply and in scale. The various errors and the methods of their determination are as follows.

Spherical Aberration. Single lenses and mirrors with spherical surfaces combine the rays traversing their borders to a nearer focus than the rays near to the axis (paraxial rays). The difference in focal length between the various zones is called the spherical aberration. It can be completely eliminated through a combination of two lenses or through an aspherical form of the surfaces between two zones, and can be kept very small for all other zones. In the case of good telescope objectives the deviation of the focal lengths from an average focal length should be smaller than 1/1000 of the latter. If the zones near the border have a shorter focal length than the central zones, then the objective or the mirror is *spherically undercorrected*; in the opposite case, it is *overcorrected*.

[15] See KOHLRAUSCH, F.: *Praktische Physik*, 19th ed., Vol. 1, p. 331. Leipzig: B. G. Teubner, 1951.

The determination of the spherical aberration is performed with *Foucault's knife-edge test*, which, however, should only be used for lenses or mirrors of more than 3 cm diameter. We require an artificial star, a hole of 0.01 to 0.05 mm diameter (G. W. RITCHEY), which is brightly illuminated. It can be produced by a carefully pointed needle in an aluminum or copper foil and brightly illuminated by acetylene, an arc, etc. This method as well as the whole procedure of the knife-edge test has been described in great detail in the literature and discussed by numerous enthusiastic mirror makers over the world. See also Hurt, B.D.: Fiber Optics for a Testor. *Sky and Telescope* **41**, 5, 315 (1971).

In the case of an *objective lens* we place beside the image of a monochromatic artificial star—on the side towards the objective at a distance of twice the focal length on the axis—a razor blade. The eye views the objective lens from just behind the razor blade. As soon as the razor is moved across to the optical axis the image becomes exactly half-occulted. If the razor blade were introduced exactly in the image-point the objective would disappear instantaneously provided its focal length is the same for all zones. Therefore, if the objective is in this way examined in single zones it is possible to find the position of the particular focus at which an instantaneous or at least uniformly progressing darkening of the observed zone of the objective lens takes place when the razor blade is inserted in the focal "point." The distance between the focal points of the various zones characterizes the spherical aberration.

In the case of a *spherical mirror* the artificial star is placed in the center of curvature, i.e., at twice the focal length. The razor blade, at the edge of which the eye looks towards the mirror, is close to the eye at the same distance from the mirror. The center of curvature is found as soon as the mirror is suddenly nearly or completely occulted when the edge of the blade is moved a very little sideways. If such an instantaneous and complete darkening of the mirror takes place it indicates that the mirror is really spherical and at the same time the error, i.e., the spherical aberration, is found as if the mirror were being used for the observation of a real star in the sky. In order to eliminate this error, the mirror is ground a little deeper at the center; the focal length of the central zone is somewhat smaller than that of the zones near to the rim. The difference *m* between the focal lengths, at the center of the mirror as compared with the zone near the rim, should be:

$$m \text{ in mm} = \frac{(\text{radius of the zone in mm})^2}{\text{focal length of this zone in mm}}. \tag{27}$$

It is necessary for the determination of the particular focal length to displace only the razor blade but not the artificial star in the direction of the axis. If the artificial star is displaced simultaneously with a razor blade, *m* is half the value given by Eq. (27). If this relationship is valid for every zone of the mirror, the latter is exactly *parabolic*, i.e., it combines the parallel beam of light which enters in the direction of the axis to a single point, even if this beam fills the whole mirror.

The spherical aberration of single lenses and mirrors is, in general, smaller if the aperture ratio is smaller. The error decreases as the square of this ratio. It is therefore possible to apply a spherical mirror to astronomical observation if its aperture ratio is small enough (see Table 2.15 on page 61).

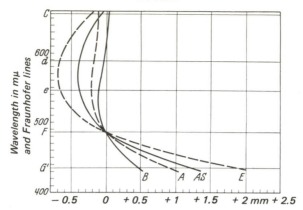

Fig. 2-10 Color aberration of an objective lens of 1000 mm focal length for light of different wavelengths. *E* is an achromat; *A* and *AS* are semiachromats; *B* is an apochromat. From *Handbuch der Astrophysik*, Vol. 1, p. 111. Berlin: Springer, 1931.

Chromatic Aberration. This occurs only by refraction and not by the reflection of the light and is caused by the fact that short wavelengths are more refracted than long wavelengths. For this reason short-wavelength light combines after passage through a lens at a shorter focal length than long-wavelength light. Through a combination of two lenses of different glass, chromatic aberration can be compensated for two wavelengths completely (in the so-called achromats), or for additional colors almost completely (in apochromatic systems). Chromatic aberration renders the star images in the telescope colored at their border (showing in the case of the achromats the "*secondary spectrum*"), and each image within the focal length (intrafocal image) or outside the focal length (extrafocal image) becomes a disk with colored border regions. Achromatic objectives for visual observation combine the wavelengths of the FRAUNHOFER lines *C* (0.000653 mm) and *F* (0.000486 mm) to one point. Objectives used for ordinary photographic plates should combine lines *D* (0.0005893) and *G′* (0.000434 mm), while the special astrographic objectives combine the wavelengths of *F* (0.000486 mm) and *G′* (0.000434 mm) (see Fig. 2-10). The smaller the aperture ratio, the smaller the chromatic aberration becomes. The error decreases with the aperture ratio (d/f). Roughly,

Focal length of a lens with inconspicuous color aberration

$$= 100 \times (\text{aperture diameter}) \quad (28)$$

The "Light Spots." These phenomena originate through single or multiple reflection of the light on lens- and mirror-surface or also on the emulsion of the photographic plate. They only become noticeable for very bright stars and are situated symmetrically to the axis, moving about in the field of view when one moves the image of a very bright object to another place in the field.

Other errors of lenses, with which we shall deal now, only occur in photographic work.

Astigmatism. This error can be present when a pencil of rays, arriving at an angle to the axis, is intercepted by the lens or reflected by a mirror. Under astigmatism even narrow parallel light beams will not combine into one point; they meet in two different perpendicular straight lines. Outside these lines they produce elliptic light disks. The farther apart these straight lines are, for the same inclination of the beam, i.e., the larger their stigmatic difference, the more pronounced is the astigmatism. It can be removed by special nonastigmatic systems. If astigmatism occurs in a telescope or a camera, the first thing to do is to test whether lenses and mirrors are properly centered and not under tension.

Curvature of the Field of View. This effect consists of the fact that parallel light beams which meet the axis under different angles are not combined with equal sharpness on a plane which is parallel to the lens or mirror, but on a spherical surface. This defect can be removed by the insertion of a simple lens (see page 43).

Distortion. This effect causes a straight line that does not pass through the axis of the lens not to be imaged as a straight line but as a curve.

The Coma. If a zone of light in the aperture meets the axis obliquely it may be subjected to an oblique radial shift so that the star images look like little comets with tails, an appearance with which astigmatism and curvature of the field of view may be combined.

2.2.3. The Aperture Diaphragm

The aperture diaphragm is as important as the objective lens or as the main mirror itself. It does not only determine which parts of the light beam coming from the objective are imaged, but is in itself sufficient for forming an image. Simple lenses and mirrors have the purpose of combining the incoming light rays more rapidly to a focus. It is for this reason that even with a pinhole camera pictures of very bright objects (Sun, Moon, bright meteors) can be obtained. The most favorable diameter of the hole of a pinhole camera, $2r$, follows from

$$2r = 1.90 \sqrt{(\text{camera-length}) \times \text{wavelength}} \tag{29}$$

If the photographic plate is at a distance of 15 cm from the diaphragm of a pinhole camera, and the particular light used is of the wavelength 0.000400 mm, which applies to the effective light of blue sensitive plates, it follows $2r = 0.47$ mm.

A parallel beam of light which passes through a circular opening produces a diffraction circle whose diameter can be calculated from formula (12).

2.2.4. Objectives for Visual and Photographic Work

2.2.4.1. Focal Length of a Single Lens and Size of the Field

Objectives which consist only of one lens can be used for visual and photographic observations. For visual observation the aperture ratio 1:20 is advisable so that the deviations from spherical shape and the color error can be kept small (see page 32). In the case of photographic observation we should keep the aperture ratio smaller than 1:12 and furthermore try to compensate the color errors by a suitable combination of filter and emulsion, e.g., by the use of a yellow filter together with a blue sensitive plate.

The *focal length* of a lens depends on the thickness g of the lens, on the refraction index n of the glass for the various wavelengths, and on the two radii r_1 and r_2 of the curvature of the surfaces of the lens. It is

$$\frac{1}{f} = (n-1)(1/r_1 - 1/r_2) + \frac{(n-1)^2}{n} \cdot g/r_1 r_2. \tag{30}$$

The radii of curvature can be measured by an optician with his spectacle spherometer, if one does not want to calculate them on the bases of the relationship

$$r = \frac{2aa'}{(a+a')},$$

in which a and a' denote the distance of the object and of the image, respectively (see page 70). The refraction index n is supplied by the manufacturer of the lens. Table 2.7 lists several indices of Jena glasses and of quartz for the

Table 2.7. Refraction Indices of Optical Glasses for the Fraunhofer Lines

Wavelengths in Å	A(O) 7608.2	B(O) 6867.2	C(H) 6562.8	D(Na) 5893.0	E(Fe) 5270.0	F(H) 4861.4	G′(H) 4340.5	G(Ca) 4307.8	H(Ca) 3968.5
Boron crown BK1	1.50491	.50674	.50762	.51002	.51302	.51567	.52017	.52050	.52457
Heavy crown SK1	1.60347	.60585	.60698	.61016	.61418	.61778	.62396	.62443	.62999
Flint F3	1.60294	.60638	.60805	.61279	.61898	.62464	.63473	.63548	.64518
Heavy flint SF4	1.73924	.74452	.74728	.75496	.76520	.77471	.79201	.79336	.81038
Quartz glass SiO$_2$	1.45443	.45604	.45682	.45886	.46140	.46358	.46731	.46758	.47091

Fraunhofer lines, referred to air at $+20°C$, of 760 mm pressure (Hg), and 10 g.m^{-3} absolute humidity. The unit length is 1 Angström, i.e., 0.0000001 mm. Table 2.7 shows that the difference of the refraction indices (the dispersion) is particularly great for short-wave light; also, the color aberration for this light becomes particularly large, and wide blue color borders are produced. These can be largely avoided by use of a yellow filter.

If a photographic plate is placed in the focal point for a particular color, the radii of the color-dispersion circle for another color on the axis can be calculated by

Diameter of the color ring = aperture ratio of the lens

              × difference in focal lengths of the two colors.   (31)

Neither simple nor composite objectives show a *"limb effect"* because the single lenses are thin and because only one of the surfaces is fully exposed to the influence of temperature changes. Also appreciable temperature differences for the different surfaces of the lenses are without importance for the quality of the image. However, the actual temperature of the objective lens determines its focal length.

In each photograph of the sky the motion of the celestial body has to be taken into account. A star moves in seconds by the angle

$$\frac{360°}{23^h56^m4{.}091} = 0{.}004/\text{sec.}$$

Since an angle of $1°$ at 100 mm, focal length is imaged as a length of

$$\frac{1° \times 100 \text{ mm}}{57{.}2958} = 1.8 \text{ mm.}$$

Thus, a star at the equator moves in one second by $0.004 \times 1.8 = 0.0072$ mm on the photographic plate. Such a small line is indistinguishable from a dot. After 4 seconds exposure time we have reached the resolving power of the stationary photographic plate (0.03 mm), and from then onwards the star records its image as a little stroke. A photograph of a planet or a comet whose motion can be considered here to be equal to that of a star becomes diffuse if, in our example, the time of 4 seconds is exceeded. If we wish to have a moving point-like celestial body (satellite, minor planet) appearing as a little stroke we must exceed the exposure time of 4 seconds many times. One can also attach the camera to a telescope in such a way that both have the same object at the same time in the field of view. During the exposure the telescope must be very carefully "guided."

It is also easily possible to calculate the size of the field of the sky which for a given focal length is imaged on a given plate-size:

Extent of the image area in degrees

$$= \frac{\text{length or width of the photographic plate}}{\text{the focal length of the objective}} \times 57{.}2958. \tag{32}$$

This relationship applies only to angles up to about $8°$. If the length of the side calculated by Eq. (32) is larger, then we have to replace this formula by

Tangent of the angular diameter of the image area of the sky

$$= \frac{\text{length or width of the photogrpahic plate}}{\text{focal length of the objective}}. \tag{33}$$

2.2.4.2. Single Lenses as Objectives

The Plane-Convex Lens. Formula (30) gives the focal length of such a lens as $f = r/(n-1)$. If its plane surface is facing toward the object at a distance and if at $a = (r-g)/n$ (the explanation of r, g, and n is given on page 36) there is a diaphragm aperture with the diameter $d < f/7$, we obtain an image free from astigmatism. The field of view gives a good image even at a size of 60°. The image-surface is curved, and the radius of curvature is n times the focal length. The beams of light which pass the diaphragm obliquely are, on the photographic plate, not convergent to a point, but appear as small spots of light. By narrowing the opening, it is possible to keep the diameters of these dispersion rings smaller than the resolution of the plate.[16]

If a shutter is to be used it must of course be placed in the plane of the diaphragm or close to it. As to the removal of field curvature, see page 43.

The Meniscus Lens. Figure 2–11 shows the positive meniscus lens which has

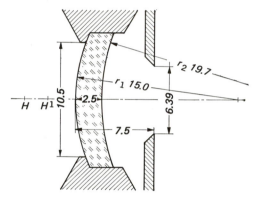

Fig. 2–11 Single lens (meniscus) as the photographic objective. With this simple lens $f/4.4$ and a field of view of 40° diameter we obtain a plane image without coma, but with 1.8% distortion and 3% astigmatism at the edge. Improvements are discussed in the text. From *Amateur Telescope Making*, Book 1, p. 29. New York: Scientific American, Inc. (1953).

been designed by J. W. SHEAN.[17] The data refer to the focal length 100, they can be changed for other focal lengths, e.g., in the case of the focal length 10 cm they can be divided by 10 and used in cm. The refractive index of glass for the particular light is $n = 1.53$. The dispersion should be kept small. The indicated diameter of the diaphragm corresponds to an aperture ratio 1:14.4. By increasing the free aperture from 10.5 to 15, and by decreasing the diaphragm aperture from 6.39 to 5.5, and by changing the distance of the aperture from 7.5 to 11.5 it is possible to improve the image. The plane of the plate is then at a distance 84 from the plane of the aperture. The imaging errors, except for the chromatic errors, are small if we restrict ourselves to a total of 40° for the field of view. The size of the plate then required is 74 × 74.

[16] GRAMATZKI, H. J.: *Hilfsbuch der astronomischen Photographie*, p. 16. Berlin, Bonn: Ferd. Dümmlers Verlag, 1930.
[17] SHEAN, J. W.: Elementary camera lenses. *Amateur Telescope Making*, Vol. 3, p. 532. New York: Scientific American, Inc. (1953).

These two examples of simple lenses show that even simple tools are able to yield to good observations. At the same time they may encourage amateur astronomers who grind their own lenses, an activity for which there exist many detailed instructions;[18] the calculation of simple and composite (achromatic) lenses is also dealt with in detail in the literature so that the amateur astronomer may proceed with it if he so wishes.[19]

2.2.4.3. The Achromatic Objectives

The preceding examples of objectives made up by single lenses have shown that all lens errors can be eliminated to a large extent—with the exception of chromatic aberration. This error, too, can be avoided if the objective is composed of at least two different kinds of glass. It is one of the great achievements of JOSEPH FRAUNHOFER (1787–1826) to have been the first to calculate in every detail such achromatic objective lenses. Such an achromat consists of a crown lens in front of a flint lens. Both have spherical surfaces and only a small air space between them. Spherical aberration and coma are not noticeable; astigmatism, distortion, and curvature of the field of view are present. The best focus contains 95% of the light; the remaining 5% forms the "secondary spectrum" (see Fig. 2–10). But this remaining color error becomes unnoticeable if the aperture ratio approaches the value 1:20. Then the achromat produces pictures so free from color that every planetary observer should use such a ratio. Further advantages of the achromatic objective lenses are insensitivity against changes of temperature and humidity and the ease with which the single lens surface, which is on the outside, can be cleaned. As a rule of thumb for achromatic objectives we may note that the secondary spectrum is sufficiently suppressed if, in centimeters,

$$\text{Focal length of an achromat} = 2 \times (\text{diameter of the aperture})^2. \qquad (34)$$

In a telescope 50/540, which contains an objective of the FRAUNHOFER type, this rule of thumb is realized. Concerning the focal length of composite lenses we refer to page 69.

Refractors of small aperture ratio, and therefore of great length of tube, can be set up in relatively small observing rooms if the path of the rays is interrupted and redirected by one or two plane mirrors. In this case, the length of the tube may be one half or one third of the focal length. The result is the so-called "BASSOON refractor," the "NEWTON refractor," or the "SCHAER refractor."[20] The image quality depends, apart from the quality of the objective,

[18] FERSON, F. B., and LENART, P.: Lens production. *Amateur Telescope Making*, Vol. 3, p. 163. New York: Scientific American, Inc. (1953).

[19] GEE, A. E.: The design of telescope objectives by the G-sum method. *Amateur Telescope Making*, Vol. 3, p. 208. New York: Scientific American, Inc. (1953).

WOODSIDE, C. L.: On computing the radii of an achromatic objective. *Amateur Telescope Making*, Vol. 3, p. 565. New York: Scientific American, Inc. (1953).

[20] We refer to *Blick ins All*, published by the Bayerische Volkssternwarte München, **9**, No. 11 (1964); and ROTH, G. D. (Ed.): *Refraktor-Selbstbau*, Munich: Uni-Verlag, 1965.

on that of the plane mirrors used. There are great requirements as to the stability of the tube and the positioning of the mirror, but the amateur astronomer will be well able to fulfill these. (See Fig. 2–12.)

Folded light beams in the refractor are frequently favored by amateur astronomers.[21]

Further developments of the achromat, which proved of importance for the amateur astronomer, are the following:

The composite apochromatic telescope objective by Zeiss, which possesses a strongly diminished secondary spectrum, is often employed. It is used in connection with the telescope 63/840 and denoted as the *AS* objective (see Fig. 2–10).

Another composite telescope objective, also consisting of two cemented parts, was developed by HARTING; it uses new Jena glasses and is completely free of coma. Its images are absolutely sharp and it is only used for relatively small apertures, probably because the cemented lenses are distorted by temperature variations.

Finally we mention an apochromat which is composed of three parts, and which ideally fulfills the demand for freedom from color effect, and which of course shows spherical errors that are quite inconspicuous, while the coma has been completely eliminated. This instrument is only produced up to a diameter of 200 mm and exceeds the requirements of an amateur astronomer (see Fig. 2–10).

Achromatic objective lenses can be supplied, for instance, by those firms advertised in *Sky and Telescope* and other astronomical periodicals. Examples of achromatic objectives consisting of two meniscus lenses cemented together are given with all necessary data for their production and application in Book 3 of the already mentioned handbook, *Amateur Telescope Making* (New York, 1953). Here we also find discussions of the influence of the position and the aperture of the diaphragm on the imaging errors, so that the amateur astronomer who is interested in the making and the use of this instrument can easily reach a decision.

An achromatic objective lens, with two or three components, is best corrected for the imaging on-axis, and of these, those instruments are best corrected that image a surface-like object in the middle of the field of view through all parts of the objective with equal size and at the same place (fulfillment of the *sine condition*). Nevertheless the field usable for photographic imaging is much larger than one usually believes. For instance, the *Paris Moon Atlas* (73 maps of the size 47×57 cm) has been produced from plates obtained with a 2-lens objective of 620/15,900 mm.

The *Carte du Ciel* (22,000 plates) has been made on the basis of an

---

[21] See PFANNENSCHMIDT, E.: Construction of a folded refracting telescope. *Sky and Telescope* **37**, 319 (1969).

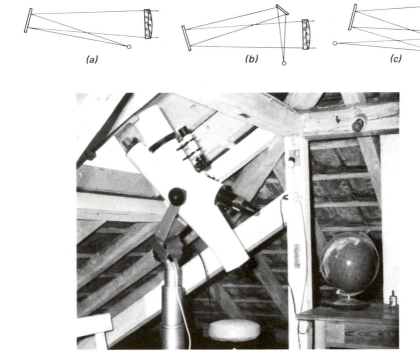

(a)   *(b)*   *(c)*

(d)

Fig. 2–12 (a) BASSOON refractor, length $\frac{1}{2}$ of the total focal length. (b) NEWTON refractor, length $\frac{1}{2}$ of the total focal length. (c) SCHAER refractor, length $\frac{1}{3}$ of the focal length. (d) Modified SCHAER refractor ("broken refractor" by M. WACHTER, Stuttgart) of 125 mm aperture and 2300 mm focal length. Length of the tube with dewcap is 850 mm. Fork mounting.

achromatic 2-lens objective of 340/3400 mm, which has the property of imaging an area of 20 square degrees on plates of 28×28 cm.

Many astronomical journals carry advertisements of suppliers of optical components, etc. A few examples follow:

Aluminising and Lens Blooming, Oaklands House, Solarton Road, Farnborough, Hampshire, England.

Brunnings (Holborn) Limited, 133 High Holborn, London W.C.1.

Cave Optical Company, 4137 E. Anaheim Street, Long Beach, California 90804.

Charles Frank Limited, 145 Queen Street, Glasgow C1.

Coulter Optical Company, 8217 Lankershim Blvd., North Hollywood, California 91605.

Edward R. Byers, 1541 W. Nancy Street, Barstow, California 92311.

Irving & Son, 258 Kingston Road, Teddington, Middlesex TW11 8JQ, England.

Manchester Telescope Centre, 34 and 36 New Brown Street, Manchester M4 2AL.

Questar Corporation, Box EE10, New Hope, Pennsylvania 18938.
Telescope House, 63 Farrington Road, London E.C.1.
Telescopics, 6565 Romaine Street, Los Angeles, California 90038.
University Optics, Inc., 2122 East Delhi Road, Ann Arbor, Michigan 48106.

Some achromats image a field of view up to 40° diameter. The refractors for visual observation possess a *field curvature* which can be compensated for by the use of curved plates or films. It is more difficult to make allowance for the fact that in a refractor only those colors are sufficiently focused in one point that are most effective to the eye, while colors that have the greatest effect on the photographic plate exhibit such a large difference in focal length that the sharpness of the image is affected. If it is possible, by using sensitized photographic plates, together with *color filters* in front of these, to achieve photochemical action only for light of the wavelengths 0.000510 mm to 0.000600 mm, an achromatic objective lens can be used for photographic work (Fig. 2–13). The loss of light through the filters must be accepted and

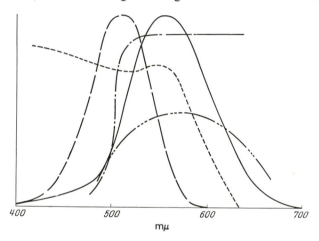

Fig. 2–13 Color sensitivity, i.e., color transparency, and color deviation of eye (solid line for observation in daytime, broken line for observation in twilight), orthochromatic photographic plate (dotted line), color filter GG 11 (—·—), and E-objective (—·· ·—). Each curve is drawn in arbitrary units.

it is particularly regrettable when we want to photograph objects which are faint and therefore require an especially long exposure time. If our aim is the determination of the brightness of stars on the basis of the size and blackening of the star images on the photographic plate, we must remember that the same star produces different images as far as size and blackening are concerned, according to whether it is at the center or nearer to the edge of the plate. Achromatic refractor objectives are most suitable for photographic records of single objects on the sky.

Limiting magnitudes and diameter of images are listed in Table 2.1.

H. E. DALL[22] has shown how it is possible to make visual observations in the ultraviolet.

While it is required that for visual objectives only the errors on the axis (spherical errors, color deviation, and deviations from the sine condition) must be sufficiently small, we demand from photographic objectives that these yield sharply defined and flat images also off-axis, i.e., that astigmatism, coma, and distortion must not exceed small values. Nearly all such objective lenses can be derived from two fundamental forms. Either their structure is *symmetrical* and consists of two front lenses and two back lenses, wherein color error, coma, and distortion must be eliminated (the prototype is the portrait objective designed by J. PETZVAL in 1880); or they are asymmetrically built with one negative lens at the center whereby the image flattening is preferably achieved by the use of special glasses (the prototype is H. D. TAYLOR's classical triplet of 1894).

The Petzval Objective. PETZVAL designed his objective to compensate for the insensitivity of the photographic plates of his day by the light power of the objective. This consists of four lenses, has a relatively large field curvature and gives good images over a field of 7°. Many firms have produced it with aperture ratios between 1:3 and 1:6 and focal lengths 25–100 cm; objective diameters up to 25 cm occur. The characteristic large distance between the front and back doublet has in some cases been decreased in order to increase the sharpness of the image off-axis. Sometimes, too, the back lens has been reversed to the disadvantage of the sharpness on the axis. A good PETZVAL objective is suitable for the amateur astronomer who wishes to take photographs of the sky. It is possible to remove the disadvantage of the field curvature (PETZVAL *curvature*).

To achieve this we measure first the amount of this curvature. We set the objective on a very distant sharp source of light on a ground glass, which appears in the middle of the field of view and at different measured distances from the middle to the edge of the ground glass. Every distance z from the center of the field of view requires another setting of the ground glass. The displacement of this glass must be measured very accurately; a mean of many repeated measurements should be taken. If now the displacement of the ground glass in its dependence on the measured distances of the images from the center of the plate is plotted we obtain a curve, representing an arc of 10°–20°.

We then need a *flattening lens* (plane concave lens), which we can also make ourselves. If n is the refracted index of the glass of this lens then the thickness of the lens at the distance z from the center of the plate becomes

Thickness of lens = $z \cdot (n/n-1)$ + thickness of the lens at the center (35)

Then we have

Radius of curvature of the flattening lens
= radius of curvature of the field of view $\times (n-1)/n$ (36)

[22] DALL, H. E.: Visual astronomy in the ultra-violet. *J. Brit. Astron. Assoc.* 77, 94 (1967).

The photographic plate is placed in contact with the plane surface of the flattening lens.

Other forms of possible flattening lenses are discussed in *Amateur Telescope Making*, Book 3, p. 310, New York: Scientific American, Inc., 1953, which also lists additional literature.

The Triplet (COOK *Lens*, TAYLOR *Lens*). This objective has aperture ratios up to 1:3, aperture diameters up to 15 cm, and angular fields up to 25°. Several well-known objectives of large observatories are derived from this type (ROSS objective, SONNEFELD's 4-lens objectives) as well as the many triplets, Tessars, and wide angle lenses that are used for the survey-photographs initiated by many observatories and that also in the hands of amateur astronomers produce remarkable celestial photographs (see, for instance, H. E. PAUL in *Amateur Telescope Making*, Book 3, p. 308).

The limiting magnitude (which refers here to stars at the plate center that are still clearly visible) of these powerful objectives reaches 14^m. This limit is, because of the long light path in the glass, the reflection on the lens-air surfaces, and the compromise needed in the focal position for the regions of the plate away from the axis, less than we would expect from the large apertures.[23]

For example (see also Table 2.1), one may mention the following much used triplets, Tessars, and wide angle lenses. The correct stops for them must be found in each case by trial and error, according to the quality desired for the given field:

Triplets (field—a few degrees) are VOIGTLANDER Heliar, and DALLMEYER Pentec.

Tessars (up to a field of view of 28°) are, for example, ZEISS Tessar, EASTMAN Ektar, SCHNEIDER Xenar, BAUSCH and LOMB Tessar, ROSS London Express, ZEISS Sonnar (1:1.4), and ZEISS Jena four-component amateur camera 70/250 mm with the field of view of 21 × 28 degrees on 9 × 12 cm plates.

Wide angle objectives (up to 90° field) are ZEISS Orthometer, BAUSCH and LOMB Metrogon, ROSS X-Press, and MAYER Double Planat.

2.2.4.4. The Optics of the Coronograph and the Prominence Telescope

A lens telescope of a special kind which can be built and used successfully by the amateur astronomer is a coronograph. This makes it possible to render the faint light of the solar corona visible. Similarly, the prominence telescope, with which prominences of the solar limb can be observed, can easily be built by the amateur astronomer from components which are even easier to obtain.

Figure 2–14(a) shows how the solar disk can be imaged on an iris diaphragm B with the help of the auxiliary lens H by an objective O_1 which is placed at the bottom of a long dewcap. This objective produces only very little scattered light (no scratches, bubbles, or dust particles). A conical diaphragm K receives the image, so that the bright disturbing light sources

---

[23] See SCHNELLER, H.: *Veröffentlichungen der Universitäts–Sternwarte zu Berlin–Babelsberg* **8**, Part 6, 7 (1930).

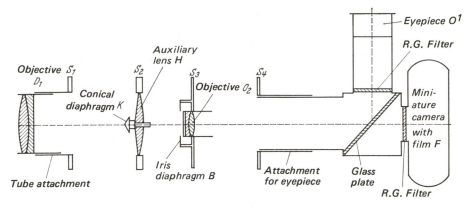

Fig. 2–14   (a) The prominence telescope of the amateur astronomer (after O. NÖGEL). Disks of light metal are denoted by S_1, S_2, S_3, S_4.

are made harmless. The image of the solar limb and of the immediate neighborhood of the Sun is produced on the rim of the conical diaphragm and is imaged with the help of a second objective O_2 onto the image plane (plate, film, eyepiece diaphragm). Since prominences glow essentially in the light of hydrogen, the bright hydrogen line $H\alpha$ (6563 Å) suffices for the observation; therefore we must eliminate all other light by corresponding red filters. It is to the merit of several amateur astronomers that they have shown how it is possible to construct by this simple means a very good prominence telescope.[24] In what follows we shall give a description of the various components and their combination according to O. NÖGEL.

The Objective O_1. We can use a cemented achromat Zeiss Mess 92.900, 25/250 mm, in order to use the telescope both for stars and nebulae (see below), or an uncoated simple lens, if we wish to observe in monochromatic light only. Clean work and optical homogeneity are of greatest importance. On the other hand the diameter of the entrance pupil and the focal length are much less important. The focal length determines the distance from the conical diaphragm K.

The Diaphragm K. Its diameter can be calculated from formulae (23) and (24) as $0.0002909 \times$ Sun's diameter in minutes of arc $\times$ focal length of O_1 in mm.

The values to be taken for the Sun's diameter are

January	1:32.'6	July	1:31.'5
February	15:32.'4	August	15:31.'6
April	1:32.'1	October	1:32.'0
May	15:31.'7	November	15:32.'4

[24] DALL, H. E.: Filter-type solar prominence-telescope for amateurs. *J. Brit. Astron. Assoc.* **77**, 99–101 (1967).
NÖGEL, O.: *Die Sterne* **28**, 135 (1952), and **31**, 1 (1955). Also in *Astro-Amateur*, p. 59. Zürich-Stuttgart: Rascher-Verlag, 1962.

The silvered or chromium-plated conical diaphragm will be provided by a clockmaker. Its base is placed inside a central opening of the auxiliary lens H which has been cemented with a special glycerine cement, or the cone is placed on top of a silver rod, which is attached with the help of a thread and a nut in the hole of the auxiliary lens.

The Auxiliary Lens H. Here we use a biconvex spectacle lens of $f = 160$ mm (6 dioptrics). Its focal length and its distance from O_1 determine distance and size of the image of the entrance pupil, according to Eqs. (38) and (39), pages 69-70. The central hole will be made by an optician or by an optical firm (e.g., SPINDLER und HOYER, Göttingen). NÖGEL recommends that the auxiliary lens be diaphragmed down to 3 times the diameter of the cone to avoid scattered light from the cone.

The Iris Diaphragm B. This diaphragm must be exactly at the place where the image of the entrance pupil O_1 is produced; therefore the objective, or better still the diaphragm, must be movable. The proper position is reached if the diaphragm, when illuminated with a torch from the eyepiece region, gives a sharp image on a transparent screen immediately in front of O_1. During observation of the prominences, the diaphragm must be enlarged sufficiently. If the diaphragm is too large it gives a bright background and weak pictures; if it is too small the images are, as a whole, too faint. The iris diaphragm B is kept in place by three long adjusting screws, with counter-sprung nuts in the same casing as the objective ring of O_2.

The Objective O_2 The Reversing Lens. Here, for instance, an objective Messers 20/150 mm or a photographic objective 50/150 mm is used. Its focal length, its distance from the conical diaphragm, and the size of the image of the solar disk produced by O_1 by the formulae (41) and (42) (see page 70), determine the image distance and the size of the solar image as produced on the film F or in front of the eyepiece O'. If we wish, for instance, that the solar image produced by O_1 should be doubled in size we have according to Eq. (42) $a' = -2a$ and thus from Eq. (41) the relation $a' = 3f$ [whereby we have in Eq. (41) to replace f' by f_2].

The Color Filter RG. The red filter RG2 (thickness 2 mm) together with a weak gray filter or a red filter RG5 (thickness 1 mm) or better still metallic interference filters by Geffcken with a semibandwidth of 3×10^{-9} to 5×10^{-9} meters (that is, 30–50 Å) and a peak transparency of 25%—all these color filters have been supplied by Schott and Genossen in Mainz or Jena. The filters are placed immediately in front of the plate, the film F, or the eyepiece diaphragm. Their minimum diameter follows from the calculated size of the solar image as produced by O_2.

All the other parts are placed on four disks of light metal (S_1, S_2, S_3, S_4) each of 45 mm thickness, which are kept together by 12 rods made of light metal, with threads and nuts. The disks S_1 and S_4 are fitted with tubes for the attachment of collars for O_1, and for a cube made of light metal, which takes a miniature camera with its metal-laminated shutter, as well as the attachment for the monitoring eyepiece O'. Inside the cube, a plane, appropriately bloomed, glass plate is placed with an inclination of 45°, or a semi-transparent mirror (aluminized glass plate) which brings the sunlight both to eyepiece and camera.

The Tube. Here, two different designs have been proposed. Either the two metal disks S_1 together with O_1 and S_2 together with the conical diaphragm K and the auxiliary lens H are in one portion of the tube, while S_3 together with O_2, S_4, and the eyepiece attachment are in a second piece of the tube (which can be displaced for focusing purposes), or the auxiliary lens with cone, the iris diaphragm, the reversing lens, and eyepiece with filter are all placed in a short tube that can be converted into a prominence telescope. This prominence eyepiece can be attached to the eyepiece collar of an ordinary reflector which can thus be converted into a prominence telescope. The centering of all parts in the tube takes place in steps, beginning with O_1 and H. Light from a small bulb enters the telescope from the front with the help of a glass plate and the observer looks through this glass into the tube. It must be possible to find a position for the eye on the extended exit of the tube from which all light reflections on the lens surfaces can be seen in a symmetrical arrangement and at equal distances around the axis of the tube. Otherwise the centering is unsatisfactory. A different construction of the prominence telescope has been proposed by ROQUES.[25] The reader is also referred to G. NEMEC, Das Protuberanzenfernrohr als Hochleistungs-instrument, a series of eight articles in *Sterne und Weltraum*, June 1971 to February 1972, which give a comprehensive discussion with many practical hints.

2.2.4.5. The Barlow Lens

The BARLOW lens[26] is the negative lens system which has the purpose of extending the focal length of a refractor objective. It is placed as a two-component achromat in front of the focus of the objective. It is possible to keep the chromatic error for stars outside the center of the field very small; the optician who calculates the BARLOW lens for a given objective will take care of this. It is easier to produce for a smaller aperture ratio of the objective and a smaller required increase of focal length. The free aperture is best determined with a transparent paper disk which is introduced at the intended position into the paths of the rays. The BARLOW lens itself should not be displaced, e.g., in order to change the total focal length, since this would diminish the quality of the picture in- and outside the optical axis. Because of the long focal length, a HUYGENS eyepiece may be used. For simple requirements we might even use a nonachromatic lens as a BARLOW lens, provided that we are satisfied with achieving a lengthening of the focal length without being concerned with the quality of the image, e.g., in guiding telescopes.

Let us denote by f_1, f_2, and f' the focal lengths of the objective, the

[25] ROQUES, C. E., and ROQUES, J. M.: *l'Astronomie* **75**, 67 (1961).
[26] P. BARLOW was an English mathematician and physicist, 1776–1862.

BARLOW lens, and the whole optical system, respectively; and by δ, s_1, and s_2, the distances from the objective to the BARLOW lens, from the BARLOW lens to the focus of the objective, and from the BARLOW lens to the focus of the whole system, respectively. We further denote with $A = f'/f_1 = s_2/s_1$, the magnification factor. It then follows from Eq. (39) with $\delta = f_1 - s_1$:

$$f' = \frac{f_1 \cdot f_2}{f_2 - s_1}, \qquad s_1 = \frac{f_2(A-1)}{A}, \qquad s_2 = f_2(A-1).$$

This last equation shows that the focus of the whole system lies in the case of a doubling of the magnification ($A = 2$) by only the small distance $s_2 = f_2$ behind the focus of the objective. It can therefore be reached without any appreciable extension of the tube or camera.[27]

2.2.4.6. Small Cameras for Amateur Astronomers

The objectives of cameras differ considerably from each other according to aperture ratio, focal length, and field of view. None of them is suitable for visual observations but all of them can be applied to the photography of certain celestial objects. The reader will take account of the general discussion on page 37.

The focal length should not be shorter than 15 cm if one wishes to photograph star fields in which the stars are sufficiently separated. The scale is usually strictly related to the size of the plate grain or the resolving power of the objective. Short focus objectives are just as suitable as long focus lenses if we are concerned with photographs of the Moon, comets, halos, Zodiacal Light, twilight phenomena, bright meteors, and the Milky Way. The size of the image on the plate or film is determined by the focal length according to formula (24). Photographs of the Sun are separately discussed on page 250.

There is no limit to the diameter of the *aperture diaphragm*. The example of the pinhole camera shows that even the smallest opening is sufficient for certain photographs, e.g., of the Sun and Moon. For small cameras the objective aperture cannot be set equal to the aperture diaphragm.

If we wish to determine the effective aperture (entrance pupil) in order to find the aperture ratio, we form the image of a pointlike light source on the axis within the image plane on transparent paper, using the full aperture. Transparent paper which is placed on the objective permits the measuring of the diameter of the bright disk, which then becomes visible and is usually somewhat smaller than the diameter of the objective lens.

The short-focal-length camera has enhanced optical errors on astronomical photographs. These camera objectives are designed for photography of landscape and portraits and can therefore produce soft pictures. They are

---

[27] Optical data of BARLOW lenses are compiled by C. R. HARTSHORN: The Barlow lens, *Amateur Telescope Making*, Book 3, p. 277. New York: Scientific American, Inc., 1953. See also BERLAUNY, S. S.: *J. Brit. Astron. Assoc.* **78**, 64–69 (1967).

therefore without consequence. The picture of a star field is, on the other hand, very suitable to disclose ruthlessly nearly all the properties of the optical performance of the particular objective.[28] Also, photographs where the stars appear as little strokes can be made useful for this purpose. Strokes of greatest blackening indicate the focus, not those of largest thickness.

The aperture ratio (f number, reciprocal of the aperture number, f ratio, speed; the light ratio is proportional to the f number) of a camera can be changed to a very large degree. A large aperture ratio easily leads to a fogging of the emulsion if intense scattered light from artificial light sources enters the objective. A bright background on the plate suppresses the faint star images and therefore reduces the limiting magnitude of the stars on the plate.

Table 2.8, calculated by WHIPPLE and RUBENSTEIN, shows the connection

Table 2.8. Limiting Magnitude and Image Sizes of Stars

Aperture ratio		1:3		1:5		1:7	
Opening in cm	Opening in inches	Image diameter in mm	Limiting magnitude	Image diameter in mm	Limiting magnitude	Image diameter in mm	Limiting magnitude
1.0	0.4	0.023	10.20	0.026	11.18	0.028	11.84
2.5	1.0	0.024	12.15	0.027	13.12	0.030	13.74
5.0	2.0	0.025	13.61	0.029	14.55	0.032	14.28
7.5	3.0	0.026	14.44	0.031	15.36	0.035	15.96
10.0	3.9	0.027	15.03	0.032	15.96	0.042	17.27
15.0	6.0	0.030	15.80	0.036	16.71	0.042	17.27
20.0	7.9	0.032	16.35	0.040	17.21	0.047	17.77

between aperture and aperture ratio on the brightness of the stars, whose diffraction disks or diffraction images stand out on the plate background because of their blackening.

Very badly disturbed air and a brightening of the sky affect considerably the values given in the table. Celestial photographs should be taken at as large an altitude as possible and at least symmetrically about the meridian in order to minimize the effect of air disturbance.

Plates should be preferred to films if there is a choice. However, only such plates should be used whose contrast and color sensitivity is accurately known (see page 112).

The record attached to the photograph should contain the following data:
Region of the sky, constellation, object.
Right ascension and declination of the plate center.
Date and time of exposure in Universal Time.
Instrument, objective aperture ratio, and diaphragm,
Type of plate, and note if sensitized.

[28] See JASCHEK, W.: Die Prüfung von photographischen Astroobjektiven. *Photographische Korrespondenz* **87**, 11 (1956).

Filter.

Seeing, transparency, scattered light, Moon, clouds, wind.

Developer, time of developing, temperature, contrast-increasing substances.

Reproduction: paper, developer.

Plate scale.

2.2.4.7. The Objectives of Small Telescopes

In the following discussion, the widely used telescopes that are immediately available for astronomical observations will be dealt with. Their application is the object of a book by R. BRANDT[29] which contains numerous instructions. Furthermore, T. HENSON has discussed in great detail the selection, construction, and care of field glasses and other small telescopes.[30]

The telescopes that we will discuss, binocular or prism telescopes, are of the type of astronomical telescope (Fig. 2–1) whereby in addition the picture is made upright by a set of PORRO-prisms (a combination of two right-angle prisms), or as in the HENSOLDT-Dialyt binocular by a special system of prisms so that in contrast to the astronomical finder, upright pictures with a correct orientation of the sides are produced. We must take this into account if we use them as finders. The objectives were formerly two-component cemented achromats of the aperture ratio 1:3.7 to 1:5, and more recently achromats have been used with a larger air space between the components—only corrected for errors on the axis (spherical and color aberration, sine condition) including corrections for the paths through the glass in the prism.[31]

Small telescopes of the Galilean type (Fig. 2–2) are used as opera glasses and as binoculars by hunters. The same optical arrangement is followed in glasses of various types. Without going into detail concerning the connection between objective diameter and the restriction of the field of view such small telescopes can really only be designed for about 4 times magnification. With increasing magnification the field of view decreases rapidly and becomes diffuse and not very uniformly illuminated toward the edge. The light power has values such as 8.3, 10.5, 11.3, and is therefore extremely large (see page 12). The objective is nearly always achromatic, as is the eyepiece in several new glasses. There exist also Galilean telescopes in which objective and eyepiece are corrected as a united system for the errors on the axis (M. VON ROHR). The eye need not be taken close to the eyepiece; the size of the field of view depends on the position of the pupil of the eye. The pictures are upright, right and left are correct, and their appearance is brilliant. Frequently the motion between the two parts of a pair of binoculars, operated by a central adjustment screw, is rather faulty.

[29] BRANDT, R.: *Himmelswunder im Feldstecher*, 7th ed. Leipzig: J. A. Barth, 1964.

[30] HENSON, T.: *Binoculars, Telescopes and Telescopic Sights*. New York: Greenberg, 1955.

[31] A special group among the commercial field glasses comprises the 11 × 80 and 22 × 80 glasses of the firm BECK in Kassel, known as "Tordalk" glasses, and the 14 × 100 "Gigant" of M. WACHTER, Bodelshaüsen, West Germany. Very handy and at the same time of fine optical quality is the newly developed glass by LEITZ, commercially known as Trinovid 10 × 40 (Ed.).

Monocular small telescopes (e.g., for terrestrial observations and for viewing distant landscapes) contain either an image-erecting system of lenses, which accounts for their greater length, or they are equipped with a four-component terrestrial telescope eyepiece as was done by FRAUNHOFER. The two real intermediate images make it possible to introduce cross wires. The light power is always smaller than for field glasses. The objective apertures are between 20 and 70 mm, the magnifications are between 15 and 60. Recently, also, terrestrial telescopes with easily changed magnification became available commercially (see page 41).

The body of a small telescope has an indication such as 6×30, 7×50, 15×50, etc.; the first number indicates the magnification, the second the free aperture of the objective in mm. According to formula (4) the division of the second number by the first gives the diameter of the exit pupil. These three parameters are important for the brightness of the image, the limiting magnitude, the resolving power, and the light power (see pages 9–20). Since 1891 an adjustment of the objective distance has been introduced to increase three-dimensional viewing (stereoscopic effect); this, however, is astronomically irrelevant. Binoculars for spectacle wearers are distinguished by the fact that the exit pupil is far enough away from the last lens apex of the eyepiece that it is possible to bring the pupil of the eye at the proper place so as to be able to survey the whole field of view. In order to test the telescope for impurity and scratches, we will reverse it and view the bright sky. Spectacle wearers with astigmatic eye lenses, i.e. with cylindrical spectacles, must wear their spectacles for all observations. Binoculars can have their performance measured by a quantity L:

$$L = \frac{\text{definition as seen by the eye with the telescope}}{\text{definition obtained with the naked eye}}.$$

Such measurements have been made in daytime, in twilight, and during the night.[32] During observations in twilight and at night, and also in astronomical work, the definition and sharpness of the images depends on the light density of the surroundings and on the magnification, so that Z_D can characterize the performance of binoculars in twilight,

$$Z_D = \sqrt{\text{magnification} \times (\text{free aperture of the objective})}.$$

We now mention some auxiliary optical equipment: colored or gray filters made of homogeneous glass for the observation of very bright objects and on the morning or evening sky. Filters for observations of the sun are discussed on page 98. All these glasses can be obtained from the manufacturers of these small telescopes and they can easily be attached to the collars which hold the eyepieces. The large SCHOTT filters can be framed appropriately and attached to the objective end of the casing of binoculars.

[32] See KÖHLER, H.: *Die Fernrohre und Entfernungsmesser*, 3rd ed., p. 100. Berlin-Göttingen-Heidelberg: Springer, 1959.

Projection screens for solar observations can, as described on page 226, be attached to the small telescope eyepiece end.

Eye shields made of soft rubber which fit close to the observer's face are useful to avoid stray light during brightness estimates.

Concerning the installation of small telescopes we have to take into account that the performance of the instrument held freely by hand is 20 to 40% smaller than if the instrument is set up in a fixed manner. Except for sporadic phenomena (meteor tails, artificial satellites), we should therefore always use these instruments on little stands, such as a photographic tripod, a ground stand or tripod, a chemical retort stand, or a homemade support made of hardwood that grips the main axis.

In order to increase the performance of binoculars we recommend supplementary telescopes. Magnification and light power of binoculars can be varied easily by being solidly connected with an additional telescope. We then obtain the combination of two telescopic images[33] (page 9). The parallel beam of light which leaves the telescope enters the additional telescope as if it comes from an infinitely distant object, and it leaves the eyepiece with a diameter that can be calculated from Eq. (3). The total magnification is then equal to the product of the single magnifications. The light power of the composite system is equal to the ratio of the two single light powers. By changing the eyepieces in the additional telescope we can change the total magnification and the total light power. How far both these quantities can be changed depends on the brightness of the observed object and on the quality of the newly obtained image. If some colored (e.g., blue) rims occur we can remove these by using colored filters (e.g., yellow filters).

As an additional telescope, we may use a monocular small field glass (e.g., "Tellup," VEB Optische Werke Jena), or a homemade telescope with a simple objective (spectacle glass) and a HUYGENS eyepiece in a metal tube which can be attached to the eyepiece end of the field glass with a tube and a clamp.

Photographic work with such a small telescope is easy and of a very varied nature. For this purpose we may either place a complete photographic camera[34] with its objective behind the small telescope (which then acts as a teleobjective) or we remove the camera objective and use the camera as a device for the projection of the solar image (see page 224). In the first case we obtain an optical combination of a telescopic and a finite imaging device. The cross section of the beam of rays which leaves the telescope determines by its cross section the size of the entrance pupil of the camera.

The size of the image can be calculated from the total focal length in

[33] See ROHR, M. VON: *Die Bilderzeugung in optischen Instrumenten*, p. 117. Berlin: Springer, 1904.
[34] A very useful combination consists of fieldglass and a Minox miniature camera which achieves outstanding definition. The manufacturers of the Minox provide a special fieldglass clamp which provides a precise attachment of the miniature camera to the eyepiece of the fieldglasses (Ed.).

Fig. 2–14 (b) A suitable monocular field glass is attached to the casing of a mirror reflex camera without optics. The connection between the field glass and the casing is provided by the special rings used with these cameras for copying. Focusing is carried out by turning the eyepiece of the fieldglass. After R. HANKE in R. BRANDT: *Himmelswunder im Feldstecher*. Leipzig: Barth, 1968.

accordance with formula (23). The total focal length is equal to the product of camera focal length times telescope magnification. The light power of the whole system is equal to the square of the ratio between the camera focal length and the exit pupil of the telescope.

Because only images near the axis are corrected (see above) the full definition remains restricted to this region. It is therefore recommended for photographing single objects only. The best procedure is to attach cameras to spectacle binoculars or to use specially-constructed photofieldglasses since the exit pupil, which for the latter is far behind the eyepiece, then lies within the objective of the camera.

> Special precautions must be taken for solar work: we should use only small apertures, if necessary by employing a diaphragm; dark glasses must be placed in front of the objective or behind the eyepiece of the telescope; cameras should have a metal shutter or the exposure should be carried out with an open shutter and a cap which can be removed without disturbance or vibration.

A second arrangement of such a telescope for the achievement of projection of solar images is shown in Fig. 2–14(b).

In both cases the maximum exposure time can be calculated according to page 253 and the whole arrangement driven by an equatorial mounting.

2.2.5. *The Mirror for Visual and Photographic Observation*

2.2.5.1. *The Limb Effect and the Focal-Length Variation*

Light is propagated in waves. If it falls on an optical surface, it suffers either a refraction or a reflection or both. In every case the direction of the *wave front* changes. If a refracting or reflecting body is irregular, the wave front undergoes a deformation. The extent of this deformation depends on the refractive index of the transparent body and on the angle of incidence of the light, and must not exceed one quarter of the wavelength of light if good image is required. This is the so-called RAYLEIGH *condition.* If only one reflecting surface yields an image then, according to this condition, the required accuracy of this surface (for light of perpendicular incidence) must be four times as great as that of a single refracting surface. We therefore see that the requirements for the grinding of reflecting surfaces are very high indeed. The FOUCAULT knife-edge test and the RONCHI test[35] check this accuracy (see pages 32 and 64). Fundamental and critical objections against the use of composite reflecting systems have been given by CROSS.[36]

Glass is a poor heat conductor. Changes in its temperature lead to deformation on its surface and to tensions inside. It is usually assumed that only nonuniform temperatures in the glass lead to deformations, although in the case of large mirrors even constant temperature does not exclude deformation. Temperature variations increase or decrease metal mirrors without changing the geometric form of the surface. The sphere remains a sphere, and the paraboloid only changes its focal length. Therefore the glass mirror changes not only its curvature (according to some uncertain law) but also the shape of the surface, and this mainly at the outer zones near the rim. This temperature effect has been called by RICCI the *"limb effect of the mirror."*

According to MAKSUTOV, mirrors made of a material for which a certain quantity Q is large probably show only a small limb effect or none at all. This quantity is defined as follows:

$$Q = \frac{\text{(modulus of elasticity)} \times \text{(thermal conductivity)}}{\text{(coefficient of thermal expansion)}}. \tag{37}$$

We quote some values of Q, if Q for crown glass is assumed to be 1.

Copper	176	Aluminum	70	Pyrex glass	2.9
Silver	156	Steel	63	Crown glass	1.0
Invar	86	Quartz glass	33	Flint glass	0.44

Table 2.9 shows that mirrors made of metal avoid the limb effect more easily than those made of glass. In spite of this, glass is preferred as a material in mirror making because it can be rapidly and easily polished. *Flint glass* is unsuitable for a glass mirror because of a small Q-value and other properties (softness and little chemical resistance). Commercially obtainable glass is sufficient for the normal mirrors with diameters below 30 cm and these are

[35] See SHERWOOD, A. A.: A quantitative analysis of the Ronchi-test in terms of ray-optics. *J. Brit. Astron. Assoc.* **68**, 180 (1958).

[36] CROSS, E. W.: Objections to compound telescopes. *The Strolling Astronomer* **21**, 70 (1968).

Table 2.9. W = Heat Coefficient; E = Elasticity Module; Q = Heat Conductivity
(see also Footnote 37)

	W	E	Q
Copper	0.000014	1000	1.15
Silver	0.00001921	7000	1.72
Invar	0.00000037	14,000	0.022
Aluminum	0.000026	7500	0.84
Steel	0.00001322	21,000	0.13
Cast iron	0.00001061	10,000	0.12
Quartz glass	0.00000057	6200	0.0083
Pyrex glass	0.00000788	6200	0.0060
Crown glass	0.00000954	7500	0.0042
Flint glass	0.00000788	5000	0.0030

much in use. *Pyrex glass* sometimes contains bubbles which can be very disturbing in the course of the grinding process. The small heat coefficient of this material requires particular care in its attachment to telescopes. The Mount Palomar mirror ($d = 5$ m) is made of pyrex. The only objection against the use of *quartz* for mirrors is its high price; probably it can only be used for mirrors with diameters smaller than 1.5 m. It is not impossible that metal mirrors which are chromium plated may again become useful in telescope optics. As to the suitability of other material MAKSUTOV and others have tabulated interesting data[37] and the Owens Illinois Company of Toledo has recently introduced Cer-Vit as a new promising material.

A difference of the temperatures between the front and the reverse side of a mirror leads to greater *deformations* the larger d is and the smaller the ratio (thickness of the mirror)/(diameter d of the mirror). On the other hand, an increase of this ratio also increases the limb effect (see page 54). We note that as a rule of thumb for glass plates that are ground to mirrors we have at best a ratio of 1:6. The justification of this rule has been confirmed by investigation of flexure. The reverse side of the mirror should be in continuous contact with air without needing ventilation. The rim should be isolated by cork (PORTER, HINDLE). In the case of solar observations the back of the heliostat mirror should receive radiation via an auxiliary mirror. Even for small temperature differences between the two surfaces of a mirror we must expect a *displacement of the focus* because of the change of the focal length. It has been proposed that parabolic mirrors of aperture ratio 1:7 be made only to 2/3 parabolic, because decreasing temperature shortens the focal length (WASSELL).

In order to get an idea of the size of the displacement of focus we list below in Table 2.10 some results obtained by MAKSUTOV for a parabolic mirror of $f = 1000$ mm.

[37] MAKSUTOV, D. D.: *Technologie der astronomischen Optik*, p. 26. Berlin: VEB Verlag Technik, 1954.
Amateur Telescope Making, Book 1, p. 325. New York: Scientific American, Inc., 1953.

Table 2.10. Focus Displacement of a Parabolic Mirror, $f = 1000$ mm

	Thickness of mirror	300 mm	100 mm	30 mm
Temperature difference	0°1	0.002 mm	0.006 mm	0.021 mm
Temperature difference	0°3	0.006 mm	0.019 mm	0.064 mm

The change in the diameter of a star disk in the image plane or on the photographic plate due to temperature variation is

2 × (aperture ratio) × (focus displacement).

The displacement of the eyepiece or of the photographic plate required to compensate the change of focus due to temperature changes of a glass mirror is a combination of the change of focal length and of the length of the tube. It is

Change of focal length = focal length × change of temperature
　　　　　　　　　× coefficient of expansion of the mirror material,

Change of length of tube = focal length × change of temperature
　　　　　　　　　× coefficient of expansion of the tube material.

A decisive *advantage in the use of a reflecting telescope* as compared with a refracting telescope is the lack of color effect (chromatic aberration). This defect, together with a spherical aberration, is the most inconvenient defect. Nearly all other advantages of a reflecting telescope derive from this, e.g., the possibility of using a larger aperture ratio and therefore smaller tube lengths at the same aperture. Consequently it is easier to mount, more convenient for observation, and the cost is less.

2.2.5.2. *Reflectors of the Newtonian, Kutter, Cassegrain, and Gregory Types*

As we have already discussed, it is possible to use a spherical mirror in spite of its various imaging defects for application to astronomical observations (see page 33). The parabolic mirror has the advantage that it produces, after only one single reflection, a color-free image. If this is received on a photographic plate, or after insertion of a *secondary mirror* or prism sideways to the outside of the tube, we have the so-called *Newtonian arrangement*. The observation then takes place at the Newtonian focus (Fig. 2–15). It is easy to transform a spherical mirror into a parabolic mirror: there is a large amount of literature available for this process and all that is required, in addition, is some patience.[38] In the Newtonian system the loss of light caused by obscuration is 7% and the loss due to two reflections 15%. Imaging errors on the axis hardly affect visual observations. Applying greater magnification and using Sonnefeld's data[39] we obtain the size of the usable field as given in Table 2.11.

---

[38] See Rohr, H.: *Das Fernrohr für Jedermann*, 4th ed. Zürich: Rascher-Verlag, 1964.
[39] Sonnefeld, A.: *Die Hohlspiegel.* Berlin: VEB Verlag Technik, and Stuttgart: Berliner Union, 1957.

Table 2.11.

Aperture ratio	1/6	1/5	1/4	1/3
Usable field of the parabolic mirror in minutes of arc	6′	5′	4′	3′

In the case of photographic observation the star images outside the axis no longer appear as circular disks but assume more and more the form of a small comet. Their size can be found in Table 2.15 (see note under this table).

Table 2.12 shows at which distance from the center of the field of view of a parabolic mirror the star image reaches a length of 0.1 mm. The transverse diameter amounts to $\frac{2}{3}$ of the length.

Table 2.12.

Aperture ratio	Distance from the center of the field of view
100/1000	3°
125/1000	2°
167/1000	1°
200/1000	0°.7

If we use a secondary mirror in a Newtonian arrangement we find that its size depends on its distance from the focus of the main mirror, the aperture ratio of the main mirror, and the size of the field of view diaphragm of the eyepiece. The secondary mirror can be placed in such a way that the focus of the main mirror lies inside the tube, while the eyepiece is outside. It is convenient to drill a central hole in the main mirror since then the Newtonian arrangement can more easily be converted into a Cassegrain or Gregory type. This hole affects only the unused central zone. Most mirror manufacturers prefer to make these holes before the grinding and then to replace the removed core immediately in its previous position and finally to remove it

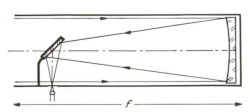

Fig. 2.15 The reflecting telescope of NEWTON. From *Handbuch der Astrophysik*, Vol. 1, p. 172. Berlin: Springer, 1931.

after polishing and parabolizing. The drilling of such a hole leads via temperature changes to a certain deformation of the main mirror (see page 54). The most favorable position of the secondary mirror and the size of the hole in the main mirror can be found by making a drawing of the main mirror and tube on an accurate scale (Fig. 2–23 on page 74).

It is possible to avoid the *loss of light* of 7% caused by the secondary mirror and also the diminished quality of the diffraction disks and thus of the definition of the image by omitting the secondary mirror altogether and making provision that the tilted main mirror throws the light immediately into the eyepiece. HERSCHEL in his design used such an arrangement. If, however, we place a secondary mirror next to the tube and arrange that the light which arrives from the tilted main mirror is reflected again in order to enter an eyepiece, which is now not very far from the main mirror, we obtain a very great focal length. This improves the quality of the image and yields a field of 30–40 minutes of arc free of defect; because of the folded paths of the rays, we still only require a relatively short tube. Such instruments are called brachytes (folded mirror systems). They represent a direct development of the Newtonian mirror.

During the last two decades, KUTTER has been successful in the construction of simple, efficient brachytes, which have been built by numerous amateur astronomers, particularly on the Continent. Here we may refer, first of all, to KUTTER's original papers[40] and, furthermore, to the discussion of this "Oblique Telescope" in *Sky and Telescope* (**44**, 190, 1972) which states: "A fine optical system for long-focus observing of bright objects, such as the moon and planets, is the oblique reflector described by ANTON KUTTER two decades ago. It is simple to construct in moderate sizes, yet it has been largely neglected by American amateurs." Full details are given (on pp. 190–191 of the above article), and reference is also made to a 21-page technical treatise by KUTTER on his telescope, issued by the Sky Publishing Corporation, Cambridge, Massachusetts 02138.

Brachytes composed of two spherical mirrors, a concave main and a convex secondary mirror, either compensate the astigmatic effects, proportional to the square of the angle of inclination and proportional to the aperture ratio (anastigmatic arrangement), or they compensate the coma error, proportional to this angle and to the square of the aperture ratio (coma-free arrangement, see Fig. 2–16). Both errors cannot be avoided by simply changing the tilt of the secondary mirror S_2 against the main mirror S_1. It is, however, possible to build absolutely error-free brachytes. KUTTER proposes for the amateur astronomer's workshop two out of four different possible types.

The anastigmatic arrangement can eliminate the remaining coma error by

(1) deformation of the secondary mirror S_2 (see *Sky and Telescope* Bulletin A, 1959),

[40] KUTTER, A.: Der Schiefspiegler, ein Spiegelteleskop für hohe Bilddefinition. Weichardt, Biberach, a.d.R., 1953, and Mein Weg zum Schiefspiegler, *Astro-Amateur*. Zürich-Stuttgart: Rascher-Verlag, 1962.

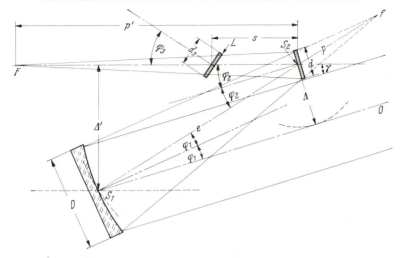

Fig. 2–16 The path of the rays in the oblique reflector. Main mirror S_1, secondary mirror S_2. The lens L only appears in the catadioptric arrangement. From A. KUTTER: Mein Weg zum Schiefspiegler, *Astro-Amateur*, p. 54. Zürich-Stuttgart: Rascher-Verlag, 1962.

(2) restricting the aperture of the main mirror to 8 cm or less, whereby the remaining coma of $1''7$ in the diffraction disk disappears [see Eq. (14)], or

(3) reducing the main-mirror aperture ratio to less than 1/12, whereby both the coma originating at the main mirror as well as the remaining coma are very much diminished (see Table 2.13).

Table 2.13. A Catadioptric Brachyte of 200 mm Aperture

Outer diameter of the secondary mirror	$d' = 100$ mm
First axial distance	$\Delta = 150$ mm
Inclination of the axis of the main mirror to the incoming rays	$\varphi_1 = 3°09'15''$
Second axial distance	$\Delta' = 425$ mm
Inclination of the secondary mirror to the incoming rays	$\varphi_2 = 9°04'22''$
Deviation of the eyepiece axis from the incoming ray bundle	$Y = 11°50'13''$
Distance of the plane surface of the correction lens from the apex of the secondary mirror	$s = 745$ mm
Correction lens made of SCHOTT BK 7 glass	$n_d = 1.517\ v_d = 64.2$
Diameter of the correction lens	$d_3 = 90$ mm
Focal length of the correction lens	$f_3 = 29$ m $\pm 3\%$
First radius of curvature of the correction lens	$r_3 = \infty$
Thickness of the correction lens	$i = 6-7$ mm
Second radius of curvature of the correction lens	$r_4 = 15$ m
Wedge error of the correction lens	$\delta = 0.05-0.06$ mm
Angle of inclination of the correction lens	$\varphi_3 = 30° \pm 8°$
The correction lens reduces p' and F to $p_1 = 1725$ mm and $F_1 = 4000$ mm, respectively	p' on $p_1 = 1725$ mm F on $F_1 = 4000$ mm

Table 2.14. Three Brachytes in an Anastigmatic Arrangement with Spherical Main and Secondary Mirrors

D	80	110	150
f_1	960	1620	2550
r_1	1920	3240	5100
e	548	919	1481
d	40	55	70
$-f_2$	1000	1750	2720
$-r_2$	2000	3500	5440
d'	50	70	82
p	412	701	1069
p'	698	1169	1781
F	1600	2700	4250
Δ	65	89	116
Δ'	144	214	277
φ_1	3°24′20″	2°47′00″	2°15′00″
φ_2	8°06′30″	6°42′50″	5°23′20″
Y	9°24′20″	7°51′40″	6°16′40″

Table 2.14 summarizes all linear and angular data for a homemade construction of cases 2 and 3 above.

Here all linear measures are in millimeters, all angles in degrees, minutes, and seconds. The meaning of the letters is obvious from Fig. 2–16 and Table 2.13. Instead of the D values of the table we can also choose some smaller values D', e.g., $D' = 60, 90, 130$. In this case all the linear values of the columns 1, 2, and 3 must be multiplied with 60/80, 90/110, and 130/150, respectively, while the angular values remain unchanged.[41]

Until recently the coma-free arrangement became catadioptric by the insertion of a small cylindrical lens placed in front of the focus F, which removed the remaining astigmatism (see page 7). In a more recent but well-tested design, the secondary mirror is tilted in such a way that both errors are present—the coma undercorrected and the astigmatism overcorrected. These two remaining errors are then removed with the help of a long-focus, plane-convex, spherical lens L. This lens makes this type of brachyte into a catadioptric system and KUTTER calls it his catadioptric arrangement. All linear and angular values for a catadioptric brachyte of the aperture $D = 200$ mm of the main mirror are compiled in Table 2.13. KUTTER remarks that, if we exceed the desired aperture $D = 22$ cm, the main mirror cannot any longer be spherical, since otherwise the system would be affected by in-admissably large remaining amounts of spherical aberration. In this case we would have to deform the main mirror to an ellipsoid. The development

---

[41] Further reference may here be made, for instance, to the advertisement pages of *Sky and Telescope*, published monthly by the Sky Publishing Corporation (49–51 Bay State Road, Cambridge, Massachussetts 02138).

Table 2.15. Summary of Errors for the Ideal Spherical Mirror $(f = 1000)^a$

$\dfrac{d}{f}$	Field of view in degrees $\dfrac{1}{2}$	$\dfrac{2}{2}$	$\dfrac{3}{2}$	$\dfrac{4}{2}$	$\dfrac{5}{2}$	$\dfrac{6}{2}$	
$\dfrac{1}{20}$	0.0039	—	—	—	—	—	B'
	0.0005	0.0019	0.0043	0.0076	0.0172	0.0172	C'
	0.0020	0.0041	0.0061	0.0082	0.0102	0.0123	F'
$\dfrac{2}{20}$	0.031	—	—	—	—	—	B'
	0.0010	0.0038	0.0086	0.0152	0.0238	0.0343	C'
	0.0082	0.0164	0.0246	0.0328	0.0410	0.0492	F'
$\dfrac{3}{20}$	0.105	—	—	—	—	—	B'
	0.0014	0.0057	0.0129	0.0229	0.0358	0.0515	C'
	0.0184	0.0369	0.0553	0.0737	0.0922	0.1106	F'
$\dfrac{4}{20}$	0.250	—	—	—	—	—	B'
	0.0019	0.0076	0.0172	0.0305	0.0476	0.0686	C'
	0.0328	0.0655	0.0983	0.1311	0.1639	0.1966	F'
$\dfrac{5}{20}$	0.488	—	—	—	—	—	B'
	0.0024	0.0095	0.0215	0.0381	0.0596	0.0858	C'
	0.0512	0.1024	0.1536	0.2048	0.2560	0.3072	F'
$\dfrac{6}{20}$	0.844	—	—	—	—	—	B'
	0.0029	0.0114	0.0257	0.0458	0.0715	0.1029	C'
	0.0737	0.1474	0.2211	0.2948	0.3685	0.4423	F'

NOTE: B' = the diameter of the dispersion circle of the spherical aberration; C' = one half the diffusion by the curvature of the image, F'-diffusion through coma; $C'+F'$ gives the meridianal extent of the dispersion figures due to image curvature and coma. "In compiling this list it has been assumed that the diaphragm is situated at the centre of the mirror surface. In practice, however, (see e.g. Fig. 2.15) it lies at or near the focal point of the mirror," (Sonnefeld). Hence the values shown above are ideal, limiting values.

^{a} SONNEFELD, A.: The Hohlspiegel. Berlin: VEB Verlag Technik, and Stuttgart: Berliner Union, 1957.

of KUTTER's brachyte, which can be reached in the homemade instruments of an amateur astronomer, have been described by KNAB[42] and BUCHROEDER.[43]

It has been demonstrated in Table 2.11 that the size of the usable field of a parabolic mirror increases when the aperture ratio decreases. It is therefore logical that an observer tries to reduce the aperture ratio of his parabolic

[42] KNAB, O. R.: An improved 4¼-inch unobstructed oblique reflector. *Sky and Telescope* **22**, 232–235 (1961).
[43] BUCHROEDER, R. A.: A new three-mirror off-axis amateur telescope. *Sky and Telescope* **38**, 418–423 (1969).

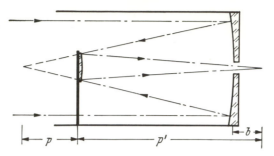

Fig. 2–17 Cassegrain reflecting telescope. From *Handbuch der Astrophysik*, Vol. 1, p. 173. Berlin: Springer, 1931.

mirror in order to achieve a larger usable field. In order to avoid the lengthening of the tube with increasing focal length two systems are recommended— both have been known for some 200 years. Either we place in front of the main mirror, with its hole, a small mirror with a convex surface (this leads to a *Cassegrain telescope*) or we place behind the focus of the same mirror a small mirror with a concave surface (this produces a *Gregorian system*). Figures 2–17 and 2–18 illustrate the resulting extensions of the focal lengths and the position of the focus behind the main mirror. Through the attachment of a second small mirror in front of the main mirror we are able to fold the beam of the rays so that the focus is now outside the tube in front of the main mirror (Nasmyth-focus). The focal length of such a composite optical system is called the *equivalent focal length*.

If f is the focal length of the main mirror and if it is intended to place the eyepiece at the distance b behind the apex of the main mirror, we obtain the position of the secondary mirror at the distance p from the focus of the main mirror from

$$p \text{ (Cassegrain)} = \frac{f+b}{A+1} \qquad A = 3$$

$$p \text{ (Gregory)} = \frac{f+b}{A-1} \qquad A = 5.$$

The radius of curvature, r, of the secondary mirror is

$$r \text{ (Cassegrain)} = \frac{2pp'}{p'-p}$$

$$r \text{ (Gregory)} \times \frac{2pp'}{p'+p}.$$

The diameter d' of the secondary mirror must not be so large that strong vignetting is caused; for this reason A has been fixed at three and five, respectively. Other A-values lead to other focal-length extensions. The diameter of d', geometrically determined, is

$$d' = \text{aperture ratio of the main mirror} \times p.$$

In fact it is necessary to make this diameter d' somewhat larger, since otherwise light rays would not be exactly reflected from the rim. As a rule an increase of 1 cm will be sufficient. If a Newtonian mirror is converted into a Gregorian system, the gain in focal length is larger than in conversion into a Cassegrain mirror. The Cassegrain focal length is equal to the Gregorian focal length minus $2p$; however, the length of the tube is also larger by $2p$.

If the main mirror is parabolic, then the Cassegrain mirror must have a hyperbolic secondary mirror, and the Gregorian telescope an elliptical secondary mirror. Furthermore, the main mirror can also deviate from the shape of a paraboloid; in this case the secondary mirror must be adapted during the process of grinding and polishing to this other form. Secondary mirrors should be somewhat polished on the reverse side to make it possible to test them from the back, just as convex mirrors are tested for uniformity, i.e., not for the correctness of the radius of curvature.

The transition from the Cassegrain focus to the Newtonian focus through the removal of the secondary mirror and the insertion of another secondary can be accomplished in a few minutes.

The rapid transition from Newtonian to the Cassegrain focus requires a very good and solid setup of the secondary mirror. Some observers give up the idea of drilling a hole into the main mirror and rather prefer the folded path of the rays. The reason is that the eyepiece is then more easily accessible and the main mirror is not so much exposed to the body heat of the observer so that the uncontrollable deformations of the main mirror during the observations can be avoided. A supplementary mirror changes the direction of the image (right-hand and left-hand sides become interchanged). A pentagon prism avoids this effect. The main, secondary, and supplementary mirrors must be very carefully supported for every mirror arrangement and equipped with adjusting screws. (See page 73, etc.).

Cassegrain mirrors can be adjusted as follows: we mark the entrance opening of the tube with two threads and place the eye at this center; the mirror image of the secondary mirror and of its support must be seen exactly at the center of

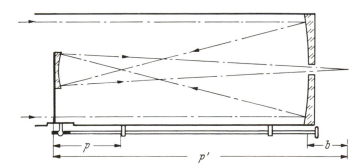

Fig. 2–18 Gregorian reflecting telescope. From *Handbuch der Astrophysik*, Vol. 1, p. 173. Berlin: Springer, 1931.

the main mirror, otherwise the position of the main mirror has to be changed accordingly with the help of an adjustment screw. The eyepieces are removed, the eye views through the eyepiece tube to check whether the center of the secondary-mirror supports really appear at the center of the secondary-mirror mounting when such a mirror is used. Now the secondary mirror is inserted, we look again through the eyepiece tube: the eye must be reflected at the center of the secondary mirror and all outlines must be seen symmetrically in these mirrors. Then the eyepiece is inserted and the final adjustment is carried out.

J. L. RICHTER[44] described a test method for Cassegrain secondary mirrors which can be carried out without additional optical components.

The use of the so-called "Cheshire squaring-on eyepiece" by COX[45] can also be recommended.

Another method for adjusting the telescope is as follows: the eyepiece is replaced by a fine line grating with 2 to 10 lines per millimeter, which is placed in the focus and behind this a parallel, brightly illuminated slit. The eye, close to it, must see parallel stripes on the secondary mirror and if not the fine adjustment must be continued. If the stripes never become parallel then the main mirror and secondary mirror are not correctly adapted to each other. This grating method (the so-called RONCHI *test*) is a very sensitive test and can also be applied to single mirrors.[46]

The light from the sky which enters the eyepiece of a Cassegrain mirror (false light) and which causes the "*day blindness*" can, according to NÖGEL, be completely obliterated if one places a diaphragm behind the eyepiece just where the image of the main mirror originates on the projection screen; aperture of diaphragm is then equal to the diameter of the image. Similarly according to NÖGEL, it is possible to avoid the *reflection images of the Moon* during observations of occultations by using a ring of the size of the secondary mirror as a cover, and to place it in the middle of the main mirror, leaving the hole free.[47]

Newtonian and Cassegrain mirrors can be obtained with or without mounting from most of the manufacturers mentioned on page 41, as well as from the numerous firms advertising in the *Journal of the British Astronomical Association*, *Sky and Telescope*, and other astronomical periodicals.

[44] RICHTER, J. L.: A test for figuring Cassegrain secondary mirrors. *Sky and Telescope* **39**, 49–53 (1970).
[45] COX, R. E.: Notes on telescope making from here and there. *Sky and Telescope* **35**, 319–324 (1968).
[46] See *Amateur Telescope Making*, Book 1, p. 264. New York: Scientific American, Inc., 1953.
[47] NÖGEL, O.: Verlustfreie Beseitigung der Tagblindheit des Cassegrain-Teleskopes. *Die Sterne* **34**, 141 (1958). Another method using an adjustable diaphragm has been recommended by G. WALTERSPIEL: Der Sternfreund und sein Instrumentarium No. 16, Monatsbeilage zu dem Nachrichtenblatt der Vereinigung der Sternfreunde, **9**, Nos. 1–2 (1960).

The imaging errors which occur in parabolic mirrors can be more or less suppressed if a *reflector corrector* is used as suggested by BAKER. It consists of a correction plate with a positive achromatic lens in its middle and is positioned between the main mirror and the focus so that we obtain a nearly flat field of view without disturbing coma or astigmatism. We also mention the lens system proposed by ROSS, which serves for the same purpose.[48]

2.2.5.3. The Mirror-Lens Instruments of Schmidt, Bouwers, and Maksutov, and Schupmann's Medial Telescope

The instruments described above are, because of their small field of view, mainly used for the observation of single objects of small angular size. They are instruments suitable for the amateur observer and are the telescopes with which the astronomers of the past centuries have made their discoveries. The reflectors described below have until recently mainly served for photographic work of large areas of the sky (star fields, cosmic clouds, and nebulae). The evaluation of such plates nearly always requires the instrumental and theoretical tools of the professional astronomers. However, the development of optical instruments makes rapid progress and we should not omit any possible stages of this development. Furthermore, celestial photography gains more and more friends and the art of telescope making becomes more and more popular among amateur astronomers, so that for this reason alone some of the following instruments deserve to be mentioned.[49]

The use of newly developed photographic cameras began in 1931 when BERNHARD SCHMIDT (1878–1935) pointed his new camera to the sky for the first time. We will now discuss some details of the so-called "Schmidt camera": very narrow parallel pencils of light, with axis going through the center of curvature of a spherical mirror, are combined after reflection in focal points whose distance from the spherical mirror are the same as half the radius of curvature r of the latter. They lie, therefore, on a spherical surface which is concentric with the reflecting spherical mirror and has a radius of curvature $r/2$. If these narrow parallel bundles are extended to finite parallel bundles, then none of them is focused in one point but all show after reflection a spherical error (see page 32). If we succeed in collecting the rays that are near the axis of the bundle, and disperse those that are further away from this axis, it should be possible to eliminate the spherical error. The problem is to admit only such bundles of light that go through the radius of curvature of the spherical mirror and to change the conversion of the rays of the finite bundle. SCHMIDT has solved this problem by inserting in the optical system a correction plate above the spherical mirror at the distance of its radius of curvature. This plate has a plane front surface and an aspheri-

[48] *Amateur Telescope Making*, Book 3, p. 1. New York: Scientific American, Inc., 1953.
[49] Material for grinding and polishing can be obtained in small quantities, e.g., from the Astronomische Arbeitsgruppe, Schaffhausen, Switzerland, or from the Schweizerische Astronomische Gesellschaft, see page 550, and we may also refer, e.g., to AUSTIN ROBERTS and Co. (73 Woodchurch Lane, Birkenhead, Cheshire, L42 9PL, England) who produce mirror-making kits.

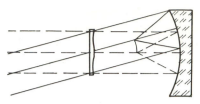

Fig. 2–19 The classical Schmidt telescope. From *Amateur Telescope Making*, Book 3, p. 359. New York: Scientific American, Inc., 1953.

cal back surface of a rather complex shape. It is plane convex in the middle, while toward the rim it is shaped like a plane-concave lens. It refracts the light and thus introduces the known imaging errors of a lens. However, all these errors are so small that they can be neglected. A Schmidt mirror can therefore justifiably be called a spherical reflector in which there are no obliquely incident bundles of light and in which the errors of spherical aberration are eliminated. (Fig. 2–19).

Its inventor has not only the merit to have found the principle of this mirror system but also to have indicated the process of the production of differently shaped correction plates and tested it. This method—grinding of the plate which is curved above a vacuum space—need not be applied any longer; recently other more convenient procedures have been described.[50]

The deviation Δ of the surface of the corrector plate of radius r, refractive index n, and zone radius x is, if R is the radius of the spherical mirror,

$$\Delta = \frac{x^4 - Rr^2x^2}{4(n-1) \cdot R^3}.$$

The constant k being taken as 1 we obtain a curve for the surface in which middle and rim are equally high; at $x = 0.707\ r$ the plate becomes plane-parallel. The angular diameter α of the field of a Schmidt telescope, having the diameter d' of the plate holder and the focal length f, is given by

$$\sin\frac{\alpha}{2} = \frac{d'}{2f}.$$

If d is the diameter of the corrector plate, the diameter of the spherical mirror must be

$$D = d + 2d',$$

if we want to achieve the greatest light-grasp.

As material for the corrector plate "Schottglas U.B.K.5" has been recommended. The testing of the plate is carried out by placing it immediately on top of the spherical mirror; we then measure the radius of curvature of each zone x with the help of the knife-edge test. If we move simultaneously blade and source of light we must have

$$\Delta R = \frac{x^2}{R}.$$

[50] *Amateur Telescope Making*, Book 3, p. 354. New York: Scientific American, Inc., 1953.

The curvature of the imaging plane can be compensated with a plane-convex lens, whose front surface has a radius of curvature $f/3$ (for $n = 1.50$); see page 43. The disadvantages of the SCHMIDT telescope are its relatively great length and the reflections which occur on the back of the corrector plate and perhaps also on the flattening lens.

The great length of the SCHMIDT telescope has been avoided in *modifications* by VÄISÄLÄ and by WRIGHT. Both of them place the corrector plates nearer to the spherical mirror and have deformed the latter in such a way that the coma due to the shortening was eliminated. The shortening can go so far that the corrector plate distance becomes equal to the focal length.[51] Then the image plane will be flat and much larger than in the classical SCHMIDT mirror. The performance of the latter as to definition of the image, however, remains unsurpassed.

> The SCHMIDT-BAKER mirror in its various modifications will not be dealt with at this time. The same applies to the modern coma-free mirror systems of SONNEFELD, SLEVOGT and RICHTER, WHIPPLE, and WYNNE and to the monocentric meniscus mirror of MAKSUTOV.[52]

Independent of each other, MAKSUTOV and BOUWERS discovered the advantageous use of meniscus lenses for *coma-free mirror systems*. In these the meniscus has the task of nearly compensating, first of all, the spherical error of the main mirror. In order to eliminate the small extent of this error the lens requires hardly any refractive properties and practically no difference in thickness. Meniscus lenses using nearly all these modifications are placed in front of the spherical mirror within the parallel beam of light, i.e., light passes through it before it arrives at the mirror. Therefore the tube is closed and air circulation is excluded. The astigmatism and coma errors can be easily compensated for, if a diaphragm is placed above the main mirror at a distance of the radius of curvature, or otherwise they can well be neglected. The image field is not flat and may lie between the spherical mirror and the meniscus, behind the holed spherical mirror, or in front of the meniscus, which is then within the nonparallel beam of light. We thus obtain mirrors which have only a superficial similarity with those of the Newtonian, Cassegrain, and Gregorian types, but are much smaller and possess only spherical surfaces.

Two of the proposed and realized systems appear here to be of importance:

The Improved Herschel Telescope. By tilting the spherical mirror it is possible to place the focus outside the tube. We obtain a telescope free from vignetting, similar to the one used by HERSCHEL. (See page 58.) HINDLE claims that his 10-inch telescope of this construction has the same quality in

---

[51] See WAINEO, T. J.: Fabrication of a Wright telescope. *Sky and Telescope* **38**, 112–118 (1969).

[52] See RIEKHER, R.: *Fernrohre und ihre Meister*. Berlin: VEB Verlag Technik, 1957.

observations of the Moon as the 24-inch telescope of the Lowell Observatory. The aperture ratio of this telescope, states Hindle, should not exceed 1/9. The most suitable aperture of the diaphragm must be found out experimentally. Only achromatic eyepieces should be used. As to the grinding of the meniscus, Hindle gives some practical instructions, for instance:[53]

Diameter of aperture	305 mm
Focal length	2745 mm
Borosilicate glass for the meniscus, with $n = 1.50$	
Diameter	63.5 mm
Thickness	9.5–11.5 mm
r (concave)	762 mm
r (convex)	757.2 mm

The Cassegrain-Maksutov Telescope. In this design the meniscus approaches the spherical mirror to within a smaller distance than the focal length, so that the light is reflected from the center of its far side. The central zone is sputtered with a reflecting coating and reflects the light through the hole in the spherical mirror so that the focus is behind this mirror.

Figure 2–20 and the following data illustrate the situation.[54]

		$f/15$ (inches)	$f/23$ (inches)	$f/2$ (mm)
D_m	Diameter of the spherical mirror	6.6	6.4	(180)
D_c	Diameter of the meniscus	6.45	6.26	(140)
T_c	Thickness of the meniscus	0.541	0.448	(14)
D_s	Diameter of the secondary mirror	1.4	1.2	
T_m	Thickness of the spherical mirror	1.1	1.1	(30)
D_h	Size of the hole in the spherical mirror	1.375	1.17	
R_1		−6.583	−8.040	(−135.6)
R_2		−6.888	−8.293	(−143.5)
R_3		−30.83	−43.39	(−583.0)
R_4		−6.888	−8.293	
C_a	Diameter of the free meniscus	6.0	6.0	(80)
B_a	Diameter of the back of the meniscus	6.3	6.11	
S	Distance from the center of the spherical mirror to the apex of the meniscus	12.867	18.541	(341)
BFL	Focal length of the secondary mirror	17.16	23.05	
	Total (effective) focal length	90.00	140.00	(280)

The design of the Maksutov camera follows similar lines. Instead of the Schmidt corrector plate the Schmidt camera is equipped with a meniscus lens with spherical surfaces. The curvature of the film is concentric with the mirror behind the meniscus. The values in parentheses in the last column of the table

---

[53] Hindle, J. S.: An improved Herschelian telescope. *Sky and Telescope* 12, 107 (1953).
[54] See Gregory, J.: A Cassegrainian-Maksutov telescope design for the amateur. *Sky and Telescope* 16, 237 (1957).

above are the data for a camera with a meniscus aperture of 140 mm. The distance from the apex of the curved film to the center of the spherical mirror is 298 mm; the meniscus lens is made of crown glass with $n = 1.516$.[55]

The firm of Dr. J. HEIDENHAIN (Traunreut/Obb., Germany) supplies a Maksutov mirror 300/3000 mm in fork mounting with observation arrangement from the side.

Finally, we mention the catadioptric system of SCHUPMANN's medial. It has been known since 1899 and has been very effective in the hands of the lunar observer PHILIP FAUTH. Recently it has again been used in the United

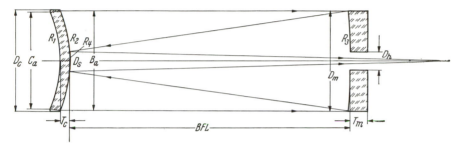

Fig. 2–20 The Cassegrain-Maksutov telescope. From *Sky and Telescope* **16,** 237 (1957).

States. It possesses a simple objective, whose errors due to deviations from the spherical shape, color deviations, and coma are compensated by having a lens in front of a mirror or by a MANGIN mirror (a spherical mirror reflecting from its back). The chromatic correction of the medials corresponds, according to SONNEFELD, to telescopes with semiachromatic objectives of the type AS.[56] There is some astigmatism, but in the case of visual observations, for which the medial is used, this is irrelevant.

2.2.6. Eyepieces

2.2.6.1. The Focal Length, Image Distance, and Image Size of a Composite System

The eyepiece and the objective lens together form a telescopic system, i.e., an optical system without focus. The light from every star leaves the eyepiece as a parallel bundle. The diameter of this light bundle determines the minimum and maximum magnification for certain purposes (see page 11). Alternatively we can say that the eyepiece serves as a magnifying lens for viewing the image of the object which has been produced in the field of view by the objective lens. The apparent angle of view of this image is

$$\text{Tangent of the angle of view} = \frac{\text{size of the image in mm}}{\text{focal length of the eyepiece in mm}}. \quad (38)$$

[55] See ORION. *Mitteilungen der Schweizerischen Astronomischen Gesellschaft* **61,** 457 (1958).
[56] See SONNEFELD, A.: *Die Hohlspiegel.* Berlin: VEB Verlag Technik, and Stuttgart: Berliner Union, 1957.

The focal length of an eyepiece, composed of a front lens (field lens) of focal length f_1 and a back lens (eye lens) of focal length f_2, separated by a distance δ, is

$$f' = \frac{f_1 f_2}{f_1 + f_2 - \delta}. \tag{39}$$

Here f' is the equivalent focal length of a two-lens system; it is equal to the focal length of the so-called equivalent-lens; δ is the distance between the cardinal points H_1, H_2. The position of the cardinal points and of the principal planes of the lenses is shown in Figs. 2–1 and 2–2. The composite eyepiece has two cardinal points H and H'. The distance a of the object must be measured from H and the distance a' of the image from H'. The position of the cardinal points H and H' can be calculated from

$$\overrightarrow{H_1 H} = \frac{f_1}{f_1 + f_2 - \delta} \qquad \overrightarrow{H_2' H'} = \frac{(f_1 - f_2) - f_1 f_2}{f_1 - f_2 - \delta}. \tag{40}$$

For a given object distance a we obtain

$$\text{The distance of the image } a' = \frac{af'}{a + f'}, \tag{41}$$

$$\text{The size of the image } b = \left(\frac{a'}{a} \right)$$

$$\times \text{ the size of the image as produced by the objective.} \tag{42}$$

For short eyepieces (orthoscopic, monocentric eyepieces) we can measure f', δ, a, and a' with rough approximation from the apex of the field or eye lens.[57]

2.2.6.2. The Various Types of Eyepieces

The Huygens Eyepiece. It was HUYGENS who found by experimenting with lenses made of the same glass that the color deviation is removed if $\delta = \frac{1}{2}(f_1 + f_2)$; see page 34. Formerly one took $f_1 = 3 \times f_2$, while today we take $f_1 = 2f_2$. The eyepiece does not produce any reflections and is free from distortion and from a color difference in the color magnification; the eye does not notice the difference of the colored images at average and small aperture ratios. The image field is strongly curved, convexly toward the eye. The usable apparent size of the field of view is therefore only 25–40°. Since the image from the objective is situated between field lens and eye lens, it is necessary to place cross wire and micrometer wires at such a place and they would consequently appear with colored borders. The eyepiece cannot be used as a magnifying lens, it is a "*negative eyepiece.*" The transparency is 93%. The eye must be very close to the eye lens and the eye lashes are easily

[57] See WÜSTEMANN, E.: Mikroskopobjektive als Fernrohrokulare. *Sterne und Weltraum* **3**, 67 (1963), and DALL, H. E.: Telescope eyepieces. *J. Brit. Astron. Assoc.* **78**, 282 (1968).

in the way. This eyepiece can easily be built, but it is not very suitable for daylight observations [Fig. 2–21(a)].

The Mittenzwey Eyepiece. This is an improved HUYGENS eyepiece. In place of the plane-convex field lens it makes use of a meniscus, so that we obtain a large field of view of 50°. It is very suitable in a normal refractor (of a focal ratio 1:12 or less) for observations of comets and large star clusters.

The Ramsden Eyepiece. This is a *"positive eyepiece"* which can be used as a magnifying glass. Cross wire and micrometer are placed in front of this eyepiece. It is free from distortion but not free from the color difference of

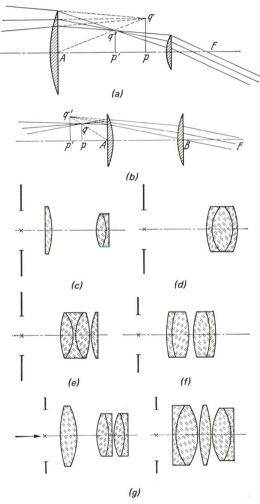

Fig. 2–21 (a) The HUYGENS eyepiece. (b) The RAMSDEN eyepiece. (c) The KELLNER eyepiece. (d) STEINHEIL's monocentric eyepiece. (e) ABBE's orthoscopic eyepiece. (f) PLÖSSL's orthoscopic eyepiece. (g) The wide-angle eyepieces designed by ERFLE with 70° field of view. From *Handbuch der Astrophysik*, Vol. 1, pp. 151–153. Berlin: Springer, 1931.

the "colored magnification." That is, image size differs, since the focal
lengths are different for the different colors. This error is not noticeable for
large aperture ratios. It could be avoided if the field lens were moved to the
position of the objective image; however, then the scratches and spots on the
field lens would be visible at the eye lens. The field of the image is slightly
curved, convexly toward the eye. The apparent size of field can be 25–40°.
The transparency is 93%. Because of the small spherical aberration, it is
suitable for micrometer measurements. It is easy to build; see Fig. 2–21(b).

  The Kellner Eyepiece. This is an improved RAMSDEN eyepiece which
removes the color errors of the latter with the help of an achromatic eye lens
made up of two components [Fig. 2–21(c)].

  The Achromatic Eyepiece. The monocentric eyepiece of STEINHEIL, the
orthoscopic eyepiece of ZEISS, the orthoscopic eyepiece of SCHIECK, and the
aplanatic eyepiece of PLÖSSL are distinguished by great purity of color and
greater distance of the exit pupil from the eye lens, as well as being free of
reflections. They are necessary for large magnification and for observation
of fine detail on planets; see Figs. 2–21(d), 2–21(e), and 2–21(f).

  The Massive Eyepieces. We also mention the massive eyepieces proposed
by C. S. HASTINGS, which consist of a very thick crown glass lens that has
been covered with a cemented thin flint glass lens. They are classed as
"positive eyepieces," have a plane field of view, and are free from reflections
and colors.[58]

  The Terrestrial Eyepieces. The terrestrial eyepieces by DOLLOND and
FRAUNHOFER consist of four plane-convex lenses which cause hardly noticeable imaging errors. The modern terrestrial eyepieces, as they are found in
most field glasses, possess instead of the two single front lenses two 2-lens
achromatic lenses. In field glasses of American make there are 3-lens eyepieces to be found. The *wide-angle eyepieces* in ZEISS binoculars increase the
diameter of the field of view as compared with the ordinary eyepieces by
nearly 40% [see Fig. 2–21(g)].

  Field glasses can be used as eyepiece systems, providing a complete erect-
ing and focusing device as has been demonstrated by FITZGERALD[59] with a
double telescope.

  The Eyepieces in the Galilean Telescope. These are two 2-lens or 3-lens
negative eyepieces that, together with a 2-lens objective, produce a telescopic
arrangement (see page 9).

  Testing of the Eyepieces. The astronomer tests his eyepieces at the telescope
by observing stars. Faults and decentering, etc., show themselves if the eyepiece is turned. Exchanging eyepieces demonstrates the quality and the
performance of any of them.

[58] *Amateur Telescope Making*, Book 1, p. 172. New York: Scientific American, Inc.,
1953.
  [59] FITZGERALD, B. L.: *Sky and Telescope* 37, 186–187 (1969).

2.2.6.3. Solar Eyepieces

Solar eyepieces must be able to absorb the very intense sunlight to such a degree that it becomes equal, at the most, to the light of the Full Moon. Since the apparent magnitude of the Sun is $-26^{m}86$ and that of the Moon $-12^{m}7$, the sunlight must be reduced by at least 14.2 magnitude classes. This is done by utilizing the reflection of light on transparent bodies. Also polarization effect at the reflection can be used, by crossing the polarizing components and obtaining a continuous weakening of the light. We must take into account the fact that each reflection leads to an inversion of the image, that is, to an interchange of right and left. It is also necessary to render harmless the sunlight which is not used for the observation.

If the sum of the angle of incidence and of the angle of refraction is 90°, then the fraction of the light which is reflected by the glass is

$$\left(\frac{n^2-1}{n^2+1}\right)^2 = 0.15,$$

for the refractive index $n = 1.50$. If we have a perpendicular incidence of the light, then the reflected fraction becomes

$$\left(\frac{n-1}{n+1}\right)^2 = 0.04.$$

In the first case 15% is reflected, in the second 96% passes through the surface of the glass. The first case leads in practice to a larger loss of light. After two reflections we achieve a reduction by 4.1 classes of magnitudes and after 4 reflections by 8.2 magnitudes since

$$m = 2.5 \times \log(0.15^2) \qquad \text{and} \qquad m = 2.5 \times \log(0.15^4).$$

This reduction of the light is increased if the objective lens is diaphragmed. If this arrangement decreases the aperture to $d/4$ the light is reduced by $3^{m}0$. A diaphragm which reduces the aperture to $d/5$ corresponds to $3^{m}5$. If we require a particularly effective weakening of the light we can sputter coats of reflection-reducing material behind the reflecting surfaces.

Concerning the various solar eyepieces which are now in use we refer to page 223.

2.2.7. The Tube and Its Accessories

2.2.7.1. The Tube

The tube of the telescope has the task of holding and of keeping in adjustment all optical parts including the photographic camera and the photoelectric cell. It must protect these components from rapid temperature variations, humidity, air disturbances, vibrations, and undesired light. Finally, it must be possible to move it in connection with the mounting so that it can be directed toward every point of the sky and be able to follow,

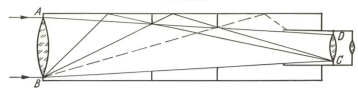

Fig. 2–22 Graphical determination of the diaphragms in a refractor (schematic). After Kolbow, in J. Plassmann's *Hevelius: Handbuch für Freunde der Astronomie*, p. 485. Berlin: Dümmler, 1922.

through the guiding of the observer, any celestial object. Sometimes we require only the observation of points of more than 15° altitude above the horizon. The *length of the tube* and its *diameter* are determined by the focal length of the objective or of the main mirror and by the size of the usable field of view. The latter is equal to the size of the photographic plate or the field of view of the eyepiece of the smallest magnification. For eyepieces that are not of the wide-angle type we can assume the field of view to be equal to one half of the focal length. The best procedure is to draw a cross section through the tube (see Figs. 2–22 and 2–23). The size of the true field of view in degrees follows from the size of the usable field of view in mm and the focal length of the objective or the mirror in mm; see formula (21).

The raw material used for the tube by the amateur astronomer ranges from boiler pipes to a wooden skeleton covered by cardboard. Metal tubes should be insulated with linen, cork, etc. Structures made of wood should be enclosed in well-glued wood in the form of a box of square cross section. This provides space for additional mounting of finder, diopter, and cameras. The tube should be strengthened so that it does not vibrate in the wind. It is suspected that air currents at the outer and inner wall of the tube affect the quality of the outer zones of the objective or mirror very unfavorably. For

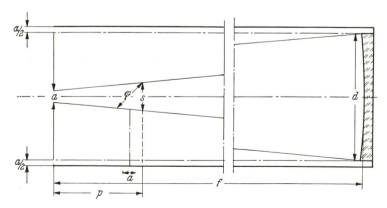

Fig. 2–23 Graphical determination of the position and size of secondary mirror or prism.

this reason the diameter of the tube should be somewhat larger than that required by the light path. The question of whether an open or a closed tube should be used is answered by different authorities differently.

As examples of strengthened but still light tubes we mention NöGEL's construction consisting of four disks of light metal, which are connected by rods of the same material, having circular holes for the insertion of the optical parts. See page 46. We also mention SCHIFFHAUER's design[60] consisting of a seamless PVC tube manufactured by Kunststoffröhren Wavin, Meppen. These tubes are manufactured up to a diameter of 300 mm, have the specific gravity 1.38, a temperature coefficient of 0.00006 m/T, and a wall thickness of 8.8 mm. In working and treating this material we have to take into account the characteristic properties of this material. In particular, reference should be made to the articles in the columns "Gleanings for ATM's" of *Sky and Telescope* (see also page 549).

The objective and main mirror must have a mounting or bearing in which they can be kept at every position of the tube without tension and displacement. This mounting or bearing should be adjustable within or outside the tube. Small mirrors are supported at three points. These lie on a circle with a diameter equal to (diameter of mirror)/$\sqrt{2}$. The mirrors are supported in a "cell" the bottom of which contains a hole (see page 57). Little pieces of leather will prevent a displacement of the mirror to the sides. The mirror cell sits on a base plate with a hole, which is attached to the tube and can be adjusted. The hole in the base plate and in the mirror cell makes it possible— together with the partly polished back of the mirror and its uncoated central front surface—to perform rapidly and safely for Schmidt mirrors the first rough adjustment.

2.2.7.2. The Dewcap

This attachment will prevent misting of the objective glass and keep off any scattered light. Various procedures have been proposed to prevent misting of the optics or to remove the condensed water:

(1) The objective must be clean, free of dust and condensation nuclei.

(2) The dewcap should be very long, cylindrical- or conical-shaped, and lined with blotting paper or with felt, cloth, etc.

(3) An electrical heating arrangement with a resistance should provide a uniform very small heating of the dewcap, the secondary and main mirrors, etc.

(4) Misted objectives should be cleared by the insertion of a warm cloth into the dewcap.

(5) Wet optics should never be wiped. Slight misting can be removed after the end of the observation by fanning, strong misting by a hair dryer or with a small bellow.

---

[60] SCHIFFHAUER, H.: Die Verwendung von Kunststoffröhren im Fernrohrbau. *Sterne und Weltraum* 7, 158 (1962).

(6) A tube with wet optics should not be closed but only covered with a moderately thick cloth. Next day the tube should be exposed in the open dome or outside to fresh air and dried in this way.

2.2.7.3. The Diaphragms and the Secondary Mirror

The position inside the diaphragms and of the secondary mirror can best be found from a drawing, as illustrated in the two examples of Fig. 2–22 and Fig. 2–23.

(1) We draw to scale the diameter of the objective, focal length, image plane, and plate size, or cross section of the weakest eyepiece, respectively. We then join the upper rim of the objective with a straight line to the upper border of the plate or the upper rim of the field lens and extend the resulting truncated cone at its upper surface by 10%. We obtain $ABCD$. The diameter of the tube should be at least equal to the diameter of the objective mounting. We now draw the first ray which enters the objective at B at an angle of 45°, and then draw from the point where this ray intersects the inner wall of the tube a connecting line to C. Where this line intersects the outer circumference of the cone, $ABCD$ must be the position at the first diaphragm, the free opening of which is determined by the diameter of the cone at this place. The second "disturbing" ray passes B at the upper rim of the opening in the first diaphragm; its point of intersection is connected with C as well. The new point of inner section gives the place and diameter for the second diaphragm, and so on, until we arrive at a connecting line above the photographic plate or the eyepiece. In the preceding discussion we have followed instructions given by H. KOLBOW.

(2) In our second example we draw the diameter of the mirror, focal lens, image plane, and plate size, or cross section of the weakest eyepiece, as we would if we were taking a photograph in the prime focus. The diameter of the tube must be of such a size that the light from the whole field of view can reach the rim of the mirror without any hindrance; see Eq. (24). The secondary mirror must have such a distance from the prime focus that the plane of the image is outside the tube; see also Fig. 2–16. Then the length of the major axis of the oval secondary mirror, or the length of the side of the hypotenuse surface of a reflecting prism, becomes

$$\varphi = \frac{p \cdot d + a(f-p)}{f}.$$

This value has to be increased by about 1 cm, so that rays arriving near the rim of the mirror can be reflected without hindrance.

The *diaphragms of the refractor* are tested by viewing the daylight sky through the objective lens toward the eyepiece. In addition, the eyepiece pupil must be visible from the rim of the objective.

2.2.8. The Finding Telescope

The finder should have a field of view of at least 3°. Furthermore, it should contain a cross wire which is visible against the sky whatever its brightness. It is for this reason that two cross wires are suitable. Some observers have added at the center of this cross a ring in order to emphasize still more the center of the field of view. Sometimes even a very thin coating of luminous paint has been applied to the wires. It is advisable to arrange the finder in such a way that, when viewing through it, we look at the same time in the correct direction to the observed celestial object. The finder eyepiece should be close to the eyepiece of the main tube so that the observer can easily move his eyes from one to the other.

In place of a finding telescope we can also use a simple sight which consists of a large ring with a cross wire and a small ring for the observer's eye, the larger the distance between the rings, the better. One half of a field glass is very useful as a finder. On the other hand field glasses have no cross wire and do not reverse the images. Therefore the relative position of the stars in the finder is very different than in the main tube.

A cross wire with adjustable brightness can be obtained by using according to BLATTNER[61] a small collimator, (a tube with a converging lens at whose focus a pointlike source of light has been placed). Between the lens and the source of light is a cross wire which is fixed on a plane-parallel glass plate. The latter is inclined at 45° to the direction of projection and to the line of vision of the observer. The latter then sees the cross wire projected onto the sky and with the help of a small adjustment can regulate the brightness of the cross wire and by coloring the source of light also the color (dark red) of the wires.

2.2.9. The Mounting and Its Accessories

2.2.9.1. The Structure of the Mounting

The main component of the whole mounting is the optical arrangement with a tube or with a camera. Both have to be selected with a view to the intended observations and the site of the instrument, i.e., whether this is fixed or movable. Thus, focal lengths and aperture, as well as dimensions, weights, and the basis of support for the tube, determine everything else: type and size of the protecting dome for a fixed instrument,[62] packing and transport for a transportable one, and finally mounting and the base. Mounting comprises all components which carry the tube and make it possible for the instrument to be directed to any required point in the sky. The mounting is fixed to the base.

Many experts in various lines of engineering (e.g., R. W. PORTER, A. STAUS, J. TEXEREAU, H. ZIEGLER) as well as many associations of amateur observers and public observatories have helped develop suitable mountings for the amateur; see also the bibliography, page 538, and Fig. 2–24. It is absolutely necessary to study some of this literature if the observer is going to build his own mounting, so that he will receive thorough instructions.

[61] BLATTNER, K.: Reflexvisier für Amateurteleskope. Orion. *Mitteilungen der Schweizerischen Astronomischen Gesellschaft* 5, 33 (1956).

[62] Reference should be made to new designs in the construction of protective domes, by the use of polyester glue as discussed by G. TEUCHERT, Ein neuer Werkstoff im Kuppelbau. *Sterne und Weltraum* 3, 67 (1964).

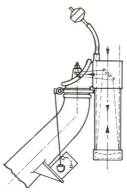

Fig. 2–24 The Springfield mounting of R. W. Porter. The observer always remains at the same place. From *Handbuch der Astrophysik*, Vol. 1, p. 178, Fig. 97. Berlin: Springer, 1931.

A systematic presentation of the various types of mountings—symmetrical kinds (cradle or yoke and fork) and the asymmetrical types (German or Fraunhofer type cross axis, torque tube, weight-stress compensation, inverted fork)—has been published by Tichenor.[63] The special torque-tube mounting, which is applicable to all sizes of instrument, has already been tested for an amateur telescope by Link.[64]

It is important to draw the mounting to scale, including the base and the location. To calculate vibrations, moments of inertia, the degree of flexure, and rigidity, the amateur observer can consult Kohlrausch's *Praktische Physik*, 19th Ed. Leipzig: B. G. Teubner, 1951.

The *base* of the telescope must be in very solid connection with a deep foundation since nearby disturbances easily propagate in layers near the surface. The foundation will be built with concrete or bricks; it will be massive or hollow with irregular substructure and filled with sand, as it can never be too heavy. It must not be in contact with the floor on which the observer moves about.

The *column*, or stand, which is the lowest part of the mounting, has the task of raising the tube to a height where there is undisturbed free vision. The column must be extremely rigid and free from vibration. This is achieved by using a large mass of low elasticity, asymmetrical cross section, or large moment of inertia of this cross section. The column should not be higher than absolutely necessary. One must consider, however, that an uncomfortable posture of the head and body of the observer affects the accuracy of observations with a refractor or a Cassegrain reflector near the zenith. This can be partly compensated for by the use of a deck chair or a zenith prism. The designer of the mounting avoids this dilemma by increasing the height of

---

[63] Tichenor, C. L.: Notes on modern telescope mounting. *Sky and Telescope* **35**, 290–295 (1968).
[64] Link, H.: An amateur torque-tube mounting. *Sky and Telescope* **38**, 258–260 (1968).

the column and by making the latter more stable. If the eyepiece of the instrument is at the zenith position 90 cm above the floor, then every observer, sitting on a little stool and using a zenith prism, can observe very comfortably. The center of gravity of the whole instrument should lie very low and over the base of the column. For this purpose the latter has to be designed to be of a considerable size; three horizontally attached feet which carry at their ends an adjusting screw for a correct setting of the polar altitude are very convenient. Two additional horizontal screws, which grip one of the feet from the floor, make the setting of azimuth possible. The center of gravity of the whole instrument should lie above the center of gravity of the triangle formed by the ends of the feet. One of these center of gravity lines should run from north to south. These requirements apply more or less to every kind of instrument, be it fixed or transportable. Meeting these requirements determines, for example, how far the photographic limiting magnitude is reduced by vibration and flexure of the column when the observer guides by hand.

The *system of the axes*, from the point of view of construction, is the most difficult part of the mounting. It consists of two mutually perpendicular axes, turning independently. The first axis rests, at least with its lower end, in a fixed bearing. The case of the bearing of the second axis is rigidly connected with the upper end of the first axis. The second axis carries the tube and is perpendicular to the axis of the tube. If the first axis is perpendicular to the ground we speak of an azimuthal mounting. If the first axis is tilted so that it points to the celestial pole (the polar altitude is equal to the geographic latitude of the observing site), and if the telescope is outside the points of intersection of both axes but is displaced on the second axis, we speak of an equatorial mounting. With such a mounting it is possible to follow the apparent course of the stars by simple turning of the first axis. This is usually done by clockwork, and the axis is appropriately called the hour axis. By turning the second axis the telescope is directed to a given declination. It is called the declination axis. Joining this declination axis with the various forms of the hour axis produces the whole variety of mountings, as illustrated in Fig. 2–25.

The system of axes, together with the tube on the camera, places the center of gravity of the whole instrument higher, the greater its own weight. To build the system of the axes in light metal therefore helps to obtain a low position of the center of gravity. Each axis of the equatorial mounting is subject to flexure. This effect is proportional to the load and the moment of the force, but inversely proportional to the modules of elasticity of the material and the moment of inertia of the cross section of the axis. These concepts should be taken into account by every amateur observer when planning his instrument. For a precise bearing, free of backlash, of the two movable axes, the amateur astronomer will require the help of an engineer and a mechanic.[65]

[65] "Reference should be made to the pages of 'Sky and Telescope' for suppliers of such mechanisms."

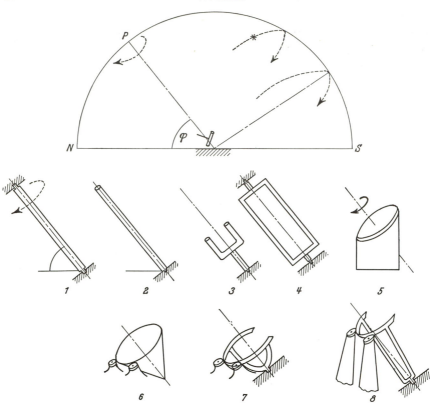

Fig. 2–25   The different forms of the polar axis. No. 5 shows the axis shortened to a disk, the so-called Springfield mounting. No. 6 is Porter's cement mounting, while No. 7 and No. 8 illustrate the split-ring mounting.

Suitable axes with the useful bearings of various dimensions, but rarely of the most suitable weight, can occasionally be found in scrap yards and car dumping grounds.

According to R. W. Porter the ends of the hour and declination axes are, besides the declination axis itself and the tube, the parts of the whole mounting which are most sensitive to flexure. These ends and their bearings must therefore be given particularly large cross sections (conical axes). The hour axis is made of steel, while the casing of the bearing is conveniently made of cast aluminium or plywood. If the shape is an open tube, some weight is saved; a piece of a tube can also be used. The casing of the hour-axis bearing is attached directly to the column, in such a way that the hour axis points to the celestial pole (see the above remarks). On the declination axis is a short tube and flange attachment. On the flange is a saddle that is adapted to the cross section of the tube and that carries the tube with the help of iron hoops. It is a characteristic of good mountings that the distance of the telescope axis from the hour axis is very small.

Hour and declination axis must be equipped with a clamp; in the simplest case this consists of two rather large wooden blocks, which can be squeezed together by ring screws or spindles in metal bushes. One of these is attached to the axis casing. See also Fig. 2–26.

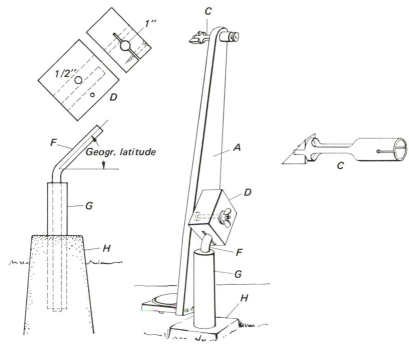

Fig. 2–26 A simple mounting without tube, drive, or protective hut. From *Amateur Telescope Making*, Book 1, p. 29. New York: Scientific American, Inc., 1953.

2.2.9.2. The Drive

Large telescopes have a setting motion (speed of about 80 minutes of arc per minute of time), a guiding motion (15′/min) and a correction drive (fine motion: 0″.1 to 15″/sec). For medium and small telescopes guiding and slow motion are nearly always necessary. The drive on the hour axis carries this axis in one sidereal day by 360° in the opposite sense to the rotation of the Earth, i.e., it turns it in 1 second by (1 revolution)/86,400. This ratio is easily achieved with the help of a worm, which is the origin of the drive, and a wormwheel. If a worm rotates once, the corresponding wormwheel advances by one tooth. It is also possible to obtain the necessary gearing by using wheels with various numbers of teeth used with different worms and wormwheels. The decomposition of 86,400 in prime factors (e.g., $2 \times 3 \times 5 \times 13 \times 13 \times 17$) provides the different number of arrangements which can be allocated to the various substages of the whole system, for instance, $2 \times 5 = 10$,

$3 \times 17 = 51$, $13 \times 13 = 169$. A large proportion should belong to the last stage, because in this way errors in the preceding stages are diminished in carrying the drive further.

A worm made of steel and a wormwheel made of lead–tin–bronze, or aluminium–bronze, must be produced by precision tools in a precision workshop. Recently, wormwheels have been manufactured from Polyvinylchloride (hard PVC, sensitive to mineral oil). The wormwheel turns the hour axis directly. To prevent damage, the observer will choose a friction clamp or better still an adjustable clamping ring. For this purpose the wormwheel is placed on a separate casing made of brass or bronze which is attached to the axis in a more or less firm manner by a clamping ring, as described by ZIEGLER.[66]

Various suitable drives have been recommended by STAUS;[67] for instance, the use of a spring, or of a weight with gramophone regulator, an electric drive with a motor for direct or alternating current.[68] The drive of a large telescope is usually a synchronous motor. Recently infinitely variable speed drives have been recommended, e.g., by W. REIMERS in Bad Homburg, Germany, and by H. HEYNAU in Munich. Homemade drives in which a weight is governed by a clock mechanism have been described in the literature. Even water drives have been used by ingenious observers, for instance by B. SCHMIDT, the inventor of the Schmidt mirror. In these designs a piston is pulled through a water-filled cylinder by a weight; the emerging water is governed and kept uniform by an adjustable orifice.

During the observation we have to take into account the fact that the drive cannot be equal at all positions of the telescope, since the latter moves more easily or with greater difficulty through the various positions. The speed of the drive must, of course, always be adapted to that of the fixed stars. The fine motion is necessary for guiding of Sun, Moon, comets, and planets, for the compensation of the apparent change in altitude of the stars due to refraction in the air, and for the correction of irregularities in the drive. Photographic work with long exposure times requires a uniform drive and a reliably fine motion. But even in the case of visual observations, the best objective lens is useless if the drive or fine motion fails at the decisive moment (e.g., at star occultations) or requires too much attention.[69]

[66] ZIEGLER, H.: Die Montierung. In H. ROHR, *Das Fernrohr für Jedermann* (see Bibliography).

[67] STAUS, A.: *Fernrohrmontierungen und ihre Schutzbauten*, 2nd ed. Munich: Unidruck, 1959.

[68] See PEEL, R.: A homemade electric drive for a 5-inch equatorial mounted telescope. *J. Brit. Astron. Assoc.* **78**, 99–104 (1968).

[69] A drive that can easily be built by amateur astronomers and that has been tested successfully has been described by H. VEHRENBERG in *Sterne und Weltraum* **2**, 19 (1963). See also ARGUE, A. N.: A quartz-crystal-controlled telescope drive. *The Observatory* **970**, 115–117 (1969).

The drive can be tested using the following technique:

There are in the field of view of the eyepiece two parallel wires in the direction of declination (hour wires), symmetrically placed to the center of the field of view. Let us assume that a star at the equator requires 6.3 seconds of time to move from one wire to the next (keeping the telescope immobile); this is to be taken as a mean of so many observations that its value does not change by more than 0.1 if more observations are added. Accordingly, we obtain for the distance of the wires $6.3 \times 15 = 94.5$ seconds of arc. If the telescope is attached to the drive our star may take 18.1 seconds to traverse the distance from left to right. The drive is too fast. The telescope has moved during this time by $18.1 \times 15 + 94.5$, instead of by 18.1×15. A tooth wormwheel with the overall ratio

$$\frac{18.1 \times 15''}{(18.1 \times 15'') + 94.''5} = \frac{100}{135.33}$$

has therefore to be used. However, we use a gear 100/135. The drive is still by 0.33% too fast. In a time interval of 1/0.0033, that is, in 300 seconds, a star at the equator still deviates by 15 seconds of arc from the middle of the cross wire. Let us assume that the focal length of the telescope is 1000 mm; then the star draws a little line of 0.075 mm on the plate. In 2 minutes of time we might therefore exceed the resolving power of a photographic plate, and the star image is then no longer circular. Whether such a deviation can be noticed with a particular guiding telescope in use depends on its focal length and the quality of the optics, the brightness of the star, and also on the quality of illumination of the cross wire used for guiding.

At present we would recommend the synchronous motors for 220 V, 50 Hz, or better still the automatically starting synchronous miniature motors for 6 or 12 V, 50 Hz, with attached reduction gear of 1/60 to give 120 rev/min. According to ZIEGLER a medium instrument (with a mirror diameter of 150 mm) will require a motor drive with a torque of 1500–300 pcm.

Motors for 220 V are connected to the ordinary lighting grid or fed by a car battery, with the help of an alternator control, with mechanical or electronic chopper. Since the resulting frequency varies by a few percent it is necessary for this reason, also, to correct the drive with the help of the fine motion. It is better to set the velocity of the drive with the help of a frequency changer whereby the frequency can be continuously adjusted within a wide range, e.g., 35–65 Hz. Such frequency changers, whose main components are transistors, transformers, condensers, and resistors, are described in the literature.[70]

The fine motion of the declination axis is, in the case of medium and small instruments, done by means of a handspindle, which is attached to a rigid extension of the fixed part of the clamp, perpendicular to it and acting

[70] Automatic Guiding is described by COVITZ, F.: An Automatic Guider for Astrophotography. *Sky and Telescope* **47**, 191 (1974). GEWECKE, J.: Elektrischer Fernrohrantrieb über Wechseltrichter mit stufenlos einstellbarer Frequenz zwischen 35 und 65 Hz. *Sterne und Weltraum* **2**, 44 (1964).

against a small attachment at the side of the tube. The slow motion of the
hour axis is conveniently done by hand via worm gears which can act directly
on the hour axis, or through a worm gear with a differential. There are count-
less modifications of slow motion, both of the types we have mentioned, as
well as of quite different types like those with Sun-and-planet gears. The
amateur astronomer has here plenty of opportunities to build up his own
homemade devices. See also Fig. 2–27.

2.2.9.3. The Circles and Their Adjustment

The setting circles are made in the form of disks or rings and are attached
on the hour or declination axis, respectively. In this case the corresponding
reading mark can be fixed on the casing of the axis as a stroke or a pointer.
Or, on the other hand, the circle can be rigidly connected to the casing of the
bearings; then the mark sits on the corresponding axis. The setting circle or
pointer must only be loosely attached (sliding fit) so that it can follow the
movement of the corresponding axis and can also be set by itself by hand.
To do this we point the telescope at the beginning of the observation to a
conspicuous star of known right ascension and declination and set the
setting circles on these known coordinates. Then all the subsequent movements of the telescope to other stars can be carried out with circles which have
been set in this manner.[71]

Older instruments have setting circles with very detailed subdivisions.
The hour circle, for instance, is subdivided to two minutes of time and, with
the help of a vernier, even two seconds of time can be read off. The divisions
of the declination circle are 20 minutes of arc, and 0'.5 can be read off. More
recent instruments have circles which are divided to 5 minutes and one
degree, respectively. The field of view of the finder requires a coarser subdivision: if the finder has a field of view of 4°, the hour circle must have
divisions of at least 15 minutes of time, and the declination circle of 4°.

These new circles require new methods of adjustment. A method devised
by KOLBOW has been described by STAUS as follows:[72]

In KOLBOW's procedure we assume that we possess for a given observing site,
the polar altitude of which must be known, a correctly adjusted equatorial
mounting. We then set the telescope exactly vertical so that its declination axis lies
horizontal while the declination circle indicates the latitude. The hour circle
will read 0^h (in the telescope position east) or 12^h (in position west), since the
telescope in its vertical position lies in the plane of the meridian.

In order to obtain this vertical position we place a plane plate of glass on

---

[71] See, e.g., LOWER, H. A.: *Telescope Drives in Amateur Telescope Making*, 11th ed.,
Vol. 2, p. 159, 1959; and NEMEC, G.: Die Einstellung nach Teilkreisen am transportablen
Amateurfernrohr. *Die Sterne* **40**, 3 (1964).

[72] See PLASSMANN, J.: *Hevelius-Handbuch für Freunde der Astronomie und der kosmischen
Physik*, p. 464. Berlin: Dümmlers Verlagbuchhandlung, 1922.

STAUS, A.: *Fernrohrmontierungen und ihre Schutzbauten*, 3rd ed., p. 62. Munich:
Uni-Druck, 1971.

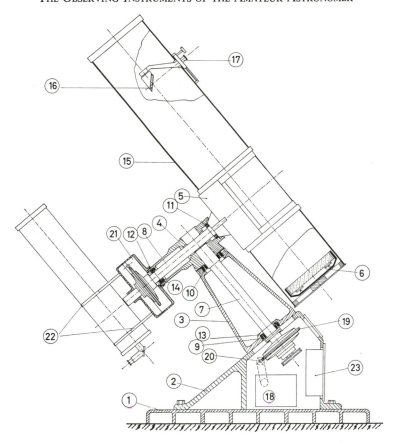

Fig. 2–27 A Newtonian telescope with a 40-cm mirror and a focal length of 2000 mm. From *Der Himmel über uns*, edited by the Astronomische Arbeitskreis Wetzlar e.V., 1965.

1 = baseplate	14 = spring plate
2 = base	15 = tube
3 = bearing cone	16 = secondary mirror
4 = cross head of axis	17 = eyepiece
5 = cradle for tube	18 = drive
6 = mirror cell with mirror	19 = wormwheel
7 = polar axis	20 = worm
8 = declination axis	21 = casing
9 = ball bearing	22 = support of the tube for Cassegrain or Schmidt telescopes
10 = ball bearing	
11 = ball bearing	23 = drive relays
12 = ball bearing	
13 = spring plate	

the objective mounting and on top of this a box level, while we can place an ordinary water level directly on the mounting, in two directions perpendicular to each other, until there are no changes if we change the direction. Also the finder can be used if it is not too small. In the case of mirror telescopes, we have in any case to rely on the finder. Of course, before the test starts we have to point it to a very distant point of the landscape or better still we look at the Pole Star so that the finder becomes exactly parallel to the line of sight of the main telescope. The further process can be demonstrated by a practical example.

Let us assume that we want to adjust the circles and to set the hour axis according to the geographic latitude, where the latter is assumed to be 48°4′. We start by setting the tube vertically in the east position. Let the reading of the declination circle be 46°38′. We then move to the west position and again set the telescope vertically. Now the declination circle may read 46°16′. We have therefore to correct the vernier to the mean of these two readings in order that we obtain the same reading for telescope east as for telescope west, i.e., in our case 46°27′. If the latitude were 46°27′ then the polar axis would be in its correct inclination. Since, however, our assumed latitude is 48°4′ we must tilt the telescope in declination with the fine motion until the vernier on the declination circle shows 48°4′. The water at this declination is of course very much out of equilibrium and it is restored by slowly changing the polar altitude. The last step is then the setting of the hour circle, at a vertical position of the tube, to 0^h or 12^h, which concludes the whole adjustment.

All this can be carried out in the workshop. A subsequent check at the observing site is, of course, essential since the floor of the workshop might have been uneven or not horizontal.

2.2.9.4. The Mounting of the Telescope in Azimuth and Latitude

A telescope will only follow the motion of a star correctly if the polar axis lies exactly above the north-south direction of the ground and has such an inclination that it points exactly to the celestial pole. Otherwise the instrument will have a *setting error in azimuth* or polar altitude, respectively. The north-south direction is fixed in accordance with the position of the Sun if the latter is exactly in the south direction, that is when our clock indicates:

Ephemeris Transit + difference in longitude of the observing site against the standard longitude in minutes of time.

The correct inclination of the polar axis can be obtained with a good approximation by the adjustment of the setting circles (see page 84).

2.2.9.5. The Setting Up of the Mounting

For a correct adjustment of the hour axis we require an eyepiece of average or short focal length equipped with a cross wire. This is turned to such a position that one of its wires, the declination wire, points to the direction of the motion of the stars when the drive is disconnected so that the stars are moving in the direction of right ascension, while the other wire (the hour wire) points in the direction of declination. The subsequent procedure has been described by SCHEINER as follows.[73]

After the instrument has been adjusted to within about 10 to 20 minutes of arc in each coordinate and the clock drive runs correctly, we view a star near the zenith and near to the meridian and bring it exactly onto the cross wire. We then find in a few minutes that the star gradually leaves the declination wire; a small turning of the instrument in azimuth is then sufficient to bring it back to the wire. However, if the star was not exactly at the zenith, then our turning will move it away from the hour wire, too, and we require the fine motion to bring it back. After a few minutes, it will be possible to see whether now the star has remained exactly on the declination wire or whether we require a new adjustment of the instrument.

If (in the reversing telescope) the star has risen above the wire, then we must shift the northern end of the hour axis from east to west. In the same way we then deal with a star near the pole (a circumpolar star), which is at the hour angle 6^h; the correction is carried out with the screw, which changes the altitude of the pole of the instrument. If the star moves apparently from the declination wire to the west, the northern end of the hour axis must be raised. The instrument is now no longer oriented with respect to the true pole but to the apparent one, as affected by refraction. In our latitudes this is only an advantage, of course.

It is advisable to repeat the adjustment of the setting circles after the completion of SCHEINER's above procedure (see page 84).

A rapid method for setting up the mounting, which is suitable for transportable instruments,[74] follows the following principle: we select from a star atlas three stars of small but nearly equal declination, whose right ascensions are very different and which in spite of this are at the same time above the horizon; for instance β Cygni 27°48′, μ Herculis 27°45′, and ε Bootis 27°20′. We now observe one of these stars, then proceed to the second and to the third and correct the deviation of the last from the declination wire by changing the azimuth of the instrument in such a way that the star then only shows half this deviation. The difference in declination of the stars must be taken into account. Then the star is moved back to the center of the cross wire and the procedure is repeated. If after several repetitions there still remains an error, we must change the polar altitude of the instruments by a small amount, which depends on the declination of the stars and on the direction of the deviation of the hour wire.

2.2.10. Auxiliary Instruments

2.2.10.1. The Eyepiece Micrometer

This instrument measures the mutual distances and the position angle of two objects in the apparent field of view.

---

[73] SCHEINER, L.: *Die Photographie der Gestirne.* Leipzig: Verlag von Wilh. Engelmann, 1897.
[74] *Amateur Telescope Making*, Book 2, p. 566. New York: Scientific American, Inc., 1954.

Simple measurements of large distances in right ascension and declination can be made without a micrometer by using setting circles and reading the latter ones with a vernier, or by estimating the intervals and forming mean values from several settings. For a more exact measurement of distances in the direction of right ascension we can use the telescope in a fixed position, whereby the single wires of the eyepiece must be exactly in right ascension and declination.

A star on the equator moves in one second Sidereal Time in the sky by 15 seconds of arc. A star of declination δ requires for this angle the time (1 Sidereal second)/cos δ. If a star of this declination requires f Sidereal seconds to move from one fixed wire to another fixed wire in the eyepiece of the arrested telescope we obtain the distance F of the wires in Sidereal seconds (if δ < 80°):

$$F_{ST}^S = f_{ST}^s \times \cos \delta, \tag{43}$$

and in seconds of arc

$$F'' = f_{ST}^s \times \cos \delta \times 15. \tag{44}$$

This approximate formula must be replaced by the correct expression if δ ≧ 80°:

$$\sin F_{ST} = \sin f_{ST} \times \cos \delta. \tag{45}$$

The amateur observer usually measures time intervals in Mean Time. The Sidereal Time that he requires for the above equations can be calculated as follows:

(Time interval in Sidereal Time) = (time interval in Mean Time)/0.997271

$$\tag{46}$$

The time interval between the transits of two objects through the same wire perpendicular to the direction of right ascension, or the time interval between the transits of an object through two such wires, is measured with the help of the "*eye-and-ear method*": the eye observes the transits with the wire and the ear listens to the ticking of the clock. The time of the transit can be estimated, after a little practice, to $\frac{1}{10}$ of a second using a clock which indicates seconds.

A pocket watch or a wrist watch can also be used. Nearly all watches of this kind perform 9000 double beats per hour, i.e., $2\frac{1}{2}$ double beats every second. Radio time signals are used for checking.

It is convenient to use, for the estimation and measurement of distances in the apparent field of view, an eyepiece of average magnification which contains not only a cross wire but also two other wires parallel to each other, which are placed halfway between the center of the field of view and the rim. Ramsden eyepieces (see page 71) are suitable.

The design of the most important part of the micrometer can be seen in Fig. 2–28. To build such an instrument we require a very accurate *measuring screw*; one with a thread of 0.5 mm and with 24 mm usable length can be easily obtained commercially. If this length is insufficient, the micrometer must have two fixed wires at larger mutual distance parallel to the movable wire. If the mutual distance of these two fixed wires has been determined according to Eq. (43), (44), or (45), measurements can be made from either of them.

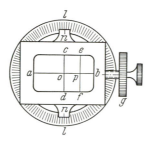

Fig. 2–28   The filar micrometer; *ab* and *cd* are the cross wire, *ef* the movable wire, *g* the drum of the micrometer screw, *l* the position circle, *n* the indices, *o* the center of the micrometer. From E. and B. STRÖMGREN: *Lehrbuch der Astronomie*, p. 52. Berlin: Springer, 1933.

A simple modification of the micrometer with perpendicularly crossed wires has been described by KIRBY.[75]

The movable thread moves by one turn of the measuring screw by an amount which corresponds in the sky to a certain distance in degrees, the so-called *screw value*.

$$\text{Screw value} = \frac{57°2958 \ (\text{thread of measuring screw in mm})}{\text{focal length of objective in mm}}. \qquad (47)$$

Taking an example, an amateur telescope 63/840 mm, with a thread of 0.5 mm gives

$$(\text{Thread value}) \times 0°034105 = 2'0463 = 122''778.$$

If the drum of the micrometer screw is divided into 50 parts it is possible to measure 2″46 and to estimate 0″246. This corresponds, according to formula (15), to the resolving power of an objective with $d = 46$ cm aperture. Since our telescope with $d = 63$ mm only resolves 1″86, we are in the position of measuring every detail shown by the objective with our micrometer.

The procedure of inserting *micrometer threads* has been described by KÖRNER (in Munich) as follows: the material for the threads is spider web or quartz fibers. The first are taken from a garden-spider cocoon. Its ends are fixed with paraffin to sufficiently wide wire forks and stretched. Then they are placed for a few hours in water. Afterwards they are placed in the grooves of

[75] KIRBY, G. J.: A simplified cross-bar micrometer. *J. Brit. Astron. Assoc.* **78**, 460–461 (1968).

the micrometer frame or micrometer sledge, stretched, and attached with shellac.

Quartz fibers must not be bent too much or they break. Their ends are fixed with sealing wax to light metal disks and placed across a frame with rounded edges. They are lifted with a suitable fork. Inside this frame is a micrometer frame or sledge which receives the threads in its grooves. If this part of the micrometer is at a suitable height the quartz fibers can be directly fixed with shellac. Quartz fibers attached to frames of a large temperature coefficient easily sag in cold weather.

For accurate measurements and setting, the wires must be seen as dark lines on a bright background. The field of view is illuminated in such a way that a little bulb of small brightness is placed in front of or behind the objective lens or the mirror, whereby the light must be prevented from falling directly into the eyepiece. If only the threads but not the field of view have to be illuminated, the source of light or the mirror must be close to the wires and screened against eyepiece and objective or mirror. It is also possible to

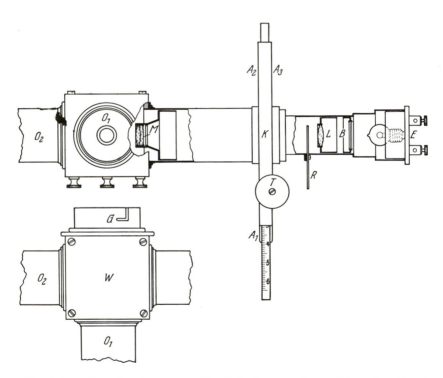

Fig. 2–29 Graff's photometer. The light from the lamp E is colored by the filter B, diaphragmed by the rotational diaphragm R, weakened by the wedge K, and imaged by the lenses L and M on a diagonal plane glass inside the tube W. The observation takes place at the eyepiece O_1 (for faint stars), or at O_2 (for bright stars). If the plane glass is replaced by a Lummer-Brodhun cube W and is removed, we obtain a surface photometer whose observer is at the eyepiece O_2. From *Zeitschrift für Instrumentenkunde* **35**, 2 (1915).

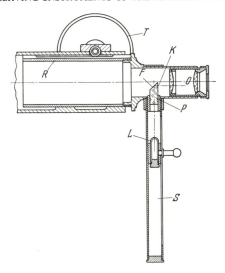

Fig. 2–30 The surface photometer of GRAMATZKI: the little tube S contains a luminous substance whose light is reflected by the surface F of the prism P and viewed by the eyepiece as a luminous surface with the edge K. With the wormwheel T, attached to the tube R, the image of a star is brought out of focus until it has the same surface brightness as F. From H. J. GRAMATZKI: *Leitfaden der astronomischen Beobachtung*, p. 100. Berlin and Bonn: Dümmler, 1928.

use homemade illumination equipment, which can be switched from "threads bright" to "threads dark."

2.2.10.2. The Photometers (Stellar and Surface Photometers)

The problems of photometry are discussed in detail on page 451. At this time, we shall only deal with the visual photometer by GRAFF and the visual surface photometer by GRAMATZKI, which are illustrated in Figs. 2–29 and 2–30. The brightness of a celestial object in the telescope is compared with the light of a pointlike source of light (artificial star) or with the energy density of a surface light source. In the first case, the light of the artificial star is weakened by a continuous gray wedge or by a polarization filter to such a degree that it appears of the same brightness as the light of the star under observation. In the second case, the energy density of the out-of-focus star image is changed by displacing the eyepiece attachment until equality with a phosphorescent luminous surface is achieved; a gray wedge, a polarization filter and eyepiece tube, must carry a scale, the unit of which is arbitrary.[76]

The position of the wedge, which is traversed by the light of the artificial star, is noted by the corresponding reading of the scale. If the light traverses the wedge at a point where its thickness is equal to the unit of our scale, and if i_0

[76] See RICHTER, N.: Radioaktive Kristallphosphore als photometrische Standard-lichtquellen. *Die Sterne*, **29**, 196 (1953).

is the incident, i the transmitted light intensity, we have $\tau = i/i_0$ for the degree of transparency at this point. Reflection can be neglected here. If s is the distance in units of the scale from the edge of the wedge with the thickness zero, then the light travels at the point s through the distance $s \cdot \tan \alpha$, where $\alpha =$ the wedge angle, and the fraction B of the transmitted light is $B = \tau^{s \cdot \tan \alpha} = \tau_k^s$. If τ_k and s are known, B can be calculated from

$$\log B = s \cdot \log \tau_k.$$

The quantity 2.5 log τ_k is usually called the *wedge constant*. If the thickness of the wedge increases linearly, then the ratio of two incident bundles of light i_1 and i_2, which are of the same brightness after they have passed the wedge, is

$$\frac{i_1}{i_2} = \tau_k^{\Delta s} \tag{48}$$

Here Δs denotes the mutual distance between the two places of transmission, expressed in scale units. It is convenient to use a linear wedge; then one needs to enter Eq. (48) only with the difference of the two-scale readings in order to obtain the ratio i_1/i_2.

The *wedge constant* 2.5 log τ_k is determined as follows: one measures with the wedge photometer the brightness of a larger number of stars of known but very different brightnesses $m_1, m_2, m_3 \ldots$ and obtains the corresponding wedge readings $s_1, s_2, s_3, \ldots$. We then have

$$2.5 \log \tau_k = \frac{\Delta m}{\Delta s}, \quad \text{or} \quad \frac{\Delta m}{\Delta s} = \text{const.} \tag{49}$$

Here Δm are the differences $m_1 - m_2$, $m_1 - m_3$, $m_1 - m_4$, $\ldots$, $m_2 - m_3$, $m_3 - m_4 \ldots$ and the values Δs are the differences $s_1 - s_2$, $s_1 - s_3$, $s_1 - s_4$, $\ldots$, $s_2 - s_3$, $s_3 - s_4 \ldots$. If, for instance, we are measuring the brightness of 5 stars we thus obtain 10 values for log τ_k. These should be nearly equal. If log τ_k depends on the brightness m of the stars, then the wedge is not linear. It is then possible, in a manner which will not be discussed in detail, to draw a curve which yields for two readings of s the corresponding Δm. If the wedge constant depends on the color of the star, then the wedge is not neutral. Usually the wedges are neutral over a sufficiently large range of color and only in the blue and red do they depart. It is then necessary to determine special wedge constants for each color.

The linearity of a gray wedge can also be tested: the wedge is copied twice in the same place of a photographic plate, with the same intensity and in diffuse light; the first time the thick end of the wedge is on the left, the second time on the right. This photograph is afterwards cut out and placed in the photometer in place of the original wedge. Then we look for a star that is just as bright as the artificial star after it has been weakened through this copy of our wedge. This equality must be maintained over the whole range of our wedge copy.

If we have found the wedge constant of the point-source photometer and if the wedge is linear, the brightness m of a star can be found from the known brightness m_1 of the comparison star from

$$m = m_1 + (s - s_1) \times 2.5 \log \tau_k. \tag{50}$$

The *surface photometer* of GRAMATZKI yields the required brightness m of a star from the brightness m_1 of a comparison star and the corresponding readings r and r_1 of the eyepiece tube (which are both to be measured from the focus position) as

$$m = m_1 - 5 \times \log r + 5 \times \log r_1. \tag{51}$$

Apart from the instrument described in Fig. 2–30, GRAMATZKI has also designed a surface photometer with which he measured energy densities on areas of major planets.[77] The comparison light source, which is attached at the side of the tube, is a ground glass uniformly illuminated by a condenser lens, which is diaphragmed by a cat's-eye diaphragm and illuminates a photometer mirror. The largest aperture of the cat's-eye diaphragm must not exceed, as seen from the position eyepiece, the apparent diameter of the objective. If the size of the latter is d, and $\bar{d}$ the largest diagonal of the cat's-eye diaphragm, e its distance from the photometer mirror, and if F and f are the focal lengths of objective and eyepiece, respectively, we have

$$dF = d(e + f).$$

Otherwise the quadratic aperture of the cat's-eye diaphragm is imaged too large, i.e., larger than the diameter of the exit pupil. The little mirror for the photometer is made from a glass tube into which a wire of platinum or nickel of 0.02 to 0.03 mm thickness is welded, by grinding it to an angle of 45°. With this instrument, GRAMATZKI has been able to measure reliably a considerable number of points on Mars and Jupiter (see also page 392, etc.).

We also refer to the simple surface photometers which have been constructed by YNTEMA and HOFFMEISTER for the photometry of rather extended regions (such as Milky Way clouds, Zodiacal Light, Gegenschein) without a telescope.[78]

A surface photometer ("Lumeter") which can be rapidly used for the photometry of aurorae, twilight phenomena, artificially illuminated night sky, etc., is sketched in Fig. 2–31.[79] The measured ratio of the energy densities is

$$\frac{i_1}{i_2} = \frac{l_1^2}{l_2^2},$$

[77] GRAMATZKI, H. J.: Zur Krisis der Amateurplanetenforschung. *Die Sterne* **28**, 95 (1952).

[78] See *Handbuch der Astrophysik*, Vol. 7, p. 110. Berlin: J. Springer, 1936.

[79] SIDGWICK, J. B.: *Amateur Astronomers' Handbook*, p. 393. London: Faber and Faber Ltd., 1954.

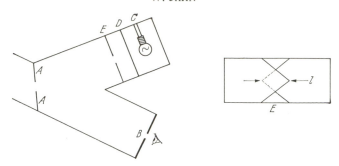

Fig. 2–31 A surface photometer for use without telescope. *A* comparison surface, *B* eye, *C* source of light, *D* ground glass, *E* cat's eye-diaphragm.

where l_1 and l_2 are the diagonals of the square opening of the cat's eye. The measured difference in magnitude is

$$m_1 - m_2 = 5 \times \log \frac{l_2}{l_1}.$$

2.2.10.3. The Clock

We demand that our clock shall have a constant rate. It is unimportant whether the clock is gaining or losing at a given moment (*clock error*). This error must be determined daily, sometimes even hourly, in order to calculate from it the changes per day or hour (daily or hourly rate). For the accurate determination of the clock error, we make a clock comparison. This can be done by listening to the radio, but not to an announcer or gong, nor by the telephone "speaking clock" but only with the help of the time signal of a radio transmitter or a special time signal station or of a station radiating on normal frequency. In Europe, for instance, it is possible to listen to time signals at every hour. A very efficient shortwave receiver receives time signals by day and night, at seconds intervals, of the following stations (Table 2.16).

Table 2.16. Time Signals at Second Intervals of Some Normal-Frequency Stations

Great Britain	MSF	0.06 MHz	0.5 kW	Czechoslovakia	OMA	2.5 MHz	1.0 kW
		2.5	50	United States	WWV	2.5	2.5
		5.0	50			5.0	10.0
		5.0	50			5.0	10.0
		10.0	50			10.0	10.0
Hawaii	WWVH	2.5	2.0			15.0	10.0
		5.0	2.0			20.0	10.0
		10.0	2.0			25.0	10.0
		15.0	2.0				

Amateur astronomers who do not have a shortwave receiver can receive time signals, also without pause and undisturbed and continuously, by listening to the longwave transmitters (in Czechoslovakia, Prague, OMA 0.05 MHz 20 kW; in Great Britain, Rugby, GBR 0.016 MHz 7.50 kHz; in Switzerland, Prangins, HBG 0.075 MHz 25 kW; in West-Germany, Mainflingen, DCF 77 0.0775 MHz 12 kW). These transmitters can be used as a clock indicating seconds with the help of a special homemade receiver which has been described by FRICK in Berne, and which uses a sound generator and a morse key so that it is possible simultaneously to record on a tape recorder time signals and observing signals (e.g., the beginning of a star occultation).[80]

The signals transmitted from the stations in Table 2.16 can replace a clock if reception is guaranteed at all times. A reliable shortwave receiver in connection with the tape recorder, a microphone, and a sound mixer permits a clock comparison and the determination of the time of an observation at any required time.

If a clock (pendulum clock, chronometer) is at a fixed place we should keep a *clock book*, which notes the time of the clock comparison, the transmitter, the error, the rate, the temperature, and if possible atmospheric pressure at the time of the clock comparison, and the correction applied.[81]

It is not necessary to have a clock indicating Sidereal Time, since the transformation of Mean Time into Sidereal Time is very easy and rapid with the help of an almanac (see page 548).

Those who can use for the observation a pendulum clock or a ship's chronometer must place it free of vibration, protect it as much as possible from temperature changes, and wind it regularly. For the amateur astronomer a reliable pocket watch will be sufficient; it has to be wound before every observation and before and after the observation compared with the time signal. Of course one will not expect a pocket watch to have a regular rate, even over a short time interval. Wrist watches have been used for observations like pocket watches.

For observation of star occultations, lunar eclipses, satellites, meteors, etc., we shall use a *stopwatch* instead of a chronometer or a pendulum clock with chronograph. Immediately after the observation the stopwatch will be compared with a reliable clock or a time signal (see page 94).

The eye-and-ear method can be applied reliably only with clocks that beat every second or every half second (see page 88).

2.2.10.4. The Photoelectric Photometer

The application of the photoelectric photometer to astronomy is discussed on page 452 in more detail. Here we shall only deal with the theoretical

[80] FRICK, M.: Ein einfacher Empfänger für den neuen Zeitzeichenempfänger HBG. *Orion, Zeitschrift der Schweizer. Astr. Gesellschaft* **11**, 185 (1966).
[81] Particular reference should be made here to a recent discussion by K. FISHER, "Quarzuhren für Amateurastronomen," in *Sterne und Weltraum* **11**, 240–242 (1972). Various types of Quartz clocks are listed in the leaflets of the Treugesell-Verlag K.G., Abt. II, Postfach 4065, D–4000 Düsseldorf 14, Germany.

Table 2.17. Selection of Some Photomultipliers

Type	M12 FS 35	931-A	150 AVP	XP 1000	K 1382	150 CVP	9558
Make	Jena	RCA	Philips	Philips	Du Mont	Valvo	EMI
No. of Stages	12	9	10	10	10	10	11
Total potential volts	1800	1000	1800	1800	1300	1800	1650
Cathode size (mm)	35	7.8–24	32	44	12	32	44
Anode sensitivity A/lum	250	24	80	3000	15	100	500
At a voltage of	1800	1000	1800	1800	1300	1800	1650
Dark current at this voltage	10	60	200	1500	1000	20	6
Sensitivity maximum at λ nm	450	450	450	450	440	800	420
Photocathode	SbCs	SbCs	SbCs	SbCs	SbCs	IR	KCsSbNa

background of the secondary electron amplifiers, i.e. the multipliers. They produce a current of 10^{-6}–10^{-9} amperes, are simple to use, possess a great bandwidth, and permit amplification with little noise. Table 2.17 lists the various types of multipliers and shows at the same time that all types must be operated at a high potential which must be kept constant within 0.1 %.

The light from the object is collected by the lens or the mirror and passes through a sufficiently large aperture, which corresponds to the cross section of the beam, in order to keep the light-sensitive cathode undisturbed by light from the neighborhood and from the sides. In order to align the beam with the center of the aperture hole and to keep it there, it should be possible to observe the front or the back of the aperture with the help of a little mirror and eyepiece. After it has gone through the aperture, the beam passes through a short-focus simple lens of high transparency. This "FABRY lens" images the aperture onto the cathode, so that a light spot of 6–10 mm diameter is produced. This imaging has the effect that the light spot remains always at the same place of the cathode, independent of the seeing, and local sensitivity differences do not have any effect. Before the light reaches the cathode it is limited to the desired color region by a color filter which is held by an eccentric rotating disk. The supply of current for the multiplier can be (1) several anode batteries in series, from which the current is led by high-tension cables to the radiation receiver at the eyepiece end of the telescope, (2) a DC converter (e.g., after H. ZIEGLER) fed by a 6-volt accumulator, or (3) a commercial "black box" which consists of a DC supply with stabilized voltage, but which is very expensive.

The output current of the multiplier is measured either directly or via an amplifier which amplifies it from 10^{-9} amperes to 10 mA.

In the first case a good mirror galvanometer can serve as a measuring

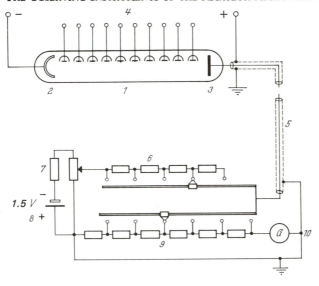

Fig. 2–32 Circuit of a photometer with galvanometer. From *Sterne und Weltraum* **3**, 69 (1965).

1 = photomultiplier	6 = resistors for dark-current compensation
2 = photocathode	7 = voltage divider
3 = anode	8 = battery
4 = dynodes	9 = decade resistance
5 = screened cable	10 = mirror galvanometer

instrument, or we can determine the capacity or voltage of a condenser to which the photocurrent is transferred.

In the second case the current is led by a shielded cable to an amplifier; the insulation (10^{11} ohm) is most important. This amplifier is preceded by an RC coupling. The resistance R and the capacity C determine the time constant of the amplifier. The time up to the final deflection of the ammeter is proportional to RC. With a larger time constant (1–4 seconds) the rapid variations of the photocurrent are more smoothed out. The amplifier must yield an output current proportional to the incoming photocurrent over the whole range of measurement (linearity). This is achieved by strong feedback. This can be so great that it eliminates differences in the characteristics of the various valves of the amplifier. Finally, to avoid nonuniformities of the exit current (which can also occur without a photocurrent, e.g., zero drift) it is necessary to employ in the first stage of the amplifier a special circuit, as has been discussed in the literature.[82] See also Fig. 2–32.

[82] KRON, G. E.: *Amateur Telescope Making*, Book 3, p. 298. See also WOOD, F. B.: *Photoelectric Astronomy for Amateurs*. New York: Macmillan Co., 1963.

2.2.10.5. The Filters

Filters are either *neutral filters*, which weaken the light in a very large range of wavelengths by the same fraction, or color filters, which act on the light of different wavelengths nonuniformly. Belonging to the last-mentioned group are filters which transmit light only in wavelength ranges from 0.0000080 mm to 0.0000005 mm (i.e., of 80 to 5 Å). The weakening of the light by filters takes place either by scattering and absorption or by polarization or by interference. If light arrives at the transparent filter, it is partly reflected. Since the amount of reflection depends on the angle of incidence it is advisable to use filters only in the particular arrangement of optical path, in which it has been tested and calibrated, for example, always with parallel beams or with the same aperture ratio as that of the particular optical arrangement. This can be achieved if the filters are calibrated in the very position in which they are being used. Usually filters are the size of a photographic plate. Larger filters are difficult to produce in uniform thickness and absolutely plane. Filters are usually placed immediately in front of the photographic plate.

If i_0 is the energy density of the incident light and i the transmitted amount of light, then the *degree of transparency* of this glass is

$$\tau = i/i_0. \tag{51'}$$

The degree of transparency is usually supplied by the manufacturer. If we place two neutral filters of the transparencies τ_1 and τ_2 behind each other, they have a total degree of transparency $\tau_1 \times \tau_2$, whereby possible mutual light reflection at normal incidence is not taken into account. The τ-values can be measured in the same way as the τ_k-values of a photometer wedge, in case they can be placed within a photometer (see page 92). For rough estimates, it is sufficient to hold the wedges in the hand and to test by how many magnitude classes bright stars are weakened by them.

Color filters show a degree of transparency which strongly depends on the wavelength of the transmitted light and frequently completely absorbs large wavelength regions or perhaps transmits them nearly completely. The manufacturers of color filters, including those used for cameras and field glasses, publish the degree of transparency in tables or in curves. We must note whether losses by reflection have been taken into account or not. Combinations of color filters and photographic plates, whose degree of transparency and color sensitivity are accurately known, or the use of color filters at visual observations with a refractor whose color correction is known, can practically suppress the light in any extended range of wavelengths, and only transmit it in a certain region range (Fig. 2–13). It is advisable to possess a whole series of neutral and color filters (which, e.g., can be obtained from Schott in Mainz or Jena), say of the size 4×4 cm.

Monochromatic filters (quartz-polaroid monochromators, Öhman, Lyot, Evans, Pettit filters, and birefringent filters) are composed of layers of double-refracting and polarizing materials, forming a composite plate. The polarizing layers are all equally thick and alternate with quartz-crystal plates each of which is twice as thick as the preceding one. The thicker the last quartz plate the narrower is the transmitted wavelength region, the so-called bandwidth. A monochromator of 1 cm thickness transmits a range of 40 Å; one of 4 cm thickness of 4 Å. These filters can be designed for different wavelengths. Of particularly frequent application are those that show the Hα-line or the K-line. If during observation they are to transmit the same wavelength very accurately they must be kept by a thermostat at a constant temperature. There exist very detailed instructions to produce homemade filters of this kind; the most expensive components are the quartz-crystal plates.[83]

Interference filters, which are not nearly as expensive and as difficult to handle, can well be used by amateur astronomers. The range of transparencies of the commercially obtainable kinds lie between 40 and 80 Å. They possess on a glass surface a sputtered reflecting layer and on top of the latter a thin homogeneous layer coated with another reflecting surface. A glass plate covers the whole filter so that it is very well protected. Additional ordinary color filters screen undesired light.

2.2.10.6. The Spectrograph and the Spectroscope

The spectrograph serves for the recording of spectra. The light is resolved by prisms, gratings, or by an interferometer. It arrives as a parallel light bundle on the refracting, diffracting, or reflecting optical parts. For astronomical spectrographs this is always the case if these parts are in front of the objective of the telescope. See also Fig. 2–33.

The deviation δ at wavelength λ by a prism with a refracting prism angle γ and the refractive index n is (see Table 2.7)

$$\delta = \alpha + \beta - \gamma,$$

where

$$\sin \beta = \sin \gamma \sqrt{n^2 - \sin^3 \alpha} - \sin \alpha \cos \gamma, \tag{52}$$

where α is the angle of incidence and β the angle of refraction.

The deviation δ of the wavelength λ by a grating with spacing g follows from

$$\sin \delta = m \frac{\lambda}{g} + \sin \alpha, \tag{53}$$

where m is the order of the spectrum.

The length of a spectrum on a plate or film, counted between two wavelengths, is determined from the difference of the angles of deviation and the focal length of the spectrum camera from the prism or the grating to the plate. The ratio (difference of angles/difference of wavelengths) is called the angular dispersion. The ratio (lengths of spectrum in mm/difference in wavelengths) is the "linear dispersion." This value is, in the case of a grating, also

[83] *Amateur Telescope Making*, Book 3, p. 640. New York: Scientific American, Inc., 1953.

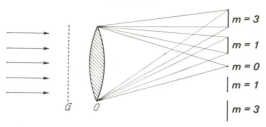

Fig. 2–33 Schematic design of an objective-grating spectrograph. The grating with equal width of rod and interspace (normal grating) is denoted by G; O is the camera objective used as a spectrograph objective; m_1 and m_3 are the diffraction spectra. (Because of the use of a normal grating there are no spectra of even order.)

equal to the focal length of the spectrograph objective multiplied by the angular dispersion. The resolving power of a spectrum is the smallest wavelength difference which is still accepted by the instrument. The ratio (smallest wavelength difference/wavelength) measures resolving power.

The material for prisms is chosen for small absorption (whereby the dispersion is also of course small) and refracting angles up to about 60° are used. Some remarks on objective-grating and objective-prism spectrographs: The gratings are made of rods or wire. Suitable thicknesses are of the order of 0.1 to 1 mm, and can be calculated from the above expression for the deviation. From this the length of the spectrum follows for a fixed focal length of the spectrograph objective. We make use of the spectra of the 1st order, in which δ is proportional to λ. The minimum length of spectra should be 0.3 mm; with a magnifying lens, they will show many details if they are 3 mm long. The direction of dispersion should coincide with the direction of declination.

The diameter of the objective lens must not exceed the diameter of the objective prism or of the objective grating. This camera can if necessary also operate without an achromatic objective, namely, if we are only concerned with narrow spectral regions, or if the film is curved. Spherical aberration should be removed. Because of this and because of the nonexisting chromatic aberration Schmidt telescopes are particularly suited as spectrograph cameras. The spectrum in the prism spectrograph is imaged on a plane that is tilted to the optical axis. Usually several spectra are taken on the same plate, facilitating comparison. During exposure, they are widened to at least 0.3 mm by operating the slow motion in right ascension, otherwise they would be difficult to analyze. The guiding and widening is usually done with the help of a star set in the guiding telescope between two eyepiece wires. Objective prism and grating can, if suitably mounted, easily be attached or removed; a counterweight is necessary.[84]

[84] See Wood, F. B.: The present and future of the telescope of moderate size. Philadelphia: University of Pennsylvania Press, 1958, and Thackeray, A. D.: *Astronomical spectroscopy*, New York: The Macmillan Co., 1961.

Disadvantages of objective-prism or objective-grating spectrographs are: light from the neighborhood and scattered light can considerably reduce the contrast if the aperture ratio of the spectrograph objective is large; in densely populated star fields many spectra overlap; air disturbances act strongly on the quality of the spectra, which become diffuse; see page 20.

For practical work in prism or grating spectroscopy and in particular in connection with the choice of the prism of the grating, the length of the spectrum, the exposure time, etc., we refer to a very useful text.[85]

2.2.10.7. Practical Notes

Contributed by Dr. W. KREIDLER, Munich

Spectroscopy opens to the amateur astronomer a new world that has received little attention previously. It is a world not only of astrophysical observing possibilities but also I would say, if I may, of mysterious beauty. Absorption and emission lines in the spectra contain an enormous amount of information about the fundamental conditions of the stars that, e.g., govern their systematic classification and scientific interpretation in spectral classes. Of course, with the restricted means at the disposal of the amateur astronomer no scientifically outstanding results can be expected (except perhaps if he has the opportunity to examine spectroscopically a bright nova). But the amateur astronomer can reproduce so to speak the rudiments of astronomical spectroscopy and thus arrive at a deeper understanding of one of today's most important astrophysical branches. See Fig. 2–34.

Relatively inexpensive are three types of instruments: (1) the straight-eyepiece spectroscope; (2) a spectrograph with objective prism; (3) the grating spectrograph. The eyepiece spectroscope (1) consists of a straight-prism system composed of three to five prisms (Amici prism). It is screwed onto the eyepiece, which should have a very large focal length, and yields a thin line as the spectrum of the observed star. A negative cylindrical lens widens this to a more or less wide band. This lens can be turned about its axis permitting a widening of the spectrum perpendicular to its length. The lens should not disperse strongly since otherwise much light is lost. It is convenient to prepare a set of about three cylindrical lenses of different (negative) focal lengths, in order to be able to adapt the width of the spectrum to the brightness of the star. In this way we can easily recognize the main spectral lines, particularly the Balmer series of the classes A0–A5 and F0, as well as the principal lines of the classes K0–K5 and M0–M3. The smaller the dispersion of the Amici prism, the larger the diameter of the objective lens, and the greater the focal length of the cylindrical lens, the brighter will be the spectrum. The observation is an outstanding aesthetic attraction.

A spectrograph with an objective prism (2) can be made from any camera

[85] GRAMATZKI, H. J.: *Hilfsbuch der astronomischen Photographie*. Berlin and Bonn: Dümmlers Verlag, 1930.

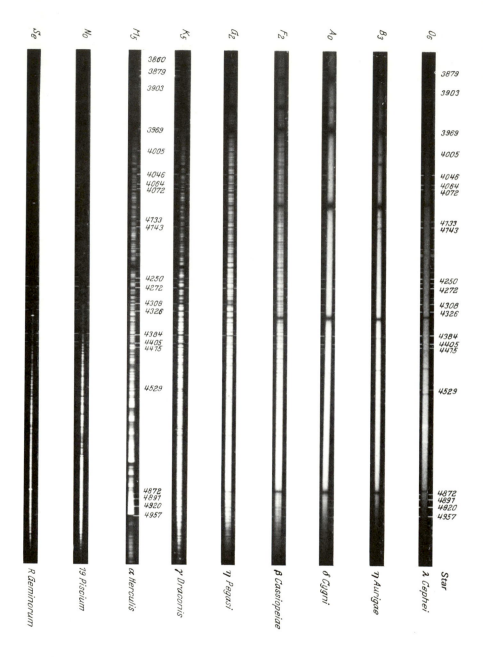

Fig. 2–34 Stellar spectra of the spectral classes O, B, A, F, G, K, M, N, and S (Mount Wilson and Palomar Observatories). The classification according to spectral type Sp and the luminosity class Lk is achieved by comparison with standard stars (see Tables 19 and 20 in the Appendix). Each observer will adapt the classification criteria of his spectra of the standard stars to his instrument.

whose focal length is not too short (say in excess of 15 cm) and with sufficiently large aperture ratio (larger than 1 : 5.6) by placing a glass prism immediately in front of the objective. It is advisable for several reasons that the prism be removable and that it can be turned and clamped around the optical axis; this is of course necessary when the focusing of the camera is done by turning the objective in the same way as most teleobjectives. The focusing is best done without the prism; the turntable mounting makes a subsequent setting of the prism parallel to the direction of the motion of the stars possible. Some patience is required in setting the spectrum correctly on the ground glass, because the angle of the light path through the prism deviates considerably from the angle of the direct line of vision. In order to lose as little light as possible we furthermore require the particular setting of the prism which causes the smallest deviation. Therefore the prism must also be adjustable as to its tilting angle with respect to the direction of the objective.

The setting is best performed on the Moon or (using the smallest diaphragm) on the Sun. First of all, we tilt the whole instrument in such a way that the color image appears on the ground glass. If we now change the tilt of the prism, the image must appear from the top on the ground glass, then reverse and disappear again at the top. If the turning point is at the center of the plane, prism and camera objectives have the correct mutual tilt toward each other (minimum deviation) and toward the star. With the help of a simple sight on the camera, it is possible to determine the inclination of the camera with respect to the direct line of sight.

As far as the widening of the produced thin spectrum is concerned, we leave this job simply to the apparent motion of the stars. Here we even do not require an equatorial mounting. The simplest case is the one when we observe the spectrum of a star at the time of its culmination, since then its direction of motion is parallel to the horizon. All that we have to do is to place the edge of the prism horizontally and to expose the plate for a few minutes (see page 407, Fig. 16–7).

It is more difficult to set on stars that do not culminate at this time if our camera is mounted in azimuth, because it is then first of all necessary to find the direction of motion of the star on the ground glass without prism. Applying a corresponding tilt of the camera sideways we can ensure that the stars move along a line on the ground glass, which has previously been drawn in parallel to the upper edge of the plate. Then, after the prism has been mounted parallel to this edge, the length-direction of the spectrum is, as required, perpendicular to the tangent to the orbit of the star. It is not necessary, in view of the short exposure times, to take account of the curvature of the stars' orbits.

An equatorial mounting of the camera has a great advantage. The telescope serves as guiding telescope if the camera is attached to it under the above-mentioned angle of vision. Once the horizontal setting of the prism and the edge of the plate has been performed, the telescope pointing to the south (i.e., parallel to the declination axis), the required setting of the prism is guaranteed for all stars. We are now also able to photograph fainter stars by superimposing several exposures through gradual forward motion in right ascension (which can be checked in the cross wire of the guiding telescope, Figs. 2–35, 2–36, and 2–37). Or, if there is a clock drive, we can slow down or accelerate the guiding by a little amount, which, in view of the technical shortcomings of the amateur astronomers' drives, usually does not require any special device. In this way it is possible to do spectroscopic work even for stars as faint as 4th magnitude.

It is very useful to use a miniature reflex camera since we are then also in the position of using a highly sensitive film such as Kodak Royal X or Agfa Rekord, as well as producing effective color slides. However, the advantage of comprising whole star fields on one exposure can be realized only on plate cameras of larger size (6 × 9 and 9 × 12 cm). Of course, such cameras can also be equipped with highly sensitive films, but perfect focusing will only be possible by pasting this plane film onto a glass plate.

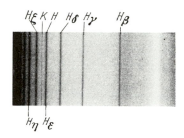

Fig. 2–35 The A0 spectrum of α Lyrae (Vega) 0ᵐ.0; azimuthal mounting without guiding; exposure time 10 minutes; magnification 5×.

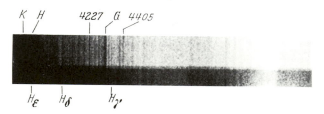

Fig. 2–36 The K_0 spectrum of α Bootis (Arcturus); −0ᵐ.1; parallactic mounting; exposure with three superimposed stages, the first and the second of 2 minutes, the third of 3 minutes duration; magnification 10×.

Fig. 2–37 The A_1 spectrum of α Coronae Borealis (Gemma); 2ᵐ.3; parallactic mounting; exposure on three superimposed stages, each of 2 minutes duration; magnification 5×. Spectrograph with flintglass prism 60r; objective Schneider Zenar 1:3.5 diaphragmed to 1:4.5; focal length 24 cm; film Agfa Rekord; developer Ultrafin.

In summary, the larger the objective diameter of the camera, the shorter the focal length, the smaller the dispersion of the prism, the greater the sensitivity of the film, the smaller the differences between clockwork and daily motion, and the larger the number of superimposed exposures, the fainter are the stars that can be examined spectroscopically. On the other hand, longer spectra, obtained with a larger dispersion of the prism and a longer focal length of the camera, provide of course a larger resolving power applied to the spectral lines.

The grating spectrograph (3) is built in two designs: as an objective-grating spectrograph or as a grating spectrograph with slit and collimator. Since the diffraction images, even those of the 1st order, are very faint, the first design requires a much-too-large diameter of the objective or mirror as well as too-large objective gratings to be still within the financial means of the observer. On the other hand a grating spectrograph with slit and collimator can yield solar spectra with a considerable resolving power. A simple construction of this kind can be found in the *VDS-Nachrichten* **15**, No. 4 (1966). As to the interpretation of the spectral lines we refer to the literature. Useful reference points are provided by the wavelengths of the Balmer series, given in Angströms:

Hα	6563	Hζ	3890
Hβ	4861	Hη	3836
Hγ	4342	Hθ	3799
Hδ	4103	Hι	3771
Hε	3971	Hκ	3751

"These wavelengths are the nearly monochromatic spectral lines of hydrogen which are prominent in the spectra of stars and nebulae. Since these wave-lengths can be produced in the laboratory, they are often used as reference points in the spectrum": (Dimitroff-Baker).

3 / Optical Radiation Receivers

R. Kühn
Revised by F. Schmeidler

3.1. Introduction

Astronomy differs from other branches of the exact sciences in certain essential ways. While physicists, chemists, geologists, and most other scientists are able not only to observe events in nature but also to interfere experimentally, experimentation is unavailable to the astronomer (if we make an exception of sending radio signals to the Moon and the planets to receive echoes). Furthermore, researchers in other fields can usually draw on results from other branches of physics and chemistry, such as in the realms of optics, mechanics, and acoustics. The only tangible object the astronomer can handle, so to speak, is—if we disregard the meteorites—the electromagnetic radiation in all its forms. Everything we know about the properties of the celestial bodies we owe to the examination of this radiation. It can reach us in very different forms—for example, as visible or invisible light, or as electromagnetic radiation in the form of radio waves. The one feature common to all these radiations is the fact that they are electromagnetic waves propagated with the velocity of light. Their wavelength, intensity, and composition, on the other hand, are very different.

The shortest waves to reach us from space and the radiations richest in energy are *cosmic radiation* and the X-rays. These radiations, however, are almost completely absorbed by the Earth's atmosphere and only very small remnants reach the Earth's surface. Practically speaking, therefore, they do not enter the observing program of the amateur astronomer.

The atmosphere becomes transparent again for *ultraviolet light*, radiation comparable to visible light but not perceived by our eye since it lies just beyond the range of its sensitivity. It can, however, be easily recorded by photographic emulsions, provided we use suitable plates (Fig. 3–1).

The spectrum of *visible light* ranges from violet through blue and green to yellow and red, or in wavelength from about 3700 Å to 6700 Å, i.e., it covers about an octave. Visible red is joined by the *infrared*, which again can only be recorded by the photographic plate, not by the eye. With infrared the photographically accessible range comes to an end and we now enter the region of heat radiation and finally that of radiowaves (see pages 107 and 126).

As already mentioned, all our insight into astronomy is obtained from the examination of this radiation; essentially there are three kinds of investigation aimed at determining (*a*) the *direction* from which the radiation reaches us, (*b*) its *quantity*, and (*c*) its *quality*.

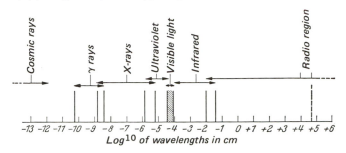

Fig. 3–1 Spectrum of the electromagnetic radiation. After WESTPHAL, *Lehrbuch der Physik*, 12th ed. Berlin: Springer, 1947.

Quantity here means the total radiation energy which our receiver records from a given celestial body. In the range of the visible light we speak of "*integral brightness.*" Quality, on the other hand, means essentially the spectral energy distribution of the radiation, that is, the composition of the total radiation as components of different wavelengths. In the visible region, for instance, color and all the spectroscopic findings come under the heading "quality of radiation." There are also investigations of the kind and intensity of any polarization that might be present.

The radiation that reaches us from the celestial bodies and is received by a radiation receiver must be converted into scientifically useful information, a process which can take place in very different ways. Two such possibilities are mentioned as examples.

If details of planetary surfaces are observed, the radiation is received by the eye; then some picture is formed by the brain and is stored in the memory. This stored picture can be reproduced and perpetuated as a drawing. During the observation of stellar brightnesses with a photoelectric instrument the radiation is received by a photocell and converted into an electric current. The latter is measured by a special recording instrument, and the end product is a measure of the brightness of the star, which can be stored and used again.

In the present chapter we are going to discuss the most important radiation receivers that can be used by the amateur astronomer. Different as these receivers may be, they have in common the ability to transform the incoming radiation into information, be it a *figure* or a *picture*.

3.2. The Human Eye

Of all the radiation receivers that interest us here the human eye is, of course, the most important. Apart from the fact that in the last instance it is the eye that processes all astronomical observations at one stage or another (e.g., by reading pointer deflections, or by judging photographic plates), it does nearly all the work at the telescope, setting the stars, guiding, finding objects, etc.

3.2.1. Structure of the Eye

We shall now deal briefly with the structure and working of the human eye (Fig. 3–2). The eye is spherical and its central part is enclosed by a tough

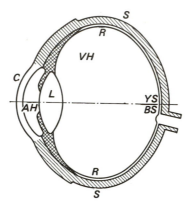

Fig. 3–2 The human eye: C = cornea; AH = aqueous humor; L = lens; VH = vitreous humor; R = retina; YS = yellow spot; BS = blind spot; S = sclera, i.e., outer tissue. After WESTPHAL, *Lehrbuch der Physik*, 12th ed. Berlin: Springer, 1947.

tissue (the sclera) which at the front consists of the curved transparent cornea. Behind the cornea is the front part of the eyeball which is filled with a clear liquid, the "aqueous humor." Toward the inside of the eye this section is closed by a lens, the front of which contains the iris. The iris fulfills in the human eye the function of an aperture that regulates the amount of light entering the eye. The opening of the iris is called the pupil. The inside of the eye is filled by a jelly-like material (the "vitreous humor"). The back of the eyeball is covered by the retina which carries the light-sensitive receivers that are connected through the optic nerve with the center of vision in the brain.

The retina contains two different kinds of light receivers, the rods (about 120 million) and the cones (about 7 million). With the less light-sensitive cones we see color in daytime and in good light. The central zone of the retina, the so-called yellow spot, contains only cones. Light that falls on this spot is therefore only recorded when it is relatively intense. At night this spot is blind.

The rods are about 10,000 times as sensitive as the cones, but it is impossible for them to distinguish different colors. In faint illumination we are so to speak color-blind. Furthermore, it should be mentioned that at the place where the nerve enters the retina there is a blind spot which we generally do not notice as long as we observe with both eyes, since the blind spots of both eyes are not exactly at the same place in the field of vision.

The blind spot of the right eye is on the right-hand side of the axis of seeing, that of the left eye, on the left-hand side of this axis. While the presence of the blind spot does not have too much practical importance for the amateur astronomer, the yellow spot within which we can record only if the light is relatively strong is very important. Many astronomical objects are faint, and we can only observe them if we do not view them directly but, so to speak, out of the corner of the eye. This *indirect vision* can, with some practice, be brought to great perfection and a good observer constantly uses it unconsciously. Indeed it is very useful to carry out some experiments without an instrument. A bright star does not disappear even if we look at it directly, but stars of the third or fourth magnitude, which we can see with the unaided eye, disappear when they lie on the axis of our eye.

As soon as an incoming light quantum hits a rod or a cone a complicated chemical process sets in, many details of which are still not clarified. But it is important to note that the rods contain a certain substance, the "visual purple." This substance is bleached by the action of light and converted into a yellow and finally a white material. In this way, the coloring matter and the carrier protein in the original visual purple are separated, and act on the ganglial cells. For us, the fact of greatest importance is that the visual purple is *regenerated* by the opposite process, when the eye is in the dark. This process requires some time, so that when we come out of a bright room we cannot see at once in the dark. This response of the eye to darkness is called adaptation. Opinions differ as to the time required by the eye to adapt completely to darkness. The time is the longer the more sensitivity we want. Generally, the largest effect occurs as soon as we have spent a few minutes in complete darkness after arrival from a bright region, but one can even state that it takes at least half an hour until the eye has reached its maximum sensitivity. Even after this half-hour interval a small increase of sensitivity still occurs. In any case, during an observing period in which one has to rely on eyes of utmost sensitivity one should avoid exposing them too much in daytime, and sunglasses should be used.

During a long observing period one should take enough rest and consume fresh vitamin-rich food. Lack of Vitamin A leads to serious night blindness. If one intends to observe for several hours it is advisable to rest for some time as soon as the preparations at the telescope are completed, and to spend the time in a dark or only faintly illuminated room. Obviously during the observation, no bright light should interfere. As soon as the eye gets used to darkness, very little illumination is required for making notes, etc. If, in spite of this, it should really be necessary during the night to switch on some bright illumination, one should close at least one eye and thus maintain its adaptation.

Through its crystalline lens the eye produces on the retina a sharp small image of the object. The vitreous humor also contributes to the production of the image. In the case of a normal eye, the focal point lies exactly on the retina when the eye is at rest. Then objects which lie at infinity are imaged sharply. If we wish to obtain sharp images from nearer objects, then the focal length of the crystal lens must be diminished. This is done through a ring muscle which envelops the lens, and by contracting increases the lens curvature. This process, which generally takes place involuntarily, is called accommodation. The ability to accommodate decreases with age.

It is important for us to set the eye for infinity during observations at the telescope so that the lens always remains without tension.

3.2.2. Defects of the Eye

It is in this context impossible to describe in detail the various defects of the eye. There are effects which originate in a disturbance of the retina or of the optic nerve, e.g., *night blindness* and *color blindness*. While night blindness does hardly disturb observations of the Sun, Moon, and planets, it is very disturbing for all other work; color blindness behaves in just the opposite manner, it actually affects only observations of planets and some solar observations, as well as work with color filters. Both kinds of disturbance can usually only be treated by the eye specialist.

Errors of the eye which are due to the optical design of the eye are not such a handicap for astronomical observation as the above-mentioned ones. Ordinary *short-sightedness* or *long-sightedness* can be compensated for by a corresponding setting of the eyepiece and even the use of spectacles at the telescope can thus be avoided. A less favorable situation arises in the case of *astigmatism*, i.e., for eye errors which cause asymmetrical distortion of the image, e.g., the transformation of a dot into a small stroke. This, too, can be compensated for by spectacles, but generally it would not be possible to achieve compensation in the eyepiece, except by a special turnable device which can be attached to the eyepiece end of the telescope and carries

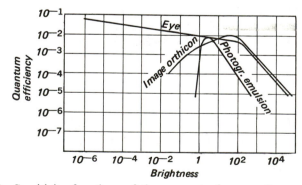

Fig. 3–3 Sensitivity functions of the eye and of some other receivers. After P. B. FELLGETT: Photoelectric devices in astronomy, in *Vistas in Astronomy* (A. BEER, ed.), Vol. 1, p. 479. London: Pergamon Press, 1955.

a suitable spectacle lens. Of course, an instrument so fitted is then only usable by the astigmatic observer.

In conclusion, we should mention the possibility of increasing the sensitivity of the eye by drugs, which are sometimes offered commercially. Frequently these are only substances which increase the opening of the pupil. However, we would like to *warn seriously* against their use. Their effect is practically nil, and the possibility of causing lasting damage is great. On the other hand, there can be no objection to the use of *vitamin preparations* for the removal of night blindness.

Of all radiation receivers the human eye is the most sensitive, the most versatile, and the most adaptable. If we increase its abilities by well-aimed and persistent training, and if we keep it in a good state by a sensible way of life, it will be possible even today to achieve outstanding scientific results in purely visual fields (see Fig. 3–3).

3.3. The Photographic Emulsion

There are very few branches in science that owe so much to the application of photographic methods, as does astronomy. One can say without exaggeration that all essential insights gained in our century have been achieved by photographic methods and indeed most of them could only be achieved in this way.

Nevertheless, it has taken a relatively long time for photography to enter the work of amateur astronomers to some degree. Even today there is a certain aversion against photographic work in these circles, sometimes even with astronomers who are keen photographers in other ways on holidays and elsewhere. This aversion is completely unfounded. Today's photographic advances allow the amateur astronomer to use excellent materials and ingenious instruments. Plates and films can be obtained quite inexpensively and permit a large number of serious and valuable astronomical activities. The amateur astronomer need no longer sensitize his plates himself as he had to do in former years.

The application of special methods will bring advantage and pleasure to the experienced amateur astronomer. However, all those who have not yet gained photographic experience are strongly advised to use material unmodified, that is, just as supplied by the photographic manufacturers.

3.3.1. Theory

3.3.1.1. The Photographic Process

The light-sensitive layer of the photographic emulsion consists of a gelatin layer in which finely divided silver bromide molecules are embedded. The emulsion is poured over a solid base, e.g., glass (plate) or celluloid (film). Then small quantities of other substances are added, which, however, are of greatest importance for the production of the image and partly represent top secrets of the photochemical factories. Silver bromide appears in the emulsion in the form of atomic lattices. The single molecules consist of a positive silver ion and a negative bromium ion. If an incoming light quantum falls onto a negative bromium ion, it causes the redundant electron to be split off. A new atom of neutral bromium and a free electron come into being. The gelatin now contains certain nuclei consisting of atomic silver, silver sulfide, and disintegration products of the gelatine, which capture the freely wandering electrons. The electrons attached to these nuclei cause the production of negatively charged electrostatic force centers. The nuclei themselves now attract again the positive silver ions and discharge them, thus producing metallic silver. In this way, the emulsion now carries an invisible image, called the latent image, which can be developed. Here the reader might like to consult the various handbooks on photography, published by Ilford, Kodak, and other manufacturers.

3.3.1.2. Developing

The purpose of developing is the intensification of the latent image to such a degree that it becomes visible. Even today not all the chemical and physical processes that take place during the developing of the latent image are completely clarified, and various theories have been proposed. The process is essentially a chemical reduction: silver salts are reduced to silver, which appears in the form of black granules. But while in the case of the latent image only a minute fraction of the silver-bromide molecules of the emulsion is used up, the amount is increased about ten million times through the developing process. Here, already reduced molecules take over the role of developing nuclei. If one continues the developing for too long, all existing granules are blackened, but in a shorter time the number of reduced silver atoms is a direct function of the number of the light quanta received by the emulsion.

3.3.1.3. Fixing

The process of fixing consists in the removal of those silver salts that have not been used up during developing; this makes the emulsion durable, "fixed." For this purpose a number of salts are available, such as sodium or potassium thiosulphate. Only after the completion of the process of fixing does the emulsion lose its light sensitivity.

3.3.1.4. Blackening and Gradation

After development, incoming light produces blackened silver granules in the layer. The more silver granules appear in a certain area of the layer, the more intense is the blackening of this spot. In general, an intense blackening also corresponds to an intense illumination of the layer, but the ideal case, that the blackening is proportional to the exposure time, is not usually fulfilled. The connection between blackening and exposure is given by a function which we call the characteristic curve; it is of utmost importance in the application of the photographic emulsion to astronomy. If we plot the blackening of the layer (e.g., the silver grains per square millimeter) as ordinates, and the exposure (product of light intensity and exposure time) as abscissae, using a logarithmic scale, we obtain the required relation between blackening and illumination. This curve is relatively flat for very small exposure times; then its gradient increases rapidly, and, in its central part,

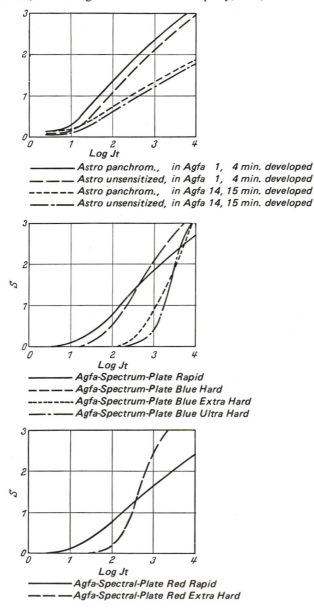

Fig. 3–4 Some examples of characteristic curves.

it remains for some time constant. It is in this part of the curve (see Fig. 3–4) that proportionality indeed exists between the logarithm of the exposure and the blackening. For large values of log It, the curve becomes less steep, and under certain circumstances it can even pass through a maximum and then decrease again. The gradient of this blackening curve is called the *gradation of the emulsion*. It is measured by the angle of the curve in its linear (middle) part and generally defined by stating the value of "gamma," which is the tangent of the angle of gradation. This value of gamma, which increases with increasing steepness of the curve, depends not only on the properties of the emulsion, but also on the development and particularly on the duration of this process.

Emulsions with a flat characteristic curve, i.e., with small gamma, are called soft. They reproduce a certain difference in the incoming light intensities as a relatively small difference in the subsequent blackening; however, they bridge a rather large range of exposure times without leading to over- or underexposure. Emulsions with a steep characteristic curve, that is, large gamma, are called hard; here, for a certain difference in the incoming light intensities the resulting difference in the blackening is relatively large. On the other hand, they only bridge a small range of exposure times. There also exist emulsions in which the linear (middle) part of the characteristic curve is not very pronounced; for these, usually, only an average gradation is quoted and the tangent of the average gradation angle is denoted by "beta."

3.3.1.5. Sensitivity

A photographic emulsion is more sensitive when less light is required to produce a certain blackening. For every particular commercial emulsion the manufacturer quotes a certain value of sensitivity. Unfortunately, no unified system has yet been introduced. Table 3.1 lists the most important systems of sensitivity rating.

Table 3.1 makes it possible to transform the various systems into each other, but we must keep in mind that the various systems are based on different methods for determining sensitivity, so that an absolute relation

Table 3.1. Measuring Systems for Photographic Sensitivity

Exposure factor	DIN	ASA and BSA	General Electric	Weston	European Scheiner	USA Scheiner	Western Scheiner	Hurter and Driffield	Kodak Speed	Ilford Group
16	$9/10°$	6	8	5	19°	14°	15°	500	20°	B
8	$12/10°$	12	16	10	22°	17°	18°	1000	23°	C
4	$15/10°$	25	32	20	25°	20°	21°	2000	26°	D
2	$18/10°$	50	64	40	28°	23°	24°	4000	29°	E
1	$21/10°$	100	125	80	31°	26°	27°	8000	32°	F
0.5	$24/10°$	200	250	160	34°	29°	30°	16,000	35°	G

between two such systems does not always exist. Finally, we must note that the sensitivity data are usually based on measurements over very short time intervals, while in astronomy we frequently have to deal with very long exposure times. This is another reason why these data have to be used with certain caution. Nevertheless, for most cases in astronomical practice they provide a very good indicator regarding the applications of the particular emulsion.

Closely connected with the sensitivity of the emulsion is its graininess. Unfortunately, sensitive emulsions are very grainy while only the insensitive ones have a fine grain. Deviations from this rule are very rare since it is based on fundamental laws of nature. We therefore have to decide to what extent we wish to use a sensitive but grainy, or a less sensitive but fine grain emulsion.[1]

An improvement of the sensitivity of the photographic emulsion for very faint objects can be achieved by cooling. The plate or film, together with the plate holder, is placed in a cool substance (ice or liquid air) and is thus kept at a very low temperature. In this way the limiting magnitude can be increased by up to one magnitude class or even more.

Another property of the photographic emulsion is of utmost importance in astronomical applications, namely its color sensitivity or to put it more precisely its *spectral sensitivity distribution*. An emulsion can under certain circumstances show the same color sensitivity as the human eye; generally, however, it will deviate from this to some extent.

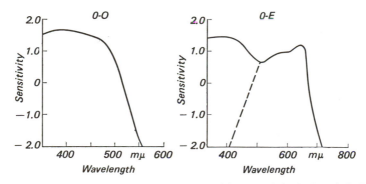

Fig. 3–5 Spectral sensitivity for the emulsions Kodak *O–O* and *O–E*.

The photochemical industry of today provides emulsions for every spectral range between the ultraviolet and the infrared. This makes it possible to take photographs in the light of only a quite definite certain color. If together with this emulsion, *color filters*, which further restrict the spectral range, are also used, it is possible to work in nearly monochromatic light and thus to enter the region between photography and spectral photometry.

[1] An instructive discussion about the usability of numerous commercially obtainable emulsions for astronomical purposes has been given by H. K. PAETZOLD: *Die Sterne* 29, 10 (1953).

Figures 3–5 and 3–6 show the spectral sensitivity distributions for a few of the most frequently used emulsions.

3.3.2. Practice

The application of photographic emulsions for practical work begins with the selection of the best usable material, and its adaptation to the available instrumental equipment. *It is a good idea to avoid extremes.* This is of particular importance to amateur astronomers who are beginners in astronomical photography. Emulsions of the greatest sensitivity as well as the most powerful developers usually give less satisfactory results than emulsions which are not at the limit of today's possibilities (Table 3.2).

This list represents only a small selection. The special emulsions listed are sufficient for the amateur astronomer, and the commercial films shown have proven very useful in practical work. Other manufacturers, Kodak, for example, supply equivalent materials.

Formerly, astronomers used plates almost exclusively; now films are employed to a large extent. Plates are still preferred for permanent record purposes if later measurements are to be made, or when the size exceeds 6×9 cm. Films are very useful if we are taking extensive series of exposures, and if we require rapid work and the size is smaller than 6×6 cm.

Solar work (see page 250) uses very insensitive plates, particularly some emulsions that are used commercially as negatives for slides. Only when we are dealing with photographs obtained with spectrographs or with strongly filtered light do we require normal emulsions of the correct color sensitivity.

Which kind of emulsion should best be used for lunar and planetary photography (see pages 279 and 385) very much depends on the instrument.

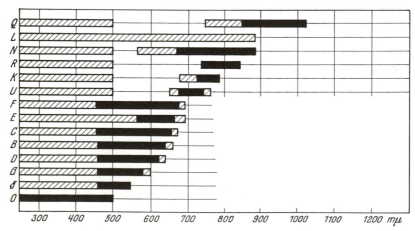

Fig. 3–6 Sensitivity ranges of the most important Kodak emulsions. Black areas denote the regions of highest sensitivity.

Table 3.2. Emulsions in Astronomical Photography

Firm	Name	Spectral range (mμ)	Sensitivity

I. Special Emulsions (Plates)

Firm	Name	Spectral range (mμ)	Sensitivity
AGFA	Astro, unsensibilized	400–500 (blue)	great
AGFA	Astro, panchromatic	550–670 (red)	great
ILFORD	Astra III, panchromatic	400–700	very great
ILFORD	Astra VII	550–650 (red)	great
KODAK	OaO (103 aO)	300–500 (blue)	very great
KODAK	OaE (103 aE)	450–680 (red)	great

II. Commercial Emulsions (Films)

Firm	Name	Spectral range	Sensitivity
AGFA	ISS	panchromatic 23/10 DIN	Very sensitive emulsions, especially for stellar photography
AGFA	Record	panchromatic 32/10 DIN	
AGFA	Ultra	panchromatic 25/10 DIN	
ILFORD	HPS	panchromatic 27/10 DIN	
PERUTZ	Peromnia 25	panchromatic 25/10 DIN	
ADOX	KB 17	panchromatic 17/10 DIN	Average sensitivity suitable for the Moon, the planets, etc.
ORWO	NP 15	panchromatic 15/10 DIN	
ADOX	KB 14	orthopanchromatic 14/10 DIN	
AGFA	Isopan FF	orthopanchromatic 13/10 DIN	
ILFORD	FP 3	panchromatic 19/10 DIN	

The general rule will again be to use sensitive, but not ultrasensitive emulsions.

The same applies, to a certain extent, also to stellar photography of all kinds. Here it must be taken into account that coarse grain, which is unfortunately the consequence of high sensitivity, does not disturb to the same extent when used with long-focus objectives as it would, for instance, with a miniature camera.

The best photographic emulsion is useless when the plate or the film is not placed accurately in the focal plane of the objective. Thus exact focusing is of utmost importance; the plate holder must be of high quality both for plates and films. Miniature film will be used with the usual film changing equipment of the particular camera. Plate holders or cassettes must keep the plate absolutely fixed, without any tension.

It is also important that we have a very reliable arrangement for the focusing of plate or film. For small instruments and normal cameras we move the objective with a worm drive. For larger instruments it is better to displace the plate holder. It is always necessary (if we do not possess an outstanding miniature camera) to have a device for correcting the inclination of the plate. Ideally this should be exactly perpendicular to the optical axis of the objective.

To achieve this, it is best to have three adjusting screws controlling the inclination of the plate holder.

The proper position of the plate holder and the best focus will be found by experimental exposure.[2] This is best done by making on the same emulsion a series of successive exposures (perhaps ten) of a few bright stars, with exposure times of the order of one minute. By adjusting the guiding telescope, it is possible to arrange that single star images appear side by side at regular intervals. Between the single exposures one changes the focusing by a certain recorded amount. Viewing the developed star images under a magnifying glass, the setting for the best focus is quickly established. In the same way the correct inclination of the plate holder can be found, except that in this case one does not choose the star at the center of the plate, but rather three stars, which are, seen from the center, in the same direction as the three adjusting screws. It can happen that the inclination of the plate can only be set satisfactorily in a number of steps, and that the best focus can only be found by several test plates. If we do not use a mirror, but a lens, then the position of the best focus depends, of course, also on the color range of the particular work. It is different for red-sensitive plates than for those in the blue. Also the temperature differences between summer and winter can cause differences in focus, and it is therefore wise to check it from time to time. For this purpose bright Full Moon nights (which are otherwise unsuitable for astronomical work) can be used.

An essential part of the photographic process is the darkroom treatment. The selection of the most suitable chemicals will have to follow the same lines as that of the emulsions. Again, reliable commercial developers, which are offered in such multitude that they will fulfill all requirements, are recommended. It is also better to use only a few types of developer and to become familiar with their operation under different conditions, instead of causing a discontinuity in the work by constant changes and new experiments. See Table 3.3.

For most astronomical work a good fine-grain developer will be most suitable. Developers which bring out the utmost sensitivity should not be used because such an increase in sensitivity is coupled with increased coarseness of the grain, which usually nullifies all other advantages. Of course, the exact temperature and developing time as given in the developer's instruction must be followed exactly.

Films are best developed in daylight containers, plates in simple dishes; or, where it is necessary to develop several plates under exactly the same conditions, in larger tanks. During development the plate or film must be kept in constant motion; however, this motion must not be allowed to damage the emulsion in any way. Afterwards it should be washed in running water before the fixing process. Fixing must be continued for a considerable time, since insufficiently fixed emulsions deteriorate very quickly. While developing

[2] ROBERTSON, R. A.: Simple photography using a knife-edge focusing technique. *J.B.A.A.* **78**, 51–52 (1967).

Table 3.3. Various Commercial Developers

Action	Agfa	Perutz	Hauff	Tetenal	Faber	Adox	May and Baker
Very strong	Methol-Hydro-chinon	MH	MH	MH			
Universal developer	Rodinal	Perinal				Adox E 10	
Fine-grain	Final	Penilin	Atofin	Leicanol	Fabolin	Mikropress	
Ultra fine-grain	Atomal	W 665	Mikrolin	Ultrafin		Super Mikro	Promikrol
Very great sensitivity				Neofin rot	Fa 19 Fa 29		Promikrol

must take place in total darkness, it is possible to use a faint darkroom illumination after about half the fixing period is over. Afterwards we have to apply again a very thorough washing. If this is done in running water, it is advisable to bathe the plate again in distilled water.

Not all amateur astronomers are in the happy position of having a special darkroom at their disposal. Usually, a kitchen or bathroom must serve for this purpose. The utmost cleanliness is of vital importance in darkroom work, and one should choose a comfortable and not too small room. Next to the so-called "wet table," which takes the work with the chemicals, there must also be a "dry table" on which material and instruments can be placed.

Finally it should be mentioned that all photographs, even apparently unimportant test plates, must be carefully numbered and their data kept in a book or card index. These include date and time of exposure of the photograph, exposure time, material, plate filter, object at the center of the plate, details about the developing, and any other remarks (see page 49). See also Fig. 3–7.

To be able to judge the results very quickly, a simple viewing frame is useful. A solidly built wooden box, closed with a not too small ground-glass plate, and housing a 40-watt lamp, will be sufficient for many purposes. If one attaches a magnifying glass or a simple microscope, it is possible to judge the quality of the pictures, to find, for instance, the best among a series of focal tests. Also the search for small planets or comets on the plate is facilitated by such a device (see Fig. 3–8).

3.3.3. Color Photography

Direct color photography of astronomical objects has only succeeded in recent times when the sensitivity of commercially obtainable color films had increased sufficiently. American telescopes have succeeded in obtaining extremely beautiful color photographs of bright nebulae. Nevertheless, direct

Fig. 3–7 Example of the sensitivity of modern emulsions: the constellation Orion (at the center, with the nebula) exposed for 10 seconds on Agfa Record Plates, without guiding, with a miniature camera Exacta Varex IIa and Steinheil Tele-Quinar 1:2.8 (f = 135 mm). Photographed by G. D. ROTH.

astronomical color photography is a difficult field, in which, at least at the beginning, disappointments are unavoidable.

Experience gained in recent years with miniature cameras equipped with good objectives (e.g., Zeiss Sonnar 1:1.5 and f = 50 mm) indicates clearly that such photography of Milky Way and starfields is very promising indeed; exposure times vary between $\frac{1}{4}$ and 1 hour.

The color-sensitive emulsion consists of three layers, each of which is sensitized for a different color. During the exposure the front layers act as a color filter for the lower layers, so that it is possible by a suitable combination of composition and arrangement of the single layers to obtain a composite picture made up of the three fundamental colors, blue, green, and red. More details can be found in the specialized literature.

In 1972 Ernst Brodkorb, in the company of Eckhart Alt and Kurt Rihm, embarked on a special astro-photographic expedition to South Africa. Using a Newtonian mirror-arrangement of 200 mm aperture and 1200 mm focal length, the outcome of this expedition was an outstandingly successful series of celestial photographs. A detailed report by Brodkorb can be found in "Sterne und Weltraum", 1972 (December), p. 347, etc.: "... Each object was photographed three times. Firstly, as a blue picture on Kodak 103 O film, which is sensitive in the blue and ultraviolet; the insertion of a special filter excluded the ultraviolet component of the radiation, which, although invisible to the human eye, would lead to an excessive blue-effect of the image. Subsequently, a green picture was obtained by the use of 103a G Kodak film; this film is sensitive to blue and green radiation, but insertion of a Schott filter KV 470 rendered the film practically exclusively green-sensitive. Finally, a red picture was achieved on a Kodak 103a E film, which as such is blue- and red-sensitive; here, the insertion of a red filter (Schott OG 590, 3 mm) suppressed the blue component. Indeed, the spectral sensitivity of these three colour components corresponds essentially to the sensitivity of the three different emulsion layers of a normal colour film . . . This very principle of colour-film photography can thus be transferred to three individual films which during the processes of exposure and developing are treated completely separately. Finally, the resulting three positive prints of the corresponding colour exposures are then superimposed and furnish in the end a perfect colour picture."

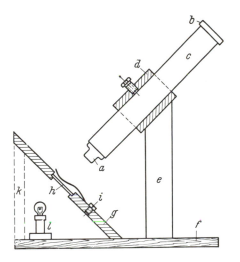

Fig. 3–8 Simple viewing device: a = objective, b = eyepiece, c = microscope, d = attachment ring, e = column, f = base, g = wooden casing, h = ground glass, i = clamp, k = back wall, l = lamp. After AHNERT, Kalender für Sternfreunde. Leipzig: VEB Barth, 1958.

The application to astronomical photography is made difficult by an imperfection common to every photographic method, namely the so-called "reciprocity error." If one wants to photograph two different objects, one of which is ten times as bright as the other, one would expect that the application of ten times the exposure will lead to a good picture of the fainter object, as compared with the single exposure time for the brighter object. However, it turns out that it is necessary to use much longer exposure times for the fainter objects than we would deduce from this simple rule of proportionality.

This explains the difficulty of color photography of astronomical objects. If, for instance, a celestial body radiates much red, but little blue light, it would be necessary to apply a much longer exposure time for the blue than for the red part of the image. Since, however, the color film is a unit, the emulsion can only be exposed with uniform exposure time and this is either too long for the red or too short for the blue light.

This difficulty is insurmountable in principle, but in practice there are a few tricks with the help of which one can partly overcome it. First of all, one will choose an exposure time which is a good compromise between the times required for the blue and the red rays. Furthermore, it is possible to smooth out the relative intensities of the various colors of the object by using a suitable filter. For example, a blue filter will be chosen for a red object, and vice versa. In the individual case all these precautions require many tests before various problems and mishaps are overcome and the correct combination is found.[3]

The amateur astronomer, too, will be pleased with the application of cooling of his colour films during exposure down to temperatures of $-60°$ to $-80°$, which eliminates the Schwarzschild effect to a large degree.

3.4. The Photoelectric Cell

Photoelectric observing methods are now becoming more and more accessible to the serious amateur astronomer. We refer to page 95 and wish to give a few more details about this kind of radiation receiver.[4]

In the same way as in a photographic emulsion the incoming light quanta release electrons from the atoms of the photoelectric layer. While the further processing of these electrons on the photographic plate takes place in a chemical manner, on photoelectric layers it is done electrically. In its simplest form we are familiar with this process in the well-known photoelectric exposure meters. Devices of a similar kind have been used originally for astronomical brightness measurements. Since, however, the astronomically available light intensities are very small indeed, these devices were only applicable to rather bright stars, even with quite large telescopes.

Since the Second World War, secondary electron multipliers have been widely used. Their light sensitivity is so great that even small telescopes are sufficient to measure the light of faint stars. The principle of the multipliers is as follows: a light quantum hits the light-sensitive layer and releases an electron. This leaves the emulsion, and enters a strong electric field in which it is accelerated and at the end of which it impacts on a new layer. Because of its great velocity, it now releases several electrons which again enter a strong accelerating field. This process repeats itself several times until the resulting electron current is so great that it can operate, via an amplifier, a measuring instrument. The value for the brightness of the star can thus be

---

[3] Further information on experiments with astronomical color photography can be found in HANS ROHR, *The Radiant Universe* (English version by ARTHUR BEER). London: Frederick Warne, 1972. See, e.g., A. A. HOAG, p. 35, etc. The columns "Observer's Page" which appear in every number of "Sky and Telescope" frequently contain (since 1973) practical advice for colour photography.

[4] WOOD, F. B.: The present and future of telescope of moderate size. Philadelphia: University of Pennsylvania Press, 1958.

read off immediately. A particularly valuable feature is that the deflections of the measuring device are, within a very wide range, proportional to the incoming intensities (see Fig. 3–9). It is, however, necessary that the voltage used for the acceleration of the secondary electrons is excellently stabilized. A good stabilizer is therefore one of the most important parts of the whole arrangement which is reproduced in our figure. A photometer equipped with a multiplier as radiation receiver can bring in very useful results, even with a telescope of only 15 or 20 cm aperture, in the field of variable star research, for instance. Here, too, it is absolutely necessary that in setting up a program and in its execution close contact is made with the staff of an observatory or with an association of amateur astronomers.

Fig. 3–9 Arrangement of the components of the tube in the 1P21 photo-multiplier. After C. M. HUFFER, Development of photoelectric photometry, in *Vistas in Astronomy* (A. BEER, ed.), Vol. 1, p. 495. London: Pergamon Press, 1955.

The experience of recent years proves that electron-multiplier cells are superior to ordinary photocells, even after the photocurrent has been amplified by a large factor with a suitable electric circuit. The main reason for this is the fact that the multiplier is much less disturbed by "noise," which occurs in all circuits.

Nevertheless, we must pay attention to some sources of error which in particular cases can appreciably restrict the good functioning of the multiplier. First of all, the multiplier possesses, as does every photocell, even the simplest one, a "dark current," which means that a faint current exists even if the cell is not illuminated at all. This is due to the fact that the gas molecules that have remained inside the cell, in spite of the vacuum, release (because of their thermal motion) occasionally electrons from the photosensitive layer. If we have to measure very faint objects, this dark current can become very disturbing. As a remedy, the multiplier can be cooled and the general rule is that a cooling by 15°C weakens the dark current by a factor of 10.

Another sort of error is the emission of electrons at various places inside the cell where great electrical field strengths exists. These release single electrons which produce, in addition to the photocurrent, another current. It is therefore advisable never to use an excessively high working voltage for a multiplier cell.

There is no remedy for another sort of error, namely the release of electrons from the material of the cell wall due to the bombardments by ions of the residual gas. Cooling of the cell helps a little, but not much. A cell which shows this fault to a disturbing degree is best replaced.

Finally, we must also mention the possibility of an electric current along the wall of the cell, which can occur if there is some humidity. The whole cell will, therefore, best be placed in a special casing, which is kept dry by some hygroscopic substance. But even in this case, experience has shown that it sometimes takes weeks before the cell functions perfectly.

Even if all these sources of error are carefully eliminated, we find that the sensitivity of many multipliers depends on temperature. This effect amounts to a few thousandths of a magnitude for each degree Celsius, so that sudden temperature changes of about 10°C can already cause conspicuous errors. Large temperature variations must therefore be avoided. If this is impossible, it is necessary to determine the temperature coefficient of the multiplier (that is the change of recorded brightness per degree C) by careful measurements and to apply this afterwards as a correction to the results.

The spectral sensitivity of various cells is very different. There are multipliers with a sensitivity maximum in the ultraviolet, and others with the maximum in the infrared. According to the requirement in hand the most suitable cell is chosen.

For many observations of a general kind, the multiplier RCA 1P21, which has been developed in the United States, will be used, as it has proved very efficient in the whole range of wavelengths between 3000 and 6000 Å. These multipliers are, of course, rather expensive, but there are also cheaper ones of the same construction, e.g., the type RCA 931–A. None of the types developed for other spectral regions can claim a special preference, and the amateur astronomer will in general always discuss the problem with an expert in this field.

3.5. Television Technique

The recent introduction of television methods into astronomy requires a considerable technical, and of course financial, effort, and will as a rule be out of the amateur's reach. We just mention it here and refer to the details in the literature.[5] Development in this field is in rapid progress, and possibilities for the future are great.

The application of television techniques leads to an enormous gain in sensitivity. Compared with the most sensitive photographic plates, a television device is superior by a factor of 20–30, and in favorable circumstances even by a factor of 100. A disadvantage is low resolution of the television picture, and as always it will be a matter of compromise whether to use this method or any other.

[5] Kühn, R., and Pilz, F.: *Sterne und Weltraum* **2**, 175 (1963). See also a complete up-to-date treatment by J. D. McGee, Image-tube astronomy, in *Vistas in Astronomy* (A. Beer, ed.), Vol. 15. Oxford–New York: Pergamon Press, 1973.

4 | Radio Astronomy for Amateur Astronomers

P. Wellmann and H. A. Schmid

4.1. Introduction

Radio astronomy, which is concerned with the analysis of high-frequency radiation arriving on the Earth from distant parts of the Universe, is still a very young branch among the astronomical sciences. Although originally discovered by K. JANSKY in 1930, the importance of this extra-terrestrial radio radiation became apparent only in the years 1940–1944, when a completely new way was opened up for the exploration of astronomical objects. Compared with the intensity we are used to in broadcasting and television, the radiation from cosmic radio sources is very weak; furthermore, special arrangements are necessary to achieve a sufficient resolution in direction. This requires in many cases very large receiving aerials and very powerful receiving instruments. Special precautions are required to achieve interference-free operation of the instruments. Practical work in radio astronomy therefore demands a large and complex instrumental outfit, and thus great financial means, quite apart from a detailed knowledge of the technique of high-frequency work. In spite of this, more and more amateur astronomers prove themselves to be well versed in electronics and high-frequency techniques. However, it has become possible to use homemade receivers with modern components that are sufficiently sensitive; many are also commercially available at quite reasonable prices. In this way, at least the strongest radio sources are now within the reach of the amateur observer. The following sections aim at giving a survey of the methods and objects of radio astronomy; Sec. 4.5 provides some indications as to the construction of simple "radio telescopes."

4.2. Fundamentals of Radio Astronomy

4.2.1. The Electromagnetic Radiation

Electromagnetic radiation, which—apart from a few exceptions—is the only link between cosmic objects and ourselves, extends over the whole range of wavelengths of the spectrum. Our eye can only receive radiation of wavelengths between 4000 and 7000 Angströms (Å). The neighboring region toward the shortwave side, down to about 200 Å, comprises ultraviolet light; with special lenses and prisms we can manipulate this light and can photograph it with suitable plates and photodetectors. This shortwave limit is further extended by the X-ray and gamma radiation, down to the shortest wavelengths recorded up to now (i.e., about 0.01 Å). These radiations, too, act on photographic plates, but because of the extremely short wavelengths they require a completely different "optics." In the direction toward the long waves we find, first of all, the infrared light. The shortwave part of the infrared is photographically still active and can be recorded with glass optics. From then on, however, we must use photocells as detectors, and the lenses and prisms must be made of material that is transparent to infrared radiation. In a transitional range, at about 0.1 to 0.5 mm, the radiation can still be received by optical means, but at the same time by electrical receivers. If the wavelengths exceed 1 mm, the effects are mainly of an electrical nature, and suitable aerials are used. Here, too, we employ different technical tools for the reception and evaluation of the radiation, we distinguish ultrashort, short, medium, and long waves, respectively. Although the radiation detectors and the "optical" devices used in the various wavelength regions are very different from each other, it is a beautiful fact that the whole complex of phenomena can be fully and accurately described by the same fundamental laws, i.e., essentially by MAXWELL'S equations.

4.2.2. The Absorption of Electromagnetic Radiation

The absorption in the Earth's atmosphere is the reason why we are unable to utilize the whole range of wavelengths, as long as we observe from the Earth's surface. The various atmospheric gases absorb in very different parts of the spectrum. The superposition of all these effects makes our atmosphere opaque for most rays; it is transparent only in two relatively narrow regions. The classical range lies between about 3000 and 10,000 Å. These limits, particularly the upper one, are not very well defined; bright objects like the Sun can still be observed between the spectral bands of water vapor until up to about 100,000 Å. The shortwave limit is due to the content of ozone in the air; in the far ultraviolet, the Rayleigh scattering is dominant, which increases inversely to the fourth power of the wavelength. In the infrared, the absorption is due mainly to water vapor and reaches far into the millimeter-wavelength range. Radio astronomy has only become possible because there exists a second transparency range in the atmosphere, namely, between about $\lambda = 1$ cm and $\lambda = 30$ m. Its upper limit is determined by the conditions in the ionosphere. This particular layer has no uniform structure, but rather consists of different layers of appreciable ionization, the lowest of which is at a height of about 100 km, but reaches up to 500 km and higher. Within these layers, the atmospheric gases (N_2, O, and N) are ionized by the ultraviolet and high-energy particles radiated from the Sun; the degree of ionization therefore varies with the intensity of this excitation. It shows daily and annual variations apart from considerable fluctuations that are correlated with solar activity, and thus also displays an 11-year cycle.

The action of the ionosphere on the high-frequency radiation can be described as a reflection effect due to the ionized layers. This effect reaches into shorter wavelengths the stronger the ionization. This is why the region of transparency in the meter range becomes narrower when solar activity is greater. This radio-astronomical "window" is restricted toward the centimeter waves by the increasing amount of water vapor in the air. These restrictions would be removed for an extraterrestrial observatory, and this is a reason for aiming at satellite and lunar observatories.

4.2.3. Thermal Radiation

The surface brightness of a radiating body can, under certain assumptions, be represented by Planck's radiation law as a function of the wavelength λ and the temperature T. The assumption is that the radiator is a so-called "black body," i.e., that it completely absorbs all incident radiation. Many astronomical objects, in particular the photospheres of the stars, fulfill this condition relatively well, and thus Planck's law can be used as a good approximation. The radiation $I(\nu)$ is given by

$$I(\nu) = \frac{2h\nu^3}{c^2} \frac{1}{e^{h\nu/kT} - 1},$$

where $\nu = c/\lambda$ is the frequency, λ is the wavelength, h is Planck's elementary quantum, k is Boltzmann's constant, and c is the velocity of light.

In radio astronomy, we deal with such low frequencies that we can replace this expression in very good approximation by the much simpler one

$$I(\nu) = \frac{2kT}{\lambda^2} = 2.77 \cdot 10^{-23} \frac{T}{\lambda^2} \text{ W/m}^2 \text{ Hz ster } (\lambda \text{ in m}) \qquad (1)$$

This is the case of the "thermal radiation." As does every other radiation, it originates in quantum-mechanical emission processes involving atoms and ions, and these processes depend on the characteristics of these particles and on their present conditions. Since Planck's equation does not contain any such factors, it follows that it represents a statistical law, which already contains averages made up of a great many individual processes.

4.2.4. Nonthermal Radiation

Frequently, however, the conditions for the validity of Planck's equation are not fulfilled. This applies particularly to very diffuse gases that are too transparent to absorb the radiation completely and can therefore not be considered as black bodies. Here the calculation of the radiation must take into account the individual radiation processes.

An important example of such a nonthermal process is the emission of synchrotron radiation. It is demonstrated theoretically and experimentally by the electrodynamics that the free electrons radiate energy as soon as they are accelerated, that is, as soon as their motion is not any longer uniform and on a straight line. An electron that moves in a circular orbit will thus radiate electromagnetic radiation. This situation occurs when the electron moves perpendicularly to a magnetic field. We may then consider the moving electron as an electric current, which is subject to a force due to the magnetic field in accordance with the fundamental laws of electrodynamics. This process produces circular orbits that are perpendicular to the magnetic field or spirals that are a combination of such a circular motion and of a uniform displacement in the direction of the lines of force. In the large electron synchrotrons, the electrons traverse a magnetic field on circular orbits; the resulting radiation has been found experimentally and has been named accordingly. Since many cosmic objects contain both fast electrons as well as sufficiently strong magnetic fields, it is often possible to trace the nonthermal radiation of both the optical and the radio range to synchrotron radiation.

There also exist other nonthermal radiation processes that are particularly active in the range of the radio frequencies. They depend on very complicated, and even today not yet completely understood, phenomena in highly ionized gases. If a gas contains many ions and free electrons, that is, many free electrically charged particles, then we observe, apart from the normal forces (like gas pressure and inner friction), additional electromagnetic forces, since the moving charge carriers produce electric fields in accordance with the laws of induction. The special scientific branch that deals with this interplay of mechanical and electrodynamic forces and their action on the motion and the condition of such a gas (a so-called plasma) is called magnetohydrodynamics. Its fundamental laws are very complex differential equations, for which general solutions have not yet been found. Even those cases that have been completely explored numerically can only be considered as very simplified models of the real situation. It is, however, now possible to state that vibrations can develop within the plasma, which in the limiting case can produce electromagnetic oscillations and thus radio radiation.

4.3. The Instruments

4.3.1. Aerials

We shall now consider the design of a device that is able to record cosmic radio radiation and the direction from where it comes. We are used to calling an instrument of this kind in optics a telescope, and the radioastronomical counterparts are therefore briefly named "radio telescopes." If we wish to emphasize the measuring process, we also speak of radiometers. If the instruments do not work in fixed wavelength regions and if λ can be changed continuously, we speak, in analogy to optical spectrographs, of radio spectrometers.

A radio telescope consists of two parts, the aerial system and the receiver. Since we are working in the range between centimeter and meter waves we should expect to use the same type of instrument as in the normal shortwave and ultrashortwave broadcasting. Indeed, such receivers have been the starting point for the development of special aerials and amplifiers.

The basic form of the aerial is an ordinary dipole [Fig. 4–1(a)]. It consists in its simplest form (as half-wave dipole) of a rod of the length $\lambda/2$, which is divided in the middle, with its inner ends connected to the receiver. The electric field strength of an incident high-frequency radiation produces within the rod periodic motions of the electrons, that is, a high-frequency alternating current, which is then applied to the receiver. Since we are here concerned with transverse oscillations, the effect will depend on the direction of the oscillation and on the direction of propagation of the emission; it is strongest when the radiation oscillates in the direction of the rod, i.e., when it arrives perpendicularly to the rod; it is zero when the wave proceeds in the direction of the rod. The dipole rod is therefore most effective and most sensitive perpendicularly to its direction. If one plots the sensitivity (which equally applies to reception as to emission) as a function of the direction of the beam (e.g., as power expressed in an arbitrary unit), we obtain the so-called aerial polar diagram. We can see immediately its essential point, if we wish to find the direction to a radiating object: the polar directivity diagram must be of such a kind that the sensitivity remains mainly restricted to an angular region that must be kept as small as possible. The simple dipole is still far off from this ideal case [see Fig. 4–1(a)].

The requirements of terrestrial broadcasting have already led to the construction of aerials with directional characteristics, the aim being to single out certain transmitters from the general radiation field. This led to the development of the well-known television aerials, and in particular to the YAGI aerials [see Fig. 4–1(b)]: a second rod is fixed parallel to the dipole

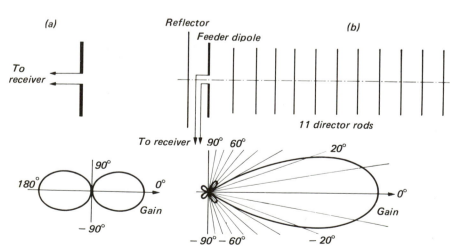

Fig. 4–1 (a) Half-wave dipole aerial, and (b) 13-element YAGI aerial with its directional polar diagram.

(on the opposite side of the transmitter) in such a way that it reflects the incident waves and thus improves the reception on the desired side while it disturbs it for the other one. In addition, we can also attach—again parallel to the dipole on the transmitter side—a sequence of (up to 20) additional rods, the so-called directors. They bundle the incident radiation provided they are fixed at the exactly calculated or tested distances, giving more gain the larger their number. They are most effective in the plane perpendicular to the dipole [Fig. 4–1(b)] and are less effective within their own plane. The dimensions are an optimum only for a given wavelength region, but the more directors used, the wider the band covered by the aerial. Frequently, to suppress branches of the polar diagram, which points backward or sideways, additional reflectors are added.

These directional diagrams can be characterized by a few data. Thus the gain G is the ratio of the maximum sensitivity to the sensitivity of an undirected aerial of the same total energy input. The angle between the direction of maximum sensitivity and the direction in which the sensitivity of the receiver is reduced by one-half is then doubled and called the half-width β of the aerial diagram [see, for example, Fig. 4–A(b), which shows β to be about 50°]. The half-width is a particularly important quantity since it is a measure of the direction-resolving power of the aerial. Two emitters must be apart by at least this angle β for the aerial to register them as separate objects.

Also other methods have been used to increase the performance of the receiver and the directivity of the aerial. Let us assume that two dipoles are parallel side by side and connected by such a circuit that both radiation waves arrive at the receiver at the same time, i.e., that they arrive "in equal phase." In the case of radiation which arrives perpendicularly to the plane of the dipole, the contributions of both part-aerials will add up in the receiver. The wave fronts of obliquely incident radiation reach the two dipoles at different times and excite them with different phases. If the phase difference of the excitation is just a half-wave, we obtain aerial currents that are opposed and cancel each other in the receiver. The angle between the directions where this happens contains the main sensitivity region of the aerial system, and its polar diagram demonstrates a corresponding half-width. Similar considerations can be applied to a larger number of dipoles, which are regularly placed side by side. The achieved resolving power β depends on the wavelength λ and on the length D of the series, and is in a plane perpendicular to the dipoles:

$$\beta \sim 50\frac{\lambda}{D} \qquad \text{(in degrees)}.$$

If now several such dipole series are placed above each other, that is, if we build a kind of "aerial wall" (Fig. 4–2), we increase the resolving power also perpendicularly to the series. The gain due to a wall consisting of m rows, of n elements in each, is:

$$G = 3.28nm.$$

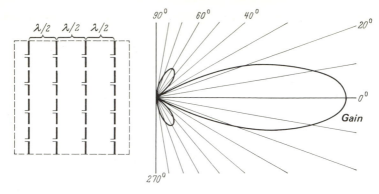

Fig. 4–2 A broadside array of 16 dipoles and its horizontal directional pattern.

Fig. 4–3 Equatorially mounted parabolic reflector (diameter 3 m, wavelengths 13 cm and 21 cm) built by SIEMENS and by amateur astronomers at their observatory in Munich.

The performance of such an arrangement is improved if the simple dipoles are replaced by YAGI aerials, or we can place a reflecting sheet one quarter of a wavelength behind the aerial. This can consist of a metal sheet, but usually a net of wire mesh will be sufficient, provided that the openings of the meshes are considerably smaller than the wavelengths to be received.

The reflector action of such a conducting layer makes it possible to build radio telescopes that correspond exactly to optical instruments. If we design a parabolic reflector aerial, which is approximately ten times larger than the wavelength, we will collect the reflected energy approximately within a single focus (Fig. 4–3). It is then only necessary to attach at this place a reception dipole. The resolving power then becomes (assuming optimum properties of the dipole) the same as the corresponding optical reflecting telescope:

$$\beta \sim 59 \frac{\lambda}{D} \quad \text{(in degrees),}$$

where λ is the wavelength and D the diameter of the reflector.

The great difficulties encountered by radio astronomy can now be easily recognized. Since the resolving power must always be proportional to the wavelength, its value for radio astronomy will be poorer than in optics (for the same given dimensions) by a factor of about one million. To compensate for this disadvantage as much as possible, radio telescopes must have extremely large dimensions. At the moment the record is about 110 m diameter for a movable telescope (Green Bank, West Virginia) and 305 m for a fixed paraboloid (Arecibo, Puerto Rico).

Nevertheless it is possible to increase the resolving power still further if we apply the well-known principle of the MICHELSON interferometer from optics to radio astronomy. Here, the basic considerations are just the same as we discussed for the dipole pair, only that we are now joining two or more complete radio telescopes that are very far apart, perhaps hundreds of meters and in some cases hundreds of kilometers (Fig. 4–4). Special methods guarantee that the theoretical resolving power can be applied. The correct phase combination of such systems offers problems that can be solved only with great technical effort, but resolving powers of much less than a minute of arc have been achieved in this way.

Another concept that is essential for the assessment of the performance of an aerial system is the reception area A. If the aerial extracts N W/Hz from the radiation flux S' W/m² Hz of a radio source and applies it to the receiver, then A is defined by

$$A = N/S'.$$

This reception area is related to the gain G by the equation

$$A = G \frac{\lambda^2}{4\pi}. \tag{2}$$

Accordingly, the effective aerial surface can be calculated for any aerial

Fig. 4–4 The 32-element interferometer of the Radiophysics Laboratory at Sydney, Australia. Its parabolic reflectors are 7 m apart, and each is of 2 m diameter; $\lambda = 21$ cm, $\beta = 3'$. (Photograph: C.S.I.R.O., Radiophysics Division, Sydney.)

whose G is known to us. For a parabolic reflector, or for an aerial wall with a not too large distance of the single elements, we obtain a value of A that is nearly equal to the geometrical surface. For a Yagi aerial the effective receiving surface is given by an effective cross-section for the radiation. We obtain, for instance, for a system of 9 directors:

$$A = 2.54\lambda^2. \tag{3}$$

If we know A, as well as the minimum performance N to which the receiver still responds, we are able to state the "brightness" of the radio sources that will be still observable with our instrument (see Sec. 4.5.3.). Since the aerials always receive radiation in only one plane of polarization they only make use of half the energy flux, that is, $S' = S/2$. Furthermore, the instrument responds only to a certain bandwidth. If we call this bandwidth B, we have

$$P = \int N dv = \int A S' dv \approx \frac{ASB}{2} \, \text{W/m}^2 \tag{4}$$

for the power gained by the aerial.

4.3.2. Receivers

It should be possible to connect the receiver with an amplifier of practically unlimited amplification and thus to achieve any amount of sensitivity. This, however, would be of little purpose since the receiver output is not only the received signal, but also comprises, equally amplified, all disturbances in the receiving system, the so-called "background noise." What actually matters is therefore not the absolute sensitivity, but the "signal to noise" ratio. We are here in a quite similar position to optics: it is of no use to increase the geometrical-optical magnification of the telescope to such a degree that in the end we only magnify the scintillation and the diffraction disks, without being able to see any more detail.

This noise level of the receiver has different causes. One of its important components is the resistance noise, i.e., the irregular fluctuations of a current consisting of a finite number of electrons. It depends on the absolute temperature of the aerial and receiver and is, in principle, unavoidable. A resistance transfers the noise,

$$P_R = kTB,$$

to another resistance of the same size. Additional noise contributions are due to the statistical fluctuations of the processes in the valves and transistors. The noise temperature is in the most favorable case given by the temperature of the receiver (i.e., about 290°K). The imperfections of the instrument increase this value (we assume that the noise that originates in the receiver can be represented by the noise at the input end of a noise-free instrument of the same output amplification). We can therefore write

$$P_R = kT_R B, \quad T_R = F \cdot 290°\text{K}. \tag{5}$$

The quantity F is a measure of the additional noise contributions.

Receivers must be as noise-free as possible. For this reason new valves were developed, masers used, certain components specially cooled, and special circuits invented (e.g., the so-called "parametric amplifier"). We shall not enter into any technical details here, the more so since they are without use to the amateur astronomer. We only mention that the optimum-noise figure, F, can reach about 0.1; the normal receiver has F-values of between 3 and 10.

We shall now discuss how the reception can be made visible. We can use an indicator with a pointer or a continuous recording device. The reaction time of the instrument, its so-called "time constant," must be of such a size that noise fluctuations are smoothed out without affecting the existing time variation of the desired signals. These time constants usually lie between 1 and 0.5 min. Irregular fluctuations in a recording can afterwards be eliminated by forming mean values. Ingenious circuits can be designed that correct for changes in the receiver sensitivity (see Sec. 4.5.2). If we use such methods and choose the bandwidths of the reception and the time constant of the recording in a suitable way, it is possible to detect the existence of signals of 1/1000 of the receiver's noise level. The amateur observer, however, will be unable to utilize such an extreme situation, and he will have to be content with a signal strength of at least about one-tenth of the noise.

4.4. The Radio-Astronomical Objects

It is impossible to list, and still more to discuss, all radio-astronomical phenomena. The following compilation, which gives the radiation flux or other brightness equivalents, should only help in assessing the possibilities for the amateur radio astronomer. For extended radiation sources we are able to relate the energy flux S to the surface brightness I of the object. If the source extends over a solid angle ω, we have

$$S = \bar{I} \cdot \omega. \tag{6}$$

Here $\bar{I}$ represents the mean value of I taken over ω.

In the case of thermal radiation, it is PLANCK's Law (1) which determines I. If the radiation is not thermal, it is usual to calculate from this equation the so-called equivalent temperature T_E, as a measure of the surface brightness.

4.4.1. The Sun

The most intense radio signals we receive originate in the Sun. These signals are not constant, as even the radiation of the "Quiet Sun," that is of a Sun free of spots and centers of activity, varies within an 11-year cycle. We have, for instance,[1]

Table 4.1.

	λ	0.03	0.2	1.0	6.0 m	
S	Minimum of activity	248	40	11	0.5	$\cdot 10^{-22}$ W/m² Hz
	Maximum of activity	270	64	16	0.6	

In addition, there is a slowly varying component which is strongly correlated with Wolf's sunspot number R, and which we may call the spot component.

---

[1] LANDOLT-BÖRNSTEIN: *Astronomy and Astrophysics*, Vol. 1, pp. 117 and 133. Heidelberg: Springer (1965).

Table 4.2.

λ	0.03	0.2	1.0	6.0 m	
ΔS R = 100	60	56	2.5	(0)	·10^{-22} W/m² Hz
R = 200	120	125	5	(0)	

This radiation is superimposed by rapid emissions from the centers of activity, for instance, by "radiation bursts" (Fig. 4–5). The main phenomenon, which lasts about 5 to 10 min, is usually followed by a slow fading, which can extend over several hours. The most intense radiation bursts in the centimeter and meter region reached intensities of $S = 10^{-18}$ W/m² Hz or more, lying generally between $S = 10^{-21}$ and 10^{-20} (in the centimeter region). The mean frequency of such a phenomenon somewhat increases with time ("frequency drift").

Furthermore, we observe for wavelengths $\lambda > 1$ m the "radio storms" ("noise storms"), which consist of a large number of radiation bursts, each lasting for a few seconds only; they can remain observable for several hours or many days. The S-values for the radiation bursts can exceed 2.10^{-20} W/m² Hz. Isolated bursts that occur in the shortwave range up to a few centimeters are usually considerably weaker; like the other bursts, they are restricted to narrow frequency regions ($\Delta\nu \sim 100$ MHz). This shows that these emissions have no thermal origin.

4.4.2. The Planets

The planets have a permanent radiation that corresponds to their surface temperature, for instance, in the case of Venus of about 700°K. Because of

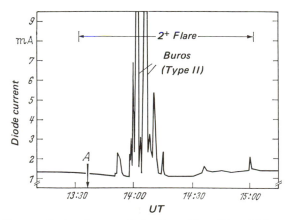

Fig. 4–5 Two radiation bursts of the Sun (Type II), which accompanied an optical flare recorded on 22 October 1963. At the point *A* preceding the flare, occurred a sudden cosmic-noise disturbance. From ALEXANDER and BROWN, *Sky and Telescope* **29**, 212 (1965).

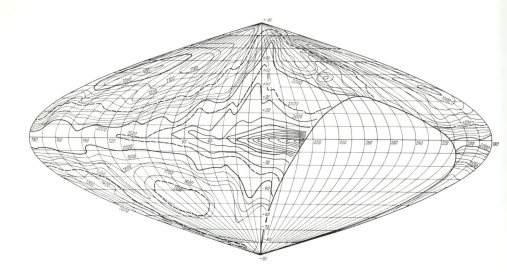

Fig. 4–6 Isophotes of the radio brightness of the Milky Way at $\lambda = 3.7$ m. The figures indicate equivalent temperatures of galactic coordinates; the plane of the Galaxy lies along the equator. After Baldwin, J. E., *Monthly Notices of the R.A.S.* **115**, 684 (1955).

the increase of the thermal radiation with decreasing wavelengths, this weak radiation is measured in the centimeter and millimeter region. For the temperature of 700°K we obtain a value of $S = 1 \cdot 10^{-24}$ W/m² Hz at $\lambda = 1$ cm, taking the distance from the Earth equal to one astronomical unit.

Jupiter, and perhaps also Venus, shows radiation bursts, which because of their narrow bandwidths are certainly not of thermal origin. In the case of Jupiter, they have been found for wavelengths between 3 and 60 cm; and at $\lambda = 60$ cm they can reach[2] $S \sim 10^{-25}$ W/m² Hz.

4.4.3. The Milky Way

The HI regions of the Milky Way show emissions due to the hydrogen line at 21 cm. These observations are of fundamental importance for the analysis of the structure of the Milky Way. The identification of these lines of relatively small bandwidth (below 1 MHz) and of small intensity relies, however, on work with expensive instruments and will therefore be mentioned here only very briefly.

[2] See Weaver, H.: *Solar System Radioastronomy* (J. Aarons, ed.), p. 391 (1965).

The Galaxy as a whole has a radiation that is nonthermal and corresponds at $\lambda = 1.5$ m in the direction to the galactic center to an equivalent temperature of $T_E = 1300°$K, whereas it decreases at the galactic equator by a few hundred degrees; at the galactic poles it is hardly $100°$K. At longer wavelengths, its intensity increases, and at $\lambda = 3.7$ m we measure a temperature of about $15,000°$K at the center, and of about $2000°$K in the opposite direction, while $800°$K is the minimum value near the poles (see Fig. 4–6 and Fig. 4–7).

4.4.4. Discrete Radio Sources

In addition, we know a large number of radio sources of a very small angular diameter, the so-called radio stars. Some can be identified with unusual galactic objects, for instance with nebulae, which are also anomalous in the optical wavelength region; others are extragalactic. Table 4.3 lists the brightest of them together with their intensities.

Nowadays, one defines "radio brightness" in a similar manner as the astronomical magnitudes by

$$m_r = -53{.}^{m}45 - 2.5 \log S \ (\lambda = 1.90 \text{ m}).$$

The Crab Nebula (Taurus A) deserves particular mention; it probably originated as a consequence of a supernova explosion, and it emits in the optical part of its spectrum nearly completely polarized light. This and other facts make it highly probable that we are dealing here with synchrotron radiation.

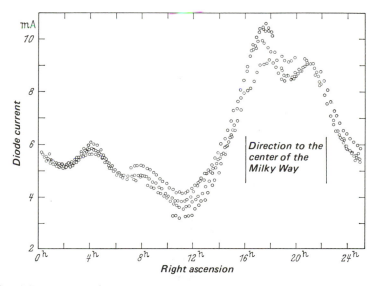

Fig. 4–7 Record of the Milky Way radio radiation at $\lambda = 10.5$ m with an antenna (4 YAGI aerials) at a fixed declination of $+5°$ to the Meridian. ALEXANDER and BROWN, *Sky and Telescope* **29**, 212 (1965).

Table 4.3.

Radio source	Photographic object	m_{Pg}	$S(\lambda_0)$ in W/m^{2}Hz	λ_0	m_r	Extent
Cas A	Galactic emission nebula	—	$220 \cdot 10^{-24}$	3.7m	—	4′
Cen A	NGC 5128	7^{m}1	18.5	3.0	2^{m}1	4′
Cyg A	Double galaxy	17.9	135	3.7	1.7	30″
Per A	NGC 1275	13.3	1.0	3.7	—	—
Pup A	Galactic emission nebula	—	35	3.0	—	30′
Tau A	NGC 1952	9.0	12.5	3.7	—	5′
Vir A	NGC 4486	9.6	10.5	3.7	4.0	5′

The radio emission of most of the more intense radio stars is approximately proportional to the wavelength and can therefore not be of thermal origin.

Recently some objects of particular interest have been discovered, the so-called "quasars," i.e., quasistellar radio sources. They are very small ($< 1″–10″$), and their apparent brightness in photographic light is only $16^m–19^m$. They display unexpectedly large red shifts (up to $\Delta\lambda/\lambda = 2$) within their line spectra, which indicates very large distances. Their absolute brightness is very great, and their masses were estimated to be of the order of about one million solar masses. They emit an intense radio radiation, but are too distant to be observable with the means available to the amateur radio astronomer.

4.5. The Instrumental Possibilities of the Amateur Astronomer

4.5.1. The Aerial and Its Site

Yagi aerials and simple dipole combinations are, as has become obvious in Sec. 4.3.1, probably the most suitable forms of aerial within the technical and financial reach of the amateur observer, and they can at once lead to interesting observations. Yagi aerials can be commercially obtained in many different forms as television aerials. We can combine them to systems, either by placing them side by side without any alteration, or by replacing their reflector rods by a common reflecting surface, as we described it previously for the dipole wall. The cable connecting the single elements must be correctly matched, i.e., the wave resistances of the aerial and of the aerial cable must be equal. Also the branching points of the current must be carefully arranged. We refer to Hyde's book.[3] Instead of the Yagi aerials, we can also use simple half-wave or whole-wave dipoles. Directions for design and circuit are also found in the literature.[4]

---

[3] Hyde, F. W.: *Radio Astronomy for Amateurs*, p. 170 (1962).

[4] See Meinke-Gundlach: *Taschenbuch der Hochfrequenztechnik*, p. 365 (1962). Also: F. W. Hyde in the preceding footnote.

The details of the technical construction do not offer any particular difficulties.[5]

The site of a "radio observatory" must be selected very carefully. It is necessary to avoid densely populated and industrial regions, as well as the neighborhood of roads with heavy traffic, electric railways, and air routes. The sparks originating from contacts, electromotors, and ignition of car engines, as well as radio and television receivers that radiate signals, the fields of high-frequency medical devices, and other electrical installations can cause considerable disturbances, even if these instruments have the usual and legally prescribed suppressors. Also the vicinity of broadcasting stations, television transmitters, and relay stations must be avoided. If the work must be done in a town, one will best settle on a high roof. Although the average noise level decreases at higher frequencies, there can be strong local maxima. In general, it is advisable to place the dipole and YAGI aerials horizontally, as in this case they accept the disturbances only within a small directional range.

4.5.2. Receiving and Measuring Devices

Radiation due to cosmic sources does not have the character of a constant or continuously variable signal of a certain frequency, but is composed of very rapidly fluctuating components or pulses that follow each other very rapidly and irregularly, usually with a wide bandwidth. The radiation has the character of noise, which is familiar to us (e.g., atmospheric disturbances or thermal noise) and is only well defined in terms of a statistical mean value of the intensity. The cosmic noise is superimposed on the normal receiver noise, and our radio telescope must be able to separate these two contributions. These conditions make it necessary to apply statistical, and thus also thermodynamical, considerations, which we cannot discuss here in detail, although we are going to apply their results.

We start out from Eq. (4), giving the radiation flux received by the aerial. If we observe a source of the temperature T_Q and the solid angle ω_Q, we obtain (7) from Eqs. (1), (6), and (4) [replacing A according to Eq. (2)]

$$P_i = \frac{\omega_Q}{4\pi} G k T_Q B. \tag{7}$$

Thermodynamics shows that it is meaningful to define an "aerial temperature" T_A by

$$T_A = \frac{\omega_Q}{4\pi} G T_Q. \tag{8}$$

This is the temperature that the aerial accepts in its interaction with the incident radiation. The limiting case of thermodynamical equilibrium, and thus of a uniform temperature of the whole range covered by the aerial,

---

[5] See, for instance, MENDE, H. G.: *Praktischer Antennenbau, Radiopraktiker-Bücherei*, **50** (1963).

would yield $T_A = T_Q$. With the help of the aerial temperature we can replace Eq. (7) by

$$P_i = kT_A B.$$

The receiver receives this flux and in addition the background noise of the instrument, corresponding to the equivalent noise temperature T_R, i.e.,

$$P_R = kT_R B.$$

The sum of these contributions produces at the output of the receiver, taking into account an amplification factor V,

$$P_a = Vk(T_A + T_R)B.$$

When the parameters V, T_R, and B of the receiver are known, we are able to calculate from the measured P_a the temperature T_A and thus the intensity S of the radio flux. The order of magnitude of T_R is given by the noise data listed in Sec. 4.3.2. In general, T_A is smaller than T_R.

Statistical considerations show, furthermore, that the measurements of a noise intensity scatter around a mean value by the amounts

$$\Delta P \sim \frac{P}{\sqrt{Bt}} \quad \text{and} \quad \Delta T \sim \frac{T}{\sqrt{Bt}}, \tag{9}$$

respectively, where t is the time that is available for averaging the single short-noise pulses, i.e., if t is equal to the time constant of the receiver (see Sec. 4.3.2). An increase of the output P_a by incident radiation can only be identified if T_A changes by more than ΔT. To ensure a safe identification, this increase should even be larger by a factor of 5 to 10. This requirement limits the performance of the receiver.

First we shall deal with the simplest case, namely a receiver that directly reproduces the received signal. It should be selected according to the range of frequencies within which we intend to receive radio radiation. There exists a whole series of commercially manufactured instruments that can form the basic unit of our radio telescope. In the range up to about 30 MHz, these are the shortwave receivers of high quality. On the other hand, their bandwidth of usually 9 kHz is very unfavorable for our purposes, and by corresponding rearrangement and rebuilding of the band-filter component this should be increased to 100 kHz or more. Ultrashortwave receivers, built for wavelengths around 3 m, are adapted for frequency modulation, but they can yield for our purposes a suitable rectified voltage at the radio detector. The advantage of these receivers is their large bandwidth (250–300 kHz). For higher frequencies (up to 800 MHz), we can use the receiver parts of television sets with one demolution stage. These have a large bandwidth, $B = 5$ MHz, and Cascode circuits, which have low noise and are often used as high-frequency input stages. The receiver is directly attached to the aerial; the measured quantity is the voltage obtained after the high-frequency rectification. A diagram illustrating the transformation of the signal is given in Fig. 4–8. The instrument becomes particularly simple if we disregard the direct-current signal and if we render the high-frequency noise (whose amplification during the transit of an object through our aerial lobe is also a measure of the incident flux) audible in a loud speaker or record it with the help of a tape recorder (see HYDE, p. 216ff) in the reference given in Footnote 3 of this chapter.

Fluctuations of the receiver amplification can simulate the occurrence of some additional radiation. It is therefore advisable to check the instrument by frequent calibration. For this purpose, we replace the aerial for a short time by a noise generator of known performance, whose output resistance is equal to the aerial resistance. For such a generator, we can use simply an ordinary Ohmic resistance of known temperature (which is normally the room temperature).

This simple auxiliary device can be replaced by a noise diode, which has the advantage that its noise level can be varied conveniently and adapted to the actual requirements. The noise produced corresponds to the temperature T_L, namely:

$$T_L = T_w + \frac{eIR}{2k} = T_w + 5.8 \cdot 10^3 \, IR. \tag{10}$$

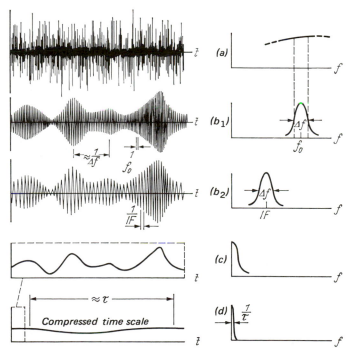

Fig. 4–8 The changes in wave form and spectrum of a signal passing through a receiver, time function on the left, frequency function on the right. (*a*) The input signal; (*b₁*) the noise-modulated wave of frequency f_0; (*b₂*) the signal converted to the intermediate frequency; (*c*) the rectified signal; (*d*) the signal *c*, after having been smoothed (now relatively constant). After PAWSEY, J. L., and BRACEWELL, R. N.: *Radio Astronomy*, p. 36, Oxford (1955).

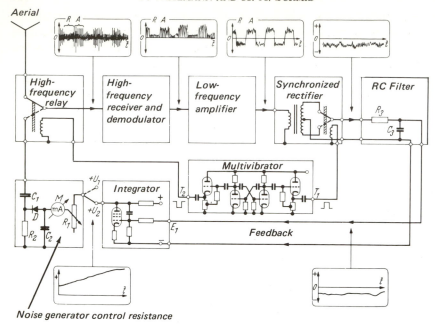

Fig. 4–9 Schematic diagram of a complete receiving system, according to Ryle-Vonberg.

Here, the output resistance R has the temperature T_w, the anode current is I, and e denotes the charge of the electron. The diode must be operated in the saturation region of its characteristic, according to the circuit given in Fig. 4–9.[6] With its help it is possible not only to test the stability of the receiver, but also to calibrate the instrument by varying I, and by relating the temperatures calculated from Eq. (10) to the deflections at the receiver output.

It is possible to use, in addition, the noise generator to make the reading of the receiver independent of variations in its amplification. This is, in principle, done by alternately connecting the aerial and the noise generator to the receiver and adjusting the noise generator until there is no difference between both sources of signals. We then require the determination only of T_L, or of the corresponding noise performance P_L, since $T_L = T_A$.

The comparison of the aerial noise with the generator noise is much improved by a low-frequency rectifier, which is synchronized with the input-switching, and yields the output-voltage zero as soon as the adjustment is complete, or a positive or negative voltage according to whether T_L is larger or smaller than T_A (see Fig. 4–9).

The switching over could be carried out at a frequency of 10–100 Hz. For this purpose we may use two relays, which may be operated with the 50 Hz-mains; the one nearer to the receiver input must be a special high-frequency relay.

[6] See, for example, Pawsey, J. L., and Bracewell, R. N.: *Radioastronomy*, p. 46 (1955).

We can still go one step further and arrange for an automatic balancing of the noise generator by using the output voltage of the synchronized rectifier for the regulation of the noise diode. We shall discuss below a few examples of the various possible types of circuit (see Fig. 4–9).

If manual operation is intended, a certain voltage is applied through U_1. The diode current is adjusted by the setting of R_1 and is read off at M. It is of great advantage to record it with one of the commercially obtainable compensation recorders with the characteristics of a milliampere meter. The noise level, which is applied through C_1 to the receiver, is calculated from the diode current.

The synchronized low-frequency rectifier is similar to a push-pull rectifier, in which the diodes are replaced by a switch operated by a relay. The pole reversal has as a consequence that both half-waves of the low-frequency amplifier output at (R_3, C_3) always will appear with the same sign: positive if P_L is larger than P_i, and vice versa. (One has to ensure, of course, that the control voltage acts with the correct sign on the noise generator.)

If the relay in our circuit is not controlled by the mains, or if one intends to use the electronic switches described below, then we must obtain two square waves of opposite phase using a multivibrator. Here, we can use the standard circuits given in many publications (see Fig. 4–9).

For automatic operation we connect this noise generator with U_2. The direct current, obtained at the RC component (R_3, C_3), the size and sign of which are a measure of $(P_i - P_L)$, is applied to the integrator E_1. As a consequence of this, the voltage at U_2 increases as long as there is a negative voltage at E_1, that is as long as $(P_i - P_L) > 0$. This increases P_L until we reach equality with P_i. (Again the same precaution mentioned above, now concerning the feed-back voltage, must be taken.)

The mechanical relays, which have been assumed in Fig. 4–9, can be replaced by electronic switches. The high-frequency relay can be replaced by the circuit in accordance with Fig. 4–10(b). The aerial and noise generator are connected with the first two grids of two heptodes, whose second control grids are coupled capacitively with the two multivibrator outputs. In this way, the valves are alternately closed and opened, and because of the different signs of the voltage on T_1 and T_2 there is always one valve open when the other is closed. The electronic switch has a great disadvantage in comparison with the mechanical relay: it increases the noise at the receiver input. In any case, both valves must produce the same additional noise. Variations in the amplification can be eliminated by a negative feedback arrangement of the cathode resistances.

Instead of the relay-controlled phase rectifier, shown in Fig. 4–9, we can also use a diode rectifier, controlled by the multivibrator [Fig. 4–10(a)]. In this case only one half-wave reaches the RC-output unit (R_3, C_3) coming from the output end of the low-frequency amplifier, while the other one is short circuited by the diodes. Thus, we always suppress the half-wave for which the outputs of the multivibrator at T_1 and T_2 are positive and negative,

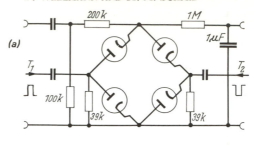

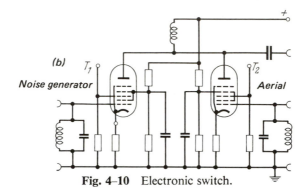

Fig. 4-10 Electronic switch.

respectively. For this connection the switch will always admit positive pulses if P_L is larger than P_i and negative pulses otherwise.

The lower limiting frequency of the low-frequency amplifier should be by a factor of 5–10 smaller than the frequency of the multivibrator.

Finally, we should also mention that the null method described above is applicable only if the aerial temperature is higher than T_W, which means for the amateur observer higher than the temperature in the vicinity of the receiver, the latter being about 290° T. Since T_W describes the minimum power of the noise generator, we would not achieve balancing at a smaller temperature T_A.

4.5.3. The Range of the Instrument

We will assume below a receiving equipment well within the possibilities of an amateur radio astronomer, and start with an estimate of the limits of its performance. The comparison with Sec. 4.4 will demonstrate which objects can be reached and which wavelengths should be chosen.

We assume a Yagi aerial with nine directors, which is simple and inexpensive, and is comparatively efficient.

According to Sec. 4.5.2, the value ΔT, which is given by Eq. (9), should exceed T_A by a factor of at least $\mu = 10$, that is,

$$T_A > \frac{\mu F T_0}{\sqrt{Bt}}.$$

With $T_0 = 290°K$, $F = 10$, $B = 100$ kHz (shortwave receiver), and $t = 1$ sec, we obtain

$$T_A > 92°K, \tag{11}$$

and for $B = 5$ MHz (television receiver) even

$$T_A > 13°K.$$

If we introduce the characteristics of the radiation source according to Eqs. (2), (3), and (8), we obtain for the first term of Eq. (12):

$$T_Q = \frac{T_A}{2.54 \cdot \omega_Q} = \frac{\omega_A}{\omega_Q} T_A. \tag{12}$$

The second term is valid if ω_A is the effective solid angle from which the aerial receives radiation; for our aerial we have $\omega_A \sim 5/\lambda^2$. For objects with small ω_Q, the limit for T_Q is much higher than that for T_A. In such cases, it is better to calculate the minimum value of S, which corresponds to the limiting temperature.

According to (1) and (6) we have

$$S = \frac{2kT_Q}{\lambda^2} \cdot \omega_Q.$$

If we relate again, according to Eq. (8), T_Q with T_A, take into account Eq. (2), and insert for the value A of our aerial the expression (3), we obtain

$$S = 1.1 \cdot 10^{-23} \frac{T_A}{\lambda^2} \text{ W/m}^2 \text{ Hz,}$$

and, according to Eq. (11), the corresponding limit

$$S > 1.0 \cdot 10^{-21} \frac{1}{\lambda^2}.$$

If we compare this result with the data given in Sec. 4.4 we can now state: It is impossible to observe the normal radio flux from the Sun, but we are able to register average or large radiation bursts or noise storms. It is also impossible to observe the radio radiation from the planets.

The two brightest radio stars can be observed with $\lambda > 3$ m. For the assessment of the Milky Way radiation, it is simpler to use Eqs. (11) and (12). Although even this extended object does not completely fill the angle of aperture of the aerial, and ω_A/ω_Q in Eq. (12) is larger than one, the observer will be able to record the Milky Way radiation on wavelengths exceeding 3 m.

Furthermore, there are possibilities to increase the performance of our instruments. We can try to increase the bandwidth to 1 MHz, and we can also increase the time constant, say up to 10 or 20 sec, and we may have possibilities to improve the aerial. Thus we should be able to reduce the limiting values by one order of ten, and this would be the outcome: The permanent radiation of the Sun would become measurable for wavelengths exceeding 20 cm. Four additional radio stars would now come within our limits. It has even been possible to measure with a similar instrumentation some radiation burst on Jupiter (at wavelengths of 10 m).[7]

This survey shows that the meter region is the most favorable one for the amateur radio astronomer. All these considerations assume that we are working on a quiet frequency band and that we have found a site for the aerial that is free of terrestrial disturbances, which increase with increasing wavelengths. In no circumstance should the observer go beyond a wavelength of 20 m, since above this limit the atmospheric disturbances increase rapidly and the ionosphere sometimes begins to weaken incident radiation. Furthermore, the aerials become then rather cumbersome.

Summarizing the preceding pages, it might appear as if the required effort is very large and the possible results—particularly in comparison with the variety of optical observations—may appear rather sparse. However, it was not the intention of our discussion to encourage every amateur astronomer to dig deep into his pocket to build up a radio observatory. But those who are by profession high-frequency experts and who concern themselves theoretically or practically as a hobby with electronics, or those who already possess by chance some suitable receivers or recorders, might perhaps nevertheless be in the position of realizing our above proposals. Since radio astronomy is a relatively young branch of science, it is not yet properly opened up for the amateur. Those who penetrate energetically into the matter have prospects of finding other solutions that might be realized more simply. The above examples have only been selected because amateur astronomers have tested them successfully, and since their installations have been discussed in the literature in numerous publications.[8]

[7] See ALEXANDER, J. K., and BROWN, L. W.: *Sky and Telescope* **29**, 212 (1965).

[8] HYDE, F. W.: *Radio Astronomy for Amateurs* (1962); HEYWOOD, J.: Radioastronomy and Radiotelescopes, *Mem. British Astronomical Associasion* **40** (1962); DOWNES, D. N.: A simple radiotelescope, *Sky and Telescope* **24**, 75 (1962); ALEXANDER, J. K., and BROWN, L. W.: A radiotelescope for amateurs, *Sky and Telescope* **29**, 212 (1965).

5 / The Terrestrial Atmosphere and Its Effects

F. Schmeidler

5.1. General Remarks Concerning the Earth's Atmosphere[1]

The atmosphere of the earth unfavorably influences the astronomical observations of the amateur astronomer and of the professional astronomer alike. In certain cases, modern techniques make it possible to compensate for these effects to a large extent (through the building of mountain observatories, observations from airplanes and satellites), but for the amateur such things are usually out of the question. He is therefore compelled to take account of all effects of the atmosphere on his observations. There are two main types of these effects, those that depend on the weather situation and are variable with time, and those that are always present. Perhaps this subdivision does not meet completely the point of view of scientifically exact systematics, but for the purpose of a presentation of practical applications it certainly appears suitable.

5.2. Weather-Dependent Phenomena

Of greatest importance, of course, is the weather situation itself, because whether an astronomical observation is possible at all depends on this. If there is a cloud cover, it is possible to observe with radio astronomical instruments, but of course not visually or photographically. However, even if the sky is clear, special phenomena might hinder astronomical observation or prevent it altogether.

5.2.1. Judgment of the Weather Situation

The question whether on a certain night we can count on the possibility of astronomical observations depends essentially on the weather situation. Unfortunately, there does not exist an unambiguous correlation with geographical distribution of the high- and low-pressure regions as are marked on the weather maps. However, certain indications can be given, although in individual cases deviations are quite possible.

[1] See also SIEDENTOPF, H.: Der Einfluss der Erdatmosphäre bei astronomischen Beobachtungen, *Die Naturwissenschaften* **35**, 289 (1948).

For instance, the present standard reference used in the German Weather Service classifies 29 typical weather situations, known as "Grosswetterlagen Europas." They are absolutely standard practice in official European publications. Eighteen types from the official list are given below.

W	West weather
BM	High-pressure bridge over central Europe
HM	Central high above middle Europe
SW	Southwest position, High in the southeast, and Low above the northern Atlantic
NW	Northwest position, High above western Europe
HN	High above the Norwegian Sea
HB	High above the British Isles
N	North position, High above the North Sea, Low above eastern Europe
TrM	Trough above central Europe
TM	Isolated Low above central Europe
TB	Isolated Low above the British Isles
TrW	Trough above western Europe
S	South position, High above eastern Europe, Low above western Europe
SE	Southeast position. High above eastern Europe, Low above the Mediterranean
HF	Isolated High above Fennoskandia
HNF	Isolated High above Fennoskandia and the Norwegian Sea
NE	Northeast position, high-pressure bridge from the Azores to Fennoscandia
Ww	Corner west position

The positions BM, HM, and HB are in central Europe always anticyclonic; the positions TrM, TM, TB, TrW, and Ww, always cyclonic. In all other positions the isobars above central Europe can be either anticyclonic or cyclonic, the different cases distinguished by attaching the letters *a* or *z*. Illustrations are given in Fig. 5–1.

A statistical analysis shows that none of these weather situations gives a guarantee for the absence of clouds; on the other hand cloudless weather is certainly not excluded a priori for any of these situations. As would be expected, the most favorable are the situations HM and BM, while the most unfavorable ones are certain cyclonic positions such as NWz and SEz. According to some counts of the author, which are of course representative

---

Fig. 5–1   Weather charts for 15 typical weather situations in central Europe. From Hess-Brezowski: Berichte des deutschen Wetterdienstes in der US-Zone, Bad Kissingen (1952).

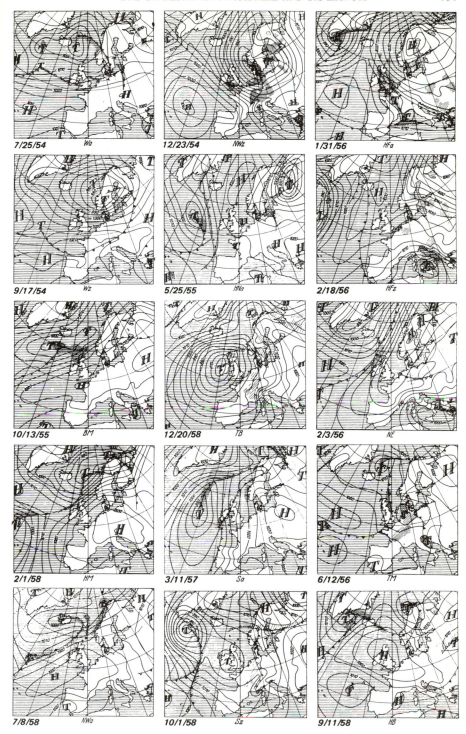

7/25/54   Wa   12/23/54   NWz   1/31/56   HFa

9/17/54   Wz   5/25/55   HNa   2/18/56   HFz

10/13/55   BM   12/20/58   TB   2/3/56   NE

2/1/58   HM   3/11/57   Sa   6/12/56   TM

7/8/58   NWa   10/1/58   Sz   9/11/58   HB

only for Munich, we might perhaps distinguish the following three groups:

Astronomically favorable: HM, BM, Wa, HNa, SWz, TB, HB, NEa
Astronomically passable: SWa, Na, SEa, HFa, HNFa, TrW, HFz, Ww
Astronomically unfavorable: All other main weather situations

Local differences are very important. For instance, Munich shows a great frequency of clear sky for certain west-weather positions (e.g., Wa, SWz, or TB), no doubt a consequence of the Föhn wind occurring in the northern part of the Alps.

The distribution of clear nights over the various seasons appears to follow, e.g., for the whole of Germany the same pattern: in late summer and autumn we find the best and in early winter the worst conditions. The total number of clear nights per year lies between 80 and 100, if we do not count nights in which gaps in the clouds occur only for very short periods; in general conditions in the northern German coastal regions are somewhat worse and in southwest Germany somewhat better.

For some purposes the above-mentioned 29 different types are reduced to about 8 groups. The bibliographical references for the German and British weather-type classifications are, respectively HESS, P., and BREZOWSKI, H.: Katalog der Grosswetterlagen Europas, in *Berichte des Deutschen Wetterdienstes*, vol. 15, No. 113, Offenbach/Main (1969); LAMB, H. H.: British Isles weather types and a register of the daily sequence of circulation patterns, in *Geophysical Memoirs* No. 116, H.M.S.O. for Meteorological Office, London (1972); see also H. H. LAMB's classification of weather types in *The English Climate*, Chap. 4, English University Press, London (1964). A full description and classification of various weather systems is included in the standard work *Atmospheric Circulation Systems* by PALMÉN, E., and NEWTON, C. W., Academic Press, New York (1969).

There are no specific circulation-pattern weather-type classifications for North America, and the ones that apply are full Northern Hemisphere classifications by Russian authors. For the Southern Hemisphere also, the only circulation-type classifications are by Russian authors and in the Russian language. Details can be found in the bibliography given in H. H. LAMB's recent book *Climate: Present, Past and Future*, vol. 1, London: Methuen, 1972.

5.2.2. Seeing and Scintillation

Even when the sky is completely clear there is a certain degree of unsteadiness of the air that causes a tremor of the pictures seen through the telescope. This tremor can be so strong that for bright stars it is conspicuous even to the naked eye. It is caused by some small local differences in density and temperature in the atmosphere, which correspond to differences in the refractive index. If in addition there is some vivid movement of the air, then the phenomenon is rapidly varying and gives rise to strongly scintillating images. In principle, this is the same phenomenon that can be observed on hot summer days above the ground, particularly strongly above heated asphalt. The optical details have been discussed in the literature.[2]

The air unsteadiness acts in two ways to bring about changes both in position and in the brightness of the stars (direction scintillation and brightness scintillation). The former shows itself either as a tremor of the star as a whole around a mean position or by the diffuse appearance of a generally steady image. According to some observers small objective apertures are supposed to favor the positional variations, whereas large apertures seem to be connected particularly with diffuseness of the image; but these indications are by no means confirmed by all observers. Indeed, experiences differ so much from each other that it is difficult to reconcile them to a simple statement.

The degree to which the unsteadiness of the air hinders astronomical observations depends on the purpose of these observations. Visual position measurements in astrometry are most affected, e.g., measurements of position angle and distance of close double stars; also observations of details on the planetary surfaces are very sensitive to unsteadiness of the air. There are nights that completely prevent observations of this kind. Less critical is the situation in the case of photometric observations, and for celestial photography the unsteadiness of air becomes prohibitive only if instruments with large focal lengths (at least two meters) are used. If, on the other hand, in the photography of the Sun, Moon, or planets, it is absolutely necessary to work with long focal lengths or with auxiliary magnification, then unsteady air is a most dangerous source of disturbance for astronomical photography. For observations without telescope (e.g., meteors, Zodiacal Light, and so on) the scintillation is almost without any practical influence.

The degree of scintillation depends on many different circumstances. First of all, it is of course larger near the horizon than in the zenith, but even this rule is not always true. There are nights when the scintillation changes only little or not at all as we go from the zenith to the horizon, while on other nights its increase is very marked. Unambiguous criteria as to when the one and the other case occurs are not yet known.

On a statistical average there is a strong dependence of the scintillation on the time of the day. It is clear that it reaches a maximum at noon, and minimum before sunrise and after sunset. We would not expect, however, the fact recorded by many observers that a secondary maximum of scintillation occurs around *midnight*, sometimes a little earlier, sometimes a little later. It is caused by a corresponding secondary maximum of the atmospheric turbulence at midnight (which is a consequence of the physical laws of mass exchange in the atmosphere). The visual observers notice that in the hours

---

[2] See, e.g., DIETZE, G.: Einführung in die Optik der Atmosphäre. Leipzig: *Akad. Verlagsgesellschaft* (1957). A popular and enthusiastically written introduction to the whole field of atmospheric-astronomical phenomena has been provided by M. MINNAERT in *The Nature of Light and Colour*, New York: Dover Publications, Inc. (1954).

after sunset the quality of the image in the telescope improves slowly only to deteriorate again one to two hours before midnight.

Considerable deviations from these mean conditions can be observed. As always in meteorological problems of this kind there are no reliable rules, but only indications, whether one can count on favorable weather or whether unfavorable conditions will prevail. As a rule, the wind even if it is only light causes an appreciable magnification of the scintillation; but even to this immediately evident correlation there exist frequent exceptions. High humidity nearly always improves the picture, even to such a degree that observations through *haze* frequently furnish very steady pictures, provided one can see the star at all.

The variation of scintillation in the different seasons differs from locality to locality. Observations in Berlin have shown that the best quality is reached in the spring. In Munich, however, it has been found that the months of August, September, and October show the smallest amount of scintillation. The reasons for these different experiences are not yet clarified; however, the differences are rather small.

Large-scale weather situations, mentioned on page 150, show a weak correlation with scintillation. As a whole, it appears that the anticyclonic weather situations are more favorable than the cyclonic ones. Seen statistically the high-pressure situations BM and HM are the most favorable ones, the latter, however, with remarkable exceptions that occur at the beginning and toward the end of a high-pressure situation lasting for some considerable time. Altogether we might accept the rather evident rule that the astronomical images are better the more stable are the large-scale conditions of the atmosphere. Additional influences are due to microclimatic situations (near a town or a forest); such influences can change the picture completely.

As a whole, we must state that the totality of all circumstances on which the quality of an image depends on a certain evening is so complex that even with the most extensive local experience, there will hardly be more than 50% of the cases that conform to a given prognosis.

5.2.3. Halos, Rainbows, and Related Phenomena

Although the phenomena that we are going to summarize in this chapter are not really of an astronomical nature, a description of their general outlines is advisable since they are so frequently conspicuous for the astronomical observer.[3] A halo is a usually ringlike light phenomenon around a bright source of light, such as the Sun or the Moon. Its origin is refraction or reflection of the light by atmospheric ice crystals. Only if such crystals exist

---

[3] Observations are analyzed in such works as the *Arbeitsgemeinschaft für Halo Beobachtungen*, Hamburg-Langenhorn-Nord, Langenhorner Chaussee 302a. The British Astronomical Association (BAA) is also very active in collecting observational data. A special section deals with material on the Sun (Director, H. Hill, Dean Brook House, 298 Orrell Road, Wigan, Lancashire WN5 8QZ, England), and one collects material on *Transient Astronomical Phenomena* (Director, H. G. Miles, Department of Mathematics and Statistics, Lanchester Polytechnic, Coventry CV1 5FB, England).

at some level in the atmosphere can a halo be produced; it is therefore linked with certain weather conditions.

According to the kind, distribution, and geometrical form of the ice crystals, we can distinguish the various types of halo. Furthermore, also the arrangement must show a preferred orientation because if all possible directions of axis were present, uniformly distributed in all directions, there would be no reason for any specific light phenomenon.

Relatively most frequent is a halo in the form of a faintly luminous ring at a distance of 22° around the Sun (or the Moon), the inner border of which is of a distinct red color. On the other hand, we might also observe the so-called large ring with a radius of 46°. Its so-called horizontal circle runs parallel to the horizon through the Sun; where it intersects the actual halo, we sometimes observe bright spots of light, the so-called auxiliary suns.

It is impossible to give here a physical explanation of these various phenomena. In the above-mentioned books by DIETZE and MINNAERT all necessary details can be found. Here we may add only that the existence of ice crystals in the atmosphere, a necessary condition for the production of halos of every type, is usually indicated by the presence of cirrus clouds, and halos are therefore most frequent if such cloud conditions exist.

Unlike the observation of halos, the phenomenon of the rainbow is much more generally known. Its bases are similar optical laws to those responsible for the halos, except that the refraction of the light is caused here by drops of water instead of by ice crystals. Since liquid water can assume only a single symmetrical shape, namely that of the sphere, the theory of the rainbow need not deal with a confusing variety of phenomena, but only with two forms: the main rainbow surrounds the counterpoint of the Sun at an angular distance of 42°, while in the case of the rather rare secondary rainbow this angle is equal to 51°. The best opportunity to see both rainbows is when the Sun is at the horizon; at higher altitudes of the Sun they become less conspicuous.

The colors of the rainbow are very distinct and have led to the phrase "the seven colors of the rainbow." For the main arc we find the red outside and the violet inside; for the secondary arc it is the other way around. Both arcs are sometimes extended by faint secondary arcs.

Of course we are only able to observe rainbows when water droplets in the atmosphere are illuminated by the Sun or the Moon. This is for instance the case when in rainy weather a small area of the celestial sphere, in which the Sun is standing, is by chance free of clouds. However, we can also see rainbows while the sky is completely free of clouds if we stand near a spraying water fountain. Rainbows caused by moonlight usually appear colorless, since on account of their low brightness the human eye is unable to distinguish the colors present.

In addition, certain favorable conditions of type, distribution, and density of the clouds give rise to the phenomena of "glories," a special halo phenomenon around the Sun or the Moon. These are colored rings of small diameter about the "anthelion point" of the Sun. Of particular interest to the reader will be M. Minnaert's splendid book *Light and Colour in the Open Air* (New York: Dover Publications Inc. 1954), which was mentioned previously.

5.3. Permanent Atmospheric Phenomena

Contrary to the weather-dependent phenomena with which we have been concerned so far, there exists a whole series of atmospheric influences which are always present in any given weather situation and in particular for a completely clear sky. Their influence must always be taken into account in dealing with astronomical observations, if this is required by the nature of the problem investigated.

5.3.1. Refraction

Just like any other physical medium, the atmosphere possesses a reflective index so that the rays of light suffer a certain, even though small, change in direction. Naturally, this phenomenon plays a role only when the direction from which the light comes, that is, the position of the celestial body on the sphere, is essential to the investigation. The amount of refraction, which always causes an apparent raising of the source of light, depends on the altitude (or zenith distance) of the object and on the color of the light. Table 3 in the Appendix lists the mean refraction according to Bessel. The fact that two objects close together in the sky, although the difference of their zenith distance is very small, suffer different amounts of refraction is called *differential refraction*. The fact that the rays from a light source of a sufficiently large zenith distance are drawn out into a spectrum because of the wavelength dependence on the atmospheric refractive coefficients is called *atmospheric dispersion*.

Of course, the general refraction of objects at the zenith is equal to zero and increases toward the horizon. The law of this increase has been the subject of extensive theoretical studies. Up to 75° zenith distance the amount of the refraction does not practically depend on the structure of the atmosphere at all, i.e. on law of the decrease of density and temperature with height (theorems of Oriani and Laplace). The amateur astronomer may find it helpful as a sufficiently good approximation for the effect of refraction to use the—of course incorrect—assumption that the atmosphere of the earth has a constant density up to a certain upper limit, and that a ray of light that arrives at this layer from outer space will only suffer one refraction. In this case, the direction of the ray after refraction could be set equal to the apparent zenith distance z, observed on the ground, while the direction before the refraction was $z + R$, where R denotes the amount of refraction. According to the law of refraction we therefore have if we denote with μ the refractive coefficient of the atmospheric air

$$\sin (z + R) = \mu \sin z.$$

Since R is a small angle we can put cos $R = 1$ and sin $R = R$ and find, since

$$\sin (z + R) = \sin z \cos R + \cos z \sin R,$$

the equation

$$R = (\mu - 1) \tan z.$$

This is the well-known law that the refraction is proportional to the tangent of the zenith distance. The constant of proportionality is very nearly $1'$ and we can thus state quite generally that the amount of the refraction expressed in minutes of arc is equal to tan z.

It is only at large zenith distances or when we need utmost accuracy that this law fails us. The exact theory proves that the refraction can be represented as a series which progresses in odd powers of tan z. The complete tables for the accurate calculation of the refraction go as far as even the thirteenth power of tan z.

The amount of refraction also depends on the pressure and on the temperature of the air, since the refractive index is a function of both these quantities. If one denotes with R_0 the "mean" refraction, that is its, value at 0°C and at 760 mm Hg ($= 1013$ mb $=$ millibars) we have

$$R = \frac{b}{1013 \cdot (1 + T/273)} R_0,$$

where b is the pressure in millibars and T the temperature in degrees Centigrade.

However, the above-mentioned development of the refraction as a series of odd powers of tan z becomes quite unusable near the horizon, where tan z approaches infinity. Thus the calculation of the horizontal refraction is a special problem in all theories in which the vertical layers of the atmosphere play a role. Observations have shown that the refraction at the horizon amounts to 34 min of arc; a star that has apparently just set is therefore in reality already more than half a degree below the horizon.

A consequence of this fact is the lengthening of the day. Beginning and end of the day are defined by the rising and setting of the upper limb of the Sun's disc. Since the apparent radius of the Sun is $16'$ we see that at the moment of sunrise the center of the Sun's disc is in reality still $34' + 16' = 50'$ below the horizon. This causes a lengthening of the day by about 10 min in central Europe; in polar regions it can be much more.

Another consequence of refraction (that is, of differential refraction) is the *apparently* elliptical shape of the Sun and of the Moon near the horizon. Since the lower limb of the disc is lower than the upper limb, it undergoes a greater lifting by refraction, so that the apparent vertical diameter of the Sun is shortened. Of course the phenomenon occurs also at any other altitude of the Sun, but only near the horizon is it pronounced enough to become conspicuous to the eye. In addition, we may just mention that for some special atmospheric conditions additional distortions of the discs of Sun and Moon can also occur.

5.3.2. Extinction

The weakening (extinction) of the light of the stars by the Earth's atmosphere has two physically very different causes; partly, the light is absorbed by the molecules of the air, i.e., transformed into heat, partly it is deviated from its direction, i.e., scattered. Thus the total loss of light is a combination of the amounts due to absorption and to scattering.

A measure of the amount of extinction is given by the so-called *extinction coefficient*, in place of which sometimes the so-called *coefficient of transmission* is used. The coefficient of extinction states the percentage of incoming radiation that is absorbed at perpendicular incidence (star at zenith); the coefficient of transmission indicates the percentage of radiation which finally arrives at the ground. Both quantities depend on the opacity of the atmosphere, and, very strongly, on the wavelength of the light. To be exact, the above definition carries a physical meaning only for monochromatic light; however, we commit only a very small error if we describe the extinction for polychromatic light simply by an average extinction or transmission coefficient taken over the whole relevant spectral region. The transmission coefficient of the atmosphere is for average atmospheric conditions and a cloudless sky, about 0.8, that is, a star at the zenith is on the average weakened by about 20% of its brightness. This value, however, is valid only for the visual spectral region while in the photographic region the extinction is usually 50 to 100% larger and also much more strongly dependent on the atmospheric conditions. Furthermore, the extinction also depends a little on the color of the star. (See Table 2 in the Appendix.)

The amount of the extinction increases, of course, toward the horizon. As a very good approximation we can put the weakening of the starlight proportional to the secant of the zenith distance. Since one does not normally correct the measured stellar brightness completely for the influence of the Earth's atmosphere but only gives the brightness that the star would have at the zenith, one has for the "reduction to the zenith" the amount

$$\Delta m = 2.5 \log p(\sec z - 1),$$

where p denotes the transmission coefficient. The factor $-2.5 \log p$ is for visual measurements about $0^m.25 - 0^m.30$, and for photographic measurements $0^m.40 - 0^m.60$. This reduction involves many difficulties, which are only rarely overcome, and it is therefore very advisable to make photometric measurements only at quite small zenith distances (if possible below 30°) and, furthermore, to arrange the measurement in such a way that two stars are if possible only compared when they are at the same zenith distance, so that only a very small difference in extinction remains to be determined.

The atmospheric extinction on high mountains is, of course, considerably smaller. The above data refer to normal altitudes within a few hundred meters of sea level; on mountain stations the extinction may decrease to a tenth of this value, and has to be specially determined for every single case.

The dependence of the extinction on the wavelength of the light (selective extinction) is important for quite a number of well-known phenomena. Generally, short-wavelength light is more weakened than long-wave radiation. That is why the Sun, Moon, and stars appear to us much redder on the horizon than in the zenith. A further consequence of this fact is the blue color of the daylight sky. It is due to the scattering of the sunlight by atmospheric dust. The red part of the direct sunlight is scattered only very little, the blue part, however, considerably; it is because of this that the light that the dust particles at large angular distance from the Sun receive and again scatter toward the observer is essentially of blue color. If there were no atmosphere, or some kind of atmosphere in which there were no scattering, the sky would be pitch black during the day, and only where the Sun or Moon were standing would we be able to see direct light.

The mathematical laws dealing with the dependence of extinction on wavelength are very different for the different physical processes involved. The simplest is RAYLEIGH's law, which states that the amount of radiation lost by scattering by the air molecules is inversely proportional to the fourth power of the wavelengths (that is, about proportional to λ^{-4}). Since, however, the atmosphere does not consist simply of air molecules but also of dust particles of very different sizes and natures, the real situation is much more complicated, and in addition, variable with time. As a whole, one can assume a proportionality to λ^{-1} to $\lambda^{-1.5}$ for the extinction caused by dust. The actual extinction is a combination of both these contributions in a manner which differs from case to case.

5.3.3. Twilight and the Brightness of the Night Sky

The phenomenon of twilight is a consequence of the scattering of the light in the atmosphere, and it can therefore be treated in conjunction with extinction. Even when twilight is over, the sky is not absolutely dark but retains a small brightness of its own.

5.3.3.1. Twilight and Related Phenomena

If there were no atmospheric scattering then the day would at sunset pass immediately into complete dark night. Indeed, such a transition takes place rapidly in the desert with its normally extremely dust-free air. The physical process consists in the fact that after sunset (the same applies, of course, at the time before sunrise) no direct solar rays reach the observer; they reach, however, the higher regions of the atmosphere and are scattered by the air molecules of these layers so that the corresponding proportion reaches the eye of the observer on the surface after having been scattered once or several times.

We must distinguish two different definitions of twilight. The *civil twilight* is the time in which it is possible to read comfortably under a cloudless sky; it ends (or starts) when the Sun is 6° below the horizon. The *astronomical twilight* ends when the solar altitude is $-18°$ and indicates the moment at which no trace of scattered sunlight can be seen any longer.

(Occasionally, one also meets the expression *nautical twilight*, which ends at the solar altitude of $-12°$). If the sky is cloudy the duration of the real twilight is of course considerably shorter.

It is within the latitude-range of $+48°5$ and $-48°5$ that we can distinguish throughout the whole year the following four sub-divisions of illumination: (1) bright daylight; (2) civil twilight between sunset and the time when the Sun's centre is $6°$ below the horizon; (3) the astronomical twilight which lasts until the Sun's centre is $18°$ below the horizon; and, finally, (4) dark night.

If the day is very clear we are able to distinguish during the evening twilight the shadow of the Earth on the eastern sky as a relatively sharp boundary between the dark and still somewhat brightened sky and correspondingly, of course, during the morning twilight on the western sky.

The duration of the civil twilight amounts at our latitudes ($+50°$) to between 30 and 40 min; it is longest in summer and winter and shortest in the intermediate seasons. At $50°$ latitude, the duration of the astronomical twilight amounts in autumn, winter, and spring rather uniformly to two hours, while the summer is characterized by very different conditions. At the latitude of $+49°$ the zone of the so-called "white nights" begins, in which there is at the time of the summer solstice an astronomical twilight lasting throughout the whole night, since the Sun is never lower than $-18°$. The more northerly the site of observation, the longer this period. It lasts:

> From 11 June to 3 July at $+49°$ latitude.
> From 21 May to 23 July at $+52°$ latitude.
> From 9 May to 5 August at $+55°$ latitude.
> From 29 April to 15 August at $+58°$ latitude.

The transition to the latitude zone of the white summer nights is so sharp that at $+48°$ (Munich and Vienna) practically nothing of midnight twilight can be seen. In some years or nights the situation can deviate from the mean conditions according to the dust content of the atmosphere.

A number of interesting color phenomena, which under favorable circumstances can be seen during twilight, are not discussed here because of their exclusively meteorological importance. Furthermore, a satisfactory explanation does not yet exist for most of them. For details we again refer to the books by G. Dietze and M. Minnaert.

Twilight is, of course, of great importance to the astronomer, indicating to him when the stars will become visible and he can begin with his observations. At which sky brightness certain observations can be carried out depends nearly exclusively on the purpose of the observation. Generally speaking, photometric measurements can be made only when the sky is absolutely dark, while astrometric measurements can start as soon as a particular star can be seen at all.

Related to all this is the question of the visibility of stars in daytime.

With the naked eye and exact knowledge of its position we are able to see in daytime the planet Venus at its greatest brightness ($-4^m.3$). Fainter stars remain invisible in daytime if no telescope is used, and the frequently made statement that an observer at the bottom of a deep well can see bright stars as well in daytime as at night is a fairytale. On the other hand, the surface brightness of the sky will disturb the visibility of the stars less, the larger the aperture of the telescope, and one can therefore see moderately bright stars in the daytime. For perfect conditions and an absolutely transparent sky a 2-in. telescope can show stars of magnitude 1^m, and a 4-in. telescope shows stars up to 3^m. Conditions are more favorable at mountain stations. However, the observation of planets with conspicuous discs (Mars, Jupiter, Saturn) during the day is almost hopeless, since the available brightness is spread over a considerable area. For Venus and Mercury the possibility of observing them with the telescope on the daylight sky very much depends on the phase, and, of course, on the magnification. We should also mention that even the Moon can be observed in the daytime with the telescope only with great difficulty.

5.3.3.2. The Brightness of the Night Sky

The faint light shown by the night sky after the end of the astronomical twilight consists of different contributions. Of course, we do not refer to the scattered light produced by terrestrial light sources (like street lighting) and assume that such light is not present. But even with this assumption the intensity of the night sky is determined by variations depending on the time and the weather situation.

The mean total brightness of the night sky corresponds to about that of a star of magnitude 22^m per second of arc squared. This sets a limit to the astronomical observation of faint surface brightnesses, a fundamental limit, which, of course, is of any practical importance only for large telescopes.

We are able to trace the brightness of the night sky to the following causes:

(a) remnants of the normal twilight
(b) recombination luminescence of air molecules
(c) Zodiacal Light
(d) faint stars and extragalactic nebulae
(e) scattered light of the sources (b), (c), and (d)
(f) from time to time, aurorae or luminous night clouds

The causes mentioned under (a), (b), (e), and (f) lie in the atmosphere itself, while (c) and (d) are of an extraterrestrial nature. The color of the night skylight is faintly reddish, but because of its faintness the eye is unable to record the color. In any case, the amateur astronomer is not handicapped by real disturbances through the night skylight.

5.3.4. Polarization of the Skylight

The diffusely scattered light of the sky is polarized to a larger or smaller degree. This means that the direction of vibration of the electromagnetic vibrations lies in the plane perpendicular to the light ray not at random but partially oriented in a certain direction. Although this fact is essentially irrelevant for astronomical observation, it should be included in a complete discussion of all optical phenomena connected with the Earth's atmosphere.

The cause of polarization of light is scattering by air molecules. The laws of optics teach us that in the process of reflection, refraction, and scattering a certain portion of the light waves becomes polarized. The percentage of the polarized light relative to the total light, the so-called degree of polarization, depends on circumstances such as angle of incidence or the size of the particles.

The human eye cannot distinguish between polarized and unpolarized light. Some creatures in nature, such as bees, have been provided with this faculty. Man, however, can as a rule observe polarization only if he views the incoming light through a polarization filter. If this is turned, the brightness of unpolarized light remains the same at any position of the filter, while for polarized light the greatest brightness corresponds to the position in which the polarization axis of the filter is identical with the direction of vibration of the incident light.

The degree of polarization of the diffuse skylight depends on the position of the particular area of the sky relative to the Sun. In a cloudless sky there are a few, usually three, points free from any polarization. All three points lie in the vertical through the Sun and are called—

1. The ARAGO point: 20° above the point opposite the Sun
2. The BABINET point: 10° above the Sun
3. The BREWSTER point: 15° below the Sun

The ARAGO point is above the horizon only if the Sun is rather low. The altitudes of the three points are mean values only and can vary according to the haziness of the atmosphere and the structure of the Earth's surface.

The largest value of the degree of polarization, some 60 to 80%, occurs at the point of the vertical through the Sun that is 90° away from the Sun. Here, too, the degree of polarization, and also to a lesser degree the place of strongest polarization, depends on the haziness of the atmosphere. Quite generally the variation of the degree of polarization can be described by the combination of two laws: (1) the degree of polarization increases with increasing distance from the Sun up to 90° and then decreases, and (2) the degree of polarization decreases toward the horizon. Details are given in MINNAERT's book.

5.3.5. The Apparent Form of the Celestial Vault

The fact that the celestial vault appears to the human eye rather flattened, not like a hemisphere, has in itself nothing to do with the atmosphere. It will be meaningful, however, to discuss this phenomenon briefly. The observer has the impression that the horizon is farther away from him than the zenith point of the sky. This, of course, is not a genuine phenomenon, since the sky as such has at no place a "distance" from the observer; at any place our view goes, in principle, to infinity. The phenomenon is of a physiological nature and is connected with the structure of the human eye. This is evident from the fact that an observer lying on the ground sees the point of the zenith of the sky as much farther away than an observer who stands upright.

An important consequence of this property of the human eye is the fact that angles of altitude are usually estimated quite incorrectly. The amount of the error is different from case to case and the following figures indicate average values only:

True altitude	0°	15°	30°	45°	60°	75°	90°
Estimated altitude	0°	30°	50°	65°	75°	84°	90°

We see that the altitude of an object is estimated up to 20° too large near the horizon. The effect is smaller at night than during the day.

Because of this overestimation of altitude angles near the horizon, the Sun and Moon at rising or setting appear larger than at the zenith. The same applies to star constellations. It is also a consequence of this property of the human eye that stars appear to be nearly at the zenith even if their true zenith distance is still 10 or 20°. For every estimate of stellar altitudes we must take into account the effect just discussed.

6 / Fundamentals of Spherical Astronomy

K. Schütte†

6.1. Introduction

The aim of this chapter is to provide a brief survey of the fundamentals of spherical astronomy. This is a field that today is often neglected, although it is in fact still the foundation of many branches of astronomy for research and practical work, even if this is not always clearly evident.

It is for this reason that the amateur will also benefit from being familiar with the foundations of astronomy, since he is unlikely to want to make merely casual and random observations. Even if he only intends to reduce his observations and make them available for further discussion, he cannot do this without a knowledge of these foundations.

It is not easy to decide where to begin and what to include in the subject of spherical astronomy. Because of the restricted space at our disposal we have had to omit many useful things. We also have had to omit derivations of any formulae; these can be found in the more detailed textbooks on spherical astronomy. We have assumed that the reader of this chapter will possess knowledge of (a) the use of logarithms, (b) the trigonometric functions, and (c) the fundamentals of error theory and of the method of least squares (see page 205). Some important tables used in spherical astronomy have been included in the Appendix of this book, and references to them appear in the text. A list of additional literature comprises Chapter 22, the bibliography of the book.

6.2. The Coordinates

6.2.1. The Coordinates on the Earth— Geographic and Geocentric Latitude

We shall begin with an explanation of the most important astronomical coordinate systems. Unless we leave the Earth with a space ship, observations are in general made from the Earth's surface, and on this surface the position of a point is determined by three coordinates:

1. Geographic latitude, φ
2. Geographic longitude, λ
3. Altitude above sea level, h

†Died September 6, 1974.

Geographic latitude is measured from 0° at the Earth's equator up to 90° at its poles, positive to the north, negative to the south. All circles parallel to the Earth's equator are called *parallels of latitude*, and all are smaller than the equator itself.

Any plane through the imaginary axis of the Earth is perpendicular to the equator and intersects the Earth along a *meridian*, which is used to measure geographic longitude (see Fig. 6–1). By international agreement, the

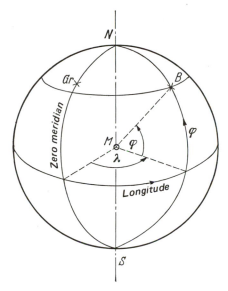

Fig. 6–1 Spherical coordinates.

meridian through the Royal Observatory, Greenwich, England, has been accepted as the *zero meridian*. Longitude is measured from 0° to 180°, positive to the west of Greenwich and negative to the east.

Determining the geographic coordinates and the shape and size of the Earth is one of the tasks of astronomy and geodesy. In reality the Earth is not exactly spherical but rather shaped somewhat like a flattened ellipsoid. For practical numerical reasons the so-called reference ellipsoid has been introduced as a reference plane. This is chosen so that the directions of its normals agree as closely as possible with the directions of a plumb line at points on the Earth's surface. Also by international agreement, the following dimensions have been accepted for this reference ellipsoid (given by HAYFORD, 1927).

a = semimajor axis = equatorial radius = 6378.388 km,

b = semiminor axis = polar radius   = 6356.912 km.

From this we obtain the oblateness

$$a = \frac{a-b}{a} = 1/297.00.$$

The polar radius is thus 21.5 km smaller than the equatorial radius. The physical (real) surface of the Earth is as a rule somewhat different from the reference ellipsoid.[1]

As a result of the ellipsoidal shape of the Earth, there is a difference between the geographic latitude φ, as determined by the plumb line, and the geocentric latitude φ' given by the direction to the center of the Earth. This difference is

$$\varphi - \varphi' = 695.''65 \cdot \sin 2\varphi - 1.''17 \cdot \sin 4\varphi. \ldots$$

Correspondingly, the distance r' of a point at sea level from the center of the Earth is given by

$$r' = a \, (0.998320 + 0.001684 \cos 2\varphi - 0.000004 \cos 4\varphi \ldots),$$

where a denotes the equatorial radius.

6.2.2. The Coordinate System of the Horizon

If we imagine a plane perpendicular to the plumb line, we notice that it intersects the celestial sphere in a circle, called the *horizon*, or, more exactly, the *apparent horizon*. Perpendicularly above the observer is the zenith, and perpendicularly downward is the *nadir* (see Fig. 6–2).

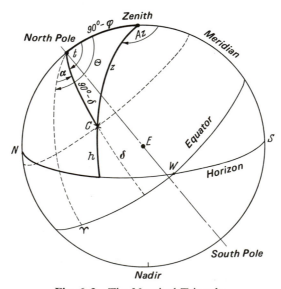

Fig. 6–2 The Nautical Triangle.

[1] On the other hand, we have the "geoid," which is a surface defined by the level of the ocean.

At any given point on the horizon, we can imagine a circle drawn perpendicular to it. Such a circle is called an *altitude circle*, and all such circles intersect at the zenith (and below the horizon at the nadir). The altitude of a star above the horizon is measured along the altitude circle from 0° at the horizon to 90° at the zenith. All altitudes above the apparent horizon are called *apparent altitudes* (*h*). Because of astronomical refraction, they are somewhat larger than the true altitudes. Frequently the *zenith distance* $z = 90° - h$ is used instead of the altitude *h*.

A plane through the center of the Earth, running parallel to the apparent horizon, intersects the celestial sphere in the *true horizon*. Altitudes referred to it are called *true altitudes*. If the celestial body is relatively close to the Earth, then its true and apparent altitudes are noticeably different; in the case of the Moon, the difference can amount to about 1°. The second coordinate of this system is given by the direction of the celestial body. This is the direction of the point where the altitude circle through the star cuts the horizon and is called the *azimuth* (Az). In astronomy it is measured south to west, north, and east, from 0° to 360°. In geodesy, on the other hand, the measurement starts from the north and proceeds in the same sense. The four points on the horizon having azimuths 0°, 90°, 180°, and 270° are called the *South Point*, *West Point*, *North Point*, and *East Point*, respectively. The altitude circle through the South Point, the zenith, and the North Point is called the local meridian; the altitude circle perpendicular to it through the West Point, the zenith, and the East Point is called the *Prime Vertical*.

Because of the rotation of the Earth, each celestial body traverses the meridian twice every day, once when it is in its highest position, *Upper Culmination* (U.C.), usually in the south, and the second time in the north in its lowest position, *Lower Culmination* (L.C.). For some stars both culminations take place above the horizon. These are called the circumpolar stars.

The condition for a star to be circumpolar is given by the relation

$$\delta \geqq 90° - \varphi,$$

where φ is the geographic latitude and δ the declination of the star.

All stars that do *not* fulfill this condition are the ones that rise and set. The positions on the horizon where the star rises and sets and the star's motion above the horizon all depend on the geographic latitude and the declination of the star.

At any point on the Earth's equator, all stars rise perpendicular to the horizon. The further the place of observation is from the equator the smaller is the angle between the horizon and the direction of the star's rising or setting. At the Earth's poles, stars move in circles parallel to the horizon and therefore never rise or set. However, the Sun, Moon, planets, and comets, because of their own motion, can also rise and set at the poles. The distances of the star's points of rising or setting from the East or West Points are called its *morning* or *evening elongations*.

6.2.3. *The Equatorial System—Spring Equinox and Sidereal Time*

The main disadvantage of having the coordinate system referred to the horizon is that the coordinates of each star change continuously in the course of the day and also that they differ from place to place. This disadvantage does not exist in the equatorial system. If one imagines an extension of the plane of the Earth's equator it intersects the celestial sphere at the *celestial equator*. The imaginary rotational axis of the Earth is perpendicular to this plane and intersects the celestial sphere at the North and South Poles. In a system defined in this way, we are able to define more permanent coordinates of the star, analogous with the geographical coordinates of a point on the Earth's surface.

Our first coordinate is obtained by measuring the distance of a star from the celestial equator toward the north or south from 0° at the celestial equator to 90° at the celestial pole; we call this quantity the *declination* (δ). It is counted positive toward the north and negative toward the south. Any great circle through the star and perpendicular to the equator is called a circle of declination.

The other coordinate is supplied by the geographic longitude on the equator. Here the zero point is the *Vernal Equinox*, also called the point of Aries (γ), which is one of the two points of intersection between the celestial equator and the plane of the solar orbit (the ecliptic). When the Sun arrives at this point, spring begins. The other point of intersection of the equator and the ecliptic is called the *Autumnal Equinox*. Starting at the Vernal Equinox we now count. The second coordinate of the star corresponds to the geographic longitude and is measured by the distance from the Vernal Equinox to the point of intersection of the circle of declination with the equator. This distance is measured from west through south to east from 0° to 360°, that is in the sense of the annual motion of the Sun, and this coordinate is called

$$\text{Right ascension} = AR = \alpha.$$

The Vernal Equinox itself therefore has the coordinates

$$\alpha = 0^h 0^m 0^s \quad \text{and} \quad \delta = 0°0'.$$

The position of a star on the sphere is uniquely determined by the two coordinates of declination and right ascension.

Because of the slow movement of the Vernal Equinox, to which we shall refer again later, the coordinates α and δ are subject also to slow changes, partly periodic and partly secular. (See pages 175–177.)

It is not difficult to find a relation between the coordinate systems referred to the horizon and to the equator, if one introduces the additional concept of the *hour angle* (t). This quantity is the angular distance between the circle of declination through the star and the upper meridian. It is measured from south through west, north, and east from 0° to 360°, or from 0^h to 24^h. At its upper culmination the star has the hour angle $0°(0^h)$, and at the lower culmination $180°(12^h)$. Since a whole revolution of $360°$ takes place in 24^h, our angles can always be expressed either in time *or* in degrees, by converting them as follows:

$$1^h = 15°$$
$$1^m = 15'$$
$$1^s = 15'' \text{ etc.}$$

Of course the Vernal Equinox, just like a star, also has a certain hour angle. This is called *Sidereal Time* (θ). When the Vernal Equinox is in Upper Culmination we have the Local Sidereal Time $\theta = 0^h0^m0^s$. Between Sidereal Time, right ascension, and hour angle t there is a very simple relationship:

$$\theta = t + \alpha \qquad \text{or} \qquad t = \theta - \alpha.$$

This relation is of greatest practical importance when, for instance, one wishes to find an object in the sky, or to set the telescope on it. If $\theta > \alpha$, the hour angle is west; if $\theta < \alpha$, the hour angle is east.

6.2.4. Transformation of Horizontal Coordinates into Equatorial Coordinates and Vice Versa

6.2.4.1. Transformation of the Horizontal System into the Equatorial System

We are *given*: $z = 90° - h$, φ, Az.
We *require to know*: t, δ
We have the following equations:

$$\sin \delta = \cos \varphi \cos z - \cos \varphi \sin z \cos \text{Az},$$
$$\cos \delta \sin t = \sin z \sin \text{Az},$$
$$\cos \delta \cos t = \cos \varphi \cos z + \sin \varphi \sin z \cos \text{Az}.$$

Or, introducing an auxiliary parameter M, defined by

$$\tan M = \cos \text{Az} \cdot \tan z,$$

we obtain expressions in a form convenient for logarithmic calculation:

$$\tan t = \tan \text{Az} \frac{\sin M}{\cos(\varphi - M)},$$

$$\tan \delta = \cos t \cdot \tan(\varphi - M).$$

6.2.4.2. Transformation of the Equatorial System into the Horizontal System

Given: φ, $t = \theta - \alpha$, δ
Required: Az, z.
Equations:

$$\cos z = \sin \varphi \sin \delta + \cos \varphi \cos \delta \cos t,$$
$$\sin z \sin \text{Az} = \cos \delta \sin t,$$
$$\sin z \cos \text{Az} = -\cos \varphi \sin \delta + \sin \varphi \cos \delta \cos t.$$

Or, if we introduce another auxiliary parameter N by

$$\tan N = \frac{\tan \delta}{\cos t},$$

we obtain, again in a form convenient for logarithmic work,

$$\tan Az = \tan t \frac{\cos N}{\sin (\varphi - N)} = \tan t \frac{\cos (\varphi - M)}{\sin M},$$

$$\cotan z = \cos Az \cdot \cotan (\varphi - N) = \cos Az \cdot \cotan M.$$

We see that only *one* parameter M is needed to carry out both transformations.

6.2.5. Other Important Coordinate Systems and Transformations

There are two other important coordinate systems that are frequently used, namely:

6.2.5.1. The System of the Ecliptic

Here the fundamental plane is the ecliptic itself. Latitudes β are measured perpendicularly from the ecliptic to the north and the south, running from $0°$ to $90°$, and ecliptical longitudes λ (which start at the Vernal Equinox and run from $0°$ to $360°$) are measured in the same sense as the right ascension.

Since the inclination of the ecliptic, the so-called obliquity ε, is known, we can easily convert equatorial to ecliptic coordinates and vice versa. The obliquity of the ecliptic changes very slowly, and its value can be calculated as follows.

$$\varepsilon = 23°27'8''.26 - 46''.84\,T - 0''.004\,T^2 + 0''.0018\,T^3,$$

where T is in Julian centuries from 1900.

Transformation is computed logarithmically with these formulae:

ecliptic → equator	equator → ecliptic
$\tan \alpha = \dfrac{\sin (P - \varepsilon)}{\sin P} \tan \lambda$	$\tan \lambda = \dfrac{\sin (Q + \varepsilon)}{\sin Q} \tan \alpha$
$\tan \delta = \cotan (P - \varepsilon) \sin \alpha$	$\tan \beta = \cotan (Q + \varepsilon) \sin \lambda$
$\tan P = \cotan \beta \sin \lambda$	$\tan Q = \cotan \delta \sin \alpha.$

6.2.5.2. The Galactic Coordinate System

For stellar astronomical investigations it is convenient to use the Milky Way as a fundamental plane. We then speak of *galactic coordinates* and measure them as follows:

Galactic latitude b runs from $0°$ to $90°$, being positive to the north and negative to the south of the galactic plane. Galactic longitude l runs from $0°$ to

360°, starting at the point of intersection (ascending node) of the galactic plane with the equator. This point has the equatorial longitude 280.°0.
The galactic pole has equatorial coordinates:

$$\alpha = 191.°3, \qquad \delta = +27.°7 \quad (1900).$$

These are generally rounded off to

$$\alpha = 190°, \qquad \delta = +28° \quad (1900).$$

The inclination of the galactic plane to the equator is taken as 62.°00, and that to the ecliptic as 60.°55.

These coordinates were in use between 1932 and 1960. In accordance with the decision of the International Astronomical Union, in 1959, we at present use the following coordinates for the galactic pole:

$$\alpha = 12^{h}49^{m} \qquad \delta = +27.°4 \quad (1950.0).$$

It was also decided that in the future the zero point of the galactic coordinates should be in the direction of the galactic center (the coordinates of which, in the old galactic systems, were $l = 327.7°$, $b = -1.4°$).

Since it is usually unnecessary to have great accuracy for the galactic coordinates we can use transformation tables that suffice for nearly all cases.[2]

6.3. Time and the Phenomena of Daily Motion

6.3.1. The Definition of Time

6.3.1.1. True Time, Mean Solar Time, and the Equation of Time

The basis for the calculation of time is the rotation of the Earth and its revolution about the Sun.

Formerly, days were divided according to the motion of the Sun. However, when observations were made of the intervals between two successive upper culminations of the Sun, it was noticed that these intervals are subject to conspicuous variations. *True Solar Time* (TST), which is given by the hour angle of the true Sun, is nonuniform and does not provide a practical way of measuring time. The reasons for this nonuniformity are the following:

1. The Earth's orbit around the Sun is not a circle, but an ellipse. Because of Kepler's second law it is therefore impossible for the Sun to traverse equal arcs in the sky in equal times.

2. Even if the Sun's motion along the ecliptic swept out equal arcs in equal intervals of time, projecting these arcs on the celestial equator, along which we measure time, would lead to different time divisions because the equator and the ecliptic are at an angle to each other.

It is for these reasons that we introduce a *fictitious* mean Sun whose position coincides with that of the true Sun (*v*) at the Vernal Equinox and

² SCHÜTTE, K.: Die Transformation beliebiger sphärischer Koordinatensysteme mit einer einzigen immerwährenden Hilfstafel. *Astr. Nachr.* **270**, 76 (1940).

then moves so that its speed along the equator is uniform. This defines *Mean Solar Time* (MST), which is a uniform measure of time.

The hour angle of this mean Sun is called *Mean Local Time* (MLT) and differs from place to place. Mean Solar Time (m) and True Solar Time (v) are therefore Local Times.

The difference

$$\text{“Mean Solar Time”} - \text{“True Solar Time”} = z$$

is called the *Equation of Time* (z). Up to the year 1930, it was used in the sense $m - v = z$. From 1930 onward, as defined by the International Astronomical Union, the sign became

$$w - m = z.$$

When using Yearbooks, one must check how the equation of time is defined, since not all yearbooks have introduced this change of sign.

At *true noon* we have

$$w = 0,$$

and therefore

$$m = -z.$$

Four times every year the Equation of Time becomes zero, namely, on 16 April, 14 June, 2 September, and 25 December. The extreme values, on the other hand, are:

Date	$w - m$
11 February	$-14^m.3$
14 May	$+\ \ 3.7$
26 July	$-\ \ 6.4$
3 November	$+16.4$

The behavior of the Equation of Time in the course of a year is shown in Fig. 6–3 and in Fig. 7–7 (on page 202).

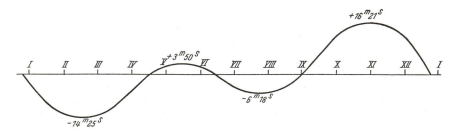

Fig. 6–3   The Equation of Time.

6.3.1.2. Sidereal Time

On page 168, we defined Sidereal Time as the hour angle of the Vernal Equinox. A day, that is, 24 hours Sidereal Time, is the interval between two upper culminations of the Vernal Equinox. Sidereal Time is also a local time, since we have

$$\theta = t + \alpha.$$

For

$$t = 0^h$$

therefore

$$\theta = \alpha.$$

This means that the right ascension of a star at its upper culmination is equal to the Sidereal Time. We can also put this another way: Sidereal Time is the right ascension of all those stars that at their upper culmination traverse the meridian at the same moment. Of course, the fictitious mean Sun also has a certain right ascension at its upper culmination, and this is called "*Sidereal Time at mean noon.*"[3]

Astronomical Almanacs generally give the Sidereal Time for 0^h Universal Time from day to day. This is valid only for the meridian of Greenwich, and since Sidereal Time is a local time, it is necessary to add a correction to obtain the Sidereal Time at mean midnight at a point on the Earth whose longitude is $\Delta\lambda$ (in hours) measured from Greenwich:

Correction of Sidereal Time $= \pm 9.8565 \cdot \Delta\lambda$ for a longitude to east or west, respectively.

6.3.2. Relation Between Sidereal Time and Mean Time

The sidereal day is somewhat shorter than a mean day, since the Sun moves toward the east from day to day, and thus every day it arrives at its upper culmination somewhat later than a star that on the previous day arrived at its upper culmination simultaneously with the Sun. If we subdivide every sidereal day into 24 hours of 60 minutes each, with each minute of 60 seconds, we find that every unit of Sidereal Time must be somewhat shorter than the corresponding unit in Mean Time. During a tropical year[4] the Sun moves from Vernal Equinox to Vernal Equinox. This requires 365.2422 mean solar days. During this time the Sun has completed one revolution; a star, therefore, has one more upper culmination per year than the Sun. Therefore, it is

---

[3] Correctly speaking, Sidereal Time is not a uniform measure of time, since it is defined by the point of the Vernal Equinox, which undergoes a small periodic variation with an amplitude of 1.05 seconds in 18.7 years, due to nutation (see page 177). However, for practical purposes Sidereal Time can be taken as a uniform measure of time.

[4] For the explanation of the concept "tropical year" see page 182.

$$\text{One mean solar day} = \frac{366.2422}{365.2422} \text{ sidereal days,}$$

$$\text{One sidereal day} = \frac{365.2422}{366.2422} \text{ mean solar days.}$$

From this it follows that

$$24^h 0^m 0^s \text{ Mean Time} = (24^h + 3^m 56^s\!.555) \text{ Sidereal Time,}$$

$$24^h 0^m 0^s \text{ Sidereal Time} = (24^h - 3^m 55^s\!.909) \text{ Mean Time.}$$

Or, per hour,

$$1^h \text{ Mean Solar Time} = (1^h + 9^s\!.856) \text{ Sidereal Time,}$$

$$1^h \text{ Sidereal Time} = (1^h - 9^s\!.829) \text{ Mean Solar Time.}$$

Tables for making this conversion are to be found as Tables 5 and 6 of the Appendix (22.1.10 and 22.1.11). For many purposes a simple approximate relation, given by Börgen (1902), is sufficient:

In every hour of Mean Time, the Sidereal Time *gains* $(10 - 1/7)^s$.
In every hour Sidereal Time the Mean Time *loses* $(10 - 1/6)^s$.

The error due to this approximation is only $0^s\!.00067$ per hour in the first case; in the second, $0^s\!.00379$.

6.3.3. The Phenomena of Diurnal Motion (Formulae)

We shall now give a summary of the more important phenomena of diurnal motion, supplemented by a few additional formulae:

Upper culmination: $z_s = (\varphi - \delta)$ for a star south of the zenith
$z_n = (\delta - \varphi)$ for a star north of the zenith

Lower culmination: $z = 180° \mp (\varphi + \delta)$ for latitudes north or south, respectively

Stars are circumpolar if

$$\delta \geqq 90° - \varphi.$$

Rising and setting (without refraction):

Hour angle: $\cos t_0 = -\tan \delta \cdot \tan \varphi,$

where t_0 is called the semidiurnal arc (see Table 4 of the Appendix, page 503).

The evening and morning elongation Az_0 is measured from the south (see Appendix, Table 1, page 502):

$$\cos Az_0 = -\frac{\sin \delta}{\cos \varphi} \quad (Az_0 \leqq 90°, \text{ if } \delta \leqq 0°).$$

Transit through the Prime Vertical (Az = 90°):

Zenith distance: $\cos z = \dfrac{\sin \delta}{\sin \varphi}$,

Hour angle: $\cos t = \tan \delta \cdot \operatorname{cotan} \varphi$.

This is above the horizon only if δ and φ have the same sign and if $|\delta| < |\varphi|$.

Maximum Azimuth: Above the horizon only, if δ and φ have the same sign, and if $|\delta| < |\varphi|$.

Zenith distance: $\cos z = \dfrac{\sin \varphi}{\sin \delta}$,

Hour angle: $\cos t = \operatorname{cotan} \delta \cdot \tan \varphi$,

Azimuth: $\sin Az = \dfrac{\cos \delta}{\cos \varphi}$,

where azimuth is measured from the north.

6.4. Changes in the Coordinates of a Star

6.4.1. *Proper Motion and Precession*

While the coordinates of a star measured in the horizon system usually undergo very rapid changes with time, the coordinates in other systems only undergo very small variation, with which the amateur astronomer will find it useful to be familiar.

6.4.1.1. *Proper Motion (PM)*

HALLEY discovered in 1718 that the stars change their position in the sky relative to each other. Stars, like the Sun, move in space and the component of their motion, which is perpendicular to the line of sight, appears to us as a change in position and is called the *proper motion* (PM of the star). Because the stars are very distant this proper motion is very small and therefore escaped detection for a long time. HALLEY was able to demonstrate the proper motions of only a very few stars; today we know more than 300,000 proper motions. The largest annual proper motion is 10.27 seconds, shown by a faint star of magnitude $9^{m}.7$. There are only a very few stars whose annual proper motions exceed 1.0 second. Thus the new *General Catalogue* by L. Boss, which contains more than 33,000 stars, shows only a hundred stars with annual proper motions exceeding 0.10 second of arc. The annual proper motion is resolved into two components, one in the direction of right ascension, the other in declination; these are denoted by μ_{α} and μ_{δ}. The amateur astronomer will, apart from a few exceptions, have no need to make allowance for the annual proper motion.

6.4.1.2. Precession

Precession is of the utmost importance since it causes relatively rapid changes in the coordinates α and δ. The pull of the Sun and the Moon on the equatorial bulge of the rotating Earth causes the polar axis to move, and in about 26,000 years it describes a complete revolution about the pole of the ecliptic. This period, which is actually 25,725 years, is also called the *Platonic Year*, and the movement of the polar axis about the pole of the ecliptic is called *precession*. This was recognized by HIPPARCHUS as early as the second century A.D., and appears as a secular retrograde movement of the spring equinox. The largest part of the precession, the so-called *lunisolar precession*, is caused by the inequality of the moments of inertia of the Earth and can only be determined empirically. For *annual lunisolar precession*, according to NEWCOMB, we have

$$p_0 = 50\rlap{.}''3708 + 0\rlap{.}''0000495 \, t,$$

where t is measured in tropical years since 1900. There is also a corresponding effect due to the planets, *annual planetary precession*:

$$p_{\text{PL}} = -0\rlap{.}''1248 - 0\rlap{.}''0001887 \, t.$$

The total effect is the so-called precession of the equinoxes:

$$p = 50\rlap{.}''2564 + 0\rlap{.}''000225 \, t.$$

In consequence of this precession the longitude of the star increases continuously.

$$P_0 \cdot \sec \varepsilon$$

is called NEWCOMB's *Constant of Precession*.

In order to evaluate the effect of precession on the coordinates α, δ, we introduce the two components m and n of the annual precession in the direction of right ascension and declination.

$$m = +46\rlap{.}''092 + 0\rlap{.}''0002797 \, t$$
$$= +3\rlap{.}^{s}0728 + 0\rlap{.}^{s}00001865 \, t,$$
$$n = +20\rlap{.}''051 - 0\rlap{.}''0000834 \, t,$$

where t is again measured in years from 1900.

For a star with coordinates α and δ, we thus obtain the annual variations as follows:

$$p_\alpha = m + n \sin \alpha \tan \delta,$$
$$p_\delta = n \cos \alpha.$$

It is evident that p_α is usually positive, while p_δ has the same sign as $\cos \alpha$. The values of p_α and p_δ calculated from these equations are given in Table 9 of the Appendix, page 509.

At the moment the celestial pole is close to the star α Ursae Minoris; it will come closest to the star in the year 2115 A.D., when its distance will be only 28'. In about 14,000 A.D. the bright star Vega (α Lyrae) will be the "Pole Star."

To calculate the effect of the precession on star positions a few years, decades, or centuries in the future, one generally uses a series expansion of the form

$$\alpha, \delta = \alpha_0, \delta_0 + \tau \, \mathrm{I} + \frac{\tau^2}{200} \, \mathrm{II} + \left(\frac{\tau}{100}\right)^3 \mathrm{III},$$

where the time τ is measured in years. The quantities I, II, and III are usually given in star catalogs and have the following names:

I = variatio annua (= annual precession + proper motion)

II = variatio saecularis

III = third term

For very large time intervals and for stars near to the pole, it is necessary, however, to use the exact formulae of spherical trigonometry.

6.4.2. Nutation and Aberration

6.4.2.1. Nutation

In 1747, the British astronomer BRADLEY discovered a short period movement of the polar axis with the period of 18.6 years, known as nutation. The celestial pole thus moves on the celestial sphere describing an ellipse with axes 18″ and 14″. The nutation is essentially caused by the fact that the lunar orbit does not lie in the ecliptic; and it therefore has the period of a revolution of the lunar nodes, namely 18.60 tropical years. The constant of nutation is 9″.207.

6.4.2.2. Annual Aberration

By 1728, BRADLEY had already discovered a phenomenon due to the fact that the orbital velocity of the Earth is a finite fraction of the velocity of light. This phenomenon is called *aberration;* because of this effect the stars appear a little displaced in the direction of the motion of the Earth. Thus a star changes its position periodically in the course of a year, describing the *aberration ellipse.*

The constant of aberration is the maximum value of the displacement,

$$k = 20''.49.$$

The coordinates of a star affected by aberration are λ' and β'. We then have

$$\lambda' - \lambda = -20''.49 \cos(\lambda - \odot) \sec \beta,$$

$$\beta' - \lambda = +20''.49 \sin(\lambda - \odot) \sin \beta,$$

where λ and β are the coordinates free of aberration, and $\odot$ is the longitude of the Sun.

The semiaxes of the aberration ellipse are

$$a = 20''.49 \quad \text{and} \quad b = 20''.49 \sin \beta.$$

At the pole of the ecliptic ($\beta = 90°$) the aberration ellipse becomes a circle, and on the ecliptic it becomes a straight line.

6.4.2.3. Diurnal Aberration

Since the Earth rotates about its axis, there is also a diurnal aberration, whose constant—corresponding to the equatorial velocity of 465 m/sec—is $0''.320$. The influence of the diurnal aberration on the coordinates α and δ is found from

$$\alpha' - \alpha = 0''.320 \cos \varphi \cos t \sec \delta,$$

$$\delta' - \delta = 0''.320 \cos \varphi \sin t \sin \delta,$$

where α' and δ' are the coordinates affected by diurnal aberration.

6.4.3. Parallax and Refraction

6.4.3.1. Diurnal and Annual Parallax

Diurnal parallax is due to the fact that angles to a star vary somewhat according to whether we measure from a point on the surface of the Earth or from the center of the Earth. In Fig. 6–4 the angle z_0 indicates the

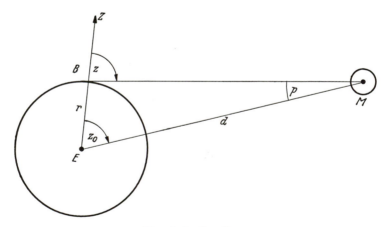

Fig. 6–4 Parallax.

zenith distance of the center of the Moon as it would appear from the center of the Earth; however, the zenith distance observed at the same time from a point B on the surface of the Earth would be z. We thus have a small angle p at the center of the Moon, which is called parallax. The greater the distance d between the centers of the celestial bodies, the smaller is p. For $z = 90°$ we obtain the maximum value of p, the *equatorial-horizontal parallax p_0''*.

This is

$$p_0'' = \frac{r}{\sin 1'' \cdot \Delta},$$

where Δ is the distance between the two bodies. For any given zenith distance z, we then have

$$p'' = p_0'' \cdot \sin z.$$

The equatorial-horizontal parallaxes of the Sun and Moon are

$$\pi_{\mathbb{C}} = 3422''.44 = 57'2''.44,$$

$$\pi_{\odot} = 8''.794.$$

In exact calculations we must also take into account the fact that the Earth is flattened at the poles. The difference between the geographic and the geocentric latitude, $\varphi - \varphi'$, and the definition of the corresponding radius r' have been given above on page 164. For all the bodies of the solar system, except the Moon and other near objects, we thus have the following approximate formulae for the correction of diurnal parallax in α and δ:

$$\alpha' - \alpha = -\frac{r' \cdot \pi \cdot \cos \varphi'}{\Delta \cos \delta} \sin t,$$

$$\delta' - \delta = -\frac{r' \pi \sin \varphi'}{\Delta} \cdot \frac{\sin(\gamma - \delta)}{\sin \gamma}$$

where $\tan \gamma = \tan \varphi' \cos t$, and π is the equatorial-horizontal parallax of the star. We have, furthermore,

α, δ, Δ the geocentric coordinates of the celestial body,

$t = \theta - \alpha$, φ', r' the geocentric coordinates of the observing site,

α, δ', Δ' the coordinates of the body referred to the observing site.

For the Moon and other near bodies such as artificial satellites, we must, however, use the exact formulae.

The *annual parallax* is defined in a similar manner, except that the baseline is now the semidiameter of the Earth's *orbit*. The annual parallax of the nearest fixed star is $0''.765$, but the amateur astronomer will hardly ever need to take this into account.

6.4.3.2. *Astronomical Refraction*

When a light ray from a star traverses the Earth's atmosphere, it is refracted, and the star therefore appears to us to be higher than it really is above the horizon. Every observation must be corrected for refraction before it can be used. Refraction has been dealt with on page 156.

6.4.4. *Reducing Mean Position to Apparent Position*

The mean position of a star is its position at the beginning of the year, including the constant part of the aberration. On the other hand, the apparent position ($\alpha_{\text{app.}}$ and $\delta_{\text{app.}}$) is the position of the star at a given moment; this apparent position is therefore still affected by proper motion, precession, and nutation. If α_0 and δ_0 denote the mean position, and $\alpha_{\text{app.}}$ and $\delta_{\text{app.}}$ the apparent position, then the reduction from the mean to the apparent position can be effected by using the following formulae:

$$\alpha_{\text{app.}} = \alpha_0 + t\mu_\alpha + \underbrace{f + 1/15g \sin (G+\alpha) \cdot \tan \delta}_{\text{precession + nutation}} + \underbrace{1/15h \sin (H+\alpha) \sec \delta}_{\text{aberration}}$$

$$\delta_{\text{app.}} = \delta_0 + t\mu_\delta + \underbrace{g \cos (G+\alpha) + h \cos (H+\alpha) \sin \delta}_{\text{precession + nutation}} + \underbrace{i \cos \delta}_{\text{aberration}}$$

Here t is the time since the beginning of the year, expressed as a decimal fraction of the year.

For very accurate calculations we must also put in the so-called short-period nutation terms, namely,

$$\text{in } \alpha : + \left[f' + \frac{1}{15} g' \sin (G'+\alpha) \tan \delta \right],$$

$$\text{in } \delta : + [g' \cos (G'+\alpha)].$$

The above form of reduction is also called the trigonometric form. The constants

$$f, \log g, G, \log h, H, \log i, i, \text{ and also } f', g', \text{ and } G'$$

were formerly given in the Berliner Astronomisches Jahrbuch for each day at 0^h Universal time; nowadays we use the Astronomical Ephemeris of London and Washington.

For the majority of the bright stars of the so-called Fundamental Catalogue most Yearbooks give apparent positions (without the short-period lunar terms) at 10-day intervals, so that for these stars a reduction to apparent position is unnecessary.

It is sometimes convenient to use the so-called algebraic form of reduction by means of the formulae

$$\alpha_{\text{app.}} = \alpha_0 + t\mu_\alpha + Aa + Bb + Cc + Dd + E + [A'a + B'b],$$

$$\delta_{\text{app.}} = \delta_0 + t\mu_\delta + Aa' + Bb' + Cc' + Dd' + [Aa' + Bb'],$$

where

$$a = m + 1/15\, n \sin \alpha \tan \delta \qquad a' = n \cos \alpha$$

$$b = 1/15 \cos \alpha \tan \delta \qquad b' = -\sin \alpha$$

$$c = 1/15 \cos \alpha \sec \delta \qquad c' = \tan \varepsilon \cos \delta - \sin \alpha \sin \delta$$

$$d = 1/15 \sin \alpha \sec \delta \qquad d' = \cos \alpha \sin \delta.$$

The quantities A, B, C, D, E, as well as A' and B', are also tabulated for each day at 12 hours Greenwich Sidereal Time. The terms denoted by brackets are again the short-period lunar terms.

6.5. Calendars, Years, Julian Dates, and Normal Times

6.5.1. The Calendar and the Measurement of Years

From the earliest times it has been apparent that difficulties are encountered in creating a useful calendar, since the year (that is, an apparent revolution of the Sun) is not an exact multiple of the length of the day or of the length of the month.

The Egyptians, whose work on a calendar goes back to the fourth millenium B.C., originally used a pure *solar year* of 360 days, starting from the annual flooding of the Nile. In later times when the deviation from the position of the Sun became noticeable, five days were added. But even then a slow displacement was found that amounted to one day in every four years. Thus the Egyptians deduced that the length of the year was 365.25 days.

In 46 B.C., Julius Caesar put an end to the arbitrary corrections made to the calendar by the priests, and introduced a new calendar, called the *Julian Calendar* after him, in which every three years of 365 days were followed by one year of 366 days. This calendar was in use up to 1582, by which time the Vernal Equinox occurred markedly before 21 March. From this it could be deduced that the time between one transit of the Sun through the (mean) Vernal Equinox to the next, the so-called *tropical year*, must be shorter than 365.25 days. Thus a new change became necessary and Pope Gregory XIII (1572–1585) decreed by a Bull of 24 February 1581 that:

1. The day following 4 October 1582 should have the date 15 October 1582 (in order to restore the beginning of spring to 21 March).
2. All years that are divisible by 4 should be leap years of 366 days, except those which coincide with the beginning of a century; the latter are leap years only if they are divisible by 400. (Thus 1700, 1800, and 1900 are not leap years, while 1600 and 2000 are.)

This calendar, called the Gregorian Calendar, therefore has (within each cycle of 400 years) three leap days fewer than the Julian calendar, and we have that the length of the mean Gregorian year is 365.2425 days; the length of the tropical year (see below) is 365.2422 days.

The difference of 0.0003 day equals 26^s, adds up to one day only after 3000 years.

> The way the years are now usually numbered was proposed by the Roman Abbot Dionysius Exiguus (in about 525 A.D.). He made an error in fixing the first year, as it appears that Christ was probably born seven years earlier than the date given by Exiguus.
>
> Later the years before the beginning of the Christian era were numbered continuously, but allowance was not made for the fact that between the year 1 A.D. and the year 1 B.C. there should have been a year with the number zero. For this reason there are differences between historical and astronomical numbering of years. There is, for instance:

$$1959 \text{ A.D. (historical)} = +1959 \text{ (astronomical)}$$
$$1 \text{ B.C. (historical)} = 0 \text{ (astronomical)}$$
$$300 \text{ B.C. (historical)} = -299 \text{ (astronomical)}$$

6.5.2. Length of the Year, Beginning of the Year, Julian Date, and the Mean Day

6.5.2.1. Different Types of Length of a Year

Since precession is not quite constant (see pages 176–177) the lengths of the years measured in different ways vary slowly. Denoting by T the time in Julian centuries since 1900 January 0.5, each century comprising 36,525 ephemeris days, we make the following definitions:

1. The *tropical year* is the interval between two transits of the mean Sun through the mean Equinox. During this time the mean longitude of the mean Sun increases by 360°. Because of the retrograde motion of the Vernal Equinox the mean Sun does not make a complete revolution. We have

$$\text{One mean tropical year} = 365\overset{d}{.}24219879 - 0\overset{d}{.}00000614\,T.$$

However, the motion of the Vernal Equinox is not quite regular, since precession slowly increases; thus the length of the tropical year decreases by 5.36 seconds every 1000 years.

2. The *sidereal year* is the time the Sun takes to return to the same star in the ecliptic. It must therefore be longer than the tropical year:

$$\text{One sidereal year} = 365\overset{d}{.}25636042 + 0\overset{d}{.}000000111\,T.$$

The sidereal year is *not* used to construct a calendar.

3. The *anomalistic* year is the time between successive transits of the Earth through the perihelion. Since the perihelion advances along the Earth's orbit, this year is about 4.5 min longer than the sidereal year.

$$\text{One anomalistic year} = 365\overset{d}{.}25964134 + 0\overset{d}{.}00000304\,T.$$

4. The *Julian year* has already been defined above:

$$\text{One Julian year} = 365\overset{d}{.}25.$$

This year is at present $11^m 14^s$ *longer* than the tropical year.

5. The *Gregorian year* or civil year has been defined above. It contains 365.2425 days and is our normal calendar year.

6.5.2.2. Beginning of the Year

The beginning of the astronomical year was fixed by BESSEL as the moment at which the right ascension of the mean Sun (including the constant part of the aberration) is

$$\text{Right ascension} = 18^h 40^m = 280°.$$

This moment coincides very closely with the beginning of our normal civil year; it is also called BESSEL's "*annus fictus.*"

The length of the annus fictus is $365\overset{d}{.}24219879 - 0\overset{d}{.}00000786\ T$ mean days. It therefore differs only very little from the tropical year. BESSEL's year is independent of the observing site and therefore starts at the same moment all over the Earth. We subdivide it in decimals, so that for instance the beginning of the BESSEL year 1959 is called 1959.0.

The difference

Beginning of the civil year − beginning of the annus fictus $= k$

which is called "*dies reductus.*"

6.5.2.3. The Julian Date and the Beginning of the Mean Day; Universal Time

It is inconvenient to express large intervals of time in terms of days. For this reason it is the *Julian Day* which is mainly used in reducing observations of variable stars. It was Joseph Justus Scaliger (1540–1609) who proposed a period of

$$19 \times 28 \times 15 \text{ years} = 7980 \text{ years,}$$

which is named after his father, Julius Cäsar Scaliger. The beginning of this Julian period was fixed at January 1.0 of the year –4712.

Up to the end of 1924, the moment of the upper culmination of the mean Sun, that is the mean noon (beginning of the astronomical mean day), was taken as the beginning of the mean day. From 1925 onward, the beginning of the mean day was moved from the upper culmination of the mean Sun to the *preceding lower culmination,* that is to mean midnight. Since 1925, the civil time at Greenwich has been called *Universal Time.*

The days of the Julian period (Julian date within the Julian period) are always numbered on the astronomical system and even after 1925 begin at mean noon, that is at 12^h Universal Time (UT).

Therefore:

1924 December $31^d, 12^h$ Mean Time Greenwich
$\qquad\qquad = 1925$ January $1, 0^h$ Universal Time;

1947 January $18^d, 3^h 5^m$ MEZ = Julian date $2432203\overset{d}{.}587$.

Table 7 of the Appendix gives the days of the Julian period between 1000 and 2000 A.D., and Table 8 shows days, hours, and minutes in decimal fractions of the Julian Year.

6.5.3. Normal Times (Time Zones), Date Line

Local Time changes from place to place and its use in practical life is therefore inconvenient. For instance, the introduction of "railway time" within countries did not help very much. In North America more than 70 different railway times were in use up to 1882. In 1883 they were replaced by 5 different zone times which differed by whole hours from the GMT, namely:

$$Pacific\ Standard\ Time\quad = GMT - 8^h,$$
$$Mountain\ Standard\ Time = GMT - 7^h,$$
$$Central\ Standard\ Time\quad = GMT - 6^h,$$
$$Eastern\ Standard\ Time\quad = GMT - 5^h,$$
$$Colonial\ Standard\ Time\ = GMT - 4^h.$$

The time zones have the Local Time of their mean meridian.

Europe followed later with:

$$Western\ European\ Time\ (WET)\quad = GMT,$$
$$Central\ European\ Time\ (CET)\quad = GMT + 1^h,$$
$$Eastern\ European\ Time\ (EET)\quad = GMT + 2^h.$$

On 1 April 1893, Germany accepted CET as the legal time; nearly all countries have now accepted some zone time.

If we go round the Earth once from east to west, then for every 15° longitude we lose one hour in time, i.e., we lose a whole day during the complete journey. If we travel in the opposite direction, that is from west to east, we gain one day. On the line of longitude 180° from Greenwich the date will therefore differ by one day from that of Greenwich. This line is called the *Date Line*. For convenience the actual Date Line does not completely coincide with the ideal one. For practical purposes we must remember that if we cross the Date Line from the east to the west we have to omit one day and if we cross from the west to the east, we have to use one day twice.

6.6. Variability of the Rotation of the Earth, Ephemeris Time, and the Definition of the Second.[5]

A few decades ago it was discovered that the rotational velocity of the Earth undergoes small variations. When extreme accuracy is required this fact must be taken into account.

The time fixed by an astronomical determination is called *Empirical Time*; it is referred to the mean Sun and to the mean Vernal Equinox. If, however, the Earth rotates irregularly, then the Empirical Time is not identical with the uniform Ephemeris or Inertial Time, based on Newtonian mechanics, which forms the basis for the Ephemerides in the Yearbooks. For an exact comparison of observations within our solar system, the difference between the variable Empirical Time and the Ephemeris Time must be taken into account.

[5] A detailed account is given in F. Gondolatsch: *Erdrotaton, Mondbewegung und das Zeitproblem der Astronomie.* Veröff. d. Astronomischen Rechen-Instituts zu Heidelberg, No. 5 (1953).

We distinguish three different kinds of changes in the rotational velocity of the Earth:

1. *A secular slowing down of the rotational velocity of the Earth by tidal friction.* This causes a continuous increase in the length of the day by 4.5×10^{-8} sec per day; and this adds up to 0.0016 sec per century. Because the effect on the mean longitude of a star is cumulative, the secular acceleration being proportional to T^2, it is of considerable importance. The theory of tidal friction is the concern of dynamic oceanography; a number of theoretical investigations by JEFFREYS and others have shown that loss of energy by tidal friction, particularly in the smaller oceans, corresponds to the secular acceleration observed in the longitude of the Moon.

2. *Irregular, positive, and negative accelerations of the rotational velocity of the Earth.* These are called "fluctuations" and their real origin is still unknown but is probably the displacement of masses in the interior of the Earth. These fluctuations have been derived from observations of the Moon and the largest deviations of the length of the day in the past 250 years have been found to be

$$-0\overset{s}{.}005 \text{ (in about 1871) and } +0\overset{s}{.}002 \text{ (in 1907)}.$$

This effect can also be cumulative and leads to noticeable time differences if we use the Earth as a clock. However, these fluctuations cannot be predicted in advance, but can only be deduced later from observations of the Moon.

3. *Seasonal variations of the rotational velocity of the Earth.* These are very small and are caused by meteorological events. These variations, which were first discovered in the years 1934–1937, when quartz clocks were used, can in general be neglected.

In astronomical work, we always use the Mean Solar Time for observations. Consequently, it is necessary to transform a given Mean Solar Time into Ephemeris Time.

At a conference on astronomical constants held in Paris in 1950, it was decided to apply the following correction Δt, to be added to the Mean Solar Time, in order to find Ephemeris Time:

$$\Delta t = t_{\text{Eph}} - t_{\text{empirical}} = +24\overset{s}{.}349 + 72\overset{s}{.}3165\,T + 29\overset{s}{.}949\,T^2 + \frac{1\overset{s}{.}821}{1''}\,B''.$$

Here T is again the number of Julian centuries, positive after 1900, negative before 1900, measured from 1900 January 0.5 Universal Time.

The term in T^2 contains the secular breaking of the Earth's rotation by tidal friction; B is in seconds of arcs and is derived afterwards from the observed fluctuations.

Table 6.1 provides the corrections Δt at 5-year intervals from 1900, where the Empirical Time is taken to be identical with Universal Time.

Since the rotation of the Earth is variable, we also require a new *definition of the second.* According to the definition by the Comité International des Poids et Mésures, Paris 1955, we have:

$$1^s = \frac{1}{31,556,925.9747} \text{ of tropical year for 1900 January 0, } 12^h \text{ Ephemeris Time}$$

Table 6.1. Correction of Time
Δt = Ephemeris Time − Universal Time

Date	Δt	Date	Δt
1900.5	− 3.79^s	1935.5	+ 23.63^s
1905.5	+ 3.26	1940.5	+ 24.30
1910.5	+ 10.28	1945.5	+ 26.57
1915.5	+ 16.39	1950.5	+ 29.42
1920.5	+ 20.48	1955.5	+ 31.59
1925.5	+ 22.55	1960.5	+ 34
1930.5	+ 23.18		

More detailed information can be found in LANDOLT-BÖRNSTEIN, Zahlenwerte und Funktionen, New Series, Group 6, *Astronomy and Astrophysics*, **1**, 74 (1965).

6.7. Spherical Trigonometry

6.7.1. The Basic Equations of Spherical Trigonometry

In a spherical triangle (Fig. 6–5) with sides a, b, c and corresponding angles α, β, γ, we have the three following fundamental relations.

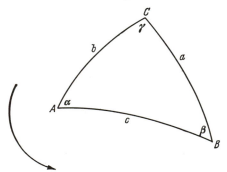

Fig. 6–5 Spherical relationships.

6.7.1.1. Sine Formula

$$\sin a : \sin b = \sin \alpha : \sin \beta,$$
$$\sin a \cdot \sin \beta = \sin b \cdot \sin \alpha,$$
$$\sin b \cdot \sin \gamma = \sin c \cdot \sin \beta,$$
$$\sin c \cdot \sin \alpha = \sin a \cdot \sin \gamma.$$

6.7.1.2. Cosine Formula

$$\cos a = \cos b \cdot \cos c + \sin b \cdot \sin c \cos \alpha,$$
$$\cos b = \cos c \cdot \cos a + \sin c \cdot \sin a \cos \beta,$$
$$\cos c = \cos a \cdot \cos b + \sin a \cdot \sin b \cos \gamma,$$

and for the polar triangle[6] whose sides are segments of Great Circles

$$\cos \alpha = -\cos \beta \cos \gamma + \sin \beta \sin \gamma \cos a,$$
$$\cos \beta = -\cos \gamma \cos \alpha + \sin \gamma \sin \alpha \cos b,$$
$$\cos \gamma = -\cos \alpha \cos \beta + \sin \alpha \sin \beta \cos c.$$

6.7.1.3. The Sine–Cosine Formula

$$\sin a \cos \beta = \cos b \cdot \sin c - \sin b \cdot \cos c \cdot \cos \alpha,$$
$$\sin b \cos \gamma = \cos c \cdot \sin a - \sin c \cdot \cos a \cdot \cos \beta,$$
$$\sin c \cos \alpha = \cos a \cdot \sin b - \sin a \cdot \cos b \cdot \cos \gamma;$$

and for the polar triangle we have

$$\sin \alpha \cos b = \cos \beta \cdot \sin \gamma + \sin \beta \cdot \cos \gamma \cdot \cos a,$$
$$\sin \beta \cos c = \cos \gamma \cdot \sin \alpha + \sin \gamma \cdot \cos \alpha \cdot \cos b,$$
$$\sin \gamma \cos a = \cos \alpha \cdot \sin \beta + \sin \alpha \cdot \cos \beta \cdot \cos c.$$

In every group the second and the third of the formulae are obtained by cyclic interchange of the letters in the given form.

6.7.2. Derived Formulae

6.7.2.1. The Cotangent Rule

If we have four successive parameters, then:

$$\cos c \cdot \cos \alpha = \sin c \cdot \cotan b - \sin \alpha \cdot \cotan \beta,$$
$$\cos a \cdot \cos \beta = \sin a \cdot \cotan c - \sin \beta \cdot \cotan \gamma,$$
$$\cos b \cdot \cos \gamma = \sin b \cdot \cotan a - \sin \gamma \cdot \cotan \alpha.$$

or, expressed in words:

The product of the cosines of the inner components is equal to the sine of the inner side times the cotangent of the outer side minus the sine of the inner angle times the cotangent of the outer angle. However, for logarithmic work the following relations are more convenient.

6.7.2.2. The Semiangle Formulae

If

$$s = \frac{1}{2}(a+b+c),$$

---

[6] A polar triangle results if we make each apex of the original triangle the pole of a great circle. The intersections of these three great circles form the apices of the polar triangle.

we have

$$\sin\frac{\alpha}{2} = \sqrt{\frac{\sin(s-b)\sin(s-c)}{\sin b \cdot \sin c}},$$

$$\sin\frac{\beta}{2} = \sqrt{\frac{\sin(s-c)\sin(s-a)}{\sin c \cdot \sin a}},$$

$$\sin\frac{\gamma}{2} = \sqrt{\frac{\sin(s-a)\sin(s-b)}{\sin a \cdot \sin b}}.$$

6.7.2.3. Semiperimeter Formulae

If

$$\sigma = \frac{1}{2}(\alpha+\beta+\gamma),$$

we have

$$\cos\frac{a}{2} = \sqrt{\frac{\cos(\sigma-\beta)\cos(\sigma-\gamma)}{\sin\beta\sin\gamma}},$$

$$\cos\frac{b}{2} = \sqrt{\frac{\cos(\sigma-\gamma)\cos(\sigma-\alpha)}{\sin\gamma\cdot\sin\alpha}},$$

$$\cos\frac{c}{2} = \sqrt{\frac{\cos(\sigma-\alpha)\cos(\sigma-\beta)}{\sin\alpha\cdot\sin\beta}}.$$

6.7.2.4. Gauss-Delambre Equations

$$\sin\frac{\alpha}{2}\cdot\sin\frac{1}{2}(b+c) = \sin\frac{a}{2}\cdot\cos\frac{1}{2}(\beta-\gamma),$$

$$\sin\frac{\alpha}{2}\cdot\cos\frac{1}{2}(b+c) = \cos\frac{a}{2}\cdot\cos\frac{1}{2}(\beta+\gamma),$$

$$\cos\frac{\alpha}{2}\cdot\sin\frac{1}{2}(b-c) = \sin\frac{a}{2}\cdot\sin\frac{1}{2}(\beta-\gamma),$$

$$\cos\frac{\alpha}{2}\cdot\cos\frac{1}{2}(b-c) = \cos\frac{a}{2}\cdot\sin\frac{1}{2}(\beta+\gamma).$$

6.7.2.5. The Napier Analogies

These follow immediately from the preceding by division:

$$\tan\frac{\beta+\gamma}{2} = \cotan\frac{\alpha}{2}\cdot\frac{\cos\frac{1}{2}(b-c)}{\cos\frac{1}{2}(b+c)},$$

$$\tan\frac{\beta-\gamma}{2} = \cotan\frac{\alpha}{2}\cdot\frac{\sin\frac{1}{2}(b-c)}{\sin\frac{1}{2}(b+c)},$$

$$\tan \frac{b+c}{2} = \tan \frac{a}{2} \cdot \frac{\cos \frac{1}{2}(\beta - \gamma)}{\cos \frac{1}{2}(\beta + \gamma)},$$

$$\tan \frac{b-c}{2} = \tan \frac{a}{2} \cdot \frac{\sin \frac{1}{2}(\beta - \gamma)}{\sin \frac{1}{2}(\beta + \gamma)}.$$

6.7.3. Application of the Equations to the Solution of Spherical Triangles

Three of the six parameters of the spherical triangle must always be given. There are thus six different cases (see Figure 6–6), which can be solved as shown in Table 6.2.

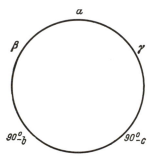

Fig. 6–6   Scheme for formulae in right-angled spherical triangles.

Table 6.2. The Solution of Spherical Triangles.

Given	Required	Formulae
a, b, c	α, β, γ	Cosine formula of the sides or semiangle formulae.
α, β, γ	a, b, c	Cosine theorem of the angles or semiperimeter formulae.
a, b, γ	α, β, c	Basic equations or GAUSS-DELAMBRE equations or NAPIER's analogies.
α, β, c	a, b, γ	Basic equations for angles or NAPIER's analogies or GAUSS-DELAMBRE equations.
a, b, α	β, γ, c	β from sine formula; c and γ from NAPIER's analogies.
α, β, a	b, c, γ	b from sine formula; c and γ from NAPIER's analogies.

6.7.4. The Right-Angled Spherical Triangle

If a is the hypotenuse, then

$$\alpha = 90°.$$

We can now use Napier's rule, omitting the right angle and replacing the two other sides by their complements as follows:

The cosine of one parameter
= the product of the *sines* of the two parameters shown furthest away from it in Fig. 6–6.

Or

The cosine of one parameter
= the product of the cotangents of the parameters shown on either side of it in Fig. 6–6.

$$\cos a = \cos b \cdot \cos c = \cotan \beta \cdot \cotan \gamma,$$
$$\sin b = \sin a \cdot \sin \gamma = \tan c \cdot \cotan \beta,$$
$$\sin c = \sin a \cdot \sin \beta = \tan b \cdot \cotan \gamma,$$
$$\cos \beta = \cos c \cdot \sin \gamma = \tan b \cdot \cotan x,$$
$$\cos \gamma = \cos b \cdot \sin \beta = \cotan a \cdot \tan c.$$

7 | Modern Sundials

K. Schütte†

7.1 Introduction

Sundials on houses and in gardens are a very popular decoration today. Their design and construction offer no difficulties to anybody who has some knowledge of spherical astronomy.

The two main components of a sundial are the rod that casts the shadow, which is called the *gnomon*, and the *dial*.

In principle the dial can be fixed on any kind of surface; however, we shall only consider some important and simple cases of plane dials.

In all cases the gnomon lies in the plane of the meridian and is inclined in such a way that the extension of its direction always points to the celestial pole. In other words, the gnomon is inclined to the horizontal plane at an angle that is equal to the geographic latitude φ.

A sundial is only able to indicate the true solar time, since the shadow of the rod is determined by the position of the true Sun. The true solar time, however, is a true local time, while the time used in daily life is always a standard time for a given zone. This zone time is the mean local time of the particular standard meridian, e.g., the mean time of Greenwich. These facts require two fundamental corrections to the actual reading of the shadow:

1. The *equation of time*, which transforms the true solar time into a mean uniform time, the so-called mean solar time.

2. The allowance for the *difference in longitude* from the standard meridian; this correction transforms the mean solar time, obtained after applying the equation of time, into the legal standard time. Contrary to the equation of time this correction for difference of longitude is, for a fixed sundial, a constant quantity that has to be applied to every shadow reading.

The above remark suggests that the longitude correction could already have been taken into account in the design and numbering of the dial. This is discussed in Sec. 7.5.5.

To determine the exact orientation of a dial, the use of a compass or an architect's plan of the building is often recommended. From my experience I would discourage the application of this method. Directions on plans are usually rather inaccurate, and the readings of a compass direction are often

†Died September 6, 1974

useless because the magnetic deviation is unknown, quite apart from the fact that some local magnetic disturbances may falsify the indication of the compass needle. Furthermore, the compass can usually not be read with the required accuracy.

For all these reasons an amateur astronomer should only use astronomical observations for the exact determination of direction.

7.2. The Equinoxial Sundial

The simplest design is the so-called equinoxial or polar dial. For its construction we can use a kind of armillary sphere, and read the shadow (of the axis which points to the celestial pole) on a ring that lies in the celestial equator. This ring carries a uniform, continuous gradation, which runs from 0^h to 24^h, whereby the point 12^h is at the place where the shadow of the axis appears at true noon, i.e., at the time when the true Sun stands in the meridian. If one takes into account the geographical longitude of the site at which the armillary sphere has been set up, we have to displace the scale corresponding-ly. Such equinoxial dials are best placed on a pillar in the garden.

7.3. Design of a Horizontal Dial and of a Vertical East–West Dial

It is also relatively simple to calculate a dial whose face lies in a horizontal plane, or on a vertical wall, which is exactly orientated in the east–west direction.

In Fig. 7–1, three planes can be found: the horizontal plane, perpendicular to it the east–west plane, and finally the plane that is parallel to the celestial

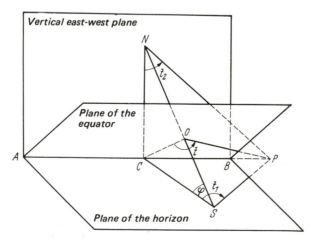

Fig. 7–1 The relationship between the shadows on equinoxial, horizontal, and vertical dials.

equator. The latter also passes through the line AB in which the two other planes intersect.

It is then quite easy to derive the equations for the gradation of the dial on the horizontal and on the vertical east–west plane.

The shadow-throwing rod NS is perpendicular to the equatorial plane which it meets at the point O. Furthermore, P may be the point at which at a given hour angle t the shadow of the rod, thrown onto the equatorial plane, meets the east–west line AB or its extension. Then SP is the shadow on the horizontal plane, which belongs to this hour angle, and NP is the shadow on the vertical east–west plane.

The inclination of the rod is equal to the geographic latitude φ, and for the hour angle $t = 0^h$ the shadow falls onto the lines SC and NC of the two principal planes. The three triangles OCP, SCP, and NCP have all their right angles at C and have the line CP in common. It is therefore

$$COP = t \quad = \text{hour angle on the equatorial plane,}$$

$$CSP = t_1 = \text{corresponding angle in the horizontal plane,}$$

$$CNP \to\ = t_2 = \text{corresponding angle in the vertical east–west plane.}$$

We then have

in the triangle OCP, $CP = OC \tan t$,

in the triangle SCP, $CP = SC \tan t_1$,

in the triangle NCP, $CP = NC \tan t_2$.

Dividing these equations we obtain

$$\tan t_1 = \frac{OC}{SC} \cdot \tan t,$$

$$\tan t_2 = \frac{OC}{NC} \cdot \tan t.$$

Furthermore, we have from the right-angle triangles COS and CON

$$\frac{OC}{SC} = \sin \varphi,$$

$$\frac{OC}{NC} = \cos \varphi.$$

Therefore

$$\tan t_1 = \sin \varphi \tan t,$$

$$\tan t_2 = \cos \varphi \tan t.$$

These are the two fundamental equations that permit us to calculate for an equinoxial dial for a given hour angle t the corresponding angle t_1 on the horizontal or t_2 on the vertical east–west clock, respectively.

7.4. The Construction of a Horizontal and a Vertical East–West Dial

For those who may prefer a graphical solution of the problem we reproduce Figs. 7–2 and 7–3. The construction is very simple.

We draw a circle with the radius $MC = 1$ and extend this radius beyond C to the point O in such a way that $OC = \sin \varphi$. We then draw a semicircle about O with the radius OC, and also draw through C the common tangent. We now mark on the smaller circle, starting at OC, the angles 15°, 30°, 45°, etc., which correspond to the full hours. Any necessary subdivision can easily be inserted, either now or later. We then draw from O the radii that belong to these angles and extend them up to their intersection with the tangent. All the resulting points of intersection are then connected with the point M and furnish on the circle drawn about M the face of the horizontal solar dial. We mark C with 12^h and then continue the numbering of the other hours in the clockwise direction. The shadow-throwing rod is set up in such a way that it is fixed in M and makes the angle φ with the 12^h-direction.

When adjusting the sundial, we first set the dial horizontally with the help of a level and place it so that it points exactly to the north at 12 o'clock. This is best done by calculating for a certain day the exact mean local time

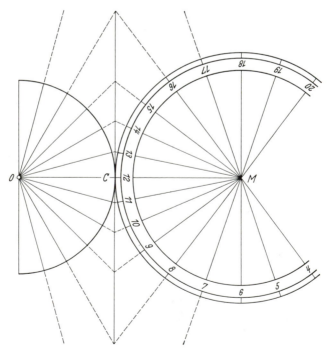

Fig. 7–2 Construction of a horizontal sundial for the geographic latitude $\varphi = 50°$.

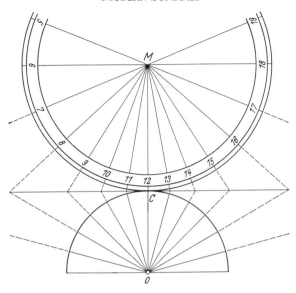

Fig. 7–3 Construction of a vertical sundial in the east–west direction for the geographic latitude $\varphi = 50°$.

which corresponds to 12^h true solar time; the whole dial is then turned accordingly. Here, the correction for longitude has not yet been taken into account. Figure 7–2 shows the construction of such a dial for the longitude $\varphi = 50°$.

If we wish to construct a vertical east–west dial, the only difference in the method is that $OC = \cos \varphi$ (see Fig. 7–3). In this case the figures run in the opposite direction. The hour markings for 6^h and 18^h are in both cases parallel to the tangent.

The setting up of the vertical solar dial in the east–west plane requires astronomical observations. There are two ways: (1) calculate the true solar time (applying the values of the equation of time and of the longitude) at any mean time, and then turn the dial about the vertical until it reads the true solar time; (2) calculate the mean solar time during the summer when the Sun is exactly east or west, and at that moment turn the dial so that the Sun's rays just graze the face of the dial.

7.5. The Vertical Declining Dial

It will not usually be possible to put a vertical sundial on an east–west plane, so we shall need to know what adjustments must be made for using a given wall, probably not accurately situated in an east–west plane. A vertical dial that is not in this plane is called a *vertical declining dial*. Our first task is then to determine the azimuth of the wall.

7.5.1. Determination of the Azimuth of the Wall

We assume for the time being that the wall is really vertical. In general, this assumption will be justified by the work of the builder. However, we have already noted that a builder's plans and compass are not sufficient for a reliable determination of the direction of the wall. It will therefore be best to find this direction by experimenting with grazing sun rays. For this purpose, we observe with a suitable dark glass and one eye the moment in which just half of the Sun's disc is eclipsed by the wall—if we are standing west or east of the extended plane of the wall. We make a note of the exact time of this moment, using a good watch which has been calibrated with a time signal.

With some experience it is possible to determine the above-mentioned moment without instruments with an accuracy of about 10^s. This observation is then repeated on different days and its accuracy improved correspondingly. We must not be surprised that the observed times differ more or less from each other. This is, of course, a consequence of the fact that the changes in the declination of the Sun and in the equation of time also change the moment of the transit through the vertical of the wall.

Each single observation furnishes the azimuth of the wall, according to the well-known formula

$$\tan \mathrm{Az}_\odot = \frac{\sin t}{\sin \varphi \cdot \cos t - \cos \varphi \tan \delta},$$

where

φ = geographic latitude,
t = hour angle of the Sun at the moment of observation,
δ = declination of the Sun at the moment of observation,
$\mathrm{Az}_\odot$ = required azimuth of the Sun, and thus of the wall, measured from the south via the west.

These observations can under favorable conditions be carried out in the morning as well as in the afternoon, and, of course, often only in the summer season.

Such observations give, if the observer has a little experience, values for the azimuth which differ amongst themselves by not more than about 20', so that we can obtain a sufficiently accurate mean value from a few repeated observations.

Of course, the shadow-throwing rod must be inserted very accurately for such a vertical declining dial as well, i.e., parallel to the rotational axis of the Earth. The way this is best done is shown in Sec. 7.5.4.

7.5.2. Calculation of the Dial for the Wall

This calculation is here, of course, considerably more complicated than in the previous cases. In order to derive the necessary formulae we consider Figure 7–4.

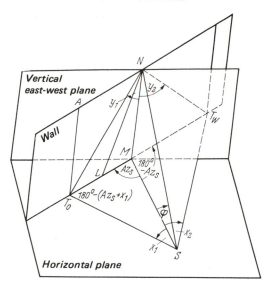

Fig. 7–4 Calculation of the dial of a "vertical declining sundial."

Let NS be again the shadow-throwing rod which meets the horizontal plane in S, and the vertical east–west plane as well as the plane of the wall in N. The figure shows, besides the horizontal plane and the plane of the wall, also the vertical east–west plane which intersects the plane of the wall in a perpendicular line passing through N. If one adds the south direction, which goes through S, we note that its extension to the north meets the two other planes in the point M. The direction of the wall is given by the line T_0M in the horizontal plane. The angle $SMT_0 = Az_\odot$ thus gives the azimuth of the wall in the same way it has been defined and determined above.

Of the two (at first arbitrary) points T_0 and T_w, we note that the former lies on the morningside (hour angle east) and the latter on the afternoon side (hour angle west). Then M is the point in which all three planes intersect. It is then

$$SMT_0 = Az_\odot \quad \text{and} \quad SMT_w = 180° - Az_\odot.$$

We further denote in the horizontal plane

$$MST_0 = x_1 \quad \text{and} \quad MST_w = x_2.$$

The angle NSM is again equal to φ.

The angles x_1 and x_2 are nothing else but the directions of the shadows of the rod in the horizontal plane which correspond to the hour angles t_1 and t_2. If we denote the angles at N in the plane of the wall as

$$MNT_0 = y_1 \quad \text{and} \quad MNT_w = y_2,$$

then they represent the corresponding angles on the wall turned by the angle

$\mathrm{Az}_\odot$. Then NT_0 and NT_w are the corresponding shadows thrown onto this wall at the hour angles t_1 and t_2. In order to find these shadows it is therefore only necessary to calculate the lengths MT_w and MT_0 or the angles y_1 and y_2.

The length MT_0 is found from the plane triangle MST_0:

$$MT_0 = \frac{MS \sin x_1}{\sin (\mathrm{Az}_\odot + x_1)} = \frac{NS \cos \varphi}{\cos \mathrm{Az} + \sin \mathrm{Az}_\odot \cdot \cotan x_1}.$$

There is, however,

$$\cotan x_1 = \frac{\cotan t_1}{\sin \varphi}$$

(just as in the case of the horizontal dial). If the above expression is inserted into the equation for MT_0 we have

$$\cotan y_1 = \frac{MN}{MT_0} = \frac{NS \sin \varphi}{NS \cos \varphi} \left[\cos \mathrm{Az}_\odot + \frac{\sin \mathrm{Az}_\odot}{\sin \varphi} \cotan t_1 \right].$$

The length of the rod NS cancels out, and we obtain

$$\cotan y_1 = \cos \mathrm{Az}_\odot \tan \varphi + \frac{\sin \mathrm{Az}_\odot}{\cos \varphi} \cdot \cotan t_1.$$

It is more convenient to write this as follows:

$$\tan y_1 = \frac{\cos \varphi}{\cos \mathrm{Az}_\odot \cdot \sin \varphi + \sin \mathrm{Az}_\odot \cdot \cotan t_1}.$$

This equation is valid for any given hour angle t_1, that is, for any true solar time in the *morning hours*.

For western hour angles t_2 we have to replace $\mathrm{Az}_\odot$ by $(180° - \mathrm{Az}_\odot)$ and x_1 by x_2. Correspondingly, we obtain

$$\tan y_2 = \frac{\cos \varphi}{-\cos \mathrm{Az}_\odot \cdot \sin \varphi + \sin \mathrm{Az}_\odot \cdot \cotan t_2}.$$

Since we had not introduced any assumptions concerning the position of the points T_0 and T_w on the line $T_0 \rightarrow M \rightarrow T_w$ (except that we required one of these points to lie on the left-hand side and the other on the right-hand side of M), we can now obtain on the horizontal line $T_0 T_w$ any reqiured point of the time scale.

Since the horizontal line $T_0 T_w$ can only have a limited length, it is advisable to move at a suitable place (e.g., in T_0 itself) onto the perpendicular to $T_0 T_w$; this may be $AT_0 = NM$. The angle ANT_0 is equal to $(90° - y_1)$. In order to find the corresponding points on the time scale on AT_0, it is therefore only necessary to make use of functions of the complementary angle.

7.5.3. Transfer of the Calculated Dial onto the Wall

Our next step involves the transfer of the calculated values of the dial onto the actual wall. First of all we determine the point of intersection for the middle of the axis of the rod, say at the point N. We then use a plumb line to determine as accurately as possible the perpendicular through N and mark it. In general, it is sufficient to use a length of exactly one meter up to the point M (as in Fig. 7–4). With the help of a good level we then determine as accurately as possible the horizontal line through M and draw it in on both sides. In this way, we obtain on both sides of the bisecting line NM two squares each of one meter, whose sides are exactly horizontal (in pairs) and exactly vertical in their orientation.

We now insert the position of the points calculated as tan y_1 and tan y_2 from the above formulae, starting from M and going horizontally to the left and to the right; correspondingly, we proceed with the values of cotan y_1 and cotan y_2 on the two perpendiculars (tan y_1 = tan y_2 = 0 lies at the point M; cotan y_1 = 0 and cotan y_2 = 0 are diagonally opposite to M).

In this way, all the calculated dial points have now been transferred to the wall. How far we wish to proceed with some subdivisions, say, for every half hour or even for every 10 minutes, depends only on the size of the clock and personal preference.

The actual dial can have a quite arbitrary shape. In order to mark on it the points of the time scale, we only need to connect all the corresponding points on our rectangular frame with the point of intersection N of the rod (which has still to be inserted). There, where these connecting lines meet our scale (or its extension outwards), we can now mark the corresponding number of the hour.

Later on, when all has been finished and the wall has been painted, the perpendicular frame with the time scale can be removed again.

7.5.4. The Positioning of the Rod

The rod must be fixed with the same care as the determination of the time scale. On the exact position of the rod depends much of the accuracy of a sundial. Two points are of particular importance:

1. During the cementing of the rod its central line must meet the wall as accurately as possible at the point N, because it is this point to which the whole time scale refers. (See Fig. 7–4.)
2. The direction of the rod must form an angle φ with the horizontal plane. Furthermore, the rod must lie exactly in the plane of the meridian.

In order to fulfill this second condition, it is best to proceed as follows. We imagine planes through the rod and consider the angles that the line of intersection of these planes with the wall make with the rod. According to the position of the turnable plane, this angle will be quite different. There is, however, one position in which this angle is a minimum, namely, that position when the turnable plane is exactly perpendicular to the wall. The line of intersection that the turnable plane then produces on the wall is called the *substylar line*; this is NL in Fig. 7–4.

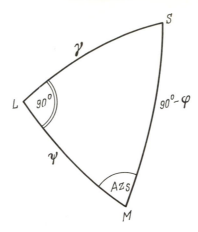

Fig. 7–5 The determination of the substylar line.

The extended directions of the line *NM*, the rod *NS*, and the substylar line *NL* go to three different points on the celestial sphere and form a spherical triangle *MSL* with the right angle at *L* (see Fig. 7–5).

In this spherical triangle the arc $MS = 90° - \varphi$, and the angle $LMS = \text{Az}_\odot$. If we now denote the arc *SL* with γ, we obtain the angle between gnomon and substylar line. In the right-angle triangle *MSL* we have, according to Napier's rule,

$$\sin \gamma = \cos \varphi \cdot \sin \text{Az}_\odot$$

and

$$\tan \psi = \cotan \varphi \cdot \cos \text{Az}_\odot.$$

Here ψ is the angle that the substylar line makes with the perpendicular *NM* in the plane of the wall (see Fig. 7–6).

If we denote with X_s the distance *ML* of the point *L* (in which the substylar line intersects the horizontal line through *M*) from the point *M*, and if we again make *NM* equal to 1.00 m, we obtain

$$X_s = \tan \psi = \cotan \varphi \cdot \cos \text{Az}_\odot.$$

If $\text{Az}_\odot < 90°$ then X_s lies on the left of *M* (to the west), while otherwise it is on the right of *M* (to the east). The position of the substylar line is therefore known from X_s.

In order to be able to place the rod exactly and correctly it is advisable to use a not too small right-angle triangle (preferably made in plywood). One of the two angles between the hypotenuse and the adjoining side is made equal to γ.

When the rod of the desired length is to be inserted, this plywood triangle is placed with one of its sides perpendicular to the substylar line—to support

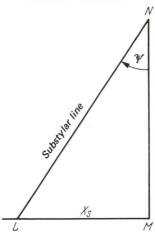

Fig. 7–6 The position of the substylar line.

the rod—whereby the angle γ comes to lie between rod and substylar line. Then the rod is placed on the hypotenuse of the wooden triangle and, if required, can be lightly attached to it. It is advisable to have this wooden triangle so large that its hypotenuse corresponds to about the length of the rod. If the rod has, as is usual, a somewhat conical shape, we can take this easily into account, be it by a corresponding change of the angle γ in the wooden triangle, or by placing a small piece of wood underneath between hypotenuse and rod. Usually the wooden triangle keeps in position by itself, due to the weight of the rod; to be safe, it can also be secured by a small nail. In any case one must take great care to make sure that the wooden triangle is exactly perpendicular to the plane of the wall.

Now the rod can be cemented in and we are certain that it is correctly placed; approximately two days later, the wooden triangle is removed.

7.5.5. *Allowance for the Geographic Longitude*

If we wish to make allowance for the geographical longitude, we can do so from the very beginning during the marking of the time scale. An example will illustrate this.

If the geographical longitude of the site at which the sundial is set up is 10° east of Greenwich, or 5° or 20 min west of the Central European Meridian, we will have to add at all relevant points of the dial 20 min to the original figures. The notation 12^h, which in the case of a vertical sundial is to be found below the point of intersection N of the rod, will now be displaced a little to the left. Where we originally had the figure 12^h, we now have 12^h20^m, because 12^h20^m is the true local time of the culmination of the true Sun on the standard meridian of Central European Time.

Our dial shows now true solar time plus longitude correction. In order to obtain the legal CET, we now only need to apply to the reading of the shadow the equation of time.

Such a sundial, to which the longitude correction has already been applied, can be recognized at a glance by the fact that the figure 12^h is not any longer exactly below the point of intersection of the rod.

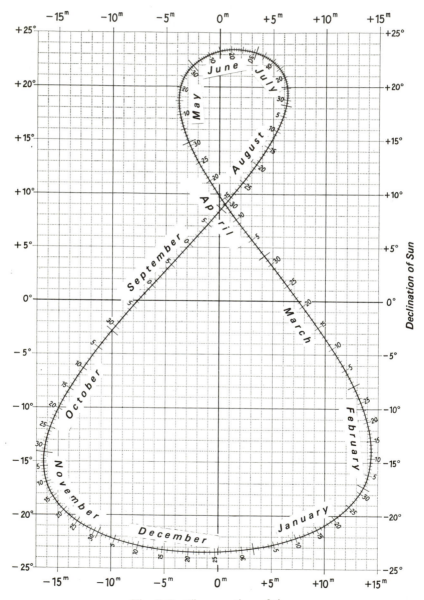

Fig. 7–7 The equation of time.

7.5.6. Allowance for the Equation of Time

Since the equation of time is only affected by very slow changes and variations, it is hardly necessary to extract its value every time from an astronomical Yearbook. The changes from year to year and from day to day are considerably smaller than the accuracy that can be achieved with our sundial, whose accuracy amounts to about 2–3 min. It is therefore possible to construct a kind of permanent diagram for the equation of time, which provides values for every day (except February 29); see Fig. 7–7. We only need to find our date on the figure-eight curve and then to read on the upper or the lower border the particular value that has to be applied (with the sign given in the diagram) to the shadow reading of the sundial to obtain CET.

Neglecting February 29 in the diagram means an error of one day in the case of leap years; the change of the equation of time on any day is at the most 30^s, namely at the end of December. This demonstrates that the error we committed by neglecting February 29 is, even under the most unfavorable circumstances, only 0.5 min.

8 / Applied Mathematics for Amateur Astronomers

F. Schmeidler

8.1. Introduction

How greatly astronomy depends on numerical calculations and mathematical formulae is shown by the fact that until late into the nineteenth century astronomy was considered to be part of mathematics. Equally today, the treatment of astronomical problems is unthinkable without mathematics, so that even the amateur astronomer frequently requires a certain minimum of mathematical knowledge. No proofs or derivations of the formulae will be given in this chapter. The reader interested in this aspect can find proofs in books of a more technical nature. Furthermore, it will be assumed that the reader is familiar with logarithms and trigonometric functions, and occasionally we assume acquaintance with the basic notions of differential calculus.

As to the accuracy of numerical calculations it is best to remember the rule of thumb which states that in five-digit calculations the final result can be expected to be accurate to about two or three seconds of arc. Thus, if we only require an accuracy of one minute of arc we only need work in four digits.

8.2. Theory of Errors

8.2.1. Direct Observations

Every measurement is subject to errors because our instruments and our senses are imperfect. We distinguish between two types of errors, systematic and accidental. Systematic errors depend in a known manner on some external circumstances and can be determined, although often only by tedious investigation. Accidental errors, on the other hand, act sometimes in one and sometimes in another direction and by their very nature cannot be predicted. It is only the latter with which the general theory of errors is concerned. Its task is to establish the law of frequency of errors and to judge the accuracy of a measurement or of a calculation.

The fundamental principle of the theory of errors was established by GAUSS, and states that the true value of a required quantity for which we possess a series of measurements has such a value that the sum of the squares of the individual errors is a minimum. This theory does not express a law of nature but rather a definition, although a very plausible one.

In the simplest case we may have a single quantity that can be measured directly. Let there be n measured values $l_1, l_2, \cdots l_n$, which would be identical if there were no errors, but of course this is never true in reality. Here the theory of errors asserts, in accordance with the principle of least squares, that the "most probable value" of the required quantity is given by the arithmetic mean

$$L = \frac{1}{n}(l_1 + l_2 + \cdots + l_n).$$

The "mean error" is expressable in terms of the differences $v_i = l_i - L$ between the individual measurements and the mean value L. Using the customary notation we denote the sum of the squares of the errors by $[vv]$. Then the mean error of a single measurement is

$$\mu = \sqrt{\frac{[vv]}{n-1}},$$

and the mean error of the mean is

$$\mu_L = \frac{\mu}{\sqrt{n}} = \sqrt{\frac{[vv]}{n(n-1)}}.$$

Thus, by forming the mean of n individual measurements, the accuracy of the results can be improved by a factor of $\sqrt{n}$.

These formulae assume that all measurements are equally reliable. In many cases this is not true. For example, an observer with a small telescope of low magnification will be able to measure the distance between the components of a double star less accurately than somebody with a large instrument. Such differences are taken care of by assigning to each measurement a "weight." The larger the degree of reliability, the larger the weight. The determination of these weight factors p_i depends in each case on an assessment, frequently only a very approximate one, of the quality of the observation. However, once the weight factors have been fixed, the formulae for the mean errors are

Mean error of unit weight: $\qquad \mu = \sqrt{\frac{[vvp]}{n-1}},$

Mean error of the mean: $\qquad \mu_L = \sqrt{\frac{[vvp]}{[p](n-1)}}.$

The mean is no longer equal to the arithmetic mean of the single measurements but is given by the expression

$$L = \frac{1}{[p]}(l_1 p_1 + l_2 p_2 + \cdots + l_n p_n) = \frac{[lp]}{[p]}.$$

Very often the required quantity cannot be measured directly, but is a known function of other quantities that can be measured. For instance, it is impossible to measure the absolute magnitude of a star directly, but it can be calculated if we know the apparent magnitude and the distance of the star. Since these two quantities can only be measured with a certain mean error, we wish to know the expected error of the computed absolute magnitude. The answer to this question is provided by the "law of the propagation of errors."

Let the required quantity x be a known mathematical function of n other quantities $x_1, x_2 \cdots x_n$, i.e.,

$$x = \varphi(x_1, x_2, \cdots, x_n).$$

If each of these quantities x_i is affected by a mean error μ_i, then the mean error of x is given by the formula

$$\mu_x^2 = \left(\frac{\partial \varphi}{\partial x_1}\right)^2 \mu_1^2 + \left(\frac{\partial \varphi}{\partial x_2}\right)^2 \mu_2^2 + \cdots + \left(\frac{\partial \varphi}{\partial x_n}\right) \mu_n^2.$$

The mean error is a measure of the accuracy of the observation. However, one must be aware of the fact that the actual uncertainty exceeds the mean error. We can then formulate the following rule of thumb: the discrepancy between a single measurement and the mean could, in unfavorable circumstances, be as much as 2.5 to 3 times the mean error. Therefore, if a previously known quantity is to be checked by new measurements and the resulting value turns out to be different, then this difference is only meaningful, i.e., real, if it exceeds the mean error of the measurement by at least a factor of 2.5; in all other cases the result of the new measurement should be interpreted as a confirmation of the old value.

8.2.2. Indirect Observations

If we are aiming at the determination of several unknown quantities whose mathematical connection with the measured values is known, we make use of a least-squares solution of indirect observations. In most cases the relation between the measured and the unknown quantities is linear, and if this is not the case we can in most cases linearize the computation by using approximations. Let us assume for the sake of simplicity that we have only three unknowns; the same principles are used for any other number of unknowns.

Let the measured quantity l be a function $\varphi(x, y, z)$ of three unknowns x, y, z. If on the basis of some plausible hypothesis we introduce the approximate values x_0, y_0, and z_0 for the unknowns and denote their true values by $(y_0 + \xi), (y_0 + \eta), (y_0 + \zeta)$, we are left with the determination of the corrections ξ, η, and ζ. According to TAYLOR's Theorem we can now set up a power series as follows:

$$f = \varphi(x_0 + \zeta, y_0 + \eta, z_0 + \zeta) = \varphi(x_0, y_0, z_0) + \xi\frac{\partial \varphi}{\partial x} + \eta\frac{\partial \varphi}{\partial y} + \zeta\frac{\partial \varphi}{\partial z} + \cdots.$$

Here the values of the derivatives are to be calculated at the point (x_0, y_0, z_0). Using the notations

$$\frac{\partial \varphi}{\partial x} = a, \qquad \frac{\partial \varphi}{\partial y} = b, \qquad \frac{\partial \varphi}{\partial z} = c, \qquad f - \varphi(x_0, y_0, z_0) = l,$$

we obtain a linear relation between the measured quantity l and the unknowns

$$l = a\xi + b\eta + c\zeta.$$

Of course, we require at least three such equations connecting the numbers l_1, l_2, l_3 with three unknowns ξ, η, and ζ. Usually, however, there are more equations than unknowns, and our task is to determine the most probable values of the unknowns ξ, η, and ζ from the measured quantities l_i, each of which carries an accidental error of measurement.

Of course, the various measurements must be performed in such a way that the values of the derivatives of the various functions φ_i differ from each other as much as possible. If we have n different measured quantities, we obtain n "equations of condition."

$$a_1\xi + b_1\eta + c_1\zeta = l_1,$$
$$a_2\xi + b_2\eta + c_2\zeta = l_2,$$
$$\dots \dots \dots \dots \dots$$
$$a_n\xi + b_n\eta + c_n\zeta = l_n. \qquad n > 3.$$

From these equations we form the "normal equations":

$$[aa]\xi + [ab]\eta + [ac]\zeta = [al],$$
$$[ba]\xi + [bb]\eta + [bc]\zeta = [bl],$$
$$[ca]\xi + [cb]\eta + [cc]\zeta = [cl].$$

The algebraic solution of these three equations yields those values of the unknowns that are most probable according to the theory of errors.

The coefficients of the normal equations are symmetrical with respect to the main diagonal of their determinant in view of $[ab] = [ba]$. This property facilitates the algebraic process of the solution. First of all we multiply the first equation by $-[ab]/[aa]$ and add the result to the second equation; we then multiply the first equation by $-[ac]/[aa]$, and add the result to the third equation. Each of these two operations leads to an equation which does not contain the first unknown ξ. We apply the same process to the two resulting equations for η and ζ and in this way we finally obtain one equation in ζ only. Knowing ζ we then can determine η and ξ from the previous equations.

To find the mean errors of the three unknowns we have to know the "mean error of unit weight" and the "weight coefficients." The mean error m of unit weight is given by

$$m^2 = \frac{[vv]}{n - \mu},$$

where μ is the number of the unknowns, in our case $\mu = 3$, and $[vv]$ the sum of the squares of the residual errors. This sum can be found either by calculating the right-hand sides of the equations of condition and then forming the differences with the observed values l_i, or from the easily verifiable equation

$$[vv] = [ll] - [al]\xi - [bl]\eta - [cl]\zeta.$$

The weight coefficients Q_{1i} follow from the normal equations by replacing the right-hand sides by 1, 0, 0:

$$[aa]Q_{11} + [ab]Q_{12} + [ac]Q_{13} = 1,$$
$$[ba]Q_{11} + [bb]Q_{12} + [bc]Q_{13} = 0,$$
$$[ca]Q_{11} + [cb]Q_{12} + [cc]Q_{13} = 0.$$

In an analogous manner we obtain the coefficients Q_{2i} by solving the normal equations with the right-hand sides 0, 1, 0, and the Q_{3i} by putting the right-hand sides equal to 0, 0, 1. Usually the greatest labor required for a complete least-squares solution is that of solving the system of normal equations three times (in the case of μ unknowns μ-times). The mean errors μ_x, μ_y, and μ_z of the three unknowns can be found from the equations

$$\mu_x^2 = m^2 Q_{11}, \qquad \mu_y^2 = m^2 Q_{22}, \qquad \mu_z^2 = m^2 Q_{33}.$$

Thus the task of the determination of the most probable values of the unknowns and of their mean errors is completely solved.

This is a subject in which familiarity with the concepts involved only comes after considerable practice with numerical cases. As an example we now give the solution for the case of two unknowns.

In the period from November 1949 to September 1950 the flexure of a meridian instrument was measured 11 times with the help of reflection observations. Table 8.1 gives the amount of flexure expressed in seconds of arc.

Table 8.1. A Sampling of the Flexure of a Meridian Instrument Throughout One Year

Date			Flexure b	Temperature $T(°C)$
1949 Nov.	18		0.39	− 0.6
Dec.	13		0.37	− 5.4
1950 Feb.	6		0.21	+ 1.2
March	27		0.60	+ 4.4
April	30		0.70	+ 9.3
May	29		0.71	+11.1
June	12		0.63	+16.3
June	29		1.12	+20.4
Aug.	15		1.32	+15.5
Aug.	28		1.13	+19.0
Sept.	30		1.41	+14.1

A superficial glance at the table shows a strong correlation between the two sets of measurements in the sense that the flexure at high temperatures is greater than at low temperatures. Thus we obtain a simple relation with two unknowns x and y:

$$x + Ty = b,$$

where T is the temperature in °C, b is the flexure, and x and y have physical interpretation as the flexure at 0°C and the change of flexure per °C change of temperature, respectively. Table 8.1 then gives the following equations of condition:

$$
\begin{aligned}
x - 0.6y &= +0.39, \\
x - 5.4y &= +0.37, \\
x + 1.2y &= +0.21, \\
x + 4.4y &= +0.60, \\
x + 9.3y &= +0.70, \\
x + 11.1y &= +0.71, \\
x + 16.3y &= +0.63, \\
x + 20.4y &= +1.12, \\
x + 15.5y &= +1.32, \\
x + 19.0y &= +1.13, \\
x + 14.1y &= +1.41.
\end{aligned}
$$

The coefficients a_i of the unknown x have in this case all the value 1, and this facilitates the numerical evaluation. We now form the sums of the products. As an example we give in detail the sum (bl):

$$
\begin{aligned}
[bl] = {}& -0.6 \cdot 0.39 - 5.4 \cdot 0.37 + 1.2 \cdot 0.21 + 4.4 \cdot 0.60 + 9.3 \cdot 0.70 + 11.1 \cdot 0.71 \\
& + 16.3 \cdot 0.63 + 20.4 \cdot 1.12 + 15.5 \cdot 1.32 + 19.0 \cdot 1.13 + 14.1 \cdot 1.41,
\end{aligned}
$$

$$[bl] = +109.98.$$

The other sums of products are calculated correspondingly; great care must be taken with the algebraic signs. (We suggest that, as a check, the observer calculates the sums twice, once starting at the beginning and the second time from the end.)

The next step is the formation of the two normal equations

$$
\begin{aligned}
+11.00x + 105.30y &= + 8.59, \qquad \| -9.5727, \\
+1741.93y &= +109.98.
\end{aligned}
$$

Here, using the usual notation, the first coefficient of the second equation need not be repeated, since by definition it is numerically equal to the coefficient given on the top right ($+105.30$). The first of these two equations is then multiplied by the factor given on the right, which is easily seen to be equal to $-105.30/11$. The first equation multiplied by this factor is then added to the second equation, which gives

$$+733.93y = +27.75.$$

Then, with the help of the first equation, x can also be evaluated. The results of the subsequent operations, which are done in an analogous manner, are as follows:

$$m^2 = 0.07, \qquad x = +0\overset{''}{.}419 \pm 0\overset{''}{.}123,$$

$$Q_{11} = 0.22, \qquad y = +0\overset{''}{.}038 \pm 0\overset{''}{.}010,$$

$$Q_{22} = 0.0014.$$

The temperature coefficient y is nearly four times as large as its mean error, so that its significance is established. The flexure has the value 0.4 second of arc at 0°C and increases per degree by $0\overset{''}{.}038$.

In all these calculations we have assumed that all equations of condition have the same weight. If the available data have different degrees of reliability, then this assumption is no longer fulfilled, and we are required to assign different weights to the equations of condition.

In this case, the coefficients of the normal equation are no longer [aa], [ab], etc., but are replaced by the sums [aap], [abp], etc. In practical work it is very easy to take these weights into account by multiplying each of the equations of condition by the factor $\sqrt{p}$, i.e., with the square root of the weight assigned to it. The resulting equations of condition are then treated as if they were all equations with the same weight, and the above-mentioned method for the calculation can be applied without change.[1]

8.3. Interpolation and Numerical Differentiation and Integration

When a mathematical function has been numerically tabulated, one frequently wants to know values for intermediate arguments. Thus, for example, we find that the coordinates of the celestial bodies in the annual almanacs are usually given from day to day for 0^h Universal Time, so that the value at any required moment can be found by "interpolation."

If we assume, with sufficient accuracy, that the graph of the function is a straight line between tabulated values, then the interpolation is very simple and is called linear interpolation. If, however, the function does not change sufficiently uniformly we form a "difference array" as follows (where w is the difference between successive entries in the table on page 211).

In this table we calculate the entries as follows. In the left-hand column are the given values for the function at the points $(a-2w)$, $(a-w)$, etc. We calculate the other columns successively from the left by putting the difference between two successive entries in one column as the entry in the column to the right and half way between the two entries. Differences are calculated by

---

[1] A number of useful hints for practical work in the theory of errors have been published by H. SCHOLZE (*Die Sterne*, 1958, p. 233) and by D. PAPERLEIN (*Die Sterne*, 1965, p. 27).

subtracting the upper entry from the lower entry. The columns f', f'', f''',

$f(a-2w)$

$$f'\left(a-\frac{3}{2}w\right)$$

$f(a-w)$ $f''(a-w)$

$$f'\left(a-\frac{1}{2}w\right)$$ $$f'''\left(a-\frac{1}{2}w\right)$$

$f(a)$ $f''(a)$

$$f'\left(a+\frac{1}{2}w\right)$$ $$f'''\left(a+\frac{1}{2}w\right)\text{ etc.}$$

$f(a+w)$ $f''(a+w)$

$$f'\left(a+\frac{3}{2}w\right)$$ $$f'''\left(a+\frac{3}{2}w\right)$$

$f(a+2w)$ $f''(a+2w)$

$$f'\left(a+\frac{5}{2}w\right)$$

$f(a+3w)$

etc., are called the first, second, third differences, etc. Thus, if $f(x)=x^2+x+1$ and $a=2$, $w=1$:

```
 1
    2
 3      2
    4       0
 7      2
    6       0
13      2
    8
21
```

If one wishes to know the value of the tabulated function f for any value $a \pm nw$ $(n<1)$ one has to use the following formulae:

$$f(a\pm nw) = f(a)\pm nf'\left(a\pm\frac{1}{2}w\right)$$

$$+ \frac{n(n-1)}{1\cdot 2}f''(a\pm w)\pm\frac{n(n-1)(n-2)}{1\cdot 2\cdot 3}f'''\left(a\pm\frac{3}{2}w\right)+\cdots \qquad \text{(NEWTON)},$$

$$f(a\pm nw) = f(a)\pm nf'(a)+\frac{n^2}{1\cdot 2}f''(a)\pm\frac{(n+1)n(n-1)}{1\cdot 2\cdot 3}f'''(a)+\cdots \qquad \text{(STIRLING)}.$$

NEWTON's formula is used if the initial value is at the beginning or at the end of the table. In other cases, STIRLING's formula is more convenient. To calculate, e.g., $f'(a)$, differences of odd order, evaluate the mean of $f'(a+\frac{1}{2}w)$ and $f'(a-\frac{1}{2}w)$; and similarly for $f'''(a)$:

$$f'(a) = \frac{1}{2}\left(f'\left(a-\frac{1}{2}w\right)+f'\left(a+\frac{1}{2}w\right)\right),$$

$$f'''(a) = \frac{1}{2}\left(f'''\left(a-\frac{1}{2}w\right)+f'''\left(a+\frac{1}{2}w\right)\right).$$

For the interpolation at the center of an interval (i.e., $n = \frac{1}{2}$) we have the simple formula

$$f\left(a+\frac{1}{2}w\right) = \frac{1}{2}\left(f(a)+f(a+w)\right)-\frac{1}{8}f''\left(a+\frac{1}{2}w\right)+\cdots,$$

according to which the value of the function for $(a+\frac{1}{2}w)$ is equal to the arithmetic mean of the two neighboring values minus one-eighth of the second difference at the same point; the error arising is considerably smaller than the fourth difference (which can nearly always be neglected). This formula can be used very advantageously if the differences are inconveniently large and we wish to reduce them by going over to the half intervals.

With the help of our difference array it is also possible to find errors in calculated values of the function. Let us assume the calculated value of $f(a)$ to be incorrect by a quantity ε, but the neighboring values to be correct. We then obtain

$$
\begin{array}{ccccc}
0 & & & & \\
 & 0 & & & \\
0 & & +\varepsilon & & \\
 & +\varepsilon & & -3\varepsilon & \\
\varepsilon & & -2\varepsilon & & +6\varepsilon \text{ etc.} \\
 & -\varepsilon & & +3\varepsilon & \\
0 & & +\varepsilon & & \\
 & 0 & & & \\
0 & & & &
\end{array}
$$

as the difference array of the error of the function.

For the higher differences the error becomes more and more conspicuous particularly in the line which contains the faulty initial value. If, therefore, particularly large jumps in the difference array occur on a certain line, a check of that particular value of the function is advisable.

This difference array also allows numerical calculation of derivatives and definite integrals of the tabulated function. The following formulae for numerical differentiation hold:

$$\frac{df(a)}{da} = \frac{1}{w}\left(f'\left(a+\frac{1}{2}w\right)-\frac{1}{2}f''(a+w)+\frac{1}{3}f'''\left(a+\frac{3}{2}w\right)+\cdots\right),$$

$$\frac{df(a)}{da} = \frac{1}{w}\left(f'(a) - \frac{1}{6}f'''(a) + \cdots\right).$$

One obtains easily the rule of thumb that the first differences are approximately equal to the first derivatives multiplied by w (the length of the interval). Corresponding formulae apply, of course, to higher derivatives.

The integral of the tabulated function can be found by forming a summation series. This is done merely by forming a further column to the left of a difference array. It is inherent in the nature of the difference array that any constant can be added to the complete column without altering the differences. If a is the lower limit of the integration, then a first approximation ${}^1f$ for the series follows from the formula

$$^1f\left(a - \frac{1}{2}w\right) = -\frac{1}{2}f(a) + \frac{1}{12}f'(a) - \frac{11}{720}f'''(a) + \cdots.$$

Then the further values of the first series follow from this by simple addition of the corresponding values of the function. The value of the integral of the tabulated function for any argument can be obtained from the formula

$$\int\limits_{a}^{a+iw} f(x)dx = w\left(^1f(a+iw) - \frac{1}{12}f'(a+iw) + \frac{11}{720}f'''(a+iw) + \cdots\right).$$

Methods of numerical integration are of greatest value in those cases where we do not know an analytical expression for the integral. Even if the analytical expression for the integral is known, its calculation is very cumbersome; however, the method can be used successfully.

Besides these formulae there exist many other expressions that are partly variants of this method, but partly independent as well. These are discussed in great detail in a publication by the Royal Observatory Greenwich, *Interpolation and Allied Tables*. The notation given there differs from that used above because it is expressly designed to be used with the aid of calculating machines. Our formulae can be used for calculations by hand which, as a rule, are more convenient for the amateur astronomer.

8.4. Photographic Astrometry

In photographic photometry we use photographic plates of a certain region of the sky to determine the coordinates of the stars. It is necessary to base this operation on a sufficient number of stars, at least three, whose coordinates on the plate are known and which can serve for the orientation of the coordinate system. TURNER has given a complete derivation of this method which can be found in the well-known book by W. M. SMART, *Textbook on Spherical Astronomy*. Here we shall only mention the most important formulae. It is assumed that all effects that change the coordinates of a star at different regions of the sky by a different amount (e.g., refraction, aberration, etc.) vary linearly within this particular region. This assumption is nearly always fulfilled and the rare deviating cases are of no importance for the amateur astronomer.

We now consider the image of a part of the celestial sphere when projected onto a plane, i.e., the plane of the photographic plate. If the center of the plate has the right ascension A and the declination D, then the rectangular coordinates X and Y of a given star with the right ascension α and the declination δ are given by the formulae

$$X = \frac{\tan(\alpha - A)\cos q}{\cos(q - D)}, \qquad Y = \tan(q - D),$$

where

$$\cotan q = \cotan \delta \cos(\alpha - A).$$

Here we have assumed that the positive Y axis points to the North Pole of the sky. These coordinates are called "standard coordinates" and can be calculated if we know the spherical coordinates of the star.

A comparison of the standard coordinates of the reference stars with the coordinates of the same stars measured on a linear scale yields the "plate constants." The measurement of the plate may give the rectangular coordinates x and y, whereby the origin of the $x, y-$ system should be very close to the center of the plate, and the direction of the positive y axis should be parallel to the north direction. If we know for a certain number of stars both the X and Y, calculated from their spherical coordinates, as well as x and y, found by direct measurement, the plate constants follow from the equations

$$X - x = ax + by + c,$$
$$Y - y = dx + ey + f.$$

Since each of these coordinates contains three constants, we require at least three reference stars. If there are more reference stars available, then the most probable values of the six plate constants can be found by the method of least squares (see page 206).

The spherical coordinates of the other stars can be easily determined as soon as we know the plate constants. We know the values of x and y of these stars from the measurement of the plate, and can therefore find the standard coordinates from

$$X = x + ax + by + c, \qquad Y = y + dy + ey + f.$$

From X and Y, the spherical coordinates can be derived by inverting the equations, namely by

$$q = D + \text{arc} \tan Y,$$
$$\tan(\alpha - A) = X \cos(q - D) \sec q,$$
$$\tan \delta = \tan q \cos(\alpha - A).$$

Of course, the calculation becomes more accurate if more reference stars are used. On the other hand, experience shows that there is usually not much gained by using more than six reference stars, because as a compensation for the greatly increased tedious numerical work we only obtain a very limited gain in accuracy. It is essential that the reference stars be uniformly distributed all over the plate.

8.5. Determination of the Position and Brightness of Planets and of the Planetographic Coordinates

Although the ephemerides of the planets are published in the various astronomical Yearbooks, it may sometimes still be desirable to calculate them directly, for instance, if we aim at a greater accuracy, or if the calculation concerns a minor planet not yet contained in the Yearbooks. For this purpose we require six orbital elements.

T = time of the transit through the perihelion of the orbit

μ = mean daily motion

e = eccentricity

Ω = ecliptical longitude of the ascending node

ω = distance of the perihelion from the node in the orbit

i = inclination of the orbit against the ecliptic.

If we require the spherical position of the planet for a given time t, we make use of the "KEPLER equation":

$$E - e \sin E = \mu(t - T) = M,$$

and solve it for the eccentric anomaly E. Since this is a transcendental equation, it can be solved only by iteration methods. Using a plausible initial value E_0 we find an improved value $E_1 = M + e \sin E_0$. If E_1 agrees with E_0 to the accuracy required, our task is completed, otherwise we use E_1 to find an improved value $E_2 = M + e \sin E_1$, and continue this procedure until we achieve agreement. Unless the eccentricity is extremely large, we can always use $E_0 = M$ as a reasonable first approximation. In practice the solution of the KEPLER equation can usually be found using only a few steps.

Using the eccentric anomaly, the radius vector r and the true anomaly v can be calculated from the formulae

$$r \cos v = a(\cos E - e),$$
$$r \sin v = a\sqrt{1 - e^2} \sin E,$$
$$\mu a^{3/2} = k = 3548''.18{,}761,$$

where a is expressed in astronomical units. From these equations there follow at once the rectangular heliocentric coordinates of the planets, referred to the ecliptic:

$$x = r(\cos \Omega \cos (v+\omega) - \sin \Omega \sin (v+\omega) \cos i),$$
$$y = r(\sin \Omega \cos (v+\omega) + \cos \Omega \sin (v+\omega) \cos i),$$
$$z = r \sin (v+\omega) \sin i.$$

To convert the heliocentric coordinates into geocentric ones, we require the heliocentric coordinates of the Earth which are the same as the geocentric coordinates of the Sun taken with the opposite algebraic sign. If we denote the ecliptic coordinates of the planet by λ and β, the ecliptic coordinates of the Sun at the time t by L and B, and the distances from the Earth to the Sun and from the Earth to the planet by R and Δ, respectively, we then have the following relationships:

$$\Delta \cos \beta \cos \lambda = x + R \cos B \cos L,$$
$$\Delta \cos \beta \sin \lambda = y + R \cos B \sin L,$$
$$\Delta \sin \beta = z + R \sin B.$$

This completes the calculation. Of course, attention must be paid to the fact that the orbital elements which characterize the position of the orbit in space (Ω, ω, i) must be referred to the same equinox as the coordinates of the Sun.

The calculation of the positions of comets having parabolic orbits is in principle quite similar, except that KEPLER's equation is now replaced by another relation which enables one to determine the true anomaly as a function of time, namely,

$$\tan \frac{v}{2} + \frac{1}{3} \tan^3 \frac{v}{2} = \frac{k(t-T)}{\sqrt{2}q^{3/2}}$$

Here, q denotes the "perihelion distance," i.e., the nearest approach that the comet makes to the Sun. It is one of the orbital elements and must be assumed as known if we embark on the calculation of ephemerides. After the true anomaly has been determined, we obtain the radius vector from

$$r = q \sec^2 \frac{v}{2},$$

and the remaining calculation is then carried out in the same way as for an elliptic orbit.

The apparent brightness of a planet changes considerably with its distance both from the Earth and from the Sun. Furthermore, because of the different illumination of the disk, the phase angle is also important. This is the angle subtended at the planet by the directions to the Sun and to the Earth, respectively, and is given by the formula

$$\tan \frac{p}{2} = \sqrt{\frac{(\sigma-r)(\sigma-\Delta)}{\sigma(\sigma-R)}},$$

$$\sigma = \tfrac{1}{2}(R+r+\Delta).$$

The changes in the brightness of a planet because of the changing distances from the Earth and the Sun follow the geometric law that the decrease in brightness is inversely proportional to the square of the distance and, after conversion to the astronomical scale, this leads to the formula

$$m = m_0 + 5 \log r + 5 \log \Delta,$$

in which m_0 is a constant that has a different value for every planet and also depends on the phase angle. Photometric measurements give for

Mercury $m_0 = +1\overset{m}{.}16 + 0\overset{m}{.}0284(p - 50°) + 0\overset{m}{.}0001023(p - 50°)^2,$
Venus $ = -4\overset{m}{.}00 + 0\overset{m}{.}0132p + 0\overset{m}{.}000000425p^3,$
Mars $ = -1\overset{m}{.}30 + 0\overset{m}{.}0149p,$
Jupiter $ = -8\overset{m}{.}93,$
Uranus $ = -6\overset{m}{.}85,$
Neptune $ = -7\overset{m}{.}05.$

For planets at greater distances than Mars, the influence of the phase angle can be neglected. Saturn has not been included in the above list, since its apparent brightness depends in a complicated manner on the position of the rings relative to the Earth.

Very often the observer is faced with the task of calculating the coordinates of a point on the observed disk of a planet relative to the equatorial planetary plane. In most cases the task can be solved with sufficient accuracy by a graphical method. However, some calculation is still necessary. We require the right ascension A and the declination D of that point on the sphere to which the northern extension of the axis of rotation of the planet points. These quantities are

Mars $\quad A = 317\overset{°}{.}8 + 0\overset{°}{.}8T, \qquad D = +54\overset{°}{.}7 + 0\overset{°}{.}3T,$
Jupiter $\quad = 268\overset{°}{.}0 + 0\overset{°}{.}1T, \qquad = +64\overset{°}{.}6 - 0\overset{°}{.}0T,$
Saturn $\quad = 38\overset{°}{.}2 + 4\overset{°}{.}0T, \qquad = +83\overset{°}{.}3 + 0\overset{°}{.}4T.$

This set of data refers to 1950, and the time T is to be counted in centuries from 1950 onwards; the changes are due to precession.

For the further treatment of the problems we need two quantities b_0 and β, which are found from the formulae

$$\sin b_0 = -\cos \delta \cos D \cos (\alpha - A) - \sin \delta \sin D,$$

$$\tan \beta = \frac{\sin (\alpha - A)}{\sin \delta \cos (\alpha - A) - \cos \delta \tan D}.$$

If we draw the planetary disk on paper as a circle with radius s, the visible pole of the planet appears at a point P, which has the distance $s \cdot \cos b_0$ from the center, and is at the position angle β if the latter is counted from the north clockwise from 0° to 360°. The quantities b_0 and β are tabulated in most astronomical Yearbooks. As to its geometrical meaning, b_0 is the planetographic latitude of the Earth above the equator of the planet.

As soon as the position of the visible pole on the planetary disk is determined, we can draw a grid of cell size according to the accuracy required for planetographic latitude and longitude. The diameter MP of the disk, which connects the center with the visible pole (see Fig. 8–1), is the projection of the central meridian at the moment of observation. All other meridians do not project as straight lines, but as ellipses which pass through the point P. If we wish to draw a meridian, for example, the longitude of which differs from the central meridian by the amount $(l-l_0)$ (Fig. 8–1), we must draw through the center M a line representing the diameter of the disk, the direction of which forms the angle γ with MP, which follows from

$$\tan \gamma = \sin b_0 \tan (l-l_0),$$

where l_0 is the length of the central meridian. This diameter is the major axis of an ellipse which also passes through the visible pole and which represents the required meridian.

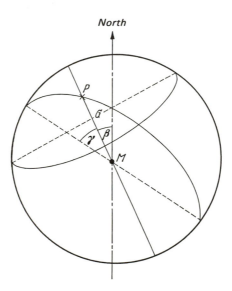

North

Fig. 8–1 Planetographic coordinates.

The latitude circles, as well, project onto the planetary disk as ellipses. If we wish to draw a circle corresponding to the planetographic latitude b then we must first find a point G on the straight line MP, which has the distance $s \cos b_0 \sin b$ from M. The major axis of that ellipse, which represents the required latitude circle, is perpendicular to MP and has the length $s \cdot \cos b$; the minor axis has length $s \cdot \cos b \cdot \sin b_0$.

If we have a sufficiently close grid of longitude and latitude circles, we can then simply place it on top of the drawing of the planetary surface, as obtained from a telescope, and then the planetographic coordinates of the various required features can be read off directly. All that is needed is the value of the planetographic longitude l_0 of the central meridian, which is given for daily intervals in tables in almanacs. The formulae for calculating it are published elsewhere.[2]

8.6. The Reduction of Star Occultations

Although as a rule it is recommended that the observers of star occultations should send their results directly to the Nautical Almanac Office (see page 548), it might be that some amateur astronomer would like to carry out the reduction of his observation himself. It is for this purpose that we summarize below the necessary formulae.

Suppose the occultation of a star is observed at the instance t (Universal Time). Then we can use Table 5 (page 505) to obtain the correction to t, which converts this into Sidereal Time t_s. If, furthermore, μ_0 is the Sidereal Time at 0^h, which is taken from the Yearbook and corrected by the amount $9^s.8565 \, \Delta \lambda$ (according to page 173), the angle h we obtain is

$$h = \mu_0 + t_s - \lambda - \alpha,$$

where λ is the geographic longitude of the observing site; places east of Greenwich count as negative. We now calculate the quantities

$$\xi = \frac{r' \cos \varphi \cdot \sin h}{k}, \qquad \eta = \frac{r' \sin \varphi \cdot \cos \delta}{k} - \frac{r' \cos \varphi \, \cos h \sin \delta}{k},$$

where $k = 0.2724953$. Here the geocentric coordinates r' and φ' of the observing site are calculated from the formulae given on page 166, while δ is the declination of the occulted star. The coordinates of the Moon are

$$x = \frac{\cos \delta' \sin (\alpha' - \alpha)}{k \sin \pi}, \qquad y = \frac{\sin \delta' \cos \delta - \cos \delta' \sin \delta \cos (\alpha' - \alpha)}{k \sin \pi},$$

where α is the right ascension of the star, while α' and δ' are the right ascension and the declination of the Moon at the moment of occultation. These quantities are taken from the hourly Ephemerides of the Moon in the Yearbook, and are then interpolated with great accuracy, using the method given on page 210. The quantity π is the parallax of the Moon, which also is interpolated in the same way.

The actual result of the observation obtained at a given site is a quantity $\Delta \sigma$ which is calculated by the formula

$$\Delta \sigma = \frac{k \sin \pi}{\sin 1''} \left(\sqrt{(x - \xi)^2 + (y - \eta)^2} - 1 \right).$$

It is equal to the distance from the center of the Moon to the star minus the radius of the apparent disk of the Moon. If the ephemerides were exactly

[2] *Handbuch der Astrophysik*, Vol. 4, p. 363, Springer, Berlin (1929).

correct, then $\Delta\sigma$ would have to be zero (apart from observational errors). Since, however, the values of the mean longitude L and latitude B of the Moon, taken from the Ephemerides (because of the incomplete knowledge of the laws of celestial mechanics), include the errors ΔL and ΔB, the quantity $\Delta\sigma$ usually differs from zero, and we have the equation

$$\Delta\sigma = \cos(\rho-\chi)\Delta L + \sin(\rho-\chi)\Delta B,$$

where χ is the position angle of the star with reference to the center of the Moon and ρ the position angle of the direction of the Moon's motion.

Thus, observations obtained at one station yield only one quantity $\Delta\sigma$ and not ΔL and ΔB, separately. If, however, we possess observations from several sites, at which the values of χ differ appreciably, then the most probable values of ΔL and ΔB can be found with the help of the method of least squares, and thus also the errors of the Ephemerides of the Moon.

9 | The Sun

R. Müller

9.1. Introduction

Observed through a darkened glass, with the eye protected by a filter, the Sun appears as a sharply defined circular disc with its brightness decreasing from the center to the limb. Only rarely, at times of great sunspot activity, can sunspots be seen by the naked eye. It is of little value to carry out such observations, which after all also require much patience from the observer. It is only the use of optical instruments that makes the study of the solar disc an extremely attractive field for the amateur astronomer.[1]

In spite of this, it appears to me that the majority of amateur astronomers are inclined rather to neglect this interesting object, the "Sun." This appears to have two reasons: somebody who is busy earning his livelihood by day can more easily follow a hobby at night. But this is not the only reason; many solar observers do not find it easy to overcome the difficulties they meet in arranging observations. It is therefore the purpose of this section to describe as clearly as possible the various tools and methods of the solar observer. If my contribution stimulates this type of work, I would consider this my best reward.

9.2. Observation of the Sun

9.2.1. The Diaphragm

The enormous luminosity of the Sun makes it necessary to use some kind of protective device. Even a careless momentary look through the eyepiece of field glasses or the telescope can do great harm to the eyes. The simplest protection is given by a sheet of glass that has been blackened over a flame and that can be made more permanent by covering it with another glass plate fixed with sellotape. Welders' goggles are also very useful. Better devices, the so-called smoked or neutral filters, are obtainable, for instance, the calibrated and parallel NG-filter produced by SCHOTT, the transparency of which is given in Table 9–1.

---

[1] See also M. G. T. MINNAERT: *Practical Work in Elementary Astronomy*. Dordrecht: Reidel, 1969, pp. 113–159.

We note that NG 2 is a filter of great heat resistance.

All filters must be used with great care because they get very hot under the influence of sunlight and so break easily. One should, therefore, use these filters only in connection with field glasses and smaller telescopes with apertures up to about 50 mm. At the Wendelstein Observatory, the author found filters that were framed in lead favorable from the point of view of heating. The lead frame can be pored into a wooden mould and then worked with a knife. It is *very* important to avoid placing such filters at the focal point or near to it; their best position is between the eye and the eyepiece.

Table 9.1. Transparency T of SCHOTT's Neutral Filter (Thickness 1 mm)

No.	T in %
NG 1	0.011
NG 2	0.36
NG 10	0.52
NG 9	3.59
NG 3	9.48
NG 7	26.6

Colored glasses, particularly red filters, should not be used because their relatively high heat transparency might lead to inflammation of the eye. We may summarize the following items for the practical observer:

1. In order to achieve an optimum in image quality, plane-parallel filters should be used if possible.

2. Since the intensity of the Sun depends on the altitude at which one observes it and thus also on the season, one should have several dark filters in readiness to provide the eye with the most convenient protection. This remark also applies to observation through cloud layers of variable density.

3. Filters that are placed between objective and eyepiece (but not near the focus) are heated less because the cone formed by the rays of the solar image has here a larger diameter. However, the absorbed heat is sufficient to produce poor seeing within the tube of the telescope and affect the quality of the image considerably. The ideal method is to place round filters in front of the objective; the only difficulty being to find suitable filters of this relatively large size.

4. The light can be weakened further by using a diaphragm to reduce the aperture of the objective. Such diaphragms are easily made out of cardboard which has been blackened with lacquer. It is advisable to use this method only for telescopes with more than 70 mm diameter.

5. The heat of the Sun causes a deterioration of the quality of the image and can also affect eyepieces made of two combined lenses. It is recommended that solar observations be interrupted from time to time and the telescope shaded so that all its parts, including the dark filters, can cool down.

9.2.2. The Helioscope for Dimming of the Sunlight

Devices that by reflection on prisms or glass surfaces or by polarization reduce sunlight to such an extent that only an additional weak neutral filter is required are called helioscopes or solar eyepieces. Figure 9–1(a) shows a simple solar eyepiece in which by the reflective prisms P_1 and P_2 (transparent glass prisms) about $1.6 \times 10^{-3} = 0.0016$ of the incoming light is being transmitted, thus achieving a considerable weakening. It is quite easy to build such a solar eyepiece, and all that is needed in addition is a weak neutral filter. We do not even need to use prisms, since plane-parallel glass plates placed at an angle of 45° fulfill a similar purpose.

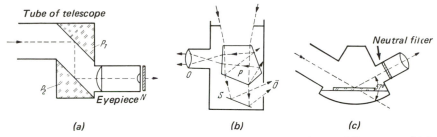

Fig. 9–1 (a) Solar eyepiece, which by reflecting prisms obliterates a high percentage of the incoming radiation. N is a weak neutral filter. After R. BRANDT. (b) Solar eyepiece with pentagonal prism; S is the mirror, O the eyepiece, and $Ö$ the aperture. After S. MAGUN. (c) Solar eyepiece with reflection and polarization properties. After R. BRANDT.

Another arrangement is the use of a pentagonal prism [Fig. 9–1(b)]. In order to produce the required weakening it is necessary to remove the metal covering from the two reflecting layers of the five-sided prism; one then obtains the factor 1.6×10^{-3} for the weakening by double-glass reflection of 4% each. The advantage of this solar eyepiece is that one obtains the same proportion at the eyepiece quite independent of the observing position. To eliminate as far as possible the heat radiation leaving the instrument at the first reflection from the prism, MAGUN [Die Sterne, 31, 228 (1955)] proposed the use of a massive roughly polished metallic mirror S built into the solar eyepiece container, to disperse the useless and harmful radiation (about 96%) through an opening O. At the second reflection the radiation entering the eyepiece container is then so small that it is practically eliminated if the outside walls are blackened. The weakening of the light is improved if one covers the reflecting surfaces with an absorbing layer.

Another method of light weakening is based on the principle of polarization. If sunlight is reflected from a glass plate at an angle of 57° [Fig. 9–1(c)], it becomes completely polarized. In this arrangement, the greater part of light and heat is removed by the reflection and polarization. The principle of a polarization helioscope, which can easily be made by the amateur astronomer,

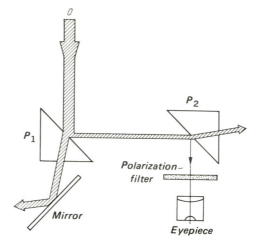

Fig. 9–2 The polarization helioscope.

has been described by GERHARD KLAUS (Fig. 9–2). Light entering the objective lens is reflected at the hypotenuse surfaces of two prisms P_1 and P_2. A large proportion of the light is already lost by deviation through the prism P_1 and is removed via a mirror. (The intensity of the light is schematically represented by the thickness of the lines.) In order to polarize the light of the glass surfaces, before it enters the eyepiece, a polarization filter is inserted which can remove practically all the remaining light or adjust it to an amount convenient for the eye.

9.2.3. The Projected Solar Image

We are now going to discuss the simplest method of solar observation—in which the solar image produced by the telescope is projected onto a screen placed behind the eyepiece of the instrument. At the focus of the objective lens the solar image is much too small and much too bright. To calculate the size of the image of the Sun we can apply the following rule: The image of the Sun is 1/100 of the focal length of our telescope; for a telescope of 100-cm focal length, we therefore obtain a focal image of only 1 cm. And one of the usual miniature cameras of 50-mm focal length provides, to the surprise of the photographer, a solar image of only 0.5 mm.

We are essentially interested in the size of the solar image obtained on the screen—let us call it S—and the distance a of the screen from the eyepiece. These data depend on the focal length f_0 of the selected eyepiece and on the diameter d of the image produced by the objective, as well as (very strongly) on the focusing of the eyepiece which we may call x. We say $x = 0$ if the focus of the eyepiece falls into the focal plane of the solar image produced by the objective. If this is not the case, that is, if we leave the primary focus by displacing the eyepiece, then the size of the image on the screen S and the distance a of the screen change rather considerably, because the above-mentioned quantities are related as follows.

$$S = \frac{f_0 d}{0}, \qquad a = \frac{f_0(f_0+x)}{x}.$$

An example: if our telescope has a focal length of 70 cm, our primary solar image has a diameter of $d = 0.7$ cm; if we now use an eyepiece of the focal length $f_0 = 1.4$ cm, a 50-fold magnification at a position which is 0.1 cm outside the primary focus, we have

$$S = \frac{1.4 \times 0.7}{0.1} = 9.8 \text{ cm} \qquad \text{and} \qquad a = \frac{1.4\,(1.4+0.1)}{0.1} = 21 \text{ cm}.$$

A small change of focus of the eyepiece, say to $x = 0.2$, would, as can easily be calculated from the above formula, reduce the size of the Sun on the screen to $S = 4.9$ cm, that is, to half that of its previous value, while the distance of the screen would now be $a = 11.2$ cm. To give another example, if we use an eyepiece of $f_0 = 28$ mm, which according to the relation $V = f/f_0 = 70/2.8$ furnishes a 25-fold magnification V, then for $x = 0.1$ we obtain $S = 19.6$ cm and the distance $a = 81.2$ cm.

These examples may suffice for the amateur astronomer to calculate for his telescope and other equipment the changes attributable to different focusing of the eyepiece. It is very important, of course, to give a good deal of thought to the size of the solar image on the screen. We recommend making the image not too large and suggest a diameter of 10–12 cm. Naturally one can project solar images to any size but it is important to take into account that the intensity of the light decreases rapidly for larger images. Even for large telescopes one should not exceed sizes of 20 cm, while for smaller instruments images from 7–9 cm might be the best ones. The eyepiece magnification, too, should not exceed 60–70.

The large amount of light around the Sun is screened by a not too small blackened diaphragm fixed around the eyepiece support. It is better still if the telescope can be placed in a darkened room with the eyepiece protruding between curtains. With a little imagination and experience one will soon find out how the most suitable darkening can be achieved. The following hint might prove useful: use a light black cloth to cover the projection screen as well as the head of the observer. If the weather is somewhat hazy or if the Sun is surrounded by a veil of cirrus clouds, a remarkable intensification of contrast is achieved.

In constructing the projection screen take into account that it should be fixed to the telescope perpendicular to the rays. It must be solidly mounted so that paper for drawings, etc., can easily be attached. Considering that the solar axis, that is, the central meridian from the North to the South Pole of the Sun, varies in the course of a year by about 53°, it is advantageous to make a device such that the circular drawing paper can be turned relative to the actual projection screen.

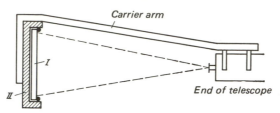

Fig. 9–3 Scheme of the construction of a projection screen. *I* = disc of 11-cm diameter; *II* = somewhat larger disc with protruding edge; ● = springy round wire for keeping the drawing paper in position.

Figure 9–3 shows some details of the construction. Let us assume that we have decided in favor of an image 10 cm in diameter. We take 2–3-mm thick plywood and cut a disc *I* of 11-cm diameter. We then take somewhat thicker plywood and cut out a plate *II* with a diameter of 11.8–12.0 cm, so that plate *I* can be placed inside and turned. We paste at its outer rim a protruding edge, or only perhaps at 8–10 places, which can take the disc *I*. A circular wire ring fixes the drawing paper inside this rim, and permits its easy removal.

The screen will, of course, be more stable if one uses two carrier arms, although this might affect the drawing of surface phenomena on the Sun, which might be a disadvantage. At the Solar Observatory Wendelstein where we observed the Sun in projection (solar diameter 15 cm) we had at the solar reflector a stable carrier arm, made of aluminum, as were the two discs which were carried on ball-bearings. In order to obtain solar images of the same size irrespective of changing solar distance, the carrier arm was adjustable in length by a screw. Of course, these are only small details because the difference between the Sun in perihelium and aphelium on the screen compared with the mean Sun of 10-cm diameter is only very small, the image being 10.17 and 9.83 cm, respectively.

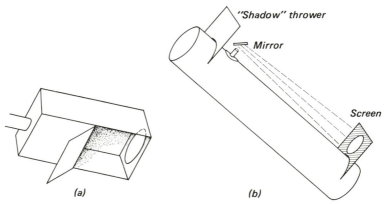

Fig. 9–4 (a) The construction of a simple projection box. (b) Projection method with an auxiliary plane mirror.

It is easy to build a kind of observing case for smaller projection, which can be equipped with a cover or sliding door [Fig. 9–4(a)]. The inner walls of this box should be painted dull black. A mirror can be attached to the projection screen as indicated in Fig. 9–4(b) (after G. KLAUS). In this case one needs an additional plane mirror and obtains a correctly orientated solar image on the projection screen.

For observations on the projection screen we advise using a kind of special "brush" that can be moved across the solar disc. We employed such a device at the Solar Observatory Wendelstein with great success. It consists of a thin kind of knife, made of wood or cardboard, one side of which is covered with a piece of semiglossy (fixed) photopaper, while the other half serves as the handle. One will be surprised how efficient contrasts can be achieved by a slight to and fro movement of this simple device.

Finally, we would like to mention the production of solar images with the help of a pinhole camera. In the section "Photography of the Sun," later in this chapter, we shall discuss this method in more detail.

9.3. Phenomena on the Sun's Disc

9.3.1. The Sunspots

9.3.1.1. Appearance and Classification of Sunspots

We now turn to the observation of the Sun's disc, in particular to that of the sunspots. These are of very different size, sometimes very small "pores" that can only be observed with larger instruments, while large sunspots can comfortably be seen through field glasses and exceptionally large ones even with the naked eye. The larger spots—that is, the nucleus surrounded by the umbra—are always surrounded by the penumbra. Anybody who is able to study the structure of the nucleus and the surrounding regions in detail, say with 70- to 100-fold magnification or in good projection, will find this a *very rewarding* task and will be surprised by the multitude of phenomena accessible to him.

It is important to note that sunspots appear in groups and that in view of its physical nature such a group actually has a structure of its own. Arranged in groups, the sunspots reveal different stages of development, which are illustrated in Fig. 9–5. This progressive evolution of a sunspot group, which is arranged approximately in the east–west direction, is an essential characteristic of the formation of two so-called major spots. By international agreement sunspots are now classified into nine groups, designated A to J, of which Fig. 9–5 gives three examples for each group. The characteristics of this classification can be summarized as follows:

Group	Characteristics
A	Small single spot or group of single spots without umbra.
B	Sunspot group with more or less distinct formation of the main spots (bipolar structure); spots without umbra.

C   Bipolar group in which one of the major spots shows formation of an umbra.

D   Bipolar group with larger major spots surrounded by umbras.

E   Larger bipolar group, the major spots of which often show an umbra of complex structure. Between the major spots we observe more single spots, sometimes surrounded by small umbra. The extent is usually larger than 10°.

F   Very large bipolar sunspot group, usually surrounded by large umbra areas, the extent of which is at least 15°.

G   Large bipolar group in which the intermediate spots have disappeared; it is still larger than 10°.

H   A group consisting of a large major spot with umbra and often surrounded by small single spots. The extent exceeds 2.5°.

J   Unipolar small spot with umbra; smaller than 2.5°.

First of all, the observer should practice the application of this classification, so that he is able to recognize the various types of spot groups and

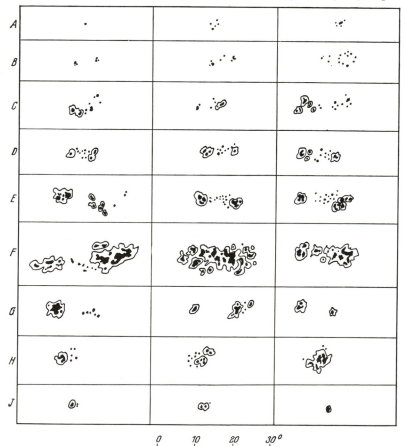

Fig. 9–5 Typical examples of the classification of sunspot groups.

to classify them correctly. Clearly, near the Sun's limb, this will usually be rather difficult because of foreshortening. Spots that appear at the east limb will reveal their true nature only after they have moved across the Sun's disc for a day or two. The difficulties of classification are, of course, greater with smaller telescopes and particularly at times of great solar activity. It is some consolation to know that even experienced solar observers arrive at different views as to the classification or subdivision in groups. The formation of sunspots is, indeed, such a complex phenomenon that it is often quite impossible to place them absolutely correctly into the scheme (see also Fig. 9–6).

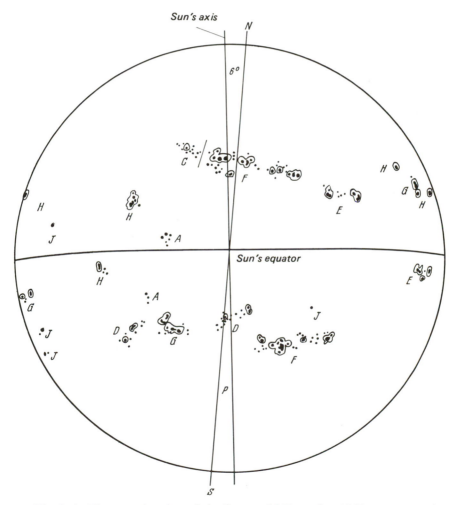

Fig. 9–6 The spot situation of the Sun on 25 December 1957, as seen on the projection screen. Groups and their classifications are sketched in. The latitude of the Sun's center was $B_0 = -2°$, and the position angle of the Sun's axis, $P = +6°$.

9.3.1.2. The Measure of Solar Activity (Relative Numbers)

At first it appears sufficient to assess the sunspot activity on a certain day simply by adding up the number of sunspots on the solar disc. However, we have seen in the preceding section that the appearance of spots in a group is an essential characteristic of their development. In order to be able to judge this evolutionary process in the sunspot activity more clearly, astronomers have taken up a proposal by R. WOLF (1816–1893), according to which not only the daily observed single spots f, but also the daily number of sunspot groups g are combined in such a way that the latter ones are given the weight 10. This measure, that is this combination of numbers of single spots and groups, of daily sunspot activity is called the relative sunspot number R. We have therefore

$$R = 10 \times g + f.$$

Let us assume we observe three sunspot groups on the disc, say, one group B consisting of nine single spots, a group C with a total of fifteen spots, and an area H with three spots. Thus we have $R = (10 \times 3) + 9 + 15 + 3 = 57$. On the other hand, if there is only a single spot on the disc, classified as A, then the relative number of the particular day becomes $R = 10 + 1 = 11$. It is a useful exercise to apply this method to a great number of daily observations and to derive the corresponding relative numbers R.

It is immediately evident that an observer X will find different R values from an observer Y who observed at the same time but at another station using a different telescope or another method. In order to coordinate observations obtained under different conditions with different instruments the daily sunspot relative numbers have to be reduced to a standard scale. A standard telescope, the FRAUNHOFER refractor of 8-cm aperture and 110-cm focal length, which has been used at the Zurich Observatory for more than 100 years, has been chosen. Data reduced to this instrument are thus given on the so-called "International Zurich Scale." The conversion factor k is obtained from long series of simultaneous observations for each observer for each station. We then have

$$R_{\text{int}} = k(10 \times g + f) = k \times R \qquad \text{and} \qquad k = \frac{R_{\text{int}}}{R}.$$

To give an example on five different days of the year in 1956, the R_{wdst} data given in Table 9.2 were recorded at the Solar Observatory Wendelstein. The relative numbers for the same days, according to the Zurich scale, R_{int}, are entered in the second column of the table. The first two columns permit the calculations of the conversion factor k, which are entered in the third column. We should emphasize, however, that such a small sampling would never suffice to determine the factor k with the necessary accuracy; for this we require a much more extensive observational series. The factor k not only depends on the method of observation and on the instrument but also on

Table 9.2. Conversion of R into R_{int}

R_{Wdst}	R_{int}	k
134	110	0.82
147	111	0.76
172	150	0.87
211	158	0.75
106	90	0.85

the atmospheric conditions (seeing, transparency), so that we need not be surprised if there is a considerable scatter between the individual values. However, it is not at all important to convert the observations into the international scale and it is advisable for one's own work to begin with $k = 1$.

Reduced to the international scale, observations can be found, for instance, in the journals *Die Sterne* and *Sky and Telescope*. In *Die Sterne* and *Sterne und Weltraum* we also find diagrams showing the general behavior of sunspot relative numbers, which can serve for comparison with the observer's own series.

Since 1946, the Fraunhofer Institute at Freiburg i. B., Germany, has published a daily "Map of the Sun," which is an extensive survey of solar activity. Figure 9–7 shows such a map. The amateur astronomer will generally

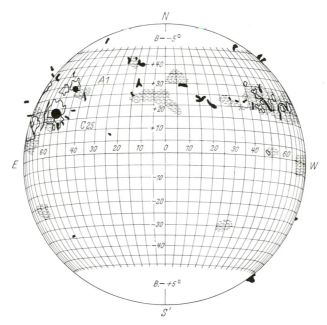

Fig. 9–7   The Sun on 15 January 1966 ($R_{int} = 32$) from a "Map of the Sun" of the Fraunhofer Institute at Freiburg.

make use of the following data shown on it: the daily (preliminary) international sunspot relative number R which corresponds, for the particular B_0 value (see page 243), to the relevant coordinate system; this is of great importance. Furthermore, we require from the map the position (center of gravity) of sunspot groups, their types, and the number of single spots. In addition (not shown in the figure) the daily map of the Sun gives data on the intensity of the monochromatic inner corona and observed solar eruptions, and a list of fields of faculae with their positions and sizes. At special observatories, which are accessible to amateur astronomers for observation of the Sun, observers will always find these maps. It is most advisable to make use of this material, which has been obtained on a wide international basis.

9.3.1.3. On the Apparent Movement of the Sunspots

Seen from the Earth, the Sun rotates in about 27 days. Since the sunspots participate in this rotation, they move from the east to the west daily by $360/27 = 13.3°$. This is quite a considerable and easily noticeable amount; if a spot appears at the east limb of the Sun, it crosses the disc in about $13\frac{1}{2}$ days and disappears at the west limb to the side that is not accessible to observation. To follow these movements of sunspots is worthwhile for every solar observer, since he can thereby derive the law of solar rotation. Indeed, the Sun does not rotate as a rigid body; there are different rotational periods for the solar equator and polar regions. If we were able, for instance, to "start" a number of spots at the central meridian on day x (Fig. 9–8) these would *not* return simultaneously to the central meridian one solar rotation later. As our figure shows, the spots near the equator would have won the race.

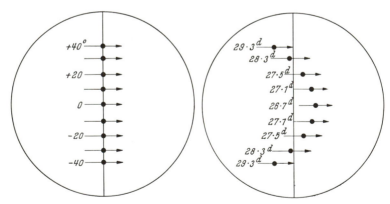

Fig. 9–8 The "race" among the sunspots: the dependence of the solar rotation on latitude is illustrated. Left, the positions of the spots on day x; right, one rotation later, that is, on day $(x+1)$.

9.3.1.4. Sunspot Period and the Distribution of the Spots in Solar Latitude

Everybody who has carried out careful sunspot observations for several years will be able to derive from his data two laws concerning the formation and position of sunspots on the solar disc. The formation of spots, that is, the solar activity, is subject to a continuous change that repeats itself about every 11 years. The increase from one sunspot minimum to maximum is steeper than the decrease to the subsequent minimum. Average figures are 4.5 years for the increase and 6.5 years for the decrease. Of course these are only approximate, just as is 11 years for the sunspot period. At times of little sunspot activity (near to the minimum) there are sometimes periods of many weeks when not a single spot is observable, while near maximum we can sometimes see more than 20 sunspot groups, among them often a really gigantic group.

Here we might add a remark about the derivation of the epochs of the sunspot minima and maxima. To smooth the scatter shown by single epochs, one forms means of 13 months, and using a special method of smoothing, the details of which do not concern us here, one obtains the epochs. [See R. Müller, *Die Sterne*, **38**, 145 (1962).] The 27-day period of solar rotation is often connected with solar-terrestrial phenomena. Since 13 months are not a multiple of a solar rotation, it has been proposed by F. Baur to apply a smoothing process which uses periods that are approximate multiples of the 27-day period. With such values Baur calculated new smoothed epochs of the sunspot minima and maxima. [See Institut für Meteorologie und Geophysik der Univ. Berlin, vol. 50, H.3, Teil I (1964); and R. Müller: *Die Sterne*, **42**, 10 (1966).] A useful summary is Table 9.3, which gives the epochs of the minima and maxima for the years 1867–1970.

A second law, which can also be discovered only after many years of observation concerns the distribution of the spots in solar latitude and is called the "zonal movement" of the sunspots. If the positions of sunspot groups are plotted over a long period, a diagram similar to Fig. 9–9 will emerge, showing that in the course of a sunspot cycle (counted from one minimum to the next), the sunspots wander from higher heliographic latitudes toward the solar equator. Even before the spots of the one cycle

Table 9.3. Epochs of Minima and Maxima

No. of Cycle	Year of minimum	Year of maximum
11	1867.2	1870.6
12	1878.9	1883.9
13	1889.6	1894.1
14	1901.7	1907.0
15	1913.6	1917.6
16	1923.6	1928.4
17	1933.8	1937.4
18	1944.2	1947.5
19	1954.3	1957.9
20	1964.6	1970.3

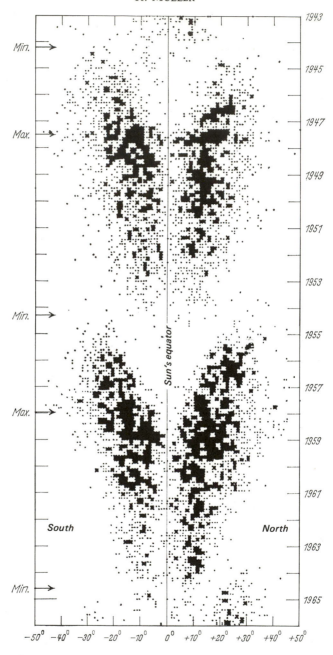

Fig. 9–9 The distribution of the sunspots in solar latitude during the period 1943–1966. When at the time of spot minimum, the last spot groups of the current cycle disappear near the equator and the first spots of the following cycle appear at high heliographic latitudes. The "butterfly diagram" is based on 8,650 single sunspot positions.

disappear near the equator (at the time of minimum), spots of the subsequent cycle appear at high latitudes.

The so-called "butterfly diagram" of Fig. 9–9 shows at once which spots belong to the new and which to the old cycle. It has been found that the separate determinations of the sunspot relative numbers for the spots of the new and the old cycle (during the time of minimum) apparently show a relationship not yet clarified. [R. G. GIOVANELLI: *Observatory*, **84**, 57 (1964); R. MÜLLER: *Die Sterne*, **41**, 23 (1965).] Although the next sunspot minimum will occur as late as 1975, we emphasize the importance of this difference.

9.3.1.5. Practical Hints for the Observer

In summary there are the following possibilities for the observation of sunspots.

Drawing of Maps. The determination of the coordinates will be dealt with later on.

Classification. Classification of the sunspots and faculae with such details as the type of the spot group, the number of single spots in the group, the intensity of the field of faculae surrounding the spot group, and the intensity of the faculae that are free of spots. (See also the section about solar faculae which follows.)

Quality. When making notes about the quality of the observation, one should judge it according to two factors, usually independent of each other, namely the steadiness r and the sharpness of the images s; both are estimated in a scale of 1 to 5:

$r = 1$: Solar limb absolutely quiet.
$r = 2$: Solar limb very little disturbed.
$r = 3$: Solar limb more disturbed.
$r = 4$: Solar limb strongly disturbed, wavy.
$r = 5$: Solar limb very strongly wavy.

$s = 1$: Solar limb very well defined, granulation well visible.
$s = 2$: Solar limb slightly diffused, granulation diffuse.
$s = 3$: Solar limb ragged, umbras of spots diffuse, granulation not visible.
$s = 4$: Solar limb "exploded," very diffuse spots.
$s = 5$: Limb completely diffuse, very diffuse spots.

Derivation. Derivation of the daily relative sunspot number R: to begin with one puts $k = 1$.

Detailed Studies. Detailed studies of the sunspots and their umbras, structure, and changes. This includes the continuous survey of *umbra* and *penumbra* of the sunspot: the nuclei of larger sunspots sometimes undergo disintegration or splitting into several parts, which may cause bright light bridges to form. This rare event usually requires several days to allow the observer to draw many pictures of the phenomenon. It is necessary to observe carefully the changing sizes of the nuclei and umbras of the spots and their relative distances in the process of any separation. The solar observers at the

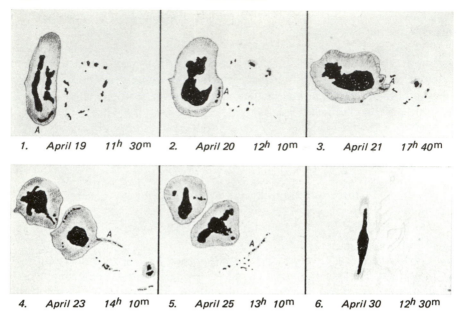

1.   April 19   11^h 30^m 2.   April 20   12^h 10^m 3.   April 21   17^h 40^m

4.   April 23   14^h 10^m 5.   April 25   13^h 10^m 6.   April 30   12^h 30^m

Fig. 9–10   The changes of a sunspot.

Budapest Observatory for amateur astronomers published impressive drawings and discussions of such observations during the previous sunspot maximum [*Die Sterne*, **35**, 205 (1959)], which are reproduced in Fig. 9–10.

Such detailed studies also include the observation of the SCHÜLEN-WILSON *phenomenon*, which is not yet completely understood. It has been found that a sunspot near the limb shows larger umbra formation toward the limb, and a smaller extent in the direction to the Sun's center. On the other hand, this is not always the case, and sometimes the phenomenon is even reversed. Unfortunately relatively little attention has been paid to these facts. This phenomena should, therefore, definitely be included in the observing program and one should make estimates in four steps (not at all, weak, strong, very strong) as proposed by H. J. NITSCHMANN [*Die Sterne*, **29**, 26 (1953)]. The situation is sketched in Fig. 9–11. Finally we would like to mention the *degree of darkening* of the spots, that is in the umbra as well as in the penumbra. This estimate is not easy and, because of the change in contrast between nucleus and umbra, is rather subjective. However, early observations by a Berlin group of observers who called themselves "Dargeso" testify that the effort is worthwhile. One may start by estimating three steps according to: umbra and penumbra dark, normal, or bright. Later on this scale can be extended by adding the estimate very dark and very bright.

Latitude Effect. Observation of the apparent wanderings of the sunspots in connection with the solar rotation.

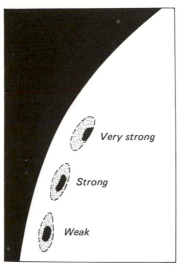

Fig. 9–11 The SCHÜLEN-WILSON phenomenon: the appearance of sunspots near the solar limb, characterized by weak, strong, or very strong asymmetry.

Sunspot Activity. Observations of the sunspot *activity* (11-year period).

The Distribution of Spots in Solar Latitude. To determine the extent of spots or spot groups—a very attractive task—one can use the following simple method: we stop the telescope drive and arrange for a spot to cross a wire in our eyepiece or a mark on the projection screen. We then turn the micrometer or the drawing board in such a way that the wire is oriented in the north–south direction. The spot should then remain exactly on a wire placed perpendular to the first wire. One starts to count seconds as soon as a front part of a spot touches the line, and stops when the end of the spot has left it. Since the time of transit depends on the declination of the Sun, which one can take from an ephemeris, we have the following relation between the length *La* of the spot, the time *T*, and the declination δ:

$$La = 15\,T\cos\delta \qquad \text{in seconds of arc.}$$

If one wishes to determine the extent of the spot in kilometers, we have to introduce the horizontal parallax of the Sun, $p = 8.8''$, and the diameter of the Earth, $D = 12{,}736$ km, leading to the formula

$$La = \frac{15 \times D}{2p}\,T\cos\delta = 10{,}855\,T\cos\delta \qquad \text{in kilometers.}$$

9.3.2. The Solar Faculae

Solar faculae are usually structures of bright lines and clouds that are only visible near the Sun's limb. One can state that all sunspots are surrounded by such faculae; the large-spot groups, in particular, are often found within large areas of bright faculae. These are frequently the forerunners of sunspot formation and deserve much attention. Quite often we observe that within a bright visual area of faculae (always near the Sun's limb) the next day reveals the position of a new spot.

A few remarks concerning the observation of such faculae are in order. First of all, the structure of the faculae must be carefully drawn, preferably with a *green pencil*. Beyond this it is of value to estimate the intensity of the bright "veins" and particularly the size of the affected area. This applies not only to the isolated faculae, i.e., those unconnected with spots, but in particular to the faculae that surround sunspot groups. Near the solar limb the spot faculae often help in the judgment of limits of the groups. If, for instance, in the case of two closely situated spot areas, a distinct border of faculae becomes observable, then the groups have to be separated.

The Fraunhofer Institute has proposed a scale of 0 to 9 for the estimation of the size of the faculae areas as given in Table 9.4.

Table 9.4. The Brightness of the Photospheric Faculae Areas

Step	0	1	2	3	4	5	6	7	8	9
Area in square degrees	0–1	2–3	4–6	7–12	13–20	21–30	31–45	46–60	61–75	> 75

For the beginner, and particularly for smaller telescopes, it is of advantage to use a restricted scale of five steps only. In a coordinate network on the Sun, the design of which is described on page 245, one draws the fields that correspond to the sizes in square degrees for these five steps. Figure 9–12 indicates such a drawing, which should be used for comparison during the observation. With some practice one will soon be able to achieve reliable estimates of the areas involved.

Let us assume that we have counted, within a spot group of type D, 15 single spots and that we have estimated the whole faculae field surrounding the group as having the brightness 4; we then note in our observing book "D 15,4." As mentioned above, the use of a smaller telescope will be better served by a less extensive scale, say 0 to 5 steps.

To take account of the intensity of the faculae one proceeds in such a way that the conspicuously bright faculae are given one step more than would correspond to the size of the actual faculae field; in the case of weak faculae one subtracts a step. It is rare that even on a spotless Sun there are no faculae at all.

Faculae have been observed near the poles even at the time of the sunspot minimum, though these were usually rather short-lived, small, pointlike structures, the observation, the drawing of which is possible only in the very

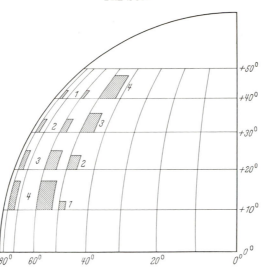

Fig. 9–12 Design of an auxiliary drawing for the estimation, in steps, of the intensity of faculae.

best atmospheric conditions. On the other hand such difficult observations are particularly useful since they can serve for the determination of the solar rotation in the neighborhood of the poles.

9.3.3. Granulation

The mosaiclike structure of the bright solar surface, called granulation, because of its grainy appearance, is visible only to the observer with a small telescope under very favorable observing conditions. On the other hand, CHRISTOPH SCHEINER (1575–1650), when describing the appearance of the Sun's disc, speaks of a "slightly tousled" surface, so that it is certainly worthwhile to consider possible observations of the granulation. This will only be successful if atmospheric conditions are excellent, that is, very transparent and calm. If one observes in projection it is advisable to make good use of the special little "knife device" (see page 227).

The essential feature revealed by observation of the granulation is that we have to think of the granulation elements as large, bubbling "cloud formations" the size of whole countries, which undergo complete changes in the course of a few minutes. The lifetime of small granules is about 3 min. These changes of granulation deserve particular attention. It is essential to note carefully all phenomena with their exact time. It is also worthwhile to follow the granulation photographically by long *series of exposures*. Apart from the relatively rapid change of structure the overall aspect of granulation, that is, the size of single granules and contrast in brightness between granules and background, remains unchanged. There is apparently no relationship with the 11-year sunspot cycle. Direct eyepiece observation is more suitable than observation on the projection screen. But for eyepiece observation also one should use the equivalent of our little knife device; that is, in this case apply some minute "shaking" to the telescope.

9.3.4. Prominences

Until quite recently observation of prominence phenomena was left to the professional astronomer. Things have changed since the French solar physicist LYOT invented his coronograph, which has become particularly useful in prominence research. Briefly, this instrument depends on masking out the bright photosphere (the luminous solar disc) by a mirror or a conical diaphragm. In this way the extended atmosphere of the Sun, in which the solar prominences rise, is accessible to view.

This sounds very simple, but it is not at all. Prominences radiate essentially in the light of hydrogen and among the visually accessible hydrogen lines particularly in the bright H_α-line (wavelength = 6563 Å). It is therefore necessary to eliminate the neighboring radiation, especially towards the yellow-green-blue part of the spectrum, and to concentrate on H_α. The solar researcher uses for this purpose rather complicated and expensive interference filters, which transmit only about 3–10 Å on each side of the hydrogen line. It was shown, however, that a good red filter (for instance, a SCHOTT RG2-filter), which absorbs visible light from the wavelengths 6300 Å toward the yellow, can be used for successful prominence observations. More useful are the interference filters designed by Dr. GEFFCKEN, produced at a reasonable price by the firm of SCHOTT in Mainz.

On the basis of such considerations, O. NÖGEL was the first to build a telescope specially for the observations of prominences by the amateur. This instrument has proved very successful and its construction by relatively simple means has been described by NÖGEL in two detailed papers [O. NÖGEL, *Die Sterne*, **28**, 135 (1952); and **31**, 1 (1955); see also page 44].

NÖGEL and others have shown that prominence observation is now by no means restricted to solar observatories; as NÖGEL says, "Our solar system can hardly offer a phenomenon more attractive than a solar prominence. Thus it is desirable for more amateur astronomers to build this relatively simple instrument. Such observations have recently been successful in Sonneberg and Stuttgart, and I would again encourage all interested amateur astronomers to new attempts in this direction; they will most certainly prove worthwhile."

9.4. Determination of Positions of Solar Phenomena

9.4.1. Coordinates of the Sun

After dealing in the preceding sections with methods of observation and with the phenomena observed on the Sun, we now proceed to the determination of the positions of the observed sunspot and other features.

Just as on the Earth, we define the poles of the Sun as those points where the solar axis "pierces" the solar sphere; we then call the north pole the point which, seen from the Earth, is directed to the northern sky. Each plane through the solar poles cuts the surface of the Sun along a meridian circle—just as for the Earth. By analogy with the coordinates by which each point on Earth is specified by geographic latitude and longitude, we speak of *heliographic latitude* (*b*) and *heliographic longitude* (*l*) on the Sun. Since there is no point on the Sun which could define a zero meridian, in the same way as Greenwich does on Earth, we must choose for every observational series a meridian from which longitudes are counted to the east (E) or west (W). For this we may choose one of three meridians:

1. The central meridian at the moment of observation.
2. The central meridian at 12^h UT ($= 13^h$ MEZ), which is favored today for global solar observations. Observations obtained before noon or in the afternoon have to then receive a small correction, which will be mentioned again below.
3. An internationally agreed zero meridian. This is the central meridian that passed through the apparent center of the solar disc on 1 January 1854 at 12^h UT. From this so-called CARRINGTON central meridian we count the heliographic longitude to the west from 0–360°. This zero meridian $L_0 = 0°$ thus always becomes established on completion of a solar rotational period (synodic period) observed from the Earth and of the mean duration of 27.2753 days.

From this date until 26 December 1974, there were 1623 solar rotations, i.e., transits through the central meridian. Table 9.5 gives the beginnings of CARRINGTON's numbering system of the solar rotations for the year 1971.

With the help of the differences listed in the table it is easy to calculate the longitude L_0 for every day of the year. For instance, on January 5.84 we had $L_0 = 0°$. Since the longitude between January 5.84 and February 2.18 changes by 360° divided by the difference $27.34^d = 13.17°$ per day, the change for 0.16^d is 2.1°, so that we obtain for January 6 (0^h UT) the longitude

$$L_0 = 360° - 2.1° = 357.9°.$$

One day later we have

$$L_0 = 357.9° - 13.2° = 344.7°.$$

Note that the differences given in the table vary in the course of the year between 27.34^d (in January) and 27.20^d (in July). This is caused by the changing velocity of the Earth in its elliptical orbit. The daily change of the zero meridian in the course of the year lies between 13.17° and 13.24°. With sufficient accuracy the mean value 13.2° can be used. Astronomical Yearbooks and calendars for amateur astronomers provide the times of the beginning of the synodic solar rotations as well as the longitudes L_0.

For geophysical purposes it is customary to use solar rotations of 27.0 days (BARTELS' rotations), which have proved very useful in comparisons of solar-terrestrial relationships. Table 9.6 therefore contains the numbering and beginning of these solar rotations.

All planes perpendicular to the solar axis intersect the solar sphere along circles of latitude. The largest of these circles through the center of the Sun

Table 9.5. Beginning of Solar Rotations for 1974

Carrington's Rotation No.	UT			Difference
1610	1974 January	5.84		
				27.34
1611		February	2.18	
				27.34
1612		March	1.52	
				27.31
1613		March	28.83	
				27.27
1614		April	25.10	
				27.23
1615		May	22.33	
				27.20
1616		June	18.53	
				27.20
1617		July	15.73	
				27.22
1618		August	11.95	
				27.25
1619		September	8.20	
				27.27
1620		October	5.47	
				27.29
1621		November	1.76	
				27.31
1622		November	29.07	
				27.33
1623		December	26.40	

Table 9.6. Beginning of Bartels' Solar Rotations for 1974

Bartels' Rotation No.	UT		
1921	1974 January	14.0	
1922		February	10.0
1923		March	9.0
1924		April	5.0
1925		May	2.0
1926		May	29.0
1927		June	25.0
1928		July	22.0
1929		August	18.0
1930		September	14.0
1931		October	11.0
1932		November	7.0
1933		December	4.0
1934		December	31.0

is the solar equator. Its inclination to the ecliptic is $7°15'$; we recall that the inclination of the terrestrial equator to the ecliptic is $23°26'$.

9.4.2. The Apparent Motion of Sunspots

Because of the inclination of the solar equator to the ecliptic, sunspots usually move not in straight lines across the Sun's disc but—as seen from the Earth—describe ellipses. Only when the Earth is at those points of its orbit where it lies in the plane of the solar equator, do they move in straight lines. At other times the solar axis is inclined toward the ecliptic by an amount varying from $+7.3°$ to $-7.3°$. Furthermore, the position of the solar axis is variable with respect to the north–south direction, that is, with respect to the great circle that passes through the celestial poles and the center of the Sun. This variation of the solar axis reaches the relatively large amount of $\pm 26°22'$.

In order to arrive at a picture demonstrating the changes of inclination and position of the solar axis, which is essential for observation of sunspots and their apparent trajectory across the Sun's disc, we have drawn pictures for six different times of year (see Fig. 9–13). For the description of these we use present-day notation: B_0 is the heliographic latitude of the center of the solar disc; B_0 varies between about $+7°$ and $-7°$. P is the position angle of the Sun's rotational axis from the north–south direction; P varies between about $\pm 26°$ (where the minus sign indicates easterly of north–south, and the plus sign westerly of north–south).

In Section I of Fig. 9–13 the situation on about January 6 is sketched; the position angle $P = 0°$ and the rotational axis coincides with the north–south direction. B_0 at this time is about $-3.6°$, and the sunspots wander at the central meridian about $4°$ above the center M of the Sun's disc; the south pole of the Sun points toward the Earth. Section II shows the situation on about April 7: P has reached its greatest eastern elongation, $P = -26.4°$; the center of the Sun has a heliographic latitude $B_0 = -6°$. In Section III, valid for about June 6, P has decreased to $-14°$; since at this time $B_0 = 0°$ the spots move in straight lines across the Sun; the solar equator divides the northern and the southern hemisphere symmetrically. Sections IV–VI show the movements of spots for positive P and positive B_0 values; in IV we have $P = 0°$, in VI $B_0 = 0°$. The values of P and B_0 are tabulated in the larger Yearbooks from day to day. Since they remain almost the same from year to year, we have listed them at five-day intervals in Table 10 of the Appendix; note also the footnote to Table 10.

Before we turn to the second portion of our discussion concerning the determination of the position of sunspots we should summarize briefly all the other data that we still require. Here we mention the heliographic latitudes, denoted with small b, which are measured on each side of the solar equator, as northerly $(+)$ and southerly $(-)$, respectively. Thus, as, for instance, in Fig. 9–7, the spot group C 25 has the coordinates $b = +18°$ and $l = 54°$ east. The heliographic longitude l is measured westerly or easterly from the central meridian toward the west, from 0–$360°$ starting from the arbitrarily

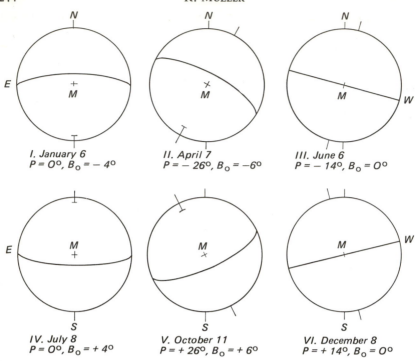

Fig. 9–13 The position of the Sun's axis and the course of sunspots at different times of the year.

defined zero meridian $L_0 = 0°$. Figure 9–13 representing the seasonal behavior of the sunspot shows the importance of the quantities P (inclination of the solar axis) and B_0 (latitude of the Sun's center). Data for both can be found in Table 10 of the Appendix.

9.4.3. The Methods for the Determination of the Solar Coordinates

9.4.3.1. Preparatory Measures During the Observation

There is no doubt that the method of observing the Sun in projection is the most suitable. Here we place our preprepared drawing paper, which has a circle of the size of the projected solar image on the particular day. A cross at the center marks the north–south and the east–west direction. The paper should be of good quality and more or less transparent. Before we begin with the drawing, classifying, or counting of the spots, it is very important to obtain the correct north–south orientation.

For this purpose we stop the drive of the telescope, so that the Sun moves steadily from the east to the west and thus marks the east–west direction. This is achieved much more accurately if we select a conspicuous sunspot and then adjust the paper in such a way that the moving spot remains exactly on our pencil line. This applies both for mountings in azimuth or for equatorial mountings. Of course it is possible to obtain a survey of the spot situation on the particular day by using an azimuthally mounted telescope, but because of the lack of exact guiding and the rapidly changing orientation in the sky in the course of the day, it is nearly impossible to obtain even approximately reliable determinations of positions. We shall therefore only deal with observations which have been made with an equatorially mounted telescope. Here we distinguish between telescopes that are mounted permanently and others that must be moved to the "observing site" from case to case. For the latter, we must take care that the instrument is placed each time in the same carefully marked position. Although the north–south or east–west orientation remains the same for an equatorially mounted telescope, it is wise to check it daily by reobserving the movement of a spot across the screen, because maladjustments occur only too frequently.

9.4.3.2. The Coordinate Grid

Larger solar observatories have special photographic grids on glass plates or film (following the Zurich procedure) for different values of B_0. These transparent films, which of course must have the same diameter as the projected solar image, are then simply placed on the drawing—in their correct orientation—and the position is read off without effort. At least eight such films are required, one for each of the values $B_0 = 0 - 7°$.[2] The construction of such grids is rather complicated and the above procedure is hardly ever used by the amateur solar observer. It is also possible to use only one grid and tabulate corrections. We shall therefore describe the design of such a coordinate grid first.

9.4.3.3. The Construction of the Fundamental Grid for the Determination of Positions

We select the grid for $B_0 = 0°$ (see Sections III and VI in Fig. 9–13). Figure 9–14 shows such a grid with a 10° interval; since there are no spots in high heliographic latitudes, the drawing has been extended only to $b = \pm 50°$. Each latitude circle is a linear distance $r \sin b$ from the equator, where r is the radius of the solar image. With $r = 50$ mm and $b = 20°$ we obtain for the distance of the required latitude circle $b_{20} = 17.10$ mm. The longitude circles l, measured from the central meridian, are at distances $r \cos b \sin l$. Therefore the coordinate $b = 40°$, $l = 20°$ lies on 32.14/13.10 (namely: $r \cos 40°$ $= 0.7660 \times 50 = 38.30$, $r \cos b \sin 20° = 0.3420 \times 38.3 = 13.10$). Experience will show whether 10° divisions are sufficient. For the less experienced observer we provide in Table 9.7 coordinates at intervals of 10° for $r = 50$. These can be easily transferred to any required size of solar image.

9.4.3.4. The First Method

We now place the $B_0 = 0$ sheet on the sunspot drawing, turning it in such a way that the north–south of the drawing coincides with the solar axis. The

[2] For negative B_0 values (-1 to $-7°$) the positive grids must be turned by 180°; see also Fig. 9.14.

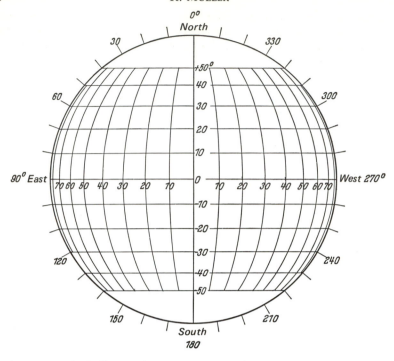

Fig. 9–14   The heliographic coordinate grid for $B_0 = 0°$. The position angles at the Sun's limb are measured from the north point of the Sun to the east from $0°$–$360°$.

Table 9.7. Coordinates of the $B_0 = 0$ Network (in mm) for a Solar Diameter of 100 mm

b＼l	0°	10°	20°	30°	40°	50°	60°	70°	80°
50°	38.30	5.58	10.99	16.07	20.66	24.62	27.83	30.20	31.65
40	32.14	6.65	13.10	19.15	24.62	29.34	33.17	35.99	37.72
30	25.00	7.52	14.81	21.65	27.83	33.17	37.50	40.68	42.64
20	17.10	8.16	16.07	23.49	30.20	35.99	40.69	44.15	46.27
10	8.68	8.55	16.84	24.62	31.65	37.72	42.64	46.27	48.49
0	0.00	8.68	17.10	25.00	32.14	38.30	43.30	46.98	49.24

position angle of the axis P can be obtained from Table 10 of the Appendix. Recall that in the case of negative P values (that is, from the beginning of January to the beginning of July) the rotation is to the east, while for the other season the solar axis (with positive P values) lies in the NW quadrant (see Fig. 9–12).

We now transfer our spot drawing on to the coordinate network and read off the coordinates. It is convenient to build a box for viewing, equipped with a lamp and a tilted glass plate on which to place network and drawing. Of course, this procedure furnishes heliograph coordinates valid only for $B_0 = 0°$. These data, b' and l', can then be converted for any required value of B_0 into the correct values b and l, using the following formulae and calculations (the logarithmic calculation is for $B_0 = +6°$, $b' = +40°$, $l' = 60°W$):

	I		II	
$\sin b =$	$\cos B_0 \sin b'$	$+ \sin B_0 \cos b' \cos l'$,		(1)
$\cot an\, l =$	$\cos B_0 \cot an\, l'$	$- \sin B_0 \tan b'\ 1/\sin l'$.		(2)

Formula (1)		Formula (2)	
lg cos 6° : 9.9976		lg cos 6° : 9.9976	
lg sin 40 : 9.8081		lg cotan 60 : 9.7614	
lg I : 9.8057		lg I : 9.7590	
I : 0.6393		I : 0.5741	
lg sin 6° : 9.0192		lg sin 6° : 9.0192	
lg cos 40 : 9.8843		lg tan 40 : 9.9238	
lg cos 60 : 9.6990		lg 1/sin 60 : 0.0625	
lg II : 8.6025		lg II : 9.0055	
II : 0.0400		II : 0.1013	
I+II : 0.6793		I−II : 0.4728	
lg I+II : 9.8321		lg I−II : 9.6747	
= lg sin b		= lg cotan l	
b : 42.8°		l : 64.7°	
$b−b'$: 2.8		$l−l'$: 4.7°	

WALDMEIER published tables for the value $b−b'$ and $l−l'$ for $B_0 = 0.5°$ to $7.5°$ [Tabellen zur heliographischen Ortsbestimmung. Basel: Birkhäuser, (1950)]. The use of these tables makes the above calculations superfluous.

9.4.3.5. On the Accuracy of the Method

In general it will be sufficient to determine the coordinates of spot groups, that is, of the center of gravity of groups. Since a group frequently extends considerably in length and may also have a width of $1°–2°$, it is quite sufficient to work with an accuracy of about $1°$. The tenths of degrees, which we have carried along in our calculation and tables, have only formal value for correct rounding-off of the results. If, however, more accurate positions are required, say of single spots, we shall usually have to use a micrometer, as described below.

9.4.3.6. The Second Method

This method determines the distance r of the spot F from the center M of the Sun's disc and the position angle that the line MF makes with the solar axis, that is, the polar coordinates of the spot. If we call ρ and ρ' the angular distances of the spot from the center of the apparent disc, as seen from the Sun and the Earth, respectively, we have the following two equations:

$$\rho' = \frac{r}{R} \cdot R', \tag{1}$$

$$\sin (\rho + \rho') = \frac{r}{R}. \tag{2}$$

Here R is the radius of the solar image as formed by the telescope, and R' the apparent diameter of the Sun (the mean value is 16.0 minutes of arc). If, for instance, we have found for r a distance of 50 mm with a solar image of $R = 75.5$ mm radius, then Eq. (1) gives, with $R' = 16' = 0.267°$, the required $\rho' = 0.18°$ and then ρ from Eq. (2) is 41.3°. For the calculation of b and l we then use the formulae

$$\sin b = \cos \rho \sin B_0 + \sin \rho \cos B_0 \cos p, \tag{3}$$

$$\sin l = \sin p \sin \rho \, 1/\cos b. \tag{4}$$

Here p denotes the above-mentioned position angle between the solar axis and the line connecting the solar center M and spot F (see Fig. 9–14).

Since it is possible to calculate the quantity ρ with sufficient accuracy (of about 0.1°) simply from the relation $\sin \rho = r/R$, the method is frequently used for the determination of spot position from spot drawings, because the measurement of polar coordinates does not require any grid. As a numerical exercise let us assume that we have measured on our drawing for the radius $R = 75.5$ mm, $r = 50$ mm, and $p = 46°$. Then with $\rho = 41.3°$ and $B_0 = +6°$, the further calculation runs as follows (Fig. 9–15 illustrates the notation):

Formula (3)		Formula (4)	
lg cos 41.3°	: 9.8758	lg sin 46.0°	: 9.8569
lg sin 6.0	: 9.0192	lg sin 41.3	: 9.8195
lg I	: 8.8950	lg l/cos 32.3	: 0.0730
lg sin 41.3	: 9.8195	lg sin l	: 9.7494
lg cos 6.0	: 9.9976	l	: 34.2°
lg cos 46.0	: 9.8418		
lg II	: 9.6589		
I	: 0.0785		
II	: 0.4559		
I + II	: 0.5344		
lg I + II			
= lg sin b	: 9.7279		
b	: 32.3°		

9.4.3.7. A Graphical Method

We also recommend an extremely simple graphical method proposed by K. SILBER of Gmunden [*Die Sterne*, **38**, 63 (1962)]. First of all, we have to orient the image on the projection screen, that is, we have to determine the east–west direction by observing a spot moving along while the telescope drive is stopped. Our Fig. 9–16(a) indicates this direction by a dashed arrow. To obtain the direction of the solar axis we enter the angle P (given for the particular day) against the north–south direction. We then draw a perpendicular from the spot F to the solar equator which meets the latter at C. The center M of the solar image is then connected with F, and at right angles to this line a line is drawn from F which meets the circle of the solar image at D. This simple construction gives the following quantities: $FC = y$, $MC = x$, $FD = m$, and $MN = r$. These four distances can be easily measured

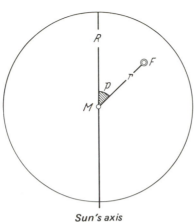

Fig. 9–15 Determination of the position of a sunspot in polar coordinates.

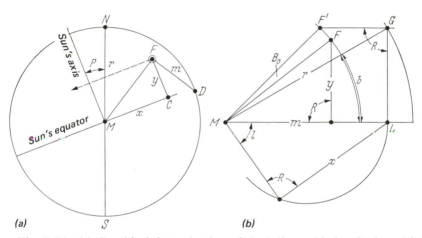

(a)   (b)

Fig. 9–16 (a) Graphical determination of the heliographic longitude and latitude of a sunspot. (b) Construction for the determination of the coordinates of a
sunspot.

on the projected image and serve, as shown in Fig. 9–15(b), as the graphical
determination of the required heliographic coordinates b (latitude) and l
(longitude). The construction with compass and ruler then proceeds as
follows: the measured quantities m and y give the point F. The line MF is
the base of the angle B_0, which meets a circle about M through F at F'. A
line parallel to the base drawn through F' cuts the circle arc drawn with the
radius r in G. The heliographic latitude b is then equal to the angle marked
by a double arrow. The perpendicular from G to the baseline gives a point L.
We then draw an auxiliary circle of diameter ML and mark on it the distance
x, starting at L. The angle l, marked on the figure. is the required heliocentric
longitude.

9.5. Photography of the Sun

9.5.1. Photography with a Pinhole Camera

I have often been told that photography of the Sun with a pinhole camera is not only a useful photographic exercise, but also a pleasant hobby. To qualify this statement, however, we should mention that in this process, images, even those of larger sunspots, become relatively diffuse. Before proceeding to the construction of a pinhole camera we must answer the following questions:

    1. What size pinhole should be chosen?
    2. What is the distance from the pinhole to the projection screen for a given size of pinhole?
    3. What is the size of the solar image?

The diameter d of the hole and the distance B of the image (both in mm) are related by

$$d = \sqrt{B \cdot 0.00127}.$$

For $B = 5000$ we thus obtain $d = 2.52$ mm.

The size D of the solar image and the distance B of the image are related by

$$D = B \cdot \tan \text{ of solar diameter} = B \cdot \tan 32' = B \cdot 0.00931.$$

Thus, for $B = 5000$ mm, we obtain $D = 47$ mm.

A graphical representation of these relationships is given in Fig. 9–17.

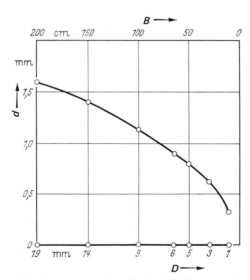

Fig. 9–17 The pinhole camera: the relationship between the diameter d of the hole, the distance B of the projection screen from the pinhole, and the size D of the solar image.

It shows, for instance, that for an image distance of 150 cm we will have to drill a hole of 1.4 mm to obtain a 14-mm solar image diameter. Experience teaches us that the most favorable distance should be a little larger than that indicated by the formulae. It is therefore advisable to prepare for a given length of the box a series of pinholes. One will soon find the best method to drill holes in different materials of the most frequently required size and how to place the carrier for plate or film in the best manner. Such a pinhole camera, equipped with a ground glass, provides a rather useful guiding telescope.

9.5.2. Photography of the Projected Sun

The ultimate goal, of course, of the amateur astronomer observing the Sun is to photograph the solar disc or single sunspots with his telescope. The projection screen must now be replaced by a camera (without a lens). If the screen is so far away from the eyepiece that the camera can be placed between, it should be made into one unit with the projection screen. With such an arrangement, we can change rapidly from photography to visual observation. Furthermore, we require a shutter to give exposure times as short as 1/200 sec. If the shutter is completely opened (T position) and the plateholder removed, everything is ready for visual observation on the screen.

As to the most suitable size of image, we must take into account that the common miniature camera cannot be recommended for solar photography, although the possibility of rapid sequences of photographs may be attractive. All we can obtain is a solar image of about 2.5-cm diameter on 35-mm film, and the great intensity of this image leads to further difficulties in giving the proper exposure time. If we aim at sections of the solar disc only, using solar images of about 4-cm diameter in the focus of the plane of the film, the miniature camera is more useful. However, in view of the normally available sizes of 6×9 or 9×12 cm for plates or films one should aim for solar images of 5 to 7 cm diameter. Since the focus of the objective and of the eyepiece is usually visually corrected we must experiment in order to obtain the best photographic focus. For this purpose it is advisable to provide the eyepiece tube with a scale that permits the setting for the sharpest image to be repeated at any time.

9.5.3. The Use of Spectacle Lenses

Spectacle lenses are long focus lenses very useful for the production of large solar images. Although they are not very good optically over the whole surface, they are still quite useful for photography if we diaphragm the border zones down to about 2 to 3 cm. If D characterizes the spectacle lens in diopters and f the focal length, we have the relation $f = 100/D$ cm. Biconvex spectacle lenses of about 0.3 to 0.5 diopters therefore give a focal length of about 330 to 200 cm.

9.5.4. The Telelens

The best solar images are obtained in solar photography with a telelens arrangement attached to the telescope. This arrangement consists of a photo-graphically corrected negative lens (BARLOW lens, see page 47) to lengthen the focal length of the solar telescope. We have to take account of the following: What is the (negative) focal length of the BARLOW lens? At what distance should the negative lens be placed from the objective and from the plane of the emulsion of our plate? What size of image can we achieve? The way to answer these questions is shown by the following example.

Let us assume our telescope has a focal length $f = 75$ cm and a negative lens of the focal length $f_n = -11$ cm. We always choose the distance d of the negative lens from the objective in such a way that d is larger than $f+f_n$, which in our example is somewhat larger than 64 cm. (The distance d should be about 1.5 cm greater than this and never less than 1 cm greater.) We thus take $d = 65.5$ cm. In order to calculate the distance B of the image (i.e., the distance of the negative lens from the plane of the film) we use the equation

$$B = \frac{f_n(f-d)}{f+f_n-d}.$$

With the above data, $B = 70$ cm. The size of the image is given by the following formula:

$$\text{Size of image} = \left(\frac{f \cdot f_n}{f+f_n-d}\right) \cdot 0.00931 = 5.1 \text{ cm.}$$

It is evident that a small change of the focal length f_n of the negative lens leads to a noticeable change of the distance B and of the size of the image. It is therefore necessary that the BARLOW lens be moved precisely in its tube so that the best focus can be found by a series of photographs with it in different positions. (See also Sec. 9.5.6.)

9.5.5. Notes on Diaphragms and Emulsions

We have already mentioned that the enormous luminosity of the Sun makes it necessary to diaphragm the telescope considerably and to use slow photographic emulsions. Neutral filters are suitable. Colored ones, such as two, three, or four red filters, SCHOTT RG1 or RG2, for example, combined parallel to each other, would be better; a yellow filter such as GG 11, better still. It is also possible to combine neutral and colored filters, but not of course red and yellow filters. Although the use of a diaphragm with the objective lens increases the depth of focus, there exist limits to this method. It is advisable to use objective diaphragms for telescopes of more than 5 to 7 cm aperture only.

As to photographic emulsions any hard kind of plate can be recom-mended, such as plates used in book printing for slides or special technical plates. These are slow and fine grained. One example would be the PERUTZ P1-plates.

9.5.6. Exposure Time and Aperture Ratio

Exposure times should be 1/100 sec or shorter. Exposure times of even 1/50 sec should be avoided to eliminate the poor seeing caused by short-lived air "bubbles." The vital quantity is

$$F = \frac{\text{focal length}}{\text{objective diameter}} = \frac{f}{d} \, .$$

As soon as we know the correct exposure time for a particular camera arrangement we might make use of the following facts for the calculation of the aperture ratios and the exposure times:

1. For photography at the prime focus we have $F = f/d$.
2. For photography in projection through the eyepiece of the telescope, which has the focal length f', we have $F = (Lf)/(df')$, where L is the distance of the eyepiece from the plate.
3. If one uses a camera with the objective of the focal length f'' focused for infinity and placed behind the eyepiece, we have $F = (f''f)/(df')$.

We note that the exposure time is proportional to the square of F; if one has found, for instance, that at $F/16$ the exposure time is 1/100 sec, it is 1/25 sec for $F/32$.

A final remark should be made concerning the determination of the correct focus. This is that the Sun is so bright that the limb and larger spots can be observed without much difficulty on the ground glass screen of the camera. But it is not always sufficiently accurate to find the best focus by eye. It is necessary to take a series of test pictures with the eyepiece tube displaced to each side of the approximate focus, and to determine the correct focus by inspection of the images with a magnifying glass. It is advisable to move the eyepiece in steps of 1 mm. If a photograph of the Sun is really good, the decrease of the brightness toward the solar limb will be distinct, and the images of the spots and umbras and outlines of the faculae will be sharp.

9.6. Conclusions

This chapter has been written for the *amateur astronomer* and therefore deals only with observational problems that can be looked into by amateurs. The professional astronomer can, of course, make more extensive observations, many of which require a large and expensive array of instruments. In order to restrict this contribution we have not described these instruments or their use, but a brief survey of them might be of interest.

Heliometer. This is an instrument in which the objective lens has been cut into halves, both of which can be measurably displaced over the other. To measure the apparent diameter of the Sun one displaces these halves until one limb of the solar image coincides with the opposite limb of the other solar image. The necessary displacement is a measure of the apparent diameter of the Sun.

Heliostat. This is a plane mirror with a driving clock, which projects the sunlight in a fixed direction.

Coronograph. The design of this instrument is described on page 44. It is used with a spectrograph or a spectroscope to investigate the inner emission corona and is usually operated at mountain stations.

Lyot filter. This expensive and complicated instrument is used with solar reflectors to explore chromospheric phenomena. It makes it possible to transmit the light of a single spectral line only 0.5–1.0 Å in width.

Magnetograph. This is a special instrument for surveying the magnetic conditions on the Sun. Special investigations of the magnetic fields of sunspots are also carried out with long focus telescopes (tower telescopes) in connection with spectrographs.

Radio telescope. These large aerials enable the radio astronomer to observe and record the quiet and disturbed solar radio radiation in the range of centimeter, decimeter, and meter waves (see also page 136).

Spectroheliograph. Here the combination of a tower telescope and a spectrograph of large dispersion allows photography of monochromatic solar images in the light of different chemical elements. Spectrohelioscopes are also used extensively for the visual observation of chromospheric phenomena.

10 / Observation of Total Solar Eclipses

W. Petri

10.1. Photography of the Solar Corona

10.1.1. The Continuous Spectrum of the Solar Corona

The most beautiful and most valuable gift of a total solar eclipse to the observer is the sight of the solar corona. From the scientific point of view, too, the corona is still the most important object of eclipse observations. When the Moon occults the observer and the atmosphere surrounding him from the brightness of the photosphere, it is then that the crown of rays around our Sun appears in all its vast extent and delicacy, much richer than even the best coronograph can reveal without an eclipse when set up on the surface of the Earth. The powerful spectrographs of large expeditions receive their light from heliostats and cover a multitude of spectral lines, whose analysis requires many months of hard work but gives us in return invaluable information about the composition and the condition of the outer atmosphere of the Sun.

The amateur astronomer with his much simpler instruments can photograph the "white corona" only. This is the continuous part of the corona spectrum, produced by light scattered by the photosphere. This scattering takes place on the free electrons of the *plasma* of the corona, its highly ionized gases. In addition, we have another part of the spectrum, caused by diffraction of the photospheric light by the particles of interplanetary matter. These are actually the same particles that at a larger angular distance from the Sun cause by reflection the Zodiacal Light. The intermediate region between the realms of diffraction and reflection is still unexplored. Here, too, eclipse photographs of long exposure times, taken at larger distance from the Sun and made possible by wide-angle optics and a very clear sky, will be most helpful.

10.1.2. The Structure of the Corona

The form of the corona as it appears on the photographs depends essentially on the phase within the cycle of the solar activity at which the eclipse takes place. The limiting cases are, on the one hand, the minimum corona that stretches along the solar equator, and on the other hand, distinct rays emerging from a more roundish maximum corona and radiating on all sides into space. Looking at small details we find that the corona has a delicate structure that is determined by the actual regions of activity (spots, prominences, faculae, etc.) near the Sun's limb and by the electromagnetic fields related to these phenomena. Thus a corona shows individual traits at each eclipse and makes it important to get full records of any eclipse. The continuum of the corona is essentially radially polarized. Photographs through polarization filters show the raylike structure conspicuously if the filter has been oriented correspondingly. One therefore uses different filter devices at different position angles. A further addition to the information obtained can be achieved by color filters with suitably corresponding emulsions. Since the degree of polarization of the corona is different in different spectral regions, photographs with a combination of polarization and color filters are advisable.

10.1.3. Evaluation of the Results

To obtain instructive pictures of the behavior of brightness within the corona, a series of enlargements on extra-hard photographic paper may be made. Although conspicuous to the eye, the rays of the corona do not come out prominently in any representation by isophotes. The thorough photometric reduction of corona photographs requires not only much skill and experience, but also the instrumental equipment of a large observatory. Those amateur astronomers who are aiming at scientifically important results should, therefore, from the very beginning make contact with an experienced professional astronomer. It can easily happen that a large expedition has been robbed of the reward of its efforts by clouds, and the leader of such an expedition would be very happy if an amateur astronomer had successfully obtained useful photographs. Usefulness includes the fact that the negatives have been calibrated. To do so there are several possibilities. It is simplest to use the uneclipsed Sun (at known zenith distance to be able to correct for extinction) as a calibration light source in order to impress a step wedge onto the original negative.

10.1.4. Photography with the Telescope

Any telescope can serve for the photography of solar eclipses. Since long focal length is more important than light-gathering power, an astrographic refractor will generally be preferred to a reflector. In order to obtain as large a field of view as possible, we give up any eyepiece magnification used for planetary photography, and use the primary focus only. Mirror reflex cameras of the single-lens type are very convenient; after removal of the objective they are clamped to the eyepiece collar. Of course, we must be careful of the heat of the solar image and use dark filters for focusing.

Guiding outside actual totality can be monitored by a simple projection finder. This is a convex lens of about one meter focal length (a spectacle lens) which is placed on the tube, near but outside the objective, and which produces next to the eyepiece a solar image that must coincide with the previously drawn circle on the projection screen.

Large instruments are usually not mounted equatorially but are set up in a horizontal position and fed with the Sun's light by a heliostat. For eclipse photography all we need is a single plane mirror that is turned by a clockwork once every 48 hours about its hour axis. In setting it up, we calculate the exact azimuth of the camera so that the solar image during eclipse will appear at the center of the field. The heliostat has to be adjusted for the geographic latitude of the observing site. Since the relatively heavy mirror will affect the clockwork differently according to its changing position, we adjust the clock rate for "eclipse position."

10.1.5. Exposure Times

The exposure time can vary within wide limits—actually it should do so. Experience shows that corona photographs are often overexposed. Because of the very steep radial brightness gradient we should make a whole series of photographs, whose exposure times differ successively by a factor of 3 to 5.

For example, if we use a refractor of the aperture ratio 1 to 10 we can well start at 1/5 sec; this would yield the series 0.2, 1, 5, 25, and 125 sec. The longish times are measured according to the beats of a metronome or with a stop watch. The filters require the application of special factors that are calculated in advance.

10.1.6. The Work of the Amateur Astronomer

Ordinary amateur cameras of the Leica or Rolleiflex type, as well as special astrographic equipment, can be used conveniently. Because of their great light power and large field of view, they can record even the outermost corona. There is no need for guiding, unless we intend to take photographs in the extreme red. If we are going to work with several short focus cameras it is advisable to attach these on a common track that may be screwed tight at the proper inclination, say, to a heavy box. This is more comfortable, and much safer in the darkness and hurry, than an array of several single photographic tripods. Easy and rapid film transport is very important. These photographs, too, should be calibrated by impressing on part of the unexposed and uncut film a series of blackening markings.

Photography with cine cameras, on the other hand, is of value only for demonstration and entertainment purposes. We must not forget to open the aperture at the second contact. Slow motion pictures, say 8 per sec, or single exposures save much film. Telelenses are necessary, but they should not restrict the angle of view to such a degree that during totality the Sun wanders out of the field of view.

Results of corona work are illustrated in Fig. 10–1.

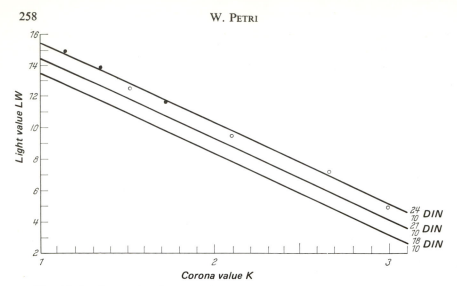

Fig. 10–1 Connection between the photographically measured diameter K of the corona (in units of solar diameters) and the light value LW. After H. GANSER, *Sterne und Weltraum*, **2**, 186 (1963). ○ Eclipse photographs of 2 October 1959; ● Eclipse photographs of 15 February 1961; (reduced to 24/10°DIN).

10.2. Special Astronomical Programs

10.2.1. The Chromosphere

During totality, prominences protrude above the solar limb beyond the disc of the Moon and, because of the H_α emission, can be seen as reddish structures either with the naked eye or better still with binoculars. They also appear on corona photographs whose exposure is short enough to give only an average blackening even at the solar limb. On the other hand it is difficult to photograph the actual chromosphere, since there are only very few seconds around the second and third contact. Spectrographic equipment for useful photography of the flash spectrum is beyond the means of the amateur astronomer. He might, however, try to observe the flash phase with its dramatic line reversal with the spectroscope, that is, visually, with a prism or grating.

10.2.2. The Times of Contact

If we know the exact position of the observing site, then the recording of the times of contact is as valuable as the registration of star occultations. However, the first and fourth contacts, when the discs of the Moon and Sun just touch from outside, cannot be recorded very sharply, not even if we know the position angles of these events from the Yearbook. At the second and third contacts the profile of the Moon's limb becomes conspicuous in the form of the "string of pearls" phenomenon, also called "BAILEY's beads." It occurs when the mountains on the Moon's limb already protrude beyond the Sun's disc: then the narrow solar crescent dissolves into various single light spots. The whole appears like a diamond ring; the ring being a prominently appearing corona, diamonds, the photospheric light reappearing through the valleys of the Moon.

10.2.3. The Partial Phase

Instead of observing the partial phase directly with binoculars, which is possible only with very dense and optically correct filters, it is better to project the image of the solar crescent onto a screen with the binoculars. This keeps us informed about the progress of the eclipse and has the additional advantage of distracting the attention of unavoidable onlookers away from the eclipse instruments.

Photographic series of the partial phase are attractive but have no scientific value. Only with a large imaging scale may they be useful to examine the law of the limb darkening of the Sun. The light scattered by the Earth's atmosphere is already much diminished through the occultation of part of the disc; furthermore, it is possible to record it well photometrically, particularly during the eclipse since the Moon's disc (apart from the negligible brightening by the light from the Earth) should appear completely black. Thus we can immediately recognize the scattered light. Since the limb darkening depends strongly on the wavelength of the light for such exposures, we use different filters.

10.2.4. The Star Field

If the total eclipse takes place in a clear sky we can also see bright stars. The stars that might become visible, for instance the inner planets Mercury and Venus, can be ascertained in advance. If one is lucky there might be the special gift of a comet. Such cases are not at all rare and are favored by the fact that comets near the Sun usually develop particularly conspicuous tails. In earlier times, astronomers searched the sky near the Sun for a planet called "Vulcan," which was suspected to move inside Mercury's orbit. Today it might even be possible to see an artificial earth satellite rushing by outside the central path of the Moon's shadow.

A special aspect of modern eclipse astronomy deals with the test of the relativistic light deflection by the gravitational field of the Sun, as predicted by Einstein. Of course, these tests require an extremely sophisticated instrumental effort, quite out of reach of the amateur astronomer. The latter, however, will like to use stars that occur on his corona photograph in order to derive the position angles of the prominences and of the rays of the corona.

10.3. Special Terrestrial Programs

10.3.1. Brightness and Color of the Sky

The strange character of the twilight that rapidly occurs immediately before the second contact usually makes a very strong impact on the observer of a total solar eclipse. The Sun sets, so to speak, in the middle of the sky while the horizon around us is brightened by partial photospheric light. There is an intense activity of prominences, together with a reddish coloring of the sky. Color photographs—of both the sky and the landscape—can then be most impressive.

The changing brightness of the sky is measured with a photoelectric cell (if necessary simply a good exposure meter), which is mounted on a rod, 6 ft or so above the ground, and directed to the zenith. The observer will switch over if he changes the sensitivity ranges. Soon after the third contact we have the subjective impression that it has become "day" again. The amateur astronomer must consult the exposure meter and not rely on his senses.

10.3.2. Flying Shadows

During the minutes when the Sun's crescent is very narrow, we have a good chance to observe the so-called "flying shadows"—narrow dark stripes that rush over the ground at the average speed of a car. They are produced by air bubbles and a slitlike source of light and are recognized if viewed at some distance from a plain by their uniform light-gray tone.

These are not to be confused with the umbra of the Moon's shadow itself, which moves at about 1 km/sec across the Earth. Its approach, particularly if there are light clouds and it is observed from a somewhat elevated point, is a thrilling spectacle.

10.3.3. Meteorological Observations

Every eclipse expedition also collects meteorological data which, in conjunction with the records for the neighboring days, provide the basis for the assessment of weather prospects of future expeditions to places with a similar climate. The rapid decrease of the solar radiation affects not only the temperature but also the wind, cloudiness, and humidity. Quite famous is the "eclipse wind," which sometimes in the very last moment literally disperses the veil before the Sun. It also increases the cooling and the demands on the stability of the instruments to a remarkable degree. Any loose piece of paper may be blown away by this sudden eclipse wind. In order to follow both the air temperature and the radiation we compare an ordinary thermometer in the shade with another whose bulb has been blackened and exposed to the direct solar radiation.

10.3.4. Biological Observations

Plants react to the twilight and night of the eclipse only as when daylight decreases normally, but animals are sometimes affected by sleep or fear. Without making detailed suggestions for biological observations, one may just mention that the behavior of domestic pets, particularly of dogs, may reflect the excitement of the observers more than the dogs' own excitement about the eclipse.

10.3.5. The Ionosphere

Finally, it has to be mentioned that the condition of the ionosphere will have various effects on an eclipse, which become evident during the eclipse. Amateur radio astronomers can find a rich field of activity here, even if they are situated outside the zone of totality. The occultation of certain active regions on the Sun by the Moon alone suffices to produce effects that may change long-range shortwave reception conspicuously.

10.4. The Observing Station

10.4.1. Devising the Program[1]

It goes without saying that a proper eclipse expedition is very expensive, if only because of the long journey that is usually necessary. In order to make full use of every second of totality, a detailed program is prepared beforehand in which all operations of every participant are precisely scheduled. Then practice begins and continues until every movement is perfect. A stopwatch is used to obtain the optimum procedure. At least one dress rehearsal should be performed in dark twilight. A preprepared tape is helpful because then a recorder can tell the observers the right moment for the beginning of the operation and continually instruct them as to the time and action. Experience shows that the end of the eclipse (the third contact) usually offers some surprise. The spatial arrangement of the instruments is also important, particularly if the same observer has to serve different instruments, for instance, short exposures with a miniature camera while the refractor is busy with long exposures.

If we intend to insert a whole pile of plateholders, it is advisable to place a cardboard marker at those places where another operation is due (e.g., a change of filter). One must be careful in the use of double plateholders to avoid getting out of step. Nothing can be too foolproof!

Exposures that require a particularly dark sky will be arranged for the middle of totality. If no time reserve in the schedule is wanted for unpredictable events, such as the slide of the plateholder jammed, or a release not wound, there will be time at the end of totality to add a few less important or repeated exposures, whose loss can more easily be borne.

Of course, the expedition will keep a complete log book, which may be supplemented by photographs.

10.4.2. Choice of Site

The choice of site is essentially prescribed by the position of the line of

---

[1] Reports on eclipse observations will be found in all astronomical journals for professionals and amateurs, as well as in the publications of the observatories that initiated the expeditions. The reports by M. WALDMEIER in the *Astron. Mitteilungen der Eidgenössischen Sternwarte Zürich*, in particular, will give many useful hints to the amateur astronomer. As a typical recent example of eclipse reports we may mention a color supplement to *Sky and Telescope*, **44**, No. 3, 173–176 (1972), which deals with the eclipse of 10 July 1972 in excellent detail.

totality, and all details will be found in the astronomical Yearbooks (see Fig. 10–2). Furthermore, there are special publications that appear often several years before each eclipse and that can be consulted in observatory libraries or by direct enquiry from the author. Accessibility and general weather prospects further restrict the choice. However, it would be wrong if everybody were to converge on the theoretically best place. Only uniform distribution of the expeditions will significantly diminish the risk of bad weather.

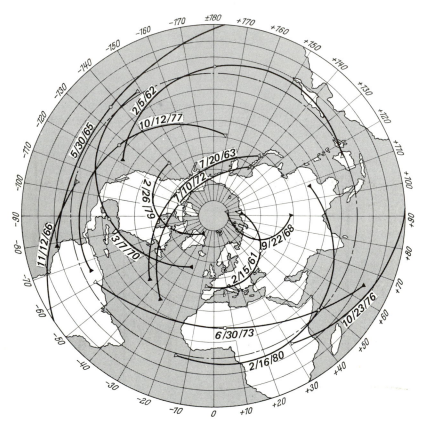

Fig. 10–2 The total solar eclipses between 1961 and 1980. The lines of totality are taken from T. von Oppolzer's *Canon der Finsternisse*, Vienna (1887).

Many considerations will go into making the final choice, such as protection against the wind, ease of guarding the station, and avoidance of any unusual and undesirable characteristic of the region, such as raininess, mosquitoes, and so on. As for the electricity supply, the observer will try to be independent of the ordinary grid. A homemade darkroom is very convenient. The joint use of professional photographic laboratories rarely works out well. However, the neighborhood might provide a refrigerator with a little space for infrared photographic material; the films and plates are put in their holders the night before the eclipse.

10.4.3. Equipment

The various items of eclipse equipment are chosen according to the circumstances. Flashlights should always be at hand, although it will rarely be so dark that the instruments cannot be operated without artificial illumination. The observer, remembering that it will get cold, will choose suitable clothing. The instruments should be protected by waterproof covers that are secured against wind, and windscreens and shields against radiation are also advisable. Nothing more need be said at this point about filters.

If one intends to count stars during totality, or to follow visually the outermost corona rays, one will use adaptation spectacles as used in X-ray laboratories. There also exist special medicines that improve the dark adaptation of the eye. A transistor radio is always convenient; and a first-aid box is definitely indispensable.

11 | The Moon

G. D. Roth

11.1. Problems and Ideas of Lunar Observations

11.1.1. The Moon as Test Object for Telescopic Work

As the nearest celestial body to our Earth, the Moon has, of course, always been a favored object for the amateur astronomer. The latter sees a celestial body with a large surface rich in various formations. Even field glasses give an idea of the multitude of visible details shown by a larger telescope. Indeed, the Moon is a nearly inexhaustible test field for the abilities of an observer and for the quality of his instrument. On the other hand, we must be aware that the large formations as such do not constitute a testimonial about the telescope and the observer: even poor lenses and uncritical eyes can always see something on the Moon. That is why the main aim of the test must be directed to very fine, small detail.

11.1.2. Special Studies of the Lunar Surface

Experienced observers concentrate on systematic investigations.[1] In our space age, the Moon has become the first space region to be explored. Space ships, of course, are solving the problems of physical selenology much more rapidly than terrestrial observers can.

Photographs of the lunar surface, showing rich detail, were already obtained by the American spacecraft Rangers VII and IX. [See the reports in *Sky and Telescope*, e.g., **29** (1965); we also mention here the series of slides "With ZEISS optics into space" produced by CARL ZEISS, Oberkochen, Germany, in conjunction with NASA.] Many problems of previous lunar maps have been solved by these photographs. The soft landing of the Russian spacecraft, Luna 9, was not only an essential step toward the manned Moon project; pictures taken from a height of 2 or 3 meters above the surface already show some fragments of rock and craters only millimeters across.

The day, 20 July 1969: "The Eagle has landed!" For the first time man had reached another celestial body. The first samples of lunar material were

---

[1] See also J. HOPMANN: Probleme der Mondforschung. *Die Sternenwelt* **4**, 157 (1952); and M. G. J. MINNAERT: *Practical Work in Elementary Astronomy*, p. 93. Dordrecht: Reidel (1969).

brought back to Earth, not to mention a multitude of instructive photographs of the lunar landscape.

In spite of this and the succeeding Apollo missions, the question as to the origin of the lunar formations has not yet been answered unambiguously. However, this is a subject where the amateur astronomer will rarely be able to contribute anything significant, as much as has been tried in this direction in the past. On the other hand, observers with sound geological knowledge will always be able to contribute original thoughts to such discussions, as testified by some telling examples [see H. KRAUSE, *Die Sterne* **28**, 199 (1952)].

There always remain tasks that if carefully attended to can lead to scientifically useful results. It is not the urgency for immediate scientific usefulness that should encourage the amateur astronomer to carry out his observations, however, but rather the value of personal experience achieved by a systematic observational series that should give him the essential impetus.

Many new aspects of lunar research will be explored by space vehicles. Optical signals are continuously received from vehicles stationed on the lunar surface. An important role is also played by the radio transmission signals, through which the terrestrial observer is able to follow with his instruments the progress of this exploration. Of course, a thorough acquaintance with lunar formations is essential.

Particular attention may be devoted to the landing sites of the original Apollo program, and it would be worthwhile to study these regions more closely through the telescope. For the choice of these sites, six requirements were considered:

1. Smooth landscape with relatively few craters and rocks.
2. Easy approach with no high mountains or large crags.
3. Low fuel consumption—a site where the landing can take place using as little fuel as possible.
4. Flexibility of timetable—a site where the time of departure can be rearranged without difficulties.
5. Within reach of the mother spaceship to help guarantee a safe return.
6. Small angle of inclination, if possible, less than 2° with reference to the landing area.

Some of the sites that fit these requirements are—

Site 1. Latitude $+2°37'54''$; longitude $34°1'31''$ East. Situated in the eastern central part of the Moon inside the southeastern Sea of Tranquillity: 100 km east from the border of the crater Maskelyne.

Site 2. Latitude $+0°43'56''$; longitude $23°38'51''$ East. Situated in the eastern central part of the Moon in the southwestern Sea of Tranquillity, about 100 km east of the border of the crater Sabine, 190 km southwest of the crater Maskelyne. This site was chosen for the landing of Apollo 11.

Site 3. Latitude $0°22'27''$; longitude $1°20'42''$ West near to the center of the visible area of the Moon in the south western part of Sinus Medii. About 40 km to the west from the center of this area and 50 km southwest of the crater Bruce.

Site 4. Latitude −3°38′34″; longitude 36°41′53″ West. Situated in the western central part of the Moon in the southeastern Ocean Procellarum. This site lies 240 km south of the crater Encke and 220 km east of the crater Flamsteed. It was chosen for the landing of Apollo 12.

Site 5. Latitude +1°46′19″; longitude 41°56′20″ West. Situated in the western central part of the visible area in the southeastern Ocean Procellarum, 210 km southwest of the crater Kepler and 190 km north-northeast of the crater Flamsteed.

Lunar observations nowadays play a role in the better understanding of solar–terrestrial relationships. The lunar surface is exposed to a permanent bombardment by the energetic ultraviolet and X-ray radiation and to the solar corpuscular radiation. To quote ZDENĚK KOPAL, "The Sun represents the *hot cathode* and the Moon the *anode* in an ionization tube of cosmic dimensions, the glass walls of which are formed by our atmosphere."

11.1.2.1. Critical Lunar Topography

There exist a number of formations that are not yet revealed in their actual structure and which might indicate some apparent or genuine changes. A comparison of the observations of well-known lunar observers (MÄDLER, LOHRMANN, NEISON, KRIEGER, FAUTH, WILKINS, and others) provides interesting material for this problem.

Discoveries of genuine changes on the lunar surface are never within the reach of the average amateur observer; they demand very extensive acquaintance with the lunar topography, achieved only through long years of telescopic work. The interplay of light and shadow has frequently become a kind of "fata morgana," as, for instance, the sensational "lunar bridge" at the eastern border of Mare Crisium between Cape Lavinium and Cape Olivium.

We have here a typical case of certain surface details and certain angles of illumination leading to a physiological illusion. This explanation has been demonstrated by H. BRENSKE, D. KIPSCH, U. KÖHLER, and H. OBERNDORFER. "We believe that the so-called lunar bridge is caused by a combination of a real terracelike fall of this region into the Mare Crisium and an optical illusion in the shape of a subjective connection of two points of light to a line of light. This illusion is favored, in addition, by the fact that the points on a dark background appear larger for physiological reasons than two dark spots would appear against a bright background" [KÖHLER and KIPSCH, *Die Sterne* 32, 124 (1956)].

Of course, all these problems are now being solved one after the other under the impact of the new photographs. A comparison of the well-known lunar maps, both the visual and the photographic ones, with the close-up pictures from lunar vehicles is most instructive. [See, for instance, the photograph of the crater Copernicus taken from an altitude of 45 km by the Lunar Orbiter 2, described in *Sterne und Weltraum* **6**, 17 (1967).]

Since critical lunar topography is nearly always devoted to studies of the smallest details, we require a telescope of at least 6 inches aperture for any meaningful work.

11.1.2.2. Representation on Maps

PAUL AHNERT of the Sonneberg Observatory describes the history of visual lunar observations as "An outstanding chapter in the history of amateur astronomy." (See "Nachrichten der Olbers-Gesellschaft Bremen." Sonderheft Mondkarten; October 1964). When man landed on the Moon in July 1969, this became again apparent, and during his broadcast in the Apollo studio of the German television service, the author of this chapter drew attention to this fact, pointing out that the topographic work of amateur astronomers over many decades had essentially helped prepare for the Moon landing. [See ERNST VON KHUON, GÜNTER SIEFARTH (eds): Mondflug in Frage und Antwort, Experten geben Auskunft, Düsseldorf, L. Schwann-Verlag (1969).]

The greatest accomplishments of topographic studies are manifested in the special maps of certain lunar landscapes and in complete presentations of the whole visible lunar surface. The complete presentations prepared by PHILIPP FAUTH and H. P. WILKINS reach a quality that will probably never be surpassed by future observers. [See H. FAUTH: PHILIPP FAUTH and the Moon. *Sky and Telescope* **19**, 20 (1959) and G. D. ROTH: Die Mondkarte von H. P. WILKINS. *Die Sterne* **31**, 180 (1955).] Nevertheless, any amateur astronomer with a 3-inch or 4-inch telescope can draw his own lunar map, provided he does so with great care and without bias. Among the special maps available, those dealing with the border zones deserve the attention of experienced observers and artists.

In presenting maps, the help of good photographs will be an important supplement to the visual results. Each map must contain the scale and its position within a grid of the lunar surface. To be of value, any work on the subject of the Moon should be represented on a topographic map. Drawings that show the interplay of light and shadow are mere contributions to discussions leading to this result. In addition to the maps based on visual observation, the last century has produced photographic lunar maps of great importance. The *Paris Atlas* is the first photographic work of this kind. In more recent years, G. P. KUIPER produced his *Photographic Lunar Atlas*, which, however, will only rarely be accessible to amateur astronomers. An attractive small photographic atlas has been produced by S. MIYAMOTO and A. HATTORI in 1964, which has proved particularly useful for the amateur observer. A remarkable publication is the *Lunar Chart*, published as a combination of the best photographs from all over the world by the National Aeronautical Space Administration (NASA); further references can be found on page 542. A good survey of the development of cartographic representations of the lunar surface, with references to the latest maps, has been given by ZDENĚK KOPAL [*The Moon*, Dordrecht: Reidel (1969)]. [See also Z. KOPAL and R. W. CARDER, *The Mapping of the Moon*: Reidel, (1974).]

11.1.2.3. Observations Relating to Physical Conditions

We now turn to lines of work devoted to the photometric behavior of certain lunar areas and single objects. The color tones of single lunar landscapes differ from each other in a characteristic manner. In particular, we note the bright rays on the lunar surface; the most conspicuous feature among these radiates from the crater Tycho.

Again it is lunar photography that supports visual observation very decisively. It makes it possible to record the intensity of difficult shadow areas; *color* and *interference filters* increase the possibilities even more. Already more than three decades ago, H. J. GRAMATZKI pointed out that it would be valuable to photograph certain lunar regions in four spectral regions—infrared, yellow-green, blue, and ultraviolet—a procedure that is well within the possibilities of a 4-inch reflecting telescope [see *Das Weltall* **39**, 145 (1939)].

Of particular interest are the peculiar luminous phenomena on the Moon, which were already observed by HERSCHEL, who thought them to be lunar volcanos. This strange luminescence is brought about by the incoming solar radiation. Some parts of the lunar surface thus become a kind of natural wavelength transformer which by luminescence turns the corpuscular radiation of the Sun into visible light. Such "lunar flares" can be so intense that they can be noticed visually without a filter. Of course, it is filter observations and photoelectric measurements that provide the most interesting results. Suspicious regions of the lunar surface are, among others, those near the craters Aristachus, Copernicus, and Kepler. The intensity varies; for instance, 1963 showed several flares, while there was calm in 1964. This is a clear indication of the connection with solar activity [see Z. KOPAL: Lumineszenz an der Mondoberfläche. *Sterne und Weltraum* **5**, 56 (1966)].

In practice it is photoelectric receivers such as image converter tubes, photocells, and photomultipliers that are employed almost exclusively for measurements of intensity, color, polarization, and spectral behavior. The amateur too has possibilities in this field; at least he can make some experiments which widen his experience, and this also applies to radio observations. Here we may refer to an important handbook for lunar observers with serious scientific aims: ZDENĚK KOPAL (ed.): *Physics and Astronomy of the Moon*, New York–London: Academic Press (1962). The recent book *The Moon* [published by Reidel, Dordrecht (1969)] brings our knowledge further up to date.

The night side of the Moon, also, under certain circumstances, shows some luminescence phenomena. Using a good 5-inch telescope and sufficient magnification (about 150-fold) the experienced lunar observer can succeed in keeping an eye on the night side—with much perseverance and self-criticism as a safeguard against illusions [see O. GÜNTER: Zur Sichtbarkeit von Einzelheiten auf der Nachtseite des Mondes. *Die Sterne* **42**, 1 (1966)].

11.1.3. Special Problems

The brightness estimates of the "earthshine" on the Moon deserve special interest. According to W. M. TSCHERNOW there exist brightness variations with an amplitude with about $0\overset{m}{.}6$ (maximum between March and May, minimum between June and August). Also the mean annual brightness is supposed to vary by about $0\overset{m}{.}8$, which is thought to be caused by atmospheric transparency changes and other effects.

Early observations of the Moon's crescent after New Moon, the so-called "New Light" observations, can be viewed from meteorological points of view as well as from astronomical-historical ones, a subject which has been discussed by such astronomers as G. SCHINDLER [*Die Himmelswelt* **46**, 204 (1936)].

The search for *Moon meteorites* has been a topical field in recent years, particularly in England and America, and is connected with speculation concerning a possible, very tenuous lunar atmosphere. On 4 July 1957, the Observatory of the town of Recklinghausen in Germany reported the observation of a phenomenon that one has tried to identify with the impact of a meteor onto the Moon.

11.2. Conditions of Visibility

11.2.1. The Moon's Phases

The visibility of the lunar formations depends on their illumination by the Sun. We speak of New Moon, First Quarter, Full Moon, and Last Quarter as the four prominent "phases." The phase is the illuminated part of the visible lunar hemisphere and an approximate characteristic angle g is given by

$$g = \lambda - \odot,$$

where λ denotes the longitude in the ecliptic for the Moon and $\odot$ the corresponding value for the Sun.

In order to determine the lunar phases for a certain date in the past or in the future we can use special tables. One such table was constructed by J. MEEUS, in which use is made of the fact that 251 synodic months are approximately equal to 269 anomalistic months: $7412\overset{d}{.}1776$ and $7412\overset{d}{.}1741$, respectively. If the Moon was once very near to its perigee it will be in the same position 251 synodic months later. If, for instance, the difference between Full Moon (opposition) and perigee was $0°0'$ at the beginning of such a cycle, then the difference will be $0°2.76'$ at the end of the cycle. The deviations between true and mean lunar phases are attributable to the eccentricity of the lunar orbit and due to repeat themselves in each cycle at the same positions, since the displacement is only very small. Simplified tables have been published by P. AHNERT in his *Astronomisch-chronologische Tafeln* (see page 540).

11.2.2. The Terminator

Of particular importance to the observer is the position of the terminator. Every feature of the Moon's surface is crossed by the terminator 25 times per year, causing at least 25 different types of illumination, corresponding to the different angles. Thus the position of the terminator constitutes an *essential* feature of every lunar observation. It is defined by the selenographic longitude at which during the moment of observation the Sun either rises (increasing Moon) or sets (decreasing Moon). Yearbooks and calendars (for instance, P. AHNERT, Kalender für Sternfreunde, Leipzig: J. A. Barth) tabulate it for every day.

The terminator can also be defined with the help of the *Colongitude* of the Sun at the center of the Moon. In the English language Yearbooks (for instance, *The Astronomical Ephemeris* or the *Handbook of the B.A.A.*) it can be found in the column *Sun's Selenographic Colongitude*. At New Moon it is nearly 270°, at the First Quarter 0°, at Full Moon 90°, and at the Last Quarter 180°.

We may give two practical examples:

1. An observer in Frankfurt drew the crater Plato on 20 February 1959 at 20^h20^m. We require the value of the colongitude. First we convert into Universal Time, i.e., 19^h20^m; we then enter the Yearbook for 1959 February $20^d0^h00^m$ UT and obtain 53.5°. The mean hourly motion of the terminator is approximately 0.51°, so that 19^h20^m corresponds to 9.8°. Thus the required value for the time of observation is 63.3°.

2. An observer in Hamburg wished to observe, at 191° colongitude at the end of May 1959, a certain lunar crater. What would be the correct observing time? The colongitude for 1959 May $30^d0^h00^m$ UT is 180.5°. The difference from the given colongitude is $+10.5°$; converting into hours at the known hourly motion of the terminator of 0.51° gives 20.6 hours, which have to be added to 0^h00^m. We thus obtain the observing time 1959 May $30^d20^h36^m$ UT.

The relation between the selenographic longitude L of the terminator and the colongitude S is:

New Moon to First Quarter: Western $L = 360 - S$ (Morning terminator)
First Quarter to Full Moon: Eastern $L = S$ (Morning terminator)
Full Moon to Last Quarter: Western $L = 180 - S$ (Evening terminator)
Last Quarter to New Moon: Eastern $L = S - 180$ (Evening terminator)

The western or eastern longitude, for the morning or evening terminator, respectively, are clearly distinguished by the sign plus or minus.

11.2.3. Libration

The Moon always turns the same face toward the Earth. Nevertheless, owing to the phenomenon of libration, it is possible to observe up to 60%

of the total lunar surface. We distinguish three types: (1) *libration in longitude* (maximum ± 7.9) caused by the ellipticity of the orbit, (2) *libration in latitude* (maximum ± 6.9) caused by the fact that the rotational axis of the Moon is not exactly perpendicular to the lunar orbital plane—it is possible to see once beyond the North Pole and then also beyond the South Pole of the Moon, and (3) *parallactic libration* (maximum $\pm 1°$), which is caused by the fact that the observer on the rotating Earth uses the Moon from changing directions. Altogether these three libration effects cause displacements of up to $10°$, which become particularly apparent to the observer at the position angles $45°$, $135°$, $225°$, and $350°$.

Observation at the lunar limb and in the libration zones, however, is made somewhat difficult by foreshortening of the perspective.

The amateur astronomer will find in the astronomical yearbooks the longitude and latitude of the apparent center of the Moon for every day, as well as the longitude and latitude of the place on the Moon facing the Earth at 1^h CET. If the longitude of the center is positive $(+)$ the observer sees more from the western limb, if it is negative $(-)$ he sees more from the eastern limb; if the latitudes are positive he sees more of the northern limb, and if they are negative more of the southern limb.[2]

11.3. The Lunar Formations

The terminology used for describing the lunar surface is not uniform. Certainly one must not try to compare the usual terms such as the Mare with the terrestrial oceans. These definitions are supposed to serve the purpose of simple description only; each one of them is, so to speak, a technical term.

It has become customary to distinguish the various types of surface detail on the Moon as follows:

Terrae—relatively bright areas at higher latitudes or island formations. Accordingly, one speaks of Terrae material (lunarite).

Mare—large dark surface formations which can be seen even without a telescope, for instance, Mare Crisium, Mare Nectaris, etc. Mare material has been called "lunabase."

Walled Plains—craters whose ringlike wall encloses a plain of diameter between 50 and 200 km. Abulfeda, Archimedes, Clavius, Fra Mauro, Grimaldi, Maurolycus, and Plato are examples.

Ring Mountains—similar to the walled plains but with higher walls and smaller extent of the lower enclosed plain, frequently with central peaks. Well-known examples are Eratosthenes, Copernicus, Petavius, Theophilus, and Tycho.

---

[2] See also the article by D. W. G. ARTHUR, Selenography, in the handbook by KUIPER and MIDDLEHURST, *The Moon, Meteorites and Comets* (The Solar System, Vol. 4), Chicago: The University of Chicago Press (1963). Here we also find all necessary formulae and calculations for the determination of libration, altitudes on the Moon, and so on.

Craters—in their proper notation, all more or less circular features that do not show any conspicuous mountainous surrounding. The diameters are smaller than those of the walled plains and ring mountains; there are very small individual craters. Photographs obtained by the lunar sondes of the Surveyor type and the observations of the first men on the Moon show that the smallest craters have diameters of only a few centimeters. Of the somewhat larger crater groups we frequently find patterns that look like a string of beads and which in a small telescope look like rills.

Chain Mountains—large mountain ranges with peaks reaching up to a few kilometers height as, for instance, the Apennines, the Carpathians, the Alps, etc.

Valleys—sharp indents in a chain mountain, of which the most conspicuous example is the valley of the Alps.

Rills—smaller indents in the lunar surface. They are long, even very long as, for example, the Ariadaeus rill, the Hyginus rill, etc. [See K. von BÜLOW, Rillen und andere lineare Elemente des Mondes. *Sterne und Weltraum* **2**, 208 (1963).]

Capes—protruding exposed blocks of mountain. They frequently indicate the transition from a mountainous landscape to a plain, for instance, Cape Heraclides and Cape Laplace.

Bright Rays—stripes and bright objects, particularly visible at Full Moon, that radiate out from a crater. Such systems of rays among others are to be found radiating out from Tycho (the most remarkable such system), Kepler, Copernicus, Olbers, etc. [See K. VON BÜLOW, Das Rätsel der hellen Strahlen, *Sterne und Weltraum* **5**, 43 (1966).]

Mare Wrinkles—or wrinkled ridges are damlike protrusions of the lunar surface, sometimes of a slightly wavy appearance.

Domes—a kind of magma-domes, rather low, of more or less circular shapes, which appear inside the Mare. Only very oblique illumination permits their observation.

The names of the various lunar formations and lunar landscapes mostly date back to the early history of visual lunar observation. It was LANGRENUS who in the Seventeenth Century was the first to select such names, which he mainly took from the Bible and biblical legends. His younger contemporary, JOHANN HEVEL (1611–1687), introduced the names of terrestrial landscapes; it was he who named the Alps and the Apennines. Most of the lunar nomography has been introduced by RICCIOLI and is still used. He chose well-known scientists and philosophers to name the craters and the mountains. The Mare received names of astrological meaning. This naming of Moon formations for personalities and geographical examples on Earth has been kept up to the present, and was also applied when the Russian astronomers named their newly discovered formations on the reverse side of the Moon (Soviet Mountain, Crater Lomonosov, Crater Joliot-Curie, etc.).

11.4. Optical Observations

11.4.1. Visual Observations

Long-focus refractors and reflectors ($f = 1:15$ to $1:20$) have been found to be the most favorable for obtaining the best definition of lunar images. The great amount of light occasionally blinds the observer and he should use neutral filters or choose hours of twilight to protect his eyes if they are particularly sensitive. Other auxiliary tools for his observations are a colored filter, a polarization filter, and for instruments of more than 6-inch aperture and if a driving clock is used, a cross-wire micrometer.

As for the best magnification to be chosen, the same remark applies to the Moon as to planetary observations. PHILIP FAUTH wrote about this point in his book [*Unser Mond.* Breslau (1936)]. In it he states, "I was most successful in drawing very fine details when using 163 mm objective with 160-fold magnification, 176 mm Apochromats with 176-fold magnification, a Medial of 300 mm aperture with magnifications 300 and 350, and at 385 mm aperture, magnifications of 350 and 430. Those who have younger and more light-sensitive eyes may increase the magnification by 50% if the seeing permits. To go further is of little use: the images become larger but also fainter and more diffuse and for observation of the Moon clarity is worth more than size."

11.4.1.1. Drawings at the Telescope

Also according to FAUTH, "One concentrates through patient efforts using every good opportunity to get familiar with a small area to such an extent that one finally grasps every small detail of its structure and sets it down on paper."

The foundation of every topographical work is familiarity with the basic features of the lunar surface. Skeleton charts serve for orientation and drawings of the outlines as a starting point. Both can be obtained by tracing reliable maps, or better still photographs. The latter, if faintly printed and enlarged, are useful "templates" for the drawings. Each drawing must show exact markings of position so that all details can be entered with their correct angles.

It is not advisable to work on a small scale. FAUTH drew his maps on a scale of $1:1$ million, corresponding to a lunar diameter of 3477 mm. This is the appropriate order of magnitude, and the observer should aim at it in the interest of a really useful survey.

The technique of the drawing depends on the ability of the observer. We distinguish between simple line drawings and fully shaded ones, which try to work bright and dark densities into the picture. FAUTH drew in "isohypses" (Fig. 11–1) and used shading that was plastically effective.

Figure 11–2 demonstrates the transition from a simple line drawing to a shaded picture, which shows shadows and light intensity impressively.

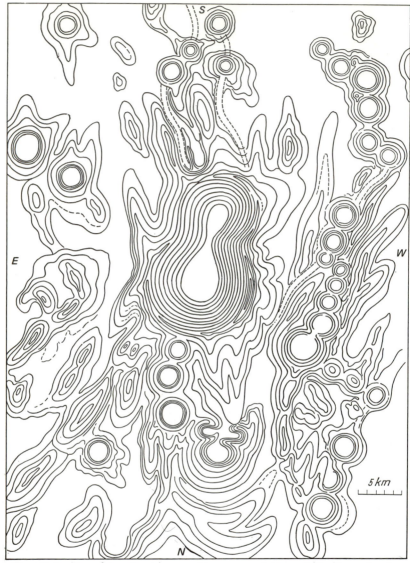

Fig. 11–1 Example of a topographic representation of the lunar landscape. The picture is taken from P. FAUTH's paper in *Die Sterne* **33**, 158 (1957) and shows the walled mountain named after him.

As an example of a plan for systematic observation, we mention several points recommended by F. BILLERBECK-GENTZ [*Die Sterne* **27**, 156 (1951)] who suggests that the observer try to answer the following questions when describing bright ray systems:

Where does the ray begin? How does it run? Where does it end? How wide is it? How bright is it? Does the brightness of the ray vary along its length and

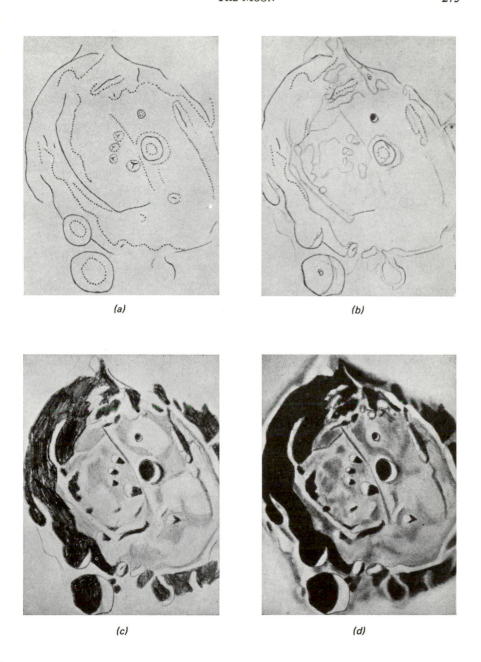

(a)

(b)

(c)

(d)

Fig. 11–2 The stages of the drawing of the walled plain Posidonius, by CLARK R. CHAPMAN (ALPO). (a) The outlines, based on a good photograph. (b) The face of the drawing now indicating the illumination and the libration. (c) Shadows added. (d) The contours and intensities in detail. From *The Strolling Astronomer* **17**, 47 (1963).

widths? If so, at which points and when? What is the general character of the ray surface? Is the structure of the ground perhaps simulated by a series of bright points, white craters, or small ridges? What is the shape of the ray? Is it straight or where does it bend; is it interrupted, and if so, by how much? Are there any parts of the ray that show a coloring at some time when the ray itself is not visible? When do these rays appear and when do they disappear?

11.4.1.2. Measurements at the Telescope[3]

The determination of absolute altitudes on the Moon meets with difficulties because of the lack of a definite zero reference. One is obliged to refer to a mean level of the Moon's crust, or to an ideal sphere, in order to derive the real distance of the surface of the crust from the geometrical center of the Moon, that is, the amount of its concave and convex departures.

This method of measurement makes use of the micrometer. For the derivation of relative altitudes on the Moon, the amateur astronomer will find measurements of the lengths of certain elevations a useful basis for his experiments. Such measurements are always *perpendicular* to the line joining the horns. The fixed horizontal wires of the micrometer (see page 87) coincide with the direction of the shadow. This gives an exact definition of the particular mountain. The accuracy of the altitude depends on the length of the shadow; the measurement of shadows which are too *short* leads to unreliable data. The altitude of the Sun during such measurements should lie below 15° but should not be smaller than 5°. If it is below 5° there is danger that the extremely long shadow intermingles with the shadow of other formations or runs into the unilluminated path of the Moon.

Concerning the theory and the numerical procedure, we refer to the publications by K. GRAFF, Formeln zur Reduktion von Mondbeobachtungen und Mondphotographien, Veröffentlichungen des Astronomischen Recheninstituts zu Berlin Nr. 14, Berlin (1901), and to the paper by G. SCHRUTKA-RECHTENSTAMM, Relative Höhenbestimmungen auf dem Monde mittels des Pariser Mondatlasses und visuelle Messungen am Fernrohr, Mitteilungen der Universitäts-Sternwarte Wien, 7, Wien (1955). SCHRUTKA-RECHTENSTAMM also discusses the determination of absolute lunar altitudes in the later publication, Mitteilungen der Universitäts-Sternwarte Wien, 8, Nr. 17, Wien (1958). A practical method suitable for the amateur has been recommended by P. AHNERT in Kalender für Sternfreunde 1952, p. 133, J. A. Barth, Leipzig (1962).

AHNERT's method makes use of some simplification. The greatest difficulty during the measurements is the determination of the terminator; furthermore, the shadows are usually observed in an uneven moonscape. Thus the observer

---

[3] Concerning the profile of the Moon, which is also important for altitude measurements, we refer the reader to the following papers: F. HAYN, Selenographische Koordinaten, III and IV. Abhandlungen der Math.-phys. Klasse der Kgl. Sächsischen Akademie der Wissenschaften, Leipzig; Vol. 30 No. 1, pp. 1–103, 1907; vol. 33 of this Leipzig Academy and "Astronomische Nachrichten," Vol. 198, pp. 125–130, 1914. See also T. WEIMER: Atlas du profiles lunaires, Paris (1952). Twelve tables 51 × 74 cm; scale of the Moon's profile: 1 mm = 0″5 for librations in longitude and latitude from 2 to 2°; Z. KOPAL, R. W. CARDER: The Mapping of the Moon, Dordrecht (1974).

obtains different values of the altitudes if a different altitude of the Sun causes the peaks of the shadows to fall into areas of different altitude. The observer must therefore measure both the length S of the shadow and the distance A of the particular mountain from the terminator. This can easily be done with an eyepiece with a cross wire. We then observe the time between the transits of the mountain peak, the end point of the shadow, and the terminator. This is best done with a stopwatch. The calculation then requires the following quantities, listed here with their symbols.

ρ Radius of the Moon in seconds of arc
β Distance of the mountain from the Moon's equator
L Longitude of the Sun in the ecliptic
l Longitude of the Moon in the ecliptic
b Latitude of the Moon in the ecliptic
P Position angle of the Moon's axis
δ Declination of the Moon
E Angle at the Earth in the plane triangle Sun-Earth-Moon
α Angle at the Sun in the plane triangle Sun-Earth-Moon
M Angle at the Moon in the plane triangle Sun-Earth-Moon
θ Angle between terminator and the line joining the peaks of the horns
ε Angle between terminator and the meridian of the mountain
φ Altitude of the Sun above the mountain
ψ Angle at the Moon's center between the directions to the peak of the mountain and the peak of the shadow
s Shadow length as fraction of the Moon's radius
h Altitude of the mountain in meters

The quantities ρ, β, L, l, b, P, and δ are obtained by the observer by using an astronomical calendar specially for the amateur astronomer and a good lunar map; S and A are measured at the telescope. The time for the above-mentioned transit is converted into seconds of arc by

$$S \text{ or } A = 14\overset{''}{.}46 \frac{D \cdot \cos \delta}{\cos P}.$$

Here D is the corresponding time of transit in seconds of time. The next step is the determination of the angle

$$\cos E = \cos b \cos (l - L),$$
$$M = 180° - E - \alpha,$$
$$\theta = M - 90° \qquad \text{for a Moon less than half illuminated,}$$
$$\theta = 90° - M \qquad \text{for a Moon more than half illuminated.}$$

The angle α amounts at the best to 9' (distance Earth–Moon as seen from the Sun). If we assume the orbits of Moon and Earth to be circular, the angle α can be expressed as a function of E as given in Table 11.1.

The unevenness of the lunar surface prevents the exact determination of the terminator but permits the following approximations:

$$\theta = 90° - (l - L) \qquad \text{for a Moon less than half illuminated,}$$
$$\theta = (l - L) - 90° \qquad \text{for a Moon more than half illuminated.}$$

Table 11.1 The Plane Triangle Sun-Earth-Moon: the Angle of the Sun (α) as a Function of the Angle E at the Earth

E	α	E
9°7—16°3	2′	163°7—170°3
16.4—23.2	3	156.8—163.6
23.3—30.5	4	149.5—156.7
30.6—38.3	5	141.7—149.4
38.4—47.2	6	132.8—141.6
47.3—57.9	7	122.1—132.7
58.0—73.9	8	106.1—122.0
74.0—90.0	9	90.0—106.0

By neglecting α and cos b in θ we commit an error of at the most 0.2°; this is usually smaller than the uncertainty in the definition of the terminator.

One now plots the position of the measured mountain on a lunar map and determines the approximate selenographic latitude. The calculation then proceeds as follows:

$$\sin(\theta \pm \varepsilon) = \sin \theta \pm \frac{A}{\rho \cos \beta}.$$

The positive sign applies for the less than half-illuminated Moon, the negative sign for the more than half-illuminated Moon. We can now determine the altitude of the Sun as seen from the peak of the mountain as

$$\sin \varphi = \sin \varepsilon \cdot \cos \beta.$$

The length of the mountain's shadow expressed in fractions of the Moon's radius becomes

$$s = \frac{S}{\rho \cos \theta}.$$

All these calculations are carried out with sufficient accuracy with four-figure logarithms. Only the final calculations may be done to six figures, namely,

$$\sin \psi = s \cdot \cos \varphi$$

$$h = \left[\frac{\cos(\varphi - \psi)}{\cos \varphi} - 1\right] \cdot 1{,}738{,}000 \; m.$$

11.4.1.3. Visual Photometry

Relatively little instrumental equipment is necessary for visual photometric work on the Moon. The beginner will certainly be interested in the measurement of relative brightnesses during Full Moon. If sufficient experience has been gained, he will turn to visual photometry in the course of the various *phases* of the Moon. For instance, it is of interest to study the changes of brightness inside the crater in connection with the position of the terminator.

All estimates are done with the help of a scale of relative intensities. The estimates themselves are done on the same lines as a step estimation of variable stars (see Sec. 19.5.2). Of course, the objects with which the scale is calibrated must be absolutely invariable during the course of a lunation.

The calibration scale given here has been based on the data provided by H. J. KLEIN and H. K. KAISER in the periodical "Die Sterne" derived from work done in the 1930's. On this scale, the blackness of the shadow corresponds to 0, while 10 corresponds to the brightest spot on the visible lunar surface.

0 = shadow on the Moon; 1.0 = the darkest part of the inside of the walled mountains Grimaldi and Riccioli; 1.5 = inner surface of De Billy; 2.0 = inner surface of Endymion and J. Caesar; 2.5 = inner surface of Pitatus and Vetruvius; 3.0 = Sinus Iridum; 3.5 = inner surface of Archimedes and Mersinius; 4.0 = inner surface of Ptolemy and Guericke; 4.5 = Sinus Medii, surface around Aristyllus; 5.0 = surface around Archimedes, walls of Landsberg and Bullialdus; 5.5 = walls of Timocharus, rays of Copernicus; 6.0 = walls of Macrobius and Kant; 6.5 = walls of Langrenus and Theaetetus; 7.0 = Kepler; 7.5 = Ukert and Euclid; 8.0 = walls of Copernicus; 8.5 = walls of Proclus; 9.0 = Censorinus; 9.5 = inner part of Aristarchus; 10.0 = central peak of crater Aristarchus.

The scale can be extended for visual estimates by linking the observed brightnesses to photographically determined photometric values of certain objects. Here a paper by K. GRAFF will be found very useful, *Zur photographischen Photometrie der Vollmondformationen*, Mitteilung der Univ.-Sternwarte Wien **4**, No. 6 (1949).

11.4.2. Photographic Observations

Both reflectors and refractors are suitable for lunar photography. We note that a telescope of 1 m focal length produces an image of the Moon of about 10 mm diameter.

As in the procedure for planetary photography (see page 385) the eyepiece collar carries either a special Moon or planet camera or a mirror reflex camera (e.g., Exakta Varex), but without optics. For photography at the focus we do not use the eyepiece of the telescope. If, however, we wish to obtain an enlarged image of the Moon we use the eyepiece to project the Moon onto the emulsion (see Fig. 11–3). The alternative between the two methods is therefore: *focal image*—bright, but relatively small image, thus short exposure time and subsequent enlargement; *projection*—darker image, but of larger dimension (see the formulae on pages 225 and 252), hence longer exposure and smaller subsequent magnification. See also Fig. 11–4.

The excellent photographs obtained by GÜNTHER NEMEC of Munich [See also, e.g. p. 283 of this book, as well as G. D. ROTH; *Handbook for Planet Observers*, London: Faber and Faber (1970)]. demonstrate the outstanding performance of the refractor as well as the achievements of the

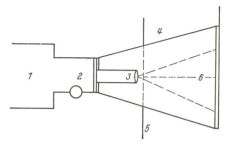

Fig. 11–3 Schematic illustration of the method of the photography of projected images: 1 = telescope, 2 = eyepiece collar, 3 = eyepiece, 4 = camera case, 5 = filter frame for the possible insertion of color filters, 6 = plane of the plate.

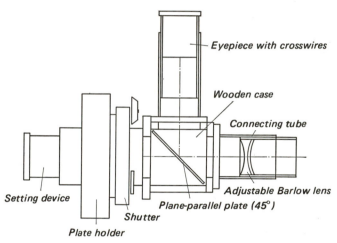

Fig. 11–4 Camera for lunar and planetary photography with Compur shutter, and checking eyepiece, according to a construction by H. OBERNDORFER of Munich.

projection method (equivalent focal lengths of 10 to 30 m at eyepiece projection and telescope apertures of 4 to 8 inches).

The differences in the apparent brightness of the Moon during the various phases pose a real problem for the photographer. The brightness does *not* increase or decrease proportionally to the extent of illuminated surface. As compared with photographs at Full Moon the exposure time for pictures obtained during the first and last quarters has to be multiplied by a factor of 4, and for the 3- or 24-day crescent, by as much as a factor of 12. It is also to be taken into consideration that the *correct* exposure succeeds only near the terminator, while regions where the light falls more steeply on the area are considerably overexposed.

While we record intensity differences for the Full Moon of the order 50:1, these differences range up to 1000:1 for other phases. To bridge such differences we have to apply quite different exposure times for the *zones of the terminator* than for the well-illuminated regions.

This can be done, for instance, for the first and the last quarter with a Compur shutter of a special shape (see Fig. 11–5). The shutter can, for example, be shaped in such a way that it provides an exposure for one-quarter of a second at the lunar limb and then with logarithmic increase up to 2 sec at the terminator. Needless to say, this procedure requires utmost accuracy in setting and guiding.

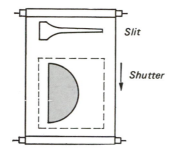

Fig. 11–5   Shutter with logarithmically shaped slit. After E. A. WHITTAKER.

There is no rule for the *exposure times*, since these depend strongly on instrument, phase, altitude of the Moon above the horizon, and kind and sensitivity of the emulsion [see also G. D. ROTH: *Taschenbuch für Planeten-beobachter*, Mannheim: Bibliographisches Institut (1966), or the English edition, *Handbook for Planet Observers*, London: Faber and Faber (1970)].

As a rough approximation we may note: Full Moon in focus (without filter) at $f = 1:15$ and 17/10 DIN emulsion requires exposure times between 1/50 and 1/10 sec. If we take a projected image of moderate size, say 10 to 15 cm, the time may be between 0.5 and 3 sec. If we use color filters we have to apply the corresponding factors. Every photograph whose exposure exceeds 1 sec demands correct guiding of the telescope with a motor drive.

The essential requirement of lunar photography is the most careful focusing. The use of a magnifying glass of 10 to $15 \times$ is very useful. The ground glass of the camera should be nearly grainless. H. J. GRAMATZKI has recommended making it by distributing on a clear glass plate a drop of diluted seccotine. A very fine opaque layer then remains behind which makes it possible to focus the lunar image, when viewed from the side, in a very reliable manner.

Every kind of disturbance or vibration of the telescope must, of course, be avoided during exposure. It is useful to release the shutter indirectly (e.g., with a cable release) or to perform the exposure with a rhythmically moved cardboard in front of the objective lens (the double tick of a wrist watch takes two-fifths of a second).

Fig. 11–6 Minox snapshot of the lunar surface in the region Bullialdus-Capuanus-Longomontanus. Refractor (1:18), of 125 mm aperture and 16 mm monocentric eyepiece. The Minox was attached to the eyepiece and set at infinity. Exposure time 0.5 sec on KB 14. No filter. Diameter of the negative 7 mm. Photographed on 30 April 1966 at 21 hours MEZ, seeing 2, by G. D. ROTH.

As to the *emulsions* to be used, we must emphasize that they must have a very fine grain (steep gradation) with the highest possible sensitivity. Sharpness and coloring can be somewhat controlled through the choice of the developer (e.g., special fine grain developing). Special developers (e.g., Neofin) act favorably with a subsequent sensitivity increase of fine grain emulsions of lesser sensitivity. Well-exposed negatives of 20-mm diameter (focal images) can yield a print corresponding to a diameter of the Full Moon of up to 50 cm. Experiments will indicate the way to the best results.

The successful photographer of the Moon and the planets, GÜNTHER NEMEC of Munich, writes, ". . . It is absolutely necessary to work very exactly, and not to be harassed by haste. It appears significant to me that photo-

Fig. 11–7 Photograph of the Moon by G. NEMEC, Munich, with an 8-inch refractor (f = 4000 mm), equivalent focal length (eyepiece projection) 16000 mm, exposure time 3 sec on 17 DIN.

graphs of the decreasing Moon succeed better than those of the increasing Moon, which culminates in the early evening when the astronomer is still occupied with his instrumental preparations and the telescope has not yet sufficiently cooled down. Atmospheric conditions usually change irregularly during the whole night and one can never say whether the second half of the night will guarantee the best seeing. Besides this scintillation it is the altitude of the object on which the definition essentially depends. Accordingly photographs taken one or two hours before or after the culmination do not yield useful results." See Figs. 11–6 and 11–7.

The photographic photometry of the Moon is also dealt with in the excellent guide by M. J. G. MINNAERT, *Practical Work in Elementary Astronomy*, pp. 99–100, Dordrecht: Reidel (1969).

11.4.2.1. Tasks for Lunar Photography

A lunar photographer with a skilled hand and good optics may make experiments whether and how his photographs can supplement or correct visual observational results; of course, he will also compare his plates with those of other observers. Good negatives can be investigated by photometric methods.

Photographs of certain lunar landscapes, say one suspected to be luminescent, can be made with color filters (e.g., with a green interference filter, 5450 Å, and a red one, 6725 Å; a good 5-inch refractor is recommended for such research.

Of special interest is the photographic determination of altitudes of lunar formations. Special instructions for this work are to be found in M. G. J. MINNAERT's book, *Practical Work in Elementary Astronomy*, on page 95 [Dordrecht: Reidel (1969)]. The photographic method has the disadvantage that shadow measurements on photographic plates are rather inaccurate owing to the influence of bad seeing, diffuseness of the images, and so on. It appears that the visual method is preferable (see page 276).

11.4.3. Photoelectric Observations

Photoelectric work offers the widest scope, but the expense of the outfit can limit the work of the amateur astronomer. However, we indicate some of the possibilities:

11.4.3.1. Photometric Investigations

Photoelectric instruments achieve much more accurate results than, for example, the photographic method. We refer the reader to the remarks on page 99 of the above-mentioned book by MINNAERT (1969).

11.4.3.2. Color Measurements

The insertion of color filters combined with photoelectric instruments of different spectral sensitivity leads to important results.

11.4.3.3. Polarization Measurements

A polarimeter connected with the photoelectric measuring instrument can examine the polarization of certain regions of the lunar surface. Extending this over a whole lunation one determines the variation of P, where

$$P = \frac{I_1 - I_2}{I_1 + I_2} \cdot 100\%.$$

Here I_1 denotes the intensity of the light which is polarized in the plane Sun-Moon-Earth and I_2 the corresponding intensity in the perpendicular plane.

11.4.3.4. Spectral Investigations

The visible spectrum of the Moon is essentially the same as the solar spectrum. There exist, however, some deviations, particularly for the absorptions in the ultraviolet region, which have been discovered with the help of color filter photography. Photoelectric spectrophotometers give better results.

11.4.3.5. Observations of Luminescence

Certain regions of the lunar surface light up when the Sun shines on them. This luminescent phenomenon can be investigated visually, by filter photography, or photoelectrically. The latter measurements are preferably carried out in three colors—green, red, and infrared.

11.4.3.6. Determination of the Structure of the Lunar Surface

It is very useful to combine the different types of photoelectric observations mentioned in the preceding sections, 11.4.3.1–5. If carried out in a great variety of conditions of illumination they can lead to a kind of "microrelief" of the lunar surface.

It is understandable that practical results of such investigation by amateur astronomers are very rare. We must mention, however, the work by K. FISCHER in Prague, who designed for his private observatory a special double refractor. At an aperture of 190 mm and a focal length of 3000 mm it has been proved suitable for work in conjunction with an infrared image tube (AEG and Tesla 22QA 41). The maximum spectral sensitivity of the image converters lies in the region of 7000–9000 Å. The purpose of his observations is to measure the difference between the image in the integrated light and the image in the infrared. The radiation at wavelengths shorter than 7500 Å is eliminated by a SCHOTT filter, RG 10 (2 mm). This instrument permits the measurement of the albedo of the Moon in the infrared and the search for lunar formations which radiate in the dark part of the Moon in the infrared region. Extensive measurements of this kind can lead to a lunar infrared map. [See K. FISCHER, Die elektro-optischen spektralen Bildwandler und ihre Anwendung in der Selenographie. *Die Sterne* **38**, 181 (1962).]

11.5. The Grid of Lunar Maps

Just as on Earth, we require a special grid for orientation on the Moon. The zero or main meridian goes through the center of the visible side of the Moon from top to bottom and connects both poles. It is cut in the middle by the equator, which connects the east point with the west point. Corresponding to the appearance of the Moon in the reversing astronomical telescope one reckons the celestial directions as follows: north at the bottom, south at the top, west on the left, and east on the right.

Seen with the naked eye or in field glasses, west is on the right and east on the left.

Maps in astronautics are oriented like terrestrial maps as follows: north at the top, south at the bottom, west on the left, and east on the right. In order to avoid confusion, the modern practice is to indicate only the directions north and south and to use for the description of positions on lunar maps the term "right" instead of east and "left" instead of west.

A positive selenographic latitude (i.e., northern latitude, below the equator) is denoted with a plus sign; the southern latitude, above the equator, is indicated by a minus sign. In the astronomical telescope the positive ($+$) selenographic longitude lies on the left, the negative ($-$) on the right.

The counting of the degrees starts at the point of intersection between the main meridian and the equator (the zero point), to both sides for the longitudes and toward the top or bottom for latitudes. In every case the 90th degree coincides with the limb of the mean visible disc of the Moon. The same procedure can be applied to the reverse side of the Moon. However, it is also possible, and it has been done in this way in the Russian publications, to continue the counting beyond the eastern and western limb. In this case the central meridian of the reverse side has the value 180°.

The quadrants of the half of the Moon which faces the Earth are counted counterclockwise as follows: *I*—northwest (left below); *II*—northeast (right below); *III*—southeast (right top); *IV*—southwest (left top).

As we see the Moon, we are immediately reminded of an *orthographic projection*. This is a parallel projection referred to a point on the intersecting line between the equatorial plane of the Moon and the plane of the zero meridian. Such an orthographic projection images all parallel circles as lines parallel to the equator, which divide the border meridian (which appears as a circle) in equal sections. The zero meridian then appears as the diameter of the border meridian which is perpendicular to the parallel circles. All the other longitude circles present themselves as ellipses whose common major axis is a projection of the zero meridian.

> The starting point for the grid is the border meridian, which is determined by the radius R chosen by the observer. The corresponding circle is then given a horizontal and a perpendicular diameter, representing the equator and the zero meridian. If, for instance, the latitude circles are drawn in every 10° one divides the periphery of the circle into 36 equal parts, starting from the equator, and draws lines parallel to the equator.
>
> One then constructs an equilateral triangle the sides of which correspond to the Moon's radius R. The selenographic longitudes are then determined geometrically, The distance of the l_nth longitude circle at a selenographic latitude b_n, measured from the projection of the zero meridian, is then given by the formula

$$R \cos b_n \cos (90° - l_n).$$

Assuming that we wish to determine the selenographic longitudes from 10 to 10° as well, one draws from the apex of the constructed triangle the radii of the *parallel circles*: $R \cos 80°$, $R \cos 70°$, ..., $R \cos 10°$ in succession. For easier readability our Fig. 11–8 shows these subdivisions from 30 to 30° only; this is sufficient, however, to elucidate the main point. The points of intersection are then joined by lines, drawn parallel to the base. The same method is applied for the base of the triangle, starting from the right or the left corner. The divisions on the base are then joined to the vertex of the triangle by lines which intersect all lines parallel to the base in the same ratio, i.e., $\cos 10 : \cos 20 : \cos 30 : \ldots \cos 80°$. It is now possible to transfer the base with the dividing points to the projection of the lunar equator, doing so from the zero meridian to the right and to the left. The first parallel line in the triangle marks on the projection the parallel circle 10°, etc.

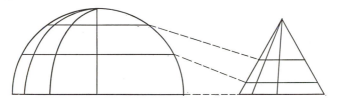

Fig. 11–8   Design of a grid for a lunar map.

Besides the spherical coordinates of the orthographic projection, which is particularly suitable for general representations of the whole surface, we also possess the rectangular coordinates of a similarly constructed grid (Fig. 11–9). For practical purposes such a rectangular grid, with ξ and η coordinates, has proved very useful. We have to imagine that it is spread in the shape of numerous squares on top of the orthographic lunar image, whereby it characterizes an unambiguous north–south direction. It retains its form only if we assume a mean libration and an exact parallel projection. Nowadays, we possess a sufficient number of reliable positions of lunar points, so that even zones near the limb (e.g., on photographs) can be used without errors of distortion.

Between the rectangular coordinates ξ and η and the selenographic longitude l and latitude b of a point on the orthographic lunar map we have, for the radius l, the following relation:

$$\xi = \sin l \cos b; \qquad \eta = \sin b.$$

The third space coordinate that gives the perpendicular distance ζ of a point from the plane of projection and that is necessary, for example, for the determination of the position of lunar formations is characterized with reference to the spherical coordinates by the relation

$$\zeta = \cos l \cos b.$$

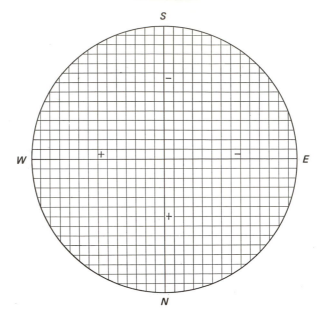

Fig. 11–9 Rectangular grid for a lunar map. Point at the western limb: $\xi = +1000$, $\eta = 0.000$. Point at the eastern limb: $\xi = -1000$, $\eta = 0.000$. Point at the northern limb: $\xi = 0.000$, $\eta = +1.000$. Point at the southern limb: $\xi = 0.000$, $\eta = 1.000$.

In the Appendix (page 510) a list of 15 fundamental craters can be found, the spherical and rectangular coordinates of which can serve as starting points for the construction of maps and measurements of positions. See also H. ROTH, Ortsbestimmung von 433 Punkten hoher Albedo auf einer Vollmondaufnahme. Mitteilungen der Univ.-Sternwarte Wien **4**, No. 7 (1949).

12 | Lunar Eclipses

F. Link

12.1. Introduction

Lunar eclipses are among the most picturesque astronomical phenomena. They have been observed since the earliest beginnings of the history of astronomy, and we find records dating back to eclipse observations made in 2282 and 2202 B.C. in Mesopotamia. For the amateur astronomer they are a rewarding object of study as we shall see.

12.1.1. Geometric Theory of Lunar Eclipses

We start from the configuration Sun-Earth-Moon, all of which during a lunar eclipse lie on a straight line. The common outer tangents to Sun and Earth, say t_1 and t_2 (see Fig. 12–1), form behind the Earth the long convergent

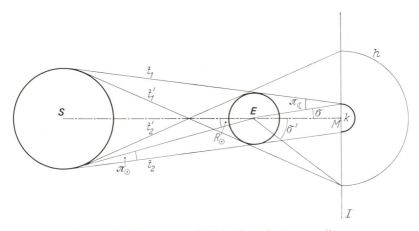

Fig. 12–1 The geometric situation of a lunar eclipse.

cone of the umbra k. On the other hand, the inner tangents, for instance, t_1' and t_2', form the divergent cone of the penumbra h. If there were no Earth's atmosphere this would be the overall course of the Sun's rays. No light would penetrate into the umbra and the Sun would be eclipsed for an observer by the Earth. An observer in the penumbra would only receive part of the sunlight, and the Sun would appear to him partially eclipsed by the Earth.

Let I be a plane perpendicular to the axis of the shadow SE at the distance of the Moon. The two shadow cones are intersected by this plane, the umbra and the penumbra forming circles corresponding to their apparent radii σ and σ' as seen from the center of the Earth. Therefore we have

Umbra

$$\sigma = \pi_{\mathbb{C}} + \pi_{\odot} - R_{\odot},\tag{1}$$

Penumbra

$$\sigma' = \pi_{\mathbb{C}} + \pi_{\odot} + R_{\odot}.\tag{1'}$$

These equations follow directly from the geometric situation shown in Fig. 12–1 where $\pi_{\mathbb{C}}$ is the horizontal parallax of the Moon, $\pi_{\odot}$ that of the Sun, and $R_{\odot}$ the apparent solar radius.

While circling the Earth, the Moon is projected onto this plane and there traverses various parts of the penumbra and umbra. If there is only a transit through the penumbra without touching the umbra, we speak of a penumbral eclipse, and usually such eclipses are listed only in the large Astronomical Ephemerides (see Fig. 12–2). In the case of a partial lunar eclipse the Moon enters the umbra only partly; in the case of a total eclipse, wholly. If the Moon's center crosses the center of shadow we speak of a central lunar eclipse (see Fig. 12–2).

The maximum of the eclipse occurs at minimum distance between the Moon and the center of the shadow. This moment is called middle of the eclipse in the Ephemerides, and the size of the eclipse in this position is expressed by the following quantity g (see Fig. 12–2).

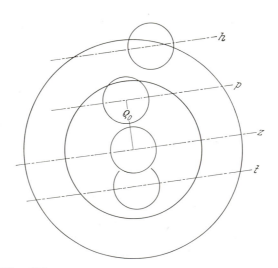

Fig. 12–2 The different types of lunar eclipses: h = penumbra, p = partial, z = central, and t = total.

$$g = \frac{\sigma - (\rho_0 - R_{\mathbb{C}})}{2R_{\mathbb{C}}}, \qquad (2)$$

in which ρ_0 is the minimum distance between the Moon and the center of the shadow.

At the beginning of a partial eclipse we have $g = 0$, during the partial phase $0 < g < 1$, at the beginning of totality $g = 1$, and during totality $g > 1$. In the course of the partial phase the quantity g denotes that fraction of the Moon's diameter that is eclipsed, but Astronomical Ephemerides give g only for the middle of the eclipse. In earlier days the size of the eclipse was given in digits, a digit being $\frac{1}{12}$ of the Moon's diameter, so that a twelve-digit eclipse was just total, one below twelve digits a partial one, and one above twelve digits a total one.

To illustrate these data we introduce numerical rounded-off values. If $\pi_{\mathbb{C}} = 57'$ and $R_{\odot} = 16'$, the radius of the penumbra is $\sigma' = 73'$ and that of the umbra $\sigma = 41'$. At a central eclipse, therefore, we have, according to Eq. (2), g about 1.8 or 21 digits.

12.1.2. Photometric Theory of Lunar Eclipses

The geometric theory neglected the presence of the Earth's atmosphere and was therefore unable to explain the visibility of the Moon in the umbra. KEPLER, at the beginning of the Seventeenth Century, showed that an explanation of this phenomenon must be sought in refraction in the Earth's atmosphere. An exact photometric theory of lunar eclipses, however, was not worked out until this century, after the necessary knowledge about the composition of the Earth's atmosphere in different layers up to the height of 100 km was acquired. This knowledge mainly concerns our understanding of the behavior of air density at different heights; from this we can then calculate the refraction, that is, the total deviation of the horizontal rays at a height h_0 above the Earth's surface, as well as the changes of the refraction with the height h_0.[1]

If we consider a beam of solar rays, which penetrates the Earth's atmosphere, we see that the refraction for the upper ray in this bundle is somewhat smaller than for the lower ray. This increases the natural and otherwise very inconspicuous divergence of the ray bundle, and the natural consequence of this is the so-called refraction attenuation of the solar rays. This effect is very noticeable, because even in high layers of the atmosphere, up to 50 km where the air density is already very small, it still causes an apparent attenuation. The attenuation by refraction is nearly neutral in color. In lower layers of the atmosphere we have, in addition, a weakening by extinction, the main cause of which is a molecular scattering of the light. The extinction depends on the traversed air mass and is distinctly selective according to RAYLEIGH's λ^{-4} law. The contribution of these two components to the total attenuation of

[1] See, e.g., PAETZOLD, H. K.: Mondfinsternisse und das Studium der Erdatmosphäre, *Die Sterne* **28**, 13 (1952).

the light also depends on the minimum height, or on the position of the particular point of the Moon's plane that is reached by the ray. The exact photometric theory must also take into account the "limb darkening," that is, the unequal distribution of brightness over the solar disk.[2]

Figure 12–3 shows some curves of the illumination within the Earth's shadow, expressed as the so-called optical density D (see Sec. 12.4.1). These give an idea of the photometric conditions under different circumstances, provided that we are dealing with an ideal RAYLEIGH atmosphere, that is, one with these characteristics:

1. A nearly neutral coloring of the shadow in its outer parts where the extinction is still small.
2. The preponderance of the red light in the inner parts of the shadow.
3. The dependence of the brightness of the shadow on the parallax of the Moon.

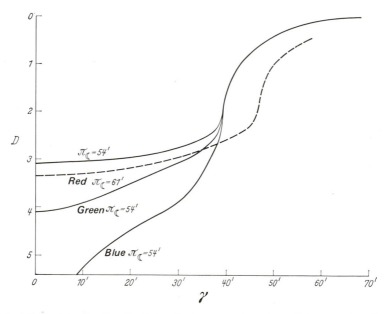

Fig. 12–3 The density D of the shadow as a function of the angular distance γ from the center of the shadow.

It is interesting, too, to analyze the situation from the point of view of an observer on the Moon, something which today, in the age of astronautics, is by no means a hypothetical case. This analysis actually also explains the photometric conditions in the penumbra. An observer in the umbra would see the Earth as a dark disc with an apparent diameter of about $2°$ ($= 2\pi_{\mathbb{C}}$).

[2] See LINK, F.: *Eclipse Phenomena in Astronomy*. New York: Springer-Verlag (1969).

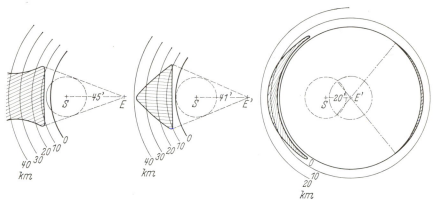

Fig. 12–4 View of a solar eclipse as seen from the Moon. The distances from the limb of the Earth (0 km) have been magnified by a factor of 100.

The geometric position of the Sun, which is covered by the Earth, is shown in Fig. 12–4 as a dotted circle. The observer would see a refraction image of the Sun produced by the Earth's atmosphere—in the shape of a very narrow crescent, which would surround part of the Earth's circumference. The radial width of this image would be very small (a fraction of a minute of arc), so that it would be seen by the naked eye as a bright, red-colored line. In reality there are two such images, as we can see in Fig. 12–4. The second image, however, since it arises in the lowest layers of the atmosphere (below 2 km), would be very faint or completely covered by clouds. The nearer the observer is to the middle of the shadow, the more would the two endpoints of the crescent approach each other; in the middle of the shadow he would see a ring that envelops the whole Earth.

An observer situated in the penumbra would see a partial eclipse of the Sun by the dark Earth. The illumination could be calculated from the ratio of the uneclipsed surface to the whole disc of the Sun, allowing for the surface distribution of the brightness on the Sun, i.e., the limb darkening. The refraction image would contribute very little to the illumination, except that portion due to the penumbra close to the border of the umbra.

Any modern photometric theory must be governed by the need for the exploration of the upper atmosphere. A comparison with the observations shows many interesting features that open up a whole complex of problems in this field and in space exploration.[3] Also the extension of the theory to artificial satellites has been completed.[4]

12.2. Prediction of Lunar Eclipses

The prediction of a lunar eclipse is very instructive for the amateur astronomer and in many respects can supplement the somewhat incomplete

---

[3] See footnotes on pages 291 and 292.
[4] LINK, F.: *Bulletin of the Astronomical Institutes of Czechoslovakia* **13**, 1 (1962).

indications in the Yearbooks. Some information on the lunar eclipses occurring through 1983 is given in Table 12.1.

Table 12.1. The Lunar Eclipses of the Period 1974–1983

(1)	(2)			(3)	(4)	(5)	(6)	(7)	(8)	(9)	(10)
4921	1974	June	4	2203	22 14	9.9	93		+ 26	− 22	Partial
4922		Nov.	29	2381	15 16	15.5	106	38	+ 128	+ 21	Total
4923	1975	May	25	2558	5 46	17.5	109	45	− 87	− 21	Total
4924		Nov.	18	2735	22 24	13.1	102	23	+ 20	+ 19	Total
4925	1976	May	13	2912	19 50	1.7	43		+ 62	− 18	Partial
		Nov.	6	3089	20 97	86%	136				Penumbral
4926	1977	April	4	3238	4 21	2.5	51		− 64	− 6	Partial
		Sept.	27	3414	8 29	93%	131				Penumbral
4927	1978	March	24	3592	16 25	17.5	109	45	+ 115	− 2	Total
4928		Sept.	26	3768	19 03	16.0	107	41	+ 73	− 3	Total
4929	1979	March	13	3946	21 10	10.5	94		+ 45	+ 3	Partial
4930		Sept.	6	4123	10 54	13.4	103	26	− 164	− 7	Total
	1980	March	1	4300	18 44	68%	122				Penumbral
		July	27	4448	17 57	28%	73				Penumbral
		Aug.	26	4478	01 42	73%	110				Penumbral
	1981	Jan.	20	4625	05 36	104%	134				Penumbral
4931		July	27	4803	04 48	6.9	80		− 71	− 21	Partial
4932	1982	Jan.	9	4979	19 56	16.2	107	42	+ 63	+ 22	Total
4933		July	6	5157	07 30	20.6	112	51	− 112	− 23	Total
4934		Dec.	30	5334	11 26	14.4	105	33	− 171	+ 23	Total
4935	1983	June	25	5511	08 25	4.1	65		− 126	− 23	Partial
		Dec.	19	5688	23 46	92%	123				Penumbral

Explanation of the columns:
(1) Current number according to TH. VON OPPOLZER in "Denkschriften der Akademie der Wissenschaften," Wien, Math.-Phys. Klasse, 52, Vienna (1887). [*Canon of Eclipses*, OWEN GINGERICH, tr., New York: Dover (1962).]
(2) Date of the maximum phase (4) in Universal Time.
(3) Day in the Julian Period diminished by 244,000 days.
(4) Universal Time of the maximum phase.
(5) Magnitude of the eclipse in digits or percentages of the Moon's diameter.
(6) Half the duration of the partial or penumbral phase.
(7) Half the duration of the total phase.
(8) Geographic longitude, counted positive to the east.
(9) Geographic latitude of the place where the center of the Earth shadow was at the zenith at the time of conjunction.
(10) Type of eclipse.

For a complete understanding of a lunar eclipse, we must first of all determine the following quantities:

1. Immersions into and emersions from the penumbra (outer contacts); beginning and end of the partial eclipse, i.e., the corresponding outer contacts with reference to the umbra; beginning and end of totality (inner contacts with the umbra); and, finally, the middle of the eclipse.

2. The graphical representation of the eclipse, i.e., the path of the Moon through both shadows in a projection onto the plane drawn perpendicularly to the axis of the shadow at the distance of the Moon.

3. The position of the terminator of the shadow on the surface of the Earth (for important phases of the eclipse or at certain particular moments).

4. The times of the transit of craters and other suitable lunar formations through the boundary of the shadow.

5. Other topocentric circumstances of the eclipse valid for the observing site, e.g., azimuth and zenith distance for important phases of the eclipse.

These data are valuable, both for the assessment of the observing conditions and for the later reduction of the measurements.

12.2.1. Numerical Determination

We start with the following data:

T = the time of the opposition in right ascension,

$\alpha_\odot$, $\alpha_\mathbb{C}$ = the corresponding right ascensions of the Sun and Moon,

$\delta_\odot$, $\delta_\mathbb{C}$ = the declinations at this time,

$\Delta\alpha_\odot$, $\Delta\alpha_\mathbb{C}$, $\Delta\delta_\odot$, $\Delta\delta_\mathbb{C}$ = the hourly changes of these coordinates,

$\pi_\odot$, $\pi_\mathbb{C}$ = the parallaxes,

$R_\odot$, $R_\mathbb{C}$ = the apparent radii.

The center of the shadow at opposition is in the position $\alpha_\odot \pm 12^h, -\delta_\odot$.

The rectangular coordinates used in our calculation are referred to a plane through the center of the shadow, perpendicular to the axis of the shadow, with the origin at 0 (the center of the shadow) and axes $+0x$ (pointing to the west) and $+0y$ (pointing to the north). At opposition the position of the Moon M_0 is given by the coordinates (see Fig. 12–5)

$$X_0 = 0 \quad \text{and} \quad Y_0 = \delta_\mathbb{C} + \delta_\odot. \tag{3}$$

One hour after the opposition the Moon is at M:

$$x' = -15 \cos \delta_\odot (\Delta\alpha_\mathbb{C} - \Delta\alpha_\odot),$$
$$y' = \delta_\mathbb{C} + \delta_\odot + \Delta\delta_\mathbb{C} + \Delta\delta_\odot \tag{4}$$

The path in the Moon's orbit after this hour will therefore be

$$\Delta s = +\sqrt{x'^2 + (y' - Y_0)^2} = -\frac{\Delta\delta_\mathbb{C} + \Delta\delta_\odot}{\sin i} =$$
$$\frac{15 \cos \delta_\odot (\Delta\alpha_\mathbb{C} - \Delta\alpha_\odot)}{\cos i}, \tag{5}$$

and its inclination to the axis $+0x$,

$$\tan i = -\frac{\Delta\delta_\mathbb{C} + \Delta\delta_\odot}{15 \cos \delta_\odot (\Delta\alpha_\mathbb{C} - \Delta\alpha_\odot)}. \tag{6}$$

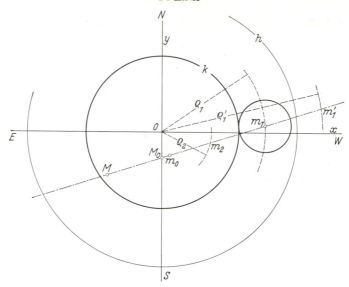

Fig. 12–5 The prediction of a lunar eclipse.

At the middle of the eclipse the distance of the Moon from the center of the shadow becomes

$$\rho_0 = (\delta_{\mathbb{C}} + \delta_{\odot}) \cos i, \tag{7}$$

and the path in the Moon's orbit between opposition and center of the eclipse is

$$\overline{m_0 M_0} = (\delta_{\mathbb{C}} + \delta_{\odot}) \sin i,$$

from which we obtain for the time of the middle of the eclipse

$$t_0 = T - \frac{(\delta_{\mathbb{C}} + \delta_{\odot}) \sin^2 i}{\Delta \delta_{\mathbb{C}} + \Delta \delta_{\odot}}. \tag{8}$$

The beginning and end of the partial phase occur when the distance of the Moon from the middle of the shadow is $\rho_1 = \sigma + R_{\mathbb{C}}$.

The length of the Moon's path is therefore

$$\overline{m_0 m_1} = + \sqrt{\rho_1^2 - \rho_0^2},$$

and the corresponding times are

$$t_{1,4} = t_0 \pm \frac{\sqrt{\rho_1^2 - \rho_0^2}}{\Delta \delta_{\mathbb{C}} + \Delta \delta_{\odot}} \sin i. \tag{9}$$

Beginning and end of totality occur correspondingly for $\rho_2 = \sigma - R_{\mathbb{C}}$ and therefore at

$$t_{2,3} = t_0 \pm \frac{\sqrt{\rho_2^2 - \rho_0^2}}{\Delta \delta_{\mathbb{C}} + \Delta \delta_{\odot}} \sin i. \tag{10}$$

Finally, we obtain for the outer contacts with the penumbra, for $\rho_1' = \sigma' + R_{\text{C}}$,

$$t_{1,4}' = t_0 \pm \frac{\sqrt{\rho_1'^2 - \rho_0^2}}{\Delta\delta_{\text{C}} + \Delta\delta_{\odot}} \sin i. \tag{11}$$

In these calculations the values σ and σ' increase by 2% for the reasons discussed in Sec. 12.3.1. The magnitude of the eclipse follows from Eq. (2).

12.2.2. Graphical Representation

The graphical representation of a lunar eclipse provides an excellent illustration of this numerical calculation. We choose the same coordinate system as above and a suitable scale, say, $1' = 5$ mm. Then we draw circles about the center, with the radii 1.02σ (umbra) and $1.02\sigma'$ (penumbra). We enter the positions M_0 and M of the Moon on the graph, that is, X_0, Y_0 and x', y'.

The line joining these two points represents the path Δs covered by the Moon in one hour. If the opposition T occurs m minutes after a full hour of Universal Time (e.g., 22 hr), the position of the Moon in its orbit at the hour is displaced backwards by $m\Delta s/60$ relative to its position in opposition. By marking the distance Δs to both sides of the point so determined we find the position of the Moon at the full hours of UT, and by further subdivision also the position after 10 min, or even after 5 min.

We now draw auxiliary circles about the middle of the shadow with the radii $1.02\sigma' + R_{\text{C}}$ and $1.02\sigma \pm R_{\text{C}}$. Their intersections with the Moon's path indicate on the latter the moment of the immersion and emersion into the penumbra, the beginning and end of the partial phase, and the beginning and end of the total phase. The position angles of the points of contact with the shadow are directly read off in angular measure. The middle of the eclipse is found at the foot of the perpendicular drawn from the center of the shadow to the Moon's path.

12.2.3. The Position of the Terminator of the Shadow on the Surface of the Earth

Sketching in the terminator of the Earth's shadow on a map—preferably on the Mercator projection—is a very interesting task for the amateur astronomer. Since the terminator is the boundary between day and night, we start from tables in the Astronomical Ephemerides that contain the times of rising and setting of the Sun for various latitudes throughout the year. If we have no almanac for the current year, one from an earlier date will suffice because the differences are small and can be neglected in the graphical representation.

The times given in such tables are converted into angular measure, and then give the geographic longitude of the places lying on the corresponding latitude at which the boundary of the shadow is situated at 0^h UT. The geographic longitude is reckoned from Greenwich, and is taken as positive to the east. For another time H (UT), the terminator moves by H^0 to the west. We are thus able to draw the position of the terminator for any moment of the eclipse.

The meteorological situation and perhaps also the geographic conditions have a certain influence on the illumination within the shadow of the Earth on the Moon. As a rule, however, this concerns only part of the whole terminator.

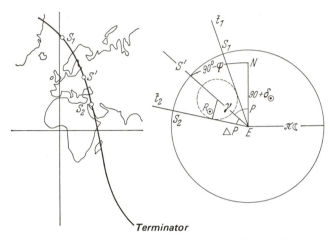

Fig. 12–6   The position of the terminator on the Earth's surface.

In order to determine this effective part of the terminator we start from a given position in the shadow defined by the distance γ from the middle of the shadow and by the position angle P reckoned from north. An observer at this point would see the Earth with radius $\pi_{\mathbb{C}}$ (see Fig. 12–6), and the geometric position of the Sun behind the Earth would be determined by the same polar coordinates, i.e., by γ and P. The terminator of the shadow lies on the disc of the Earth and its effective part is subtended by the tangents t_1 and t_2. It only remains to calculate the geographic latitude of the halfway point S' on the terminator, using the equation

$$\sin \varphi = \cos \delta_{\odot} \cos P. \tag{12}$$

The geographic latitudes of both endpoints of the effective arc of the terminator are

$$\sin \varphi_{1,2} = \cos \delta_{\odot} \cos (P \pm \Delta P), \tag{13}$$

in which

$$\sin \Delta P = \frac{R_{\odot}}{\gamma}. \tag{14}$$

It is possible to determine in this way the position of the effective terminator of the shadow for any phase of the eclipse, e.g., for the beginning and end of the partial or total phase, or also for the middle of the eclipse. To some extent the choice of the point on the Moon plays a role. In the case of contacts with the shadow such points will usually lie on the Moon's limb. In every case in which the distance of the supposed point goes beyond the limit $\gamma \leq R_\odot$, the whole 40,000-km-long terminator comes into play.

12.2.4. Graphical Determination of the Transit of Craters

To determine the approximate situation during the transit of a crater through the boundary of the shadow we make use of the same sketch as in Sec. 12.2.3 (Fig. 12–6). We draw on tracing paper the disc of the Moon at the same scale as in the previous figure and make its center the origin of a system of selenographic coordinates ξ and η, and further indicate the direction of the Moon's axis to the north point, which is given in the Ephemerides by the position angle p. Furthermore, we mark on the Moon's disc the position of the lunar formation as given in the Berlin system (see Table 12 in the Appendix), or of other suitable points. This would be the view of the Moon as seen from the Sun, if during eclipse its selenographic longitude $\lambda_\odot$ and latitude $\beta_\odot$ were zero. However, owing to the influence of the libration this is not the case, particularly not for $\lambda_\odot$, while $\beta_\odot$ during an eclipse is very small (according to CASSINI's law) and can therefore be neglected within the accuracy required. There remains therefore only the corrections of the coordinates, according to the equations

$$\xi' = \cos b \sin (l - \lambda_\odot),$$
$$\eta' = \eta. \tag{15}$$

We now shift the transparent picture of the Moon, properly orientated to the north, along the Moon's orbit and determine the times of transit of the various craters through the boundary of the shadow (1.02σ).

Although this method gives only approximate results, it serves for the arrangement of the observing program which aims at the determination of the enlargement of the shadow (see Sec. 12.3). Greater accuracy could even be harmful because it might prejudice the observer.

12.3. Enlargement of the Earth's Shadow

12.3.1. General Considerations

The magnification of the Earth's shadow during lunar eclipses has been well known for about three centuries. LAHIRE (1707) determined its approximate value on the basis of his observations as 1/41 (2.5%), and was followed later by MAEDLER, BROSINSKY, HARTMANN, and others. Theoreticians like SEELIGER and HEPPERGER made great efforts toward the end of the last century to find an explanation of this phenomenon and devised the first photometric theories of lunar eclipses. These efforts, however, were premature because the structure of the Earth's atmosphere was not sufficiently well known at that time, so that the necessary basis of an adequate theory and its comparison with observations were still lacking.

The problem of the enlargement of the Earth's shadow is concerned with the difference between the theoretical value of the radius σ of the umbra, according to the HIPPARCHUS Eq. (1), and the value σ_1, which has been determined from the transits of craters and other lunar formations, and is about 2% larger. For this reason an increased value of the radius of the umbra is used in the calculation of the Ephemerides. Further studies then yield the accurate value σ_1, which differs from one lunar eclipse to another, and perhaps even along the circumference of the boundary of the shadow of a given eclipse.

From the physical point of view, the concept of the boundary of a shadow and the value σ are of no importance, since the density D of the shadow varies as a continuous function of the distance γ from the center, and does so from the border of the penumbra ($D = 0$) up to the middle of the shadow ($D \approx 4$). In the neighborhood of σ, the density varies rapidly. As the contrast theory of KÜHL demonstrates, a line of separation is seen in such cases at approximately the distance where the curve $D = f(\gamma)$ shows a point of inflexion, whose position appears to be connected with the layer of meteoric dust at an altitude of about 120–150 km.[5]

12.3.2. Determination of the Enlargement of the Shadow by the Observation of the Transit of the Craters through the Terminator

We determine during a lunar eclipse the moment at which certain craters or other suitable lunar formations pass through the boundary of the shadow. We know their position with reference to the center of the Moon, and we can calculate the position of the Moon's center relative to the center of the shadow. In this way we determine the distance of the crater from the middle of the shadow of the time of its transit, i.e., the quantity σ_1. Its comparison with the geometric value σ yields the enlargement of the shadow.

In order to guarantee a successful outcome of these observations, certain technical conditions must be fulfilled apart from the allowance made for weather conditions:

1. A small or average telescope (aperture 5–15 cm), and a magnification not exceeding 30–100, will suffice to produce a sufficiently clear image, with a sufficiently large field of view to make a survey and an appropriate selection of the craters.

2. An assistant who keeps records, reading the clock with an accuracy of 1–2 sec, and noting the times of transit, as well as any additional remarks of the observer. The latter is therefore relieved of the writing, is not disturbed by light, and can concentrate completely on the observations. In view of the relative rarity of lunar eclipses, the help of such an assistant will be invaluable.

3. Complete Ephemerides for the orientation during the transit of the crater prepared according to Sec. 12.2.4. This facilitates the observations,

---

[5] KÜHL, A.: *Physikalische Zeitschrift* **29**, 1 (1928).

provides the observer with the times, and familiarizes him with the position of the relevant craters. Generally, an accuracy of about one minute will be sufficient to prevent bias on the part of the observer. The announcement of the expected transit is also made by the assistant.

The required accuracy of the time measurements is nowadays attained by any high-quality wrist watch with a large second hand. A correction of the time is carried out before and after the eclipse with the help of the radio time signals. The actual observing accuracy of time determinations during eclipse is usually smaller. In the case of smaller and brighter craters, we shall be satisfied with a single determination of the transit through the Earth's shadow. For larger craters, if at all included in our observing program, we determine the transit of both rims and also of the estimated center, which frequently is characterized by a central peak.

The actual determination of the transit is done by estimation, which to a certain degree is, of course, subjective. The accuracy and the homogeneity of the observational series are affected if we make great efforts to follow the crater too far into the shadow. HOFFMEISTER[6] proposes to determine for every crater three moments: T_I, when the crater begins to disappear; T_{II}, when it most probably traverses the boundary of the shadow; and T_{III}, when it is certain that its transit has been completed. We then determine the weighted mean of these three observations as follows:

$$T = \frac{T_I + 2T_{II} + T_{III}}{4}. \tag{15'}$$

12.3.3. Reduction of the Measurements of the Enlargement of the Shadow

If we do not want to leave the reduction of the measurements to an observatory, we can perform it ourselves with the help of KOZIK's[7] method, which is adapted to the modern Ephemerides of the Sun and the Moon.

KOZIK chooses the following rectangular system of spatial coordinates, placing their origin in the center of the Earth: the axis $+0z$ is directed along the line joining Sun and Earth, the axis $+0y$ to the celestial North Pole, and the axis $+0x$ lies in the equatorial plane and in the direction of the Moon's motion around the Earth. The unit of length is the Earth's radius. In this system the Moon has the following coordinates:

$$x_{\leftmoon} = \frac{\cos \delta_{\leftmoon} \sin (\alpha_{\leftmoon} - \alpha_1)}{\sin \pi},$$

$$y_{\leftmoon} = \frac{\sin (\delta_{\leftmoon} - \delta_1)}{\sin \pi} + 0.008726(\alpha_{\leftmoon}^\circ - \alpha_1^\circ)x_{\leftmoon} \sin \delta_1, \tag{16}$$

[6] HOFFMEISTER, C.: Zur Beobachtung der Mondfinsternisse. *Die Sterne* **29**, 166 (1953).
[7] KOZIK, S. M.: *Bulletin Tashkent Observatory* **2**, 79 (1940); or Z. KOPAL, ed.: *Physics and Astronomy of the Moon*, p. 167, New York: Academic Press (1962).

where α_1 and δ_1 are the equatorial coordinates

$$\alpha_1 = \alpha_{\odot} + 12^h \quad \text{and} \quad \delta_1 = -\delta_{\odot}.$$

The selenographic coordinates of the lunar craters are either spherical or rectangular ones and are given by the equations

$$\begin{aligned}
x_0 &= r_{\mathbb{C}} \cos \beta \sin \lambda, \\
y_0 &= r_{\mathbb{C}} \sin \beta, \\
z_0 &= r_{\mathbb{C}} \cos \beta \cos \lambda.
\end{aligned} \tag{17}$$

with

$$m_{\mathbb{C}} = 0.27252$$

We transfer these coordinates into KOZIK's system by rotating the axes and displacing the origin, which leads to the equations

$$\begin{aligned}
x &= x_{\mathbb{C}} + a_x x_0 + b_x y_0 + c_x z_0, \\
y &= y_{\mathbb{C}} + a_y x_0 + b_y y_0 + c_y z_0.
\end{aligned} \tag{18}$$

We now introduce the auxiliary quantities

$$\begin{aligned}
a_x &= -\cos \lambda_{\odot} \cos p - \sin \lambda_{\odot} \sin p \sin \beta_{\odot}, \\
b_x &= \sin p \cos \beta_{\odot}, \\
c_x &= \sin \lambda_{\odot} \cos p - \cos \lambda_{\odot} \sin p \sin \beta_{\odot}, \\
a_y &= \cos \lambda_{\odot} \sin p - \sin \lambda_{\odot} \cos p \sin \beta_{\odot}, \\
b_y &= \cos p \cdot \cos \beta_{\odot}, \\
c_y &= -\sin \lambda_{\odot} \sin \beta - \cos \lambda_{\odot} \cos p \sin \beta_{\odot},
\end{aligned} \tag{19}$$

in which $\lambda_{\odot}$ and $\beta_{\odot}$ are the selenographic coordinates of the Sun, while p is the position angle of the rotational axis of the Moon, projected onto the plane drawn through the center of the Moon perpendicular to the axis $0z$.

All quantities in the preceding equations are given in hours of Universal Time, at intervals of whole hours; if they are not, they must be interpolated. In this way we determine the coordinates of the observed craters for whole hours and by interpolation also their values for the observed moment of transit through the boundary of the shadow. The distance from the center of the shadow then follows from the equation

$$r_0 = \sqrt{x^2 + y^2}, \tag{20}$$

and the position angle, reckoned from the equator

$$\tan \psi = \frac{y}{|x|}. \tag{21}$$

The theoretical radius of the shadow results from the dimensions of the geoid as

$$r_0 = 1 - \operatorname{cosec} \pi_{\mathbb{C}} \tan (R_{\odot} - \pi_{\odot}) - 3.376 \cdot 10^{-3} \cdot \cos^2 \delta_{\odot} \sin^2 \psi. \tag{22}$$

A comparison with the value r yields the enlargement of the shadow. A large number of accurate determinations, combined in groups at intervals of, say, $10°$ of the angle ψ, indicate a distinct flattening of the shadow which exceeds the theoretical value. We therefore make a graphical representation for r_0 and ψ, from which the values m and n can be determined from the equation

$$r_0 = m - n \sin^2 \psi \qquad (23)$$

These constants (m, n) characterize the size of the shadow and its flattening for the particular lunar eclipse.

12.4. Photometry of Lunar Eclipses

12.4.1. General Problems in the Photometry of Lunar Eclipses

In the ideal case, the photometry of lunar eclipses would aim at the complete determination of the behavior of the density of the Earth's shadow in all those parts of the Moon's plane through which the Moon passes in the course of the eclipse, preferably by plotting isophotes. The practical application of this ideal therefore consists in the measurement of the brightness of certain lunar details during the different phases of the eclipse and outside of it. The density of the shadow can be calculated from the ratio between the brightness i during eclipse and of the value I outside eclipse, given by the equation

$$D = -\log_{10} \frac{i}{I}. \qquad (24)$$

The quantity D is sometimes also expressed in stellar magnitudes by the relation

$$\Delta m = 2.5 D. \qquad (25)$$

The numerical relationships between D in ordinary logarithms or stellar magnitudes and the corresponding brightness i (when $I = 1$) are given in the following table:

D	0	1	2	3	4	5	6
Δm	0	2.5	5.0	7.5	10.0	12.5	15.0
$i \ (I = 1)$	1	0.1	0.01	0.001	0.0001	0.00001	0.000001

The study of the theory (see Sec. 12.1.2) has shown, and the measurements have confirmed it, that we may expect shadow densities up to $D = 5$. This means that our measurements ought to bridge intensity ratios of about $1:100,000$ in a reliable manner, and this is indeed a very delicate problem. It demands an exact calibration of the photometric instruments and an adaptation of their sensitivity over this range.

The measurements of the two quantities i and I [see Eq. (24)] cannot be made simultaneously, but are an hour or more apart. During this time the

Moon changes its position in the sky; thus the extinction also changes, which, moreover, can also be disturbed in its normal course by changes in the meteorological conditions (see Sec. 5.3.2). The elimination of the influence of the extinction is therefore an important problem for every photometric method applied to lunar eclipses.

The brightness measurements of i and I are also affected by the scattered light which originates in the optics of the instrument and in the Earth's atmosphere. Suppose we measure immediately after the beginning of the partial phase a point on the Moon that lies in the umbra. Then the other part of the Moon, the light of which is weakened relatively little by the penumbra, illuminates the optical surface of the instrument (dust, diffraction, etc.), as well as the air column between Moon and observer. In this way a parasitic scattered light is present, which is superimposed on the relatively faint light of the measured details in the umbra, and thus apparently diminishes the density of the shadow.

The density at the border of the umbra changes rapidly with the distance from the center, and therefore also with time. Thus, we must measure a small surface area of the Moon (smaller than one square minute of arc) relatively rapidly; photographic methods are therefore excluded since they require longer exposure times (exceeding one min).

Finally, it is necessary to make the measurements in sufficiently mono-chromatic light with the help of glass, gelatin, or interference filters, since the density of the shadow depends strongly on the color of the light, as the theory shows (Sec. 12.1.2). In particular, one must be cautious of blue filters that transmit some red light, and which at first glance appear harmless enough for normal photometry or even for photometry in the penumbra. In the umbra, however, where the blue light is weakened by a factor of about 100 relative to the red light, such traces of red light transmitted by the blue filter can affect the measurements appreciably.

We have made it a point to discuss here in detail all the basic problems of the photometry of lunar eclipses, in order to show the difficulties that can be overcome only by special methods. Persevering and accurate measurements of the shadow density should therefore be attempted only by an experienced and well-equipped amateur astronomer.

12.4.2. Estimates by Danjon's Method

Early records show us that there are bright or dark, grey or colored lunar eclipses. There are even records that indicate that the Moon disappeared completely for shorter or longer intervals. In the time of KEPLER, at the beginning of the Seventeenth Century, no fewer than four such eclipses were reported, and other similar cases were recorded in other centuries. Even one of the recent lunar eclipses (24 June 1964) belongs to this category. On the contrary, sometimes such clear eclipses occurred that the observer doubted the reality of the phenomenon.

When DANJON was occupied with the collection and the reduction of all eclipse observations, with special reference to brightness and coloring, he introduced a scale of four steps for the classification of these eclipses.[8]

Step L	Description
0	Very dark eclipse, Moon nearly invisible, particularly at the middle of totality
1	Dark eclipse, grey or brownish coloring, details difficult to distinguish
2	Dark-red or rust-colored eclipse with a dark spot at the center of the shadow, limb regions of the shadow rather bright
3	Brick-red eclipse, the shadow frequently surrounded by a bright or yellowish border
4	Copper-red or orange-red, very bright eclipse with a bluish and very bright border zone.

We see that the scale is based on the simultaneous gradation of brightness and coloring, which relies both on the observational material and on theoretical findings. Theoretically (see Sec. 12.1.2) we should expect a conspicuously red-colored Moon, assuming an ideal Earth atmosphere. Every pollution of the atmosphere by fog or volcanic or meteoritic dust causes further diminution of the sunlight. Unlike the ideal atmosphere, however, where the diminution is strongly selective, the effect on the sunlight should be practically neutral. Extremely dark eclipses will therefore be less colored, especially so since the eye at low brightnesses does not distinguish colors very well.

The experiences made with DANJON's classification of the observing material, covering the past three centuries, are very satisfactory, and so are those of DE VAUCOULEURS over the past 50 years. DANJON was able to find the relation between the brightness of the lunar eclipse and the solar activity in the course of its 11-year cycle.[9] At the beginning of the cycle the lunar eclipses are darker ($L = 0$ to 1), then they brighten up ($L = 1$ to 2), and shortly before the new minimum and the beginning of another cycle, they are usually brightest ($L = 3$ to 4). The minimum of solar activity corresponds in lunar eclipses to a sudden drop of brightness (see Fig. 12–7), while the maximum does not seem to affect the brightness curve at all. The 20 most recent lunar eclipses in which it was possible to measure the shadow density in different colors have been tested on the Danjon scale; the results were generally good, particularly with respect to the coloring.

This normal behavior of the brightness of lunar eclipses is, however, sometimes disturbed, for example, by eruptions of volcanoes, which pollute the atmosphere with a large quantity of fine volcanic ash. This is carried up into the stratosphere where it floats for a long time—sometimes for several months—and is spread over the whole Earth by atmospheric circulation. For instance, the above-mentioned darkness of the lunar eclipse on 24 June 1964 took place 15 months after the catastrophic eruption of the volcano

[8] DANJON, A.: *C.R. Academy of Sciences*, Paris, **171**, 127 (1920).
[9] DANJON, A.: *C.R. Academy of Sciences*, Paris, **171**, 1207 (1920).

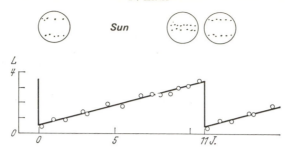

Fig. 12–7 DANJON's relationship. The top part of the figure shows the distribution of the sunspots.

Agunk on the island of Bali in Indonesia; this darkness was probably caused by the eruption as such, or by a combination of the volcanic effect with the minimum of solar activity that took place in 1964.

The meteor activity, too, has considerable influence on the brightness of lunar eclipses, as has been shown by ŠVESTKA.[10] The pollution of the atmosphere by meteoritic dust following the maximum of the large meteor swarms (Perseids, Lyrids, Geminids) causes a decrease of brightness by 1.2 steps. This lasts from one to three months, until the atmosphere is cleared again by the slow sinking of the meteoritic particles.

An explanation of DANJON's relation between brightness of lunar eclipses and solar activity probably can be found in the luminescence of the Moon. In the umbra, where the direct light from the Sun is very weak, the lunar surface shines by luminescence that is due to the corpuscular solar radiation, the so-called solar wind. Owing to the lack of a lunar atmosphere, corpuscles are unhindered in bombarding the lunar surface and, since this radiation is not propagated in a straight line like the light, it can also penetrate into the umbra. The solar corpuscles are mainly emitted from the active zones of the Sun, that is, where there are sunspots. According to SPÖRER's law these zones move during the 11-year cycle in such a way that they appear after the minimum in the high heliographic latitudes, and thus the corpuscles have then only little chance to reach our Earth and the Moon. In the later course of events these zones (on both solar hemispheres) move toward the equator and shortly before the minimum are nearest to it. This increases the number of corpuscles that fall onto the Moon and thus also increases the luminescence of its surface. At the minimum the zone of spots suddenly moves up again into high heliographic latitudes.

These discussions show that an observation of the brightness of lunar eclipses with the help of DANJON's classification will always be important for further study of the relationships mentioned above (Sun, meteors, volcanoes) and that, in addition, its simplicity makes it particularly suitable for the observing work of the amateur astronomer. The estimates would be carried

[10] ŠVESTKA, Z.: *Bulletin of the Astronomical Institutes of Czechoslovakia* **2**, 41 (1950).

out with a low-power telescope, perhaps binoculars. In the course of the lunar eclipse DANJON's scale will be particularly valuable for the study of the total phase, which is not symmetrical with respect to the middle of the eclipse. We therefore recommend continuing the estimates during the whole total phase. During the partial phase the light from that part of the Moon that is still in the penumbra and that therefore will be many times brighter than the illumination of the rest is rather disturbing. Here, in the partial phase, estimates from $L = 0$ to 2 need not be real. On the other hand, estimates of $L = 3$ to 4 have great importance, since the disturbing effect here acts in the opposite sense. Indeed lunar eclipses of this type are, in reality, very bright.

12.4.3. Observation of the Visibility of Lunar Formations in the Umbra

Within the umbra the illumination by the Sun is reduced to a small fraction (about $\frac{1}{10,000}$) of the original value at Full Moon. The majority of details on the Moon's surface are therefore indistinct or quite invisible. Nevertheless, we can see during many eclipses the outlines of the Mare and a few brighter objects, such as Aristarchus, Copernicus, Tycho, and others. However, these are not all equally visible, and there are cases of anomalous visibility which actually led even experienced observers to the assumption of lunar volcanoes. Until recently such observations were refuted, and relegated into the realm of optical illusions and subjective impressions of the observers.

More recently, however, it became possible to provide at least a partial explanation, taking into account the lunar luminescence. Most of the terrestrial minerals exhibit a visible luminescence radiation when they are subjected to shortwave ultraviolet light, X-rays, or corpuscular radiation. The lunar surface is exposed to the solar radiation without any protective atmosphere, a favorable condition for luminescence radiation. Luminescence on the Moon has actually been discovered at some eclipses,[11] and also during noneclipse it was possible to study it as a residual intensity of the core of the strong FRAUNHOFER lines in the lunar spectrum.[12]

In the inner parts of the penumbra, where the direct solar illumination is reduced, e.g., to 0.1, the penumbra should have the density 1. In reality, however, we measure an essentially smaller density there, frequently as low as 0.3, which is twice the value of the illumination expected by the theory. Such a discrepancy could not be explained in the normal way and it became necessary to involve luminescence. The latter is excited by rays from the solar corona and the chromosphere, which at this particular optical phase (partial eclipse of the Sun by the Earth) is relatively less occulted than the solar disc. The contribution of the luminescence to the lunar radiation, which at Full Moon outside an eclipse reaches a few percent only, gains more importance

[11] See LINK, F.: C.R. Academy of Sciences, Paris, **223**, 976 (1946).
[12] See DUBOIS, J.: Rozpravy ČSR Akad. Věd. Prague **69**, No. 6 (1959).

in the penumbra and can cause the above-mentioned decrease in the density of the shadow. The corpuscular radiation, which because of terrestrial magnetism does not propagate in a straight line, can also penetrate to a considerable degree into the penumbra, perhaps even into the umbra. Those lunar formations, which show stronger luminescence than their surroundings because of their composition, become in this way relatively more plainly visible in the penumbra and umbra.

A careful comparison of the visibility of single formations during the different phases of the eclipse therefore can supply information about their composition; this is the more important since a detailed measurement of the density of the shadow on a large scale is difficult to obtain. During such a study it is necessary to take into account the relative visibility with respect to the distance from the middle of the shadow. A complete method for this has not yet been worked out, but nevertheless the amateur astronomer can obtain valuable results in this field.

Confronting the intensity of the observed phenomena with the available excitation energy of solar radiations, together with the quantum efficiency of lunar materials brought by Apollo missions, a serious conflict arose because the assumed luminescence on the Moon is several orders too high as compared with laboratory tests. Therefore the explanation is still pending [cf. The Moon 5, 265–285, 1972].

12.4.4. Integral Photometry in General

The term integral photometry indicates the measurement of the illumination received by the Moon at an observing site on a plane that is perpendicular to the incident rays. The amount of illumination depends essentially on the density of the shadow in that region in which the Moon is found at the given time; it represents an average value within this region of the shadow. If the Moon's disc had everywhere the same albedo, and if the isophotes of the Earth's shadow were circular, then the light curve representing the behavior of the total brightness would be symmetrical with respect to the middle of the eclipse, as is demonstrated in Fig. 12–8(a).

In reality, the distribution of the lunar Mare and of the bright continents

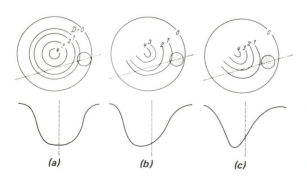

Fig. 12–8 The total photometry of lunar eclipses. At the top are shown the isophotes in the plane of the Moon; below are the corresponding light curves.

(a)   (b)   (c)

is very irregular. We note that in the northern hemisphere the dark spots dominate as opposed to the southern formations (Tycho), and this circumstance alone causes a definite, predictable asymmetry of the light curve. Furthermore, the isophotes of the shadow are usually not circular, but show previously unknown deformations, which is another cause of the asymmetry of the light curve. There are two possibilities. If it happens that the light curve is asymmetrical owing to the distribution of the albedo on the disc, as shown in Fig. 12–8(b), we cannot of course draw any conclusions as to the form of the isophotes, and the case is uninteresting. The other case, however, deserves much more attention, that is, when the observed asymmetry of the light curve is in the opposite sense to the distribution of the albedo and the whole phenomenon shows a very conspicuous deformation of the isophotes of the Earth's shadow as seen in Fig. 12–8(c).

This discussion shows that the total photometry, although not so powerful as the photometry of details in the shadow, is nevertheless able to provide interesting information in certain cases—with procedures that are well within the reach of amateur astronomers. It is, however, important to extend the measurements to the advanced partial phase and to continue them as far as the sensitivity of the method permits, during totality. ZIMMERMANN[13] proposed measuring the total brightness, particularly at the beginning and at the end (inner contacts of the Moon with the umbra), to obtain a general characterization of the lunar eclipse. If such a brightness is determined directly, i.e., without extrapolation, this procedure is justified and characterizes essentially the outer part of the umbra.

12.4.5. Visual Total Photometry

To carry out visual total photometry we may use an instrument designed by RICHTER,[14] the silver-sphere photometer. Here the image of the Moon appears in a convex mirror practically as a point source and can be compared directly with brighter stars. The mirror, with a radius of curvature R and the distance r from the eye of the observer, shows a virtual image of the source of light which is weakened in the ratio

$$k\,\frac{R^2}{r^2},$$

in which k is a constant that depends on the nature of the mirror surface.

If we now observe the image of the Moon on this spherical surface at the distance r, so that its brightness is equal to that of the directly observed star, we have the following relation between the magnitude of the Moon and that of the star:

$$m_{\mathbb{C}} = m_* + 5\log_{10}\frac{R}{r} + k + \Delta\,\text{ext}, \tag{26}$$

[13] ZIMMERMANN, G.: Astronomische Nachrichten **247**, 211 (1933).
[14] RICHTER, N.: Zeitschrift für Astrophysik **21**, 249 (1942).

where the term Δ ext denotes the difference between star and Moon caused by the effect of extinction.

The practical application of this method offers several possibilities. RICHTER used, as far as possible, faultless and colorless spherical Christmas tree decorations. Ball bearings can be used, particularly when they are freshly chromium plated and polished. We observe this sphere through an aperture of, say, 1 cm, whose position defines one end of the distance r, while the other end is positioned at the center of the sphere. This distance can, according to SCHUBERT,[15] best be determined with a measuring tape. The sphere is for this purpose attached to a vertical beam, about one foot above eye level. The observer's position is such that he sees the image of the Moon and the comparison star close to each other. Sometimes on humid nights the sphere may become misted over, which can be reduced by warming it with a bright lamp or an electric heater.

The application of Eq. (26) requires a knowledge of the constant k. Since, however, we are mainly concerned with the relative shape of the light curve during the eclipse—referred to the magnitude of the Moon before and after eclipse—it is not necessary to determine this value. The magnitudes of the comparison stars (visual or photovisual) can be found in a catalog. If we choose only stars that are as close as possible to the same altitude above the horizon as the Moon, we can neglect the determination of the extinction.

12.4.6. Photoelectric Photometry

The sensitivity of some electrical exposure meters used in photography is sufficient to give measurable values during a lunar eclipse. For this purpose, we bring the sensitive surface of the exposure meter near the exit pupil of an astronomical telescope, which is equipped with an eyepiece of the lowest magnification, to ensure that the whole image of the Moon appears within the field of view. The deflection of the exposure, if this has a linear scale, corresponds to the brightness of the Moon. The sensitivity can be reduced by diaphragming the objective and by setting the values of the diaphragm directly, according to the scale of the exposure meter.

This method has not yet been tested, but it appears to contain all essentials for obtaining usable results, and as far as objectivity is concerned, it far exceeds visual estimates.

12.5. Photography of Lunar Eclipses

12.5.1. Black and White Photography

If we possess one of the familiar small cameras we might try to obtain a series of photographs, perhaps a continuous record of the phenomenon, by keeping the camera in a fixed position. It is essential here to determine the correct direction in which the camera is to point, in order not to lose the Moon from the field of view during the whole duration of one hour or

[15] SCHUBERT, M.: *Die Sterne* **27**, 40 (1951).

more. Furthermore, it is important to have the camera set up rigidly, so that it does not move during the repeated exposures and thus disturb the regular sequence of the Moon's images on the film. If the particular camera does not allow the coupling between shutter and film movement to be disconnected, we must expose the film by carefully lifting the lens cover, arranging the diaphragm in such a way that (1) exposures are not too short, (2) they can be carried out by hand, and (3) they can be reliably repeated. Plate cameras are therefore more suitable, the automatic release at the objective lens being used repeatedly.

The intervals between the exposures must be chosen so that the single images do not overlap. The Moon moves through a whole lunar diameter in about 2 min, and an interval from 3 to 5 min is therefore quite sufficient to give a reliable separation of the resulting images. As far as the aesthetic effect is concerned, we may choose any suitable interval, preferably about 5 min. Such a series is taken only at partial eclipses, or in the case of total eclipses only in their partial phase.

Total lunar eclipses are studied by continuous photography. Then the picture shows a connected strip of images whose widths correspond to the diameter of the Moon, and whose density changes with the position of the Moon within the shadow. If we expect some overexposure at the beginning and the end of totality, we use a diaphragm and note the time of its application. Such a strip gives a good survey of the behavior of the shadow density at different distances and position angles, particularly when the Moon is high above the horizon during the eclipse, so that low-altitude extinction is absent. If we have two equal cameras or a stereo camera at our disposal, we use panchromatic material with two different color filters, preferably a blue and an orange.

Continuous photography is also successful in the penumbra, for which color filters are of no importance. However, it is necessary to apply diaphragms that can be tested at another occasion at Full Moon.

It is also possible to use cameras of long focal length or even projection behind the eyepiece; these achieve considerably larger but less bright images. If it is known, for instance, that at an aperture ratio $f = 1:15$ and photographic material of the sensitivity 17/10 DIN (40 ASA) the correct exposure for the Full Moon is 1/25 sec, then in the case of the eclipse, for a shadow density $D = 4$, we must choose the exposure time at least 10,000 times longer, i.e., make it at least 7 to 10 min. If we are going to use a very sensitive emulsion, e.g., 32/10 DIN, the exposure time shortens to about 1 min.

Long exposure times naturally result in somewhat blurred images. If we follow the Moon with the camera, the movement of the umbra on the Moon's disc causes photometric blurring. If we succeed (and this is difficult) in following the umbra itself with the camera, the details of the Moon become blurred and the photometric blurring is replaced by a lunar blurring.

If we consider the large differences in illumination between penumbra and umbra, we realize that it is normally impossible to obtain both shadows in one picture. Here we are helped by a little trick: we place a neutral filter in the focal plane immediately in front of the photographic emulsion. This filter has been cut from a gelatin film of the density of about 2.5, in which we have made, with the point of a compass, an opening of the size of the umbra at the focus of the camera ($r = f \sin \sigma$). At the beginning of the eclipse we position the image of the Moon in such a way that the part of the Moon's disc that lies in the umbra appears in the opening of the film, the boundary of the shadow running along the rim of the opening. It is then possible, provided the exposure time is long enough, to photograph simultaneously both parts of the Moon, i.e., the one in the umbra and the one in the penumbra. This arrangement is particularly important for photographic photometry of the Moon, where it has been used successfully.[16] If we make alternate exposures with a blue and with an orange filter, we obtain a conspicuously selective reduction of brightness in the umbra relative to a neutral reduction in the penumbra. This is best seen on positive prints which have been brought to the same density by suitable exposure and development. Experimenting here leads to interesting results. The work is facilitated if the finder telescope is equipped with a similar focal device as the camera.

12.5.2. Color Photography

The value of color photographs is not very great if we remember that the color of the image strongly depends on the exposure time, although some authors have achieved quite satisfactory results.[17] Here, too, the use of the above-mentioned focal diaphragm can provide a good compromise between both shadows. It is important, however, to test the exposure time beforehand, during another Full Moon.

12.5.3. Cinephotography

With a film camera equipped with a suitable teleobjective ($f > 15$ cm and size of image > 1 mm), and capable of taking single exposures, we can approach the problem of taking a slow-motion film of a lunar eclipse. Preferably the camera is attached to an equatorially mounted telescope in such a way that the position of the image in the field of view of the camera cannot change at the same setting in the guiding telescope.

If we expose at intervals of about 10 sec (the much shorter exposure time during the partial phase having previously been determined at Full Moon), one hour of this phase furnishes us with 360 pictures. At a projection speed of 15 pictures per second we thus obtain a film of the eclipse, which is accelerated by the factor 150 and therefore lasts a little less than half a minute. Here, too, the amateur astronomer is confronted with a whole range of experimental possibilities. We advise the help of an assistant for the guiding of the telescope and for the routine work of the exposures.

[16] See LINK, F., and GUTH, V.: *Journal des Observateurs* **19**, 129 (1936).
[17] See WALDMEIER, M.: *Die Himmelswelt* **52**, 89 (1942). Also GRAMATZKI, J. H.: *Die Himmelswelt* **52**, 90 (1942).

13 | The Observation of Star Occultations

A. Güttler

Revised by *W. D. Heintz*

13.1. General Remarks

The visual observation of star occultations by the Moon involves the determination of the time of immersion and emersion and requires very few auxiliary devices. Not only is the impressive experience of the motion of our Moon a rich reward in itself, but every observer has here the opportunity of participating in a program that is of great importance for positional astronomy. The combination of many observations leads to an accurate knowledge of the position of the Moon, and thus to the determination of *Ephemeris Time*, that is, an improved Sidereal Time, in which the small irregular fluctuations in the Earth's rotation, which serves as a clock, can be eliminated by a comparison with the motion of the Moon. For a detailed discussion of this time problem we refer particularly to special publications.[1]

There are also photographic methods for measuring the Moon's position with respect to the stars, and MARKOWITZ[2] has constructed a special camera. The photoelectric registration of occultations has been successfully employed at several observatories, but it will not be included in the present chapter, owing to the high technical requirements.

13.2. Predictions

In order to predict the time at which the disappearance or reappearance of a star becomes observable at a certain place we may use graphical methods. The exact calculation leads to a system of transcendental equations, which can only be solved by trial and error. The Nautical Almanac Office of the Royal Greenwich Observatory makes extensive predictions, the first approximation being obtained in an optical-mechanical manner with the help of the

---

[1] SADLER, D. H.: Ephemeris Time, *Occasional Notes (Royal Astronomical Society)* **3**, 103 (1954); Temps des éphémerides, *Annales Françaises de Chronométrie* (2) **25**, 141 (1955); CLEMENCE, G. M.: The practical use of Ephemeris Time, *Sky and Telescope* **19**, 148 (1960).

[2] MARKOWITZ, W.: Photographic determination of the moon's position and application to the measure of time, rotation of the earth, and geodesy. *Astron. J.* **59**, 69 (1954).

so-called *occultation machine*.[3] The publication of these results will enable many observers to participate in the general program of Ephemeris Time determination; see literature on page 542.

The predictions mentioned are supplied by the Royal Greenwich Observatory for a wide network of about 90 standard geographic positions, usually large cities. For example:

	Longitude (λ_0)	*Latitude* (φ_0)
Greenwich	0°000	+51°477
Edinburgh	+3°175	+55°925
Washington, D.C.	+77°065	+38°920
Denver	+104°950	+39°677
Toronto	+79°400	+43°663
Vancouver	+123°100	+49°500

Data for the British stations and for Sydney, Melbourne, Dunedin, and Wellington are given each year in the *Handbook of the British Astronomical Association*. Predictions for 15 stations in Canada and the United States (including those given above) are published annually in a special supplement to *Sky and Telescope*.

For these stations the predictions include all stars down to 7^m5, but with certain limitations, which depend on the phase of the Moon and the daylight brightness of the sky. The basis for these selected stars is provided by J. ROBERTSON's *Catalog of 3539 Zodiacal Stars for the Equinox 1950.0*, the so-called *New Zodiacal Catalog*, abbreviated NZC; see also page 548. The catalog number of the star and its usual designation are given in the lists of predictions (see page 313), together with the calculated times t of contact, to the nearest tenth of a minute, the apparent visual brightness of the star, the elongation of the Moon in degrees, and the position angle P. The latter is referred to the center of the Moon, and indicates the position at which the star disappears or reappears; it is always measured counterclockwise from north through east. Disappearance and reappearance are generally marked, e.g., by D and R, or by 1 and 2.

The values of P and t change appreciably from place to place as a comparison of the predictions shows.

Observers who are not at a standard station may obtain the times of contact with sufficient accuracy by taking from the lists two coefficients a and b. These enable the observer, in longitude λ and latitude φ, to very easily convert the contact times t_0 at the standard station (λ_0, φ_0) to his own observing site. He obtains the contact time t by applying to t_0 the correction

---

[3] Descriptions and illustrations are given in: KOPFF, A.: Neue Hilfsmittel zur Vorausberechnung und Reduktion von Sternbedeckungen, *Die Sterne* **17**, 241 (1937); and McBAIN-SADLER, F.: Predicting and observing lunar occultations, *Sky and Telescope* **19**, 84 (1959).

$$t - t_0 = a(\lambda - \lambda_0) + b(\varphi - \varphi_0). \tag{1}$$

In this formula λ and φ are in degrees, eastern longitudes being reckoned negative, so that the correction is obtained in minutes of time. The accuracy of t is generally better than ± 1 min for distances of less than 300 km from the standard station, and up to 450 km distance it is still better than ± 2 min.

Although a much greater accuracy of λ and φ is required for the reduction of the results (see page 320), it is sufficient for the present purpose of prediction to use values to an accuracy of only 0.01 degree, sometimes even to only 0.1 degree; these can be taken from a good map, even from one of a small scale. For large towns, however, such average coordinates are generally not good enough. On the other hand, we can neglect differences in altitude as far as our required accuracy is concerned.

If the prediction lists do not contain the coefficients a and b, the occultation will be grazing and its observation depends too strongly on the site of observation to permit the use of the conversion given above.[4]

13.3. Optical Requirements

The observation of stars down to the visual magnitude $7^m\!\!.5$ in the neighborhood of the Moon presents no difficulties provided the sky is transparent. Binoculars with a good support will suffice. All this changes very much, however, if the sky is less transparent owing to fog, haze, or cloudiness, for then even the brightest star may become obliterated in the scattered moonlight.

The difference between the limiting magnitude for different instruments (different magnification, aperture, and optical quality) can be read off from the equation on page 30 that applies to observation with both eyes. It has been shown that the limiting magnitude, and thus also the possibility of finding a star in scattered moonlight, rapidly increases with increasing aperture, but also noticeably with increasing magnification. Because of their small magnification and small aperture, binoculars are then very unfavorable, but this does not detract from their suitability in a transparent sky. Every successful observation can well compete with the results obtained with larger instruments.

Although it is possible to offset the effect of atmospheric scattered light by using higher powers, one should not go too far here. Increasing magnification also increases the scintillation due to the atmospheric turbulence, and can thus psychologically affect the accuracy.

The observation of the times of reappearance involves another restriction. The estimate of the position angle which is required demands that the observer is able to see the whole of the Moon's disc in his field of view. If

---

[4] Special predictions are given for grazing occultations in the publications mentioned above.

the field is about 45° in angular measure, a magnification of about 45× is the upper limit.

The observer is unlikely to have a filar micrometer to set the expected position angle of reappearance, but a simple estimate will be quite good enough. A cross wire in the field of view of the eyepiece is very useful, or according to WRIGHT,[5] also a single transverse wire. The construction of such a device is discussed on page 89. Turning the eyepiece moves the wire into such a position that with a fixed telescope a star or some detail of the Moon's surface moves through the field of view parallel to the wire. The wire then indicates the east–west direction ($P = \pm 90°$). MAK[6] suggests fitting the freely turnable cross wires (or single wire) with a position-angle scale, which enables a rapid setting of the wires in the expected direction. A simple scale at intervals of 10° is drawn on cardboard. To attach it in the correct position to the eyepiece, we determine the east–west direction as before, keeping the telescope in a fixed position. The reading should then show 90°. The eyepiece is then turned again, but is slightly offset so that the wire does not interfere with the observation. The same author also emphasizes the possibility of inserting in the field of view of some eyepieces a little homemade circular gradation. In any case we can manage quite well without these auxiliary devices.

An equatorial mounting with automatic guiding is of course a great help for the observation of star occultations, but it is by no means indispensable. If the telescope is not equipped that way, we may quite safely observe by keeping the instrument in a fixed position. This avoids continual correcting of the position while waiting for the event, and removes the danger that we may just miss the vital moment: according to JAHN[7] we first determine the travel time of the star, or of one of its neighbors, or of a detail of the Moon's surface across the field of view; we then take half this travel time, subtract it from the predicted time of contact, and set the telescope in such a way that at the calculated time the star just appears in the east side of the field of view (on the right, therefore, in the case of an astronomical telescope, if we are observing in the Northern Hemisphere). Then the occultation will take place near the center of the field of view. If the diameter d of the latter is given in degrees, we can also calculate the travel time t from the declination δ of the star or the Moon, since

$$t = \frac{d}{\cos \delta} \cdot 4 \text{ min.} \tag{2}$$

This quantity depends on the magnification V, since (if we call σ the angular size of the field of view, usually 60° or a little more) we have $d = \sigma/V$.

[5] WRIGHT, H. N. D.: Lunar occultations: A method of observing reappearances, *J. Brit. Astron. Assoc.* **66**, 265 (1956).

[6] MAK, A.: De bedekking van de Pleiaden op 30 September 1950. *Hemel en Dampkring* **48**, 156 (1950).

[7] JAHN, W.: Wir beobachten Sternbedeckungen. *Sternenwelt* **2**, 46 (1950).

13.4. Time Measurement

For the determination of the time of contact t we can use one of three methods: the *stopwatch*, *chronograph*, or *eye-ear method*. A reliable clock with a clearly visible second hand that is checked regularly (see page 94) is indispensable for the last two measures, but even for the stopwatch procedure it is very desirable. Only the chronograph requires a special clock for astronomical purposes, which must be equipped with a seconds contact.

13.4.1. Stopwatch

This method is by far the simplest and will therefore be described first, introducing us at the same time to further details of the observation.

The commercially obtainable stopwatches which give $\frac{1}{5}$ or $\frac{1}{10}$ seconds involve an accidental error, owing to the design of the toothed wheels, somewhat below 0.2 or 0.1 second. The $\frac{1}{10}$-second stopwatch will be preferable. The performance of the stopwatch, that is, its rate in sec/min, can be determined by measuring a lengthy time interval of 10 to 60 minutes, which can easily be compared with a normal clock with a second hand. Since the rate of the watch changes because of the hardening of the oil, we should repeat this procedure at intervals of about one month. The stopwatch should always be fully wound, but it should be made to run down after the observation, in order to release the spring and to prevent the hardening of the oil.

At all visual observations of sudden events the observer must be aware of his "personal equation" (*PE*) and that of his instrument, that is, the time that elapses between the actual event and the observer's response to it. It depends on the kind of stimulus received by the eye or the ear, on the physical and mental state of the observer (such as his alertness or fatigue), as well as on the required reaction, and on the type of measuring instrument, e.g., the operation of a stopwatch or a key of a chronograph. Defined in this way the personal equation becomes:

$PE = $ (reaction time of the observer)

+ (systematic delay of the time-measuring instrument).

If the personal equation is to be measured, it must therefore always be done with the same stopwatch used for the observation and is best done immediately afterwards. The method of stopping must always be the same: all attention will be concentrated on the forthcoming event, and as soon as this occurs the stopwatch will be started. Or the observer will wait for the event with the thumb "ready on the trigger," in full concentration on his aim to stop as rapidly as possible. This method yields a smaller personal equation and is preferred by the author.

The determination of the personal equation can be made in the following simple manner: the observer covers half the well-illuminated dial of the stopwatch with a piece of paper, in such a way that the edge of the paper is exactly above a selected reading on the dial, while the moving hand of the watch remains invisible. As soon as the hand appears below the paper, the watch is stopped. The extra time interval covered by the hand can then immediately be read off and furnishes the personal equation. This simple experiment may be repeated three times for different dial readings, and the mean taken. This method gives values between 0.2 and 0.3 second.

SIDGWICK[8] gives much smaller values. He also describes the dependence of the personal equation on the peculiarities of the occultation, the immersion and emersion, the brightnesses of the star and the Moon's limb, etc. All the resulting differences are far below 0.1 second and can therefore be neglected.

Differing from SIDGWICK's small values RUIZ[9] found from laboratory experiments with photoelectrically controlled model-occultation values for the personal equation of up to 0.9 second, and on the average of about 0.4 second.

The waiting time at the eyepiece before the beginning or end of the occultation must be long enough to permit sufficient concentration of the observer, and to prevent missing the event in case it deviates a little from the predicted time. The author prefers looking through the eyepiece about $1\frac{1}{2}$ minutes before the predicted time, the telescope having been set beforehand.

If a clock that fulfills the indicated requirements is available, the observer will start his stopwatch at the observation of disappearance or reappearance, will then walk to his clock, and, concentrating on the movement of the hands of the clock, will stop the watch at the very moment at which the second hand has reached one half or a whole minute. This clock time we call the signal time t_s, and it is noted together with the time read off on the stopwatch, Δt. All this may be followed by the above-mentioned determination of the personal equation. The value t_s must then be corrected for the rate of the clock, which is obtained by interpolation of the daily rate between two clock comparisons (see page 94). In addition, Δt must also be corrected, using the known rate of the stopwatch.

The following example of a disappearance observation indicates the procedure. The stopwatch was a little too slow, corresponding to $+0.2$ second per minute (the plus sign refers to a clock that is running slow). On 9 October 1954 the author observed at the Munich University Observatory the disappearance of 6 G. Piscium as follows:

Signal time t_s	$22^h48^m0^s\!.0$ Central European Time
Clock correction	$+0.4$
Watch reading Δt	-52.6
Watch correction	-0.2
Personal equation	-0.3
Total	$22^h47^m7^s\!.3$ Central European Time

[8] SIDGWICK, J. B.: *Amateur Astronomer's Handbook*, p. 80. London: Faber & Faber Ltd., 1957.

[9] RUIZ, J. J.: An artificial occultation experiment. *Sky and Telescope* **12**, 275 (1953); MEURERS, J.: *Astronomische Experimente*, p. 29. Berlin: Akademie-Verlag (1956); HERRMANN, J.: Die Beobachtung von Sternbedeckungen und ihre Bedeutung für die theoretische Astronomie, I., *Nachrichtenblatt der Vereinigung der Sternfreunde* **7**, 47 (1958).

If we refer our stopwatch immediately to the nearest preceding or following time signal of a shortwave radio transmitter (in the first case the stopwatch will be started well before the occultation), we do not require any other clock. At places where we can receive one of the regular time signals (see Table 2.16) this procedure is certainly advisable. On the other hand, radio time signals, such as those broadcast by the B.B.C., do not give smaller errors, since there may be a long interval before the signal is given, and the watch rate may alter in this time. The time information of the telephone service is a possible alternative.

The accuracy of the timings can be improved by adopting the TAYLOR method. The observer starts the watch at the occultation as before, and then listens to a series of seconds signals while looking at the watch face. This allows him to determine the few extra tenths of a second that the watch is registering at each signal, and this quantity is noted—say 0.4 sec. The watch is then stopped at the next convenient beginning of a minute signal. Suppose this is at $22^h37^m00^s$, and the watch then reads 43.6 sec. The watch has clearly stopped a few tenths of a second late, but this is ignored because the true reading at each second is already known to be 0.4 sec. The observed time is therefore $22^h37^m00^s - 43^s.4 = 22^h36^m16^s.6$.

13.4.2. Chronograph

The operation of the chronograph requires a precision clock with special contact, two polarized intermediate relays, as well as a hand key and a suitable power supply, say 12 Volts and 12 Watts DC, which can usually be provided by a selenium rectifier fed by alternating current mains.

The method of observation resembles the stopwatch reference to a clock. Here we are able to determine the personal equation simply by the second beats of the signal clock itself. Because of the self-induction of the coils and the mechanical inertia of the switches, the chronograph, etc., the personal equation is appreciably larger than in the case of the preceding method; see SIDGWICK.[10] We must remember that the two springs of the chronograph in general do not react with the same beat, so that their tasks cannot be interchanged. Furthermore, we must make allowance for the possible errors in the gear wheel, which may account for irregular intervals between the recorded second signals. On the other hand, these wheel contacts cannot be replaced by pendulum contacts which are never regular enough.

Definite advantages of this method are the higher accuracy and the documentary evidence provided by the supply of a written record. However, there are considerable demands in expense and instrumental care, and it will be advisable to use a chronograph for occultation observations only when such an instrument is already available for other purposes.

Similar remarks apply to a proposal to use a tape recorder; its performance is about the same as that of a chronograph; see HERRMANN.[11]

[10] SIDGWICK, J. B.: *Amateur Astronomer's Handbook*, p. 81. London: Faber & Faber Ltd., 1957.
[11] HERRMANN, J.: Tonbandgeräte bei astronomischen Beobachtungen. *Nachrichtenblatt der Vereinigung der Sternfreunde* **8**, 9 (1956).

13.4.3. Eye-Ear Method

For the application of this method we require only a chronometer or some other reliable and carefully checked clock that possesses a clearly visible second hand and whose half-second beat is clearly audible. If necessary, we can intensify the ticking by placing the clock on a suitable wooden box that provides good resonance. We have then to estimate the moment of contact by interpolating it between the beats that follow each other every half second, immediately afterwards counting the full seconds up to the signal time t_s read off from the clock. We first note the beat of the clock by counting (in our thoughts) 0 and 0 and 0, etc. As soon as immersion or emersion takes place, we count on 1 and 2 and 3, etc. At the same time we turn to the clock, stop counting at, say, the arrival of the second hand at a full ten-second mark, and note the last number counted and the time read off at this moment. In addition, we must estimate the fraction of the first second (which we counted as "0 and 1") at which the observation took place. If, for instance, this was shortly before the beat corresponding to $0\overset{s}{.}5$ (the "and" beat), say at 0.3 second, and if at the number 8 the clock showed $4^h25^m40^s$, then the moment of observation was $4^h25^m40^s + 0\overset{s}{.}3 - 8^s = 4^h25^m32\overset{s}{.}3$. Experienced observers will be able to estimate ± 0.1 second. There is no Δt error, and as to the personal equation, only the very small difference in the reaction time of the eye and the ear remains. Sidgwick,[12] nevertheless, indicates for this difference in the mean 0.08 second. In this way the accuracy of this method should exceed that of the stopwatch method and should approach that of the chronograph. Three observers at the Cape Observatory found a mean error as large as ± 0.15 second.[13] It must be emphasized that the eye-ear method makes great demands on the concentration and the presence of mind of the observer.

The differences in accuracy of the methods described will only be of importance if, in reducing the observations, the profile of the Moon's limb is taken into account. This correction is now included in the reductions made by H.M. Nautical Almanac Office. We recall that the Moon travels every second by about 1 km in its orbit. Thus, a change of 100 m in the height of the Moon's limb, measured parallel to the Moon's orbit, affects the observed instant of a star occultation by about 0.1 second.

13.5. Evaluation of the Observational Results

The results of occultation observations are reported quarterly or annually to *H.M. Nautical Almanac Office, Royal Greenwich Observatory (Herstmonceux Castle, Hailsham, Sussex, England)* not later than six months after the end of the observing year. The Nautical Almanac Office employs automatic reduction and hence prefers to receive the unreduced data, so there is no need for the observer to attempt a reduction.

---

[12] See Footnote 10.

[13] See Evans, D. S.: Timing of occultations. *The Observatory* **71**, 155 (1951).

The report should give complete results, name and address of the observer, and geographic coordinates and altitude of the observing site. Since we aim at an accuracy of 0.1 second for the times of contact, we must take care that the true observing site differs from the reported one by less than 33 m in horizontal direction. Since the amateur observer will hardly be able to obtain a sufficiently accurate astronomical determination of his site, he will try to proceed along different lines:

If there is nearby a triangulation point of the geodetic survey whose geographic coordinates are available in the offices of the geodetic survey, the observer can make reference to this point, if he can state how far north, south, east, or west he is situated. One second of arc in geographic latitude corresponds in the direction of the meridian to a distance of 31 m, and 1 second of arc in the geographic longitude at the geographic latitude φ to a distance of 31 m/cos φ in the east–west direction. With the help of a tape it is possible to measure distances of this order with sufficient accuracy.

In many cases the observer will use the commercially obtainable official maps of the Ordnance Survey which, on a scale 1:25,000, permit an easy determination of geographic coordinates. The required accuracy of ± 33 m in nature is represented on such a map by ± 1.3 mm; it is easy to measure this sufficiently accurately, and the errors of the printed map are below ± 1 mm. We must be careful that the map is not "generalized" at the observing site, a procedure sometimes used by the cartographer to achieve greater clarity, e.g., streets are drawn in their correct position, but considerably widened, so that all details near their borders are somewhat displaced. While in such a case the user of the map will easily recognize the generalization which has taken place and can eliminate it, this will not be possible within the borders of closed villages where the generalization can exceed the indicated tolerance. In Great Britain the more accurate maps, such as the ungeneralized ones on a scale of 1:5000, may not give geographic coordinates, but only National Grid coordinates. The necessary conversion into geographic coordinates will be carried out by H.M. Nautical Almanac Office.

Less difficult is the determination of the altitude of the observing site, for which an accuracy of ± 50 m is adequate. For this purpose the contour lines of the Ordnance Survey maps are quite sufficient.

Summarizing, the observer's report should contain the following data:

1. Occulted star.
2. Date and observed contact time in Universal Time to the nearest 0.1 sec. It is not necessary to apply the personal equation, but if this is done, the fact and the amount should be stated.
3. Observing site with geographic longitude and latitude to the nearest full second of arc, as well as the altitude above sea level to ± 50 m, with an indication as to the sources of these data.

4. Characteristic data concerning the observing instruments (e.g., 2-inch refractor of $f = 75$ cm and $45 \times$).

5. Short description of the observing method (e.g., references with a $0\overset{s}{.}1$ stopwatch to the DIZ time signal at 20^h00^m Universal Time).

6. Notes concerning the observing conditions and, perhaps, a personal assessment of the quality of the observation (e.g. very good transparency, unsteady images, etc.).

7. Any special remarks (e.g., flickering disappearance of the star).

8. Name and address of the observer.

13.6. Occultations of the Planets

Although not of the same scientific importance as the occultations of stars, occultations of the planets are much more impressive sights (see also Sec. 15.3). Owing to the considerable size of the planetary discs, immersion can here extend over several minutes. Thus the beginning or the end of a nearly grazing occultation by the Moon can last, in the case of Jupiter, between the first and second, or the third and fourth contacts, respectively, as long as 10 minutes; for Mars, 21 minutes; and for Venus, even 35 minutes. In view of the Moon's profile, which has been neglected in giving these figures, these can only indicate the order of magnitude. The observer will therefore require not only predicted times for disappearance and reappearance, but also an accurate knowledge of the durations involved.

To determine these data we take from the list of predictions the time of contact, and the position angle p of the occultation; and for the time of contact the following quantities are taken from the Astronomical or American Ephemeris:

$\Delta\alpha_{\mathbb{C}}$, $\Delta\delta_{\mathbb{C}}$: The hourly changes of the apparent position of the Moon in seconds of time or arc, respectively (to be taken from the table "Moon for each hour of Ephemeris Time," to be found on the right of each column of "Apparent Right Ascension" and "Apparent Declination").

$\Delta\alpha$, $\Delta\delta$: The daily changes of the apparent position of the planet in seconds of time or arc, respectively (to be found on the right in the corresponding columns of the planetary tables "For 0^h Ephemeris Time").

$\delta_{\mathbb{C}} \approx \delta$: The apparent declination of the Moon and the planet, respectively, which is given there as well.

ρ: The apparent radius of the planet in seconds of arc (to be found in the planetary tables under the heading "semidiameter" or "polar S.D.").

It is possible to make allowance for the flattening of the planets and the peculiar shape of Saturn's rings with the help of a good drawing or other illustration (as we shall see further below in the example of Saturn), but the effort is hardly worthwhile. The observer will prefer to take a permissible increase of the calculated duration into account: for the two strongly flattened major planets he will obtain the maximum radius of Jupiter and of Saturn's rings by multiplying the smaller tabular values for the polar radius by 1.066 or 2.542, respectively.

From these data we first of all calculate the apparent relative motions of the two planets in seconds of arcs per minute of time by using

$$\mu_\alpha = \left(\frac{\Delta\alpha_{\mathbb{C}}}{4} - \frac{\Delta\alpha}{96}\right) \cdot \cos\delta,$$

$$\mu_\delta = \frac{\Delta\delta_{\mathbb{C}}}{60} - \frac{\Delta\delta}{1440},$$

(3)

and therefore obtain the required time interval in minutes by

$$\Delta t = \frac{2\rho}{|\mu_\alpha \sin p - \mu_\delta \cos p|}.$$

(4)

It will therefore suffice to start the actual observation a little earlier than $\frac{1}{2}\Delta t$ minutes before the predicted time. No account has been taken of the Moon's parallax, which can be neglected.

It is useful, although not necessary, to have a drawing of the conjunction of the Moon and planet on the correct scale; frequently such drawings can be found in Yearbooks specially devoted to the work of amateur astronomers, where these phenomena are regularly announced and described.[14]

It is also possible to sketch the event correctly for the observing site as follows: The Astronomical or American Ephemeris furnishes for some hours before and after the calculated or assumed time of contact the apparent place ($\alpha_{\mathbb{C}}$, $\delta_{\mathbb{C}}$) of the "Moon for each hour of Ephemeris Time." Furthermore, we obtain from these tables the interpolated apparent place of the planet (α, δ) under the heading "for 0^h Ephemeris Time." A table for the parallaxes of the Moon provides the corrections $\Delta'\alpha_{\mathbb{C}}$ and $\Delta'\delta_{\mathbb{C}}$ as required for the reduction. We now calculate the coordinates

$$x = -(\alpha - \alpha_{\mathbb{C}} - \Delta'\alpha_{\mathbb{C}}) \cdot \cos\delta_{\mathbb{C}},$$

(3')

$$y = \delta - \delta_{\mathbb{C}} - \Delta'\delta_{\mathbb{C}}$$

(4')

at equal time intervals and plot these values (both in minutes of arc), using a rectangular coordinate system (positive x-axis to the right, and y-axis upwards) on squared paper. The resulting curve $y(x)$ is nearly a straight line; it represents the apparent path of the planet relative to the nearby Moon. The origin of this coordinate system is the center of the Moon; the latter can be drawn as a circle with the radius $\rho_{\mathbb{C}}$ in minutes of arc; the values are taken from the table headed "semidiameter" on the pages of "Moon for 0^h and 12^h Ephemeris Time" in the Ephemeris. We can now read off immediately the position angles p_1 and p_2 for the immersion and the emersion (referred to the $+y$ axis), as well as the corresponding times t_1 and t_2.

[14] See, for example, *The Handbook of the British Astronomical Association*; and R. NAEF: *Der Sternenhimmel*; see page 548.

To draw in the terminator we extract from the table "Moon: Ephemeris for physical observations" of the Ephemeris the position angle of the bright limb, which we plot with reference to the $+y$ axis counterclockwise. Perpendicular to the diameter drawn in this direction lie the endpoints of the terminator on the Moon's limb. Finally, we find in the same table the "fraction illuminated," f, and mark off along the diameter which we have drawn to the bright limb, starting at the latter, the distance $2f\rho_{\mathfrak{C}}$. The arc of an ellipse passing through this mark and the two endpoints of the terminator represents the complete curve of the terminator.

In the case of planets whose figure much deviates from a sphere, Saturn, for instance, it is useful to indicate on our drawing the correct orientation of the planet, even if this means increasing the scale of the drawing. Here again the Ephemeris provides the necessary data; for instance, for the ring system of Saturn we find in the table "Rings of Saturn" the major and the minor axes of the outer edge of the outer ring, as well as the position angle P of the minor axis. Furthermore, we also find here the reduction factors, with which we can derive all the other details of the ring system from this ring ellipse.

The observer using a power of, say, about $150\times$ notes all important times to the nearest second, which can later be rounded off to tenths of minutes. HOFFMEISTER's proposed method of time determination can be applied (see Sec. 12.3.2); for instance, in the case of Saturn we may use the first contact of the ring system with the Moon's limb, later that of the planetary disc, and later still the two instants when planet and rings disappear behind the Moon.

14 | Artificial Earth Satellites

W. Petri

14.1. Nature and Purpose of the Satellites

It was in 1955 that both the United States and the Soviet Union declared that they would in the course of the International Geophysical Year, 1957–58, send artificial Earth satellites with measuring instruments into space. Originally, when this was announced at the Dublin meeting of the International Astronomical Union, most of the assembled astronomers had their doubts as to the practicability of this project. However, by the end of the Geophysical Year both parties had kept their promise. Soon after the historic date, 4 October 1957, when the first "Sputnik" began to circle the Earth, news of the launch of a new satellite was no longer a sensation. Not only the general public, but also the astronomers had become used to the existence of these strange objects. Only a few of the large observatories were able to specialize in satellite observations, but amateur astronomers have found a rich and fruitful field in this work and very soon united in organized cooperation. They have contributed with their modest means very useful results, and in compensation have received joy, stimulation, and enrichment of their knowledge. Thus in a modern handbook for amateur astronomers a special chapter dealing with artificial satellites must certainly be included.

The original scientific publications on the subject appeared in the leading journals, *Nature*, *Priroda*, and specialist journals for astronautics, geophysics, astrophysics, and radio technology, which need not be listed here. In Germany the *Naturwissenschaftliche Rundschau* and *Umschau* published relevant reviews. Of particular interest to the amateur astronomer are the Publications of the Smithsonian Astrophysical Observatory; since summer 1956 a series of *Bulletins for Visual Observers of Satellites* has been published as a supplement to "Sky and Telescope," as well as a special *Phototrack Bulletin*. As far as Russian publications are concerned the observer will find many contributions in the *Astronomical Circulars* (particularly since No. 186, 1957) and in the *Astronomical Journal* (particularly since Volume 35, 1958). Fundamental contributions concerning satellite techniques and observations are contained in the *Proceedings of the International Astronautical Federation* (IAF), particularly in Volume VIII (Barcelona, 1958) and Volume IX (Amsterdam, 1959), which appeared at Springer-Verlag in Vienna, which also publish the "Astronautica Acta." Finally we may mention the astronautical journals in the various countries, for example, *Spaceflight*, the journal of the British Interplanetary Society; see pages 544 and 549.

14.1.1. The Satellite as a Celestial Body

A body subjected to gravitational attraction by the Earth and orbiting the latter according to the laws of celestial mechanics must be termed a "celestial body." To call it an "artificial moon" is somewhat sensational but not wrong, particularly because as early as the beginning of 1959 we were already presented with an "artificial planet," a man-made object which escaped from the Earth's gravitational attraction and went into orbit round the Sun.

However, in comparison with the classical lunar theory the motion of an Earth satellite offers important practical differences. On the one hand, we can generally neglect the influence of other celestial bodies as well as the intrinsic gravitation of the satellite. On the other hand, however, the deviation of the Earth from a true spherical shape and the irregular composition of its interior, which in certain regions causes anomalies of gravity, produce, in the course of time, very significant effects. To this we have to add the braking action of the Earth's atmosphere and even that of the terrestrial magnetic field and the interplanetary plasma, not to mention the influence of solar radiation pressure and certain relativistic effects that are not yet completely clarified.

An artificial satellite can have any direction of motion and any orbital inclination. Originally the eccentricities of the orbits were very small because the aim was to arrange an orbit in such a way that the satellite was brought around the Earth without moving too far away from it, for reasons of both energy saving at the launch and better observing conditions. In what follows we shall deal only with such near-Earth satellites.

14.1.2. The Satellite as a Space Probe

Nearly all satellite payloads include measuring instruments and small radio transmitters that telemeter the measured data to Earth. Often these data are stored until the Earth station orders their transmission. There are also many passive satellites, such as discarded rocket stages, fragments of instrumental parts, or the inflated silvered plastic spheres such as the much-observed Echo 1 and Echo 2 satellites, which because of their great reflective ability have been useful for passive communication transmission. They also make it possible to determine the density of the high atmosphere rather accurately, since they offer large air resistance for a small mass. Other satellites have been used to carry out special tasks like probing the radiation belt surrounding the Earth (which was only discovered by satellite measurements), or carrying out extensive communication work on ultrashortwaves.

14.1.3. The Satellite as a Spaceship

The artificial Earth satellites occupy a peculiar intermediate position between science and technology. They are at the same time research tools and means of transport of a quite unusual kind—one which has heralded a new epoch in human civilization: the epoch of space travel. As soon as it was possible to retrieve a large-scale satellite, bringing it safely back to Earth, men have come forward to entrust themselves to this means of transportation with the aim of reaching the Moon and establishing permanent stations in space to serve a multitude of scientific and practical purposes. To make possible the tremendous achievements of these recent years it was necessary to study carefully the secrets and dangers of outer space beyond the protective cover of our atmosphere. These studies were performed by the small, unmanned satellites of the early pioneering days. Thus every observer in a satellite program is in a way a pioneer of space travel. He himself gains a treasure of knowledge and experience, which carries its own reward, but which under certain circumstances can also achieve results of real practical value to science: active experts are needed everywhere, and certainly in the field of astronautics with its immense future.

14.2. Conditions of Visibility

A satellite becomes observable from the moment it rises above the horizon of the observer, but it must have a certain minimum brightness to impress the eye or the photographic emulsion to an appreciable extent. It must appear in sufficient contrast against the celestial background, so that it is not only the apparent brightness of the satellite but also that of the sky which counts. To find the satellite we, of course require that approximate times and directions be calculated in advance. Before we turn to the actual technique of observation we shall therefore deal with a few general problems concerned with the visibility of satellites.

14.2.1. The Satellite as an Illuminated Sphere

Let us assume for simplicity that the satellite is a sphere that primarily receives its light directly from the Sun. (Self-luminous satellites have remained a rare exception, and the light reflected from the Moon or the Earth is very weak indeed.) Let the intensity of illumination by the Sun be E, the radius of the sphere r meters, and its reflectivity R. We then have two limiting cases. In the case of mirror reflection the brightness of the satellite I_{sp} becomes $ERr^2/4$. This quantity is independent of the direction; see Fig. 14–1. Conditions are different in the case of diffuse reflection. Here the phase angle α becomes important, that is, the angle between Sun and observer measured at the place of the satellite. We have

$$I_{diff} = ERr^2 \cdot \frac{2}{3}\varphi(\alpha),$$

where

$$\varphi(\alpha) = \frac{\sin \alpha + (\pi - \alpha) \cdot \cos \alpha}{\pi}$$

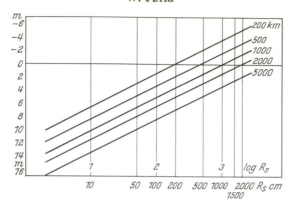

Fig. 14–1 Apparent visual brightness *m* for reflecting spherical satellites of radius R_s at different distances, valid for the albedo $A = 0.8$. For $A = 0.95$ the brightness increases by 0.1 magnitudes. After U. GÜNTZEL-LINGNER, *Sterne und Weltraum* **3**, 228 (1964).

represents the phase law; α is to be taken in radians. For New Moon we have $\alpha = \pi$ and $\varphi = 0$; at Full Moon $\alpha = 0$ and $\varphi = 1$; in quadrature $\alpha = \pi/2$ and $\varphi = 1/\pi = 0.3183$.

Up to the time of quadrature the diffuse sphere is therefore brighter than the totally reflecting sphere, and in the conditions of Full Moon by a factor 8/3. Numerically this means that for $E = 150,000$ lux, $R = 1$, and $r = 0.30$ we have $I_{sp} = 3375$ cd and $I_{diff(\alpha = 0)} = 9000$ cd. In order to exceed the brightness of a pure white, diffusely reflecting satellite, we would have to renounce the uniform distribution of the light received in space from the Sun and cover the surface of the satellite with plane mirrors, which would light up from time to time momentarily and intensely. Irregular reflection becomes particularly pronounced if the outline of the satellite strongly deviates from a spherical shape.

14.2.2. Calculation of the Phase Angle

In order to calculate the phase angle α we must know the spherical coordinates of the Sun and satellite in the horizon system of the observer; we then have $\alpha = 90° + h - \zeta$, where h is the altitude of the Sun (negative, when the Sun is below the horizon) and ζ an auxiliary angle to be found from the formula $\tan \zeta = \tan z \cdot \cos d$. Here we take z, the zenith distance of the satellite, always positive; d is the difference in azimuth between the Sun and satellite (always positive, therefore between 0° and 180°); the sign of ζ is taken to be positive if $d > 90°$, otherwise negative. Thus we imagine at the observing site a plane perpendicular to the direction to the Sun and passing through the satellite, intersecting the surface of the Earth. The angle between this direction and the line to the zenith of the observer has been denoted by ζ.

We distinguish two elementary cases. (1) The satellite moves perpendicularly to the direction to the Sun: $d = 90°$, $\cos d = 0$, $\zeta = 0$, $\alpha = 90° + h$;

the phase angle thus depends only on the solar altitude so that it remains rather constant during the transit of the satellite. (2) The satellite moves exactly in the direction of the Sun's azimuth and passes through the zenith of the observer: $d = 0°$ or $180°$, $\cos d = \pm 1$, $\zeta = \pm z$, $\alpha = 90+h \pm z$. We see that the phase angle represents fully the zenith distance of the satellite, and does so with different sign, according to whether the satellite is on the side toward the zenith or the side away from it.

14.2.3. Position of the Earth's Shadow

Since in the case of diffuse reflection the phase angle should be as small as possible, it is favorable if the Sun is very low at the time of the observation (h strongly negative) and the satellite is far away from it on the opposite side of the sky ($d > 90°$, z large). A low solar altitude has the additional advantage that twilight has then progressed sufficiently and that the sky background is dark enough (see Fig. 14–2). This, however, is only useful as long as the

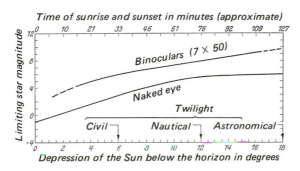

Fig. 14–2   Limits of visibility for a faint star–like object in twilight. After R. H. EMMONS, *Sky and Telescope* **16**, 171 (1957).

satellite itself is still illuminated by the rays from the Sun. It is therefore necessary to take into account the position of the Earth's shadow; it will then be possible to estimate the visibility of the satellite on the basis of the predicted orbit, or to calculate the height of the orbit if one has actually observed the disappearance of the satellite into the Earth's shadow.

The boundary of the Earth's shadow is in reality of a conical shape, and is diffuse because of the finite angular size of the Sun. We restrict ourselves here to the approximation that it is a plane, and determine in the first instance the height H at which this plane intersects the line from the observer to its zenith: we then have $H = R\,(\sec h - 1)$. Here R is the Earth's radius, and h as before the altitude of the Sun. Since the altitude H is as a rule higher than 100 km, we can neglect the actual profile of the Earth and the altitude of the observer above sea level. Let us now imagine, at the observing site, a plane perpendicular to the direction of the solar azimuth; then this plane will be cut by the Earth's shadow at the angle h with respect to the horizontal. We are interested, however, in the angle at which the boundary of the shadow appears at the azimuth of the satellite. We call this angle β and make use of the previously defined azimuth difference d. We then have $\tan \beta = \tan |h| \cdot \cos d$. Thus, the angle is always smaller than h and becomes zero if the satellite is perpendicular to the direction to the Sun.

We now calculate the orbital height H' which is the minimum requirement for a satellite to be illuminated by the solar rays:

$$H' = H\{\tfrac{1}{2}[1 + \cos(\beta + z) \cdot \sec(\beta - z)]\}.$$

This formula is valid only near to the observing site. In reality H' can even be a little larger since the line along which the shadow boundary intersects the plane perpendicular to the observing site and the satellite azimuth is usually less curved than the Earth's surface. Only for $d = 90°$ are both these curvatures equal.

14.2.4. Zenith Distance and Satellite Distance

Also of importance for the assessment of the visibility of a satellite are its zenith distance z and its distance E from the observer. At large zenith distances we must allow for the atmospheric extinction (see Sec. 5.3.2). For this reason the satellite in a clear sky at $z = 45°$ appears fainter by $0\overset{m}{.}1$, and at $z = 65°$ it appears by $0\overset{m}{.}3$ fainter than it would at the zenith.

The brightness of the satellite varies inversely as the square of the distance, so that the apparent brightness for twice the value of E amounts to only one quarter of the initial value; this corresponds to a difference of $1\overset{m}{.}5$. If we know m_E, that is, the apparent brightness at the distance E, we can easily convert this value to other distances by the relation $m_{E'} = m_E + 5 \log E'/E$.

We now ask how big are the zenith distance and the satellite distance if we know that the satellite at a given time is at a height of H km above a place that is at a distance D km from the observer. We first convert D to the geographic arc γ, by $\gamma = D/111.2$. We then calculate an auxiliary quantity: $F = H - R(\sec \gamma - 1)$. Here we again denote the Earth's radius by R. The final formulae become

$$\tan z = \frac{R \tan \gamma + F \sin \gamma}{F \cos \gamma}$$

and

$$E = F \cos \gamma \sec z.$$

These relationships can also be expressed by a diagram (Fig. 14–3). This is most simply done by drawing the basic geometric relations on a suitably chosen scale. To save space, of course, we restrict ourselves in the final drawing to only a portion of the complete diagram.

14.2.5. Apparent Angular Velocity

Finally we are interested in the speed with which the satellite crosses the sky of the observer. Its measure is the apparent angular velocity W (Fig.

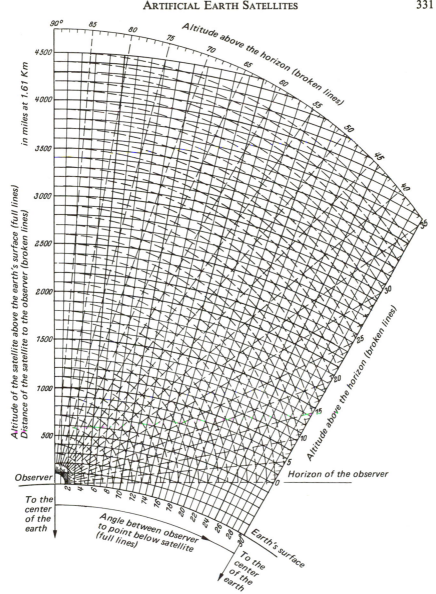

Fig. 14–3 Altitude angle and distance of a satellite for an observer on the Earth's surface. After L. N. CORMIER, N. GOODWIN, and R. K. SQUIRES: Simplified Satellite Prediction from Modified Orbital Elements, p. 46. Washington: National Academy of Sciences (1958).

14–4), which is $W = V/E \cdot 57.3$, expressed in degrees per second. Here, V is the linear relative velocity of the satellite in km/sec and E again its distance from the observer in km. The factor 57.3 provides the conversion from radians to degrees.

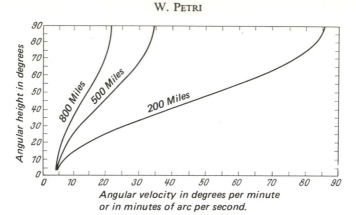

Fig. 14-4 Apparent angular velocity of the satellite as a function of height of the orbit and height angle. After R. H. EMMONS: *Sky and Telescope* **16**, 171 (1957).

In practice W always remains below $2°/\text{sec}$ and usually amounts only to a few tenths of a degree per second. Nevertheless, these are values exceeded only by meteors and make special demands on the alertness of the observer and the sensitivity of the photographic material.

14.3. Optical Observations

14.3.1. Survey Instruments

Meteor observers find by experience that a moving object must be about 2 magnitudes brighter than the faintest still visible star if we are to see it clearly. Since the satellite is usually fainter than 5^m, we generally require a telescope. Here brightness and large field of view are decisive. American and Russian observers at satellite patrol stations use small monocular field glasses of 50 to 60 mm objective aperture, 6- to 7-fold magnification, and 10 to 12° diameter of the field of view. The eyepiece must have a wide-angle character, and to prevent loss of light its exit pupil must not be larger than the pupil of the eye adapted to twilight.

Generally we know neither time nor zenith distance of the satellite accurately in advance. It is for this reason that observers work in groups in such a way that one or two great circles on the sky are covered by a corresponding number of instruments with overlapping fields of view. These "optical fences" are preferably arranged in the direction of the meridian, or the First Vertical, or across the expected satellite orbit. If we omit zenith distances beyond 70°, each such fence requires at least a dozen observers.

The instruments are set at the desired azimuth and then turned to the corresponding zenith distance about a point which should preferably be near the eyepiece end; see Fig. 14-5. We then observe with fixed telescopes. To facilitate the positioning of the head we use a sufficiently large plane mirror, placed in front of the objective; this can also be turnable but usually it is

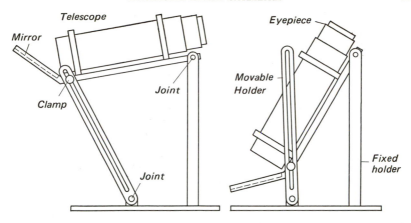

Fig. 14–5 Mounting of a Moonwatch telescope. After G. VAN BIESBROEK: Bulletin for Visual Observers of Satellites, No. 5, Cambridge, Mass. (1957).

fixed at an angle of 45° to the optical axis of the telescope. Attention must be paid to the reversal of the images, if we compare these with star maps. The field of view is best determined by pairs of stars of known distance, for instance:

$$\alpha - \beta \ UMa \ 5°.4; \qquad \alpha - \delta \ UMa \ 10°.1.$$

14.3.2. Special Visual Instruments

Occasionally satellites also appear in a larger instrument. Thus the carrier rocket of Sputnik 1 appeared in a three-inch telescope at a magnification of $80\times$, clearly visible as a stroke of 5 minutes of arc in length. Vanguard 1 was seen in a 10-inch reflector at a distance of about 4000 km, at an apparent brightness of nearly 14^m. But these are accidental events. It soon became necessary to design special instruments for continuous observation of faint satellites at their greatest distance from us. For this purpose the observers in the United States replace the 50 mm objective of a kind of coudé telescope, obtained from military surplus stores, by an objective of 120 mm aperture. The magnification is then 21.5 and the diameter of the field of view $2°.3$. This seems rather small, but is sufficiently large since the satellites at apogee move slowly; and in addition their apparent angular velocity decreases with increasing distance. These instruments have azimuthal mounting, altitude circles, and an arrangement for rapid clamping. An insulated hot-wire heater protects the objective against misting.

Sometimes a guiding telescope of moderate magnification is rigidly connected with the tube of a theodolite, both axes being parallel. This arrangement combines the advantage of accurate circles with that of a large, bright field of view.

To observe the moment of transit with the naked eye, prisms have been used which consist of two plane mirrors, forming with each other an angle exceeding 90° together with a small unsilvered glass plate placed above this arrangement. When set up correctly we now see two images of the satellite that move toward each other; their coincidence is very conspicuous.

14.3.3. Determination of the Apparent Orbit

The most important results of a satellite observation are three data: right ascension, declination, and time. The determination of the time will be dealt with later. The equatorial coordinates are topocentric, that is, they refer to the observing site. To convert them into geocentric values we must make use of the geographic coordinates of the station. The following other data are also desirable: apparent brightness, position angle of the satellite orbit on the sphere, apparent angular velocity, and meteorological conditions or other indicators of the quality of the observation.

In principle we possess two methods for solving an astrometric problem such as the reduction of satellite observations. On the one hand we can obtain some direct reference to the system of the observing site, or, on the other hand, obtain a relative position with the help of reference stars. In practice both methods are frequently used at the same time. The use of reference stars is preferred if the observations are carried out with the naked eye, and of course always in photographic work. The visual observer at the telescope will be well advised to study beforehand the conspicuous stars in the field of view. He will be greatly assisted by copies from the Atlas by Bečvář,[1] which includes stars down to $7^m.75$. Immediately following the transit of the satellite its orbit is drawn and marked with the exact time.

The method of direct reference requires the location of a great circle through the zenith—in general the meridian—in the field of view. The U.S. observers proposed using a wooden beam attached to a pole about 10 m high, carrying at its end small light bulbs to facilitate the setting. These are of course switched off before the actual observation. All the observers sit in one row. Because of the finite distance of the beam, however, we have to expect diffuse contours and a parallax effect in the field of view. Preference should therefore be given to the use of threads in the eyepiece.

To mark the reference circle in the field of view we employ a thin wire that is fixed in the telescope at a suitable place and that can be turned with the eyepiece or by a special setting device. To check the position of the wire it is preferable to use star transits.

Particularly well equipped are the apogee telescopes described above. Their eyepiece carries a little plate with short strokes and marked at its center by a small circle; see Fig. 14-6. The measuring line is subdivided by short perpendicular lines into sections of $0°.25$ and is placed eccentrically by $0°.4$, since faint objects are usually noticed only when they are so near the center of the field of view that a reliable measurement in the central position is not always possible. This plate is faintly illuminated with red light, which does not disturb the eye's adaptation to the dark.

[1] BEČVÁŘ, A.: *Atlas Coeli*, 1950.0. 2nd ed., Prague, 1956.

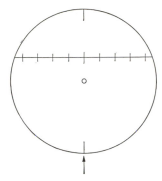

Fig. 14–6 Grid plate of a telescope for apogee observations of satellites. After A. BOGGESS III, I. S. GULLEDGE, M. J. KOOMEN, and R. TOUSEY: Bulletin for Visual Observers of Satellites, No. 9, Cambridge, Mass. (1958).

14.3.4. Photographic Results

With some luck a bright satellite can be caught by any amateur's camera, provided he uses full aperture and a highly sensitive film. More difficult is the photography of fainter objects with larger equipment which permits a more reliable astrometric reduction. Since the satellite moves very fast we cannot follow it, which would make possible longer exposures to increase the available limiting brightness (an exception is a special American camera which costs $70,000!). We thus obtain a trace of the satellite only if the following condition is fulfilled:

$$\frac{I}{WE^2} \cdot \frac{L^2}{f \cdot d} \geq C.$$

Here I, W, and E are again brightness, apparent angular velocity, and linear distance of the satellite, respectively. The quantity L is the diameter of the entrance pupil, f the focal length, and d the diameter of a sharp star image on the emulsion. The quantity C, which has to be determined empirically, contains the transmission of the instrument and the sensitivity of the photographic material.

The formula shows that a powerful optical system giving the best possible definition is the right choice for moderate focal lengths. In practice, cameras used for aerial photography, having $L = 10$ cm and $f = 25$ cm, have been found very useful. At $W = 1°/\text{sec}$ they reach a brightness of 3^m. Also popular are objectives up to $f = 50$ cm and $L = 12.5$ cm. Larger focal lengths are not advisable since then the field of view and aperture ratio become too small.

Preliminary exposures will establish the perfect setup and focusing of the optics; however, both require constant checking. Cameras of the old type with expanding bellows, etc., are unusable. More suitable are large box cameras with adjusting screws at the base. A diopter finder is necessary. At a fixed position of the camera we may expose 10 to 20 seconds. Time markings can be superimposed on the satellite trace if the objective is covered for a very short moment at the given times. The azimuthal mounting should be as stable as possible. The experienced amateur astronomer will of course adapt his instrument to the outside temperature, and will use a dewcap that does not interfere with the field of view.

The light of the night sky after the end of astronomical twilight can also considerably influence the limiting magnitude. An experimental series showed that the loss of light was 0".1 for a short exposure and increased to more than 0".2 at 15 minutes' and to 0".5 at 30 minutes' exposure time.

14.3.5. Light Variations Caused by Rotation of the Satellite

Some satellites show rotational light variations in a very conspicuous manner of the type familiar to us in the case of some minor planets, e.g., Eros. It is possible to set the tops of the payloads in unmanned satellite rockets into rapid rotation about their longitudinal axes at the time of launching. This stabilization process keeps the axis constant in space for some time. In this case the satellite presents the same side to us after a complete rotation of the Earth. If it is of an elongated shape it thus appears (under otherwise equal conditions) with varying brightness. More impressive, however, is the rotational light variation when a large unstabilized rocket stage rapidly tumbles over. This usually leads to a double wave in the brightness variation, in which the amplitudes differ alternately a little from each other.

Of special scientific interest are changes of the period, since they permit conclusions as to the condition of the high atmosphere and of the magnetic fields in space. For instance, the rotational period of 1957 β increased from 150 seconds in December 1957 to 240 seconds in January 1958; for the satellite 1958 δ_1 it increased on the average of 1 second per month: from 15.0 seconds on 2 June 1958 to 18.4 seconds on 2 October 1958. In this case it could be shown that the carrier rocket of Sputnik 3 made its somersaults in the reverse direction to the orbit.

The evaluation of all these effects is still in progress, and we may still expect many surprises. Satellites with a long life may exhibit secular brightness variations, such as when an originally reflecting surface becomes scratched and dull through the impact of micrometeorites, or if the reflective properties of the envelope vary through the absorption of interplanetary matter.

Photographic photometry of the satellite trace can be carried out with some effort. Even photoelectric measurements have already been successful where a special night-sky photometer was placed on the tube of a cine-theodolite. More suitable for the amateur astronomer, and at the same time astonishingly accurate, is the method of visual brightness estimates (see page 289). It consists of the comparison of the brightness of the satellite with that of a star that is somewhat brighter and another a little fainter, using a scale of about 10 steps. The observer must act very quickly and it is advisable to use simultaneously a tape recorder for the timing and recording. Otherwise one should have an assistant.

The photometry of satellites soon became a special field of keen groups of amateur astronomers, working together in close cooperation, since simultaneous observations (to fractions of a second at different places) are particularly useful for a unified evaluation of the results. For instance, it was once agreed to pay particular attention to every second "flaring up" of the satellite in each even minute on days of even calendar date. Of course, such observations will also include large zenith distances, since "one observer's horizon is the other's zenith. . . ."

14.3.6. Changes in Color

When a satellite enters the shadow of the Earth, it experiences a solar eclipse (for the terrestrial observer this is like a lunar eclipse). Accordingly we observe here, too, the same kind of color phenomena that we noted in the colored boundaries of the Earth's shadow during lunar eclipses (see Chapter 12). As a rule, the color of the satellite changes from yellowish-white to reddish. It is preceded by a short green phase, which is caused by the ozone layer at the upper boundary of the stratosphere. If heavy clouds pass through the shadow limit near the Earth's surface we sometimes do not reach the reddish phase, or the satellite flickers before it disappears completely.

These phenomena are most interesting from the meteorological point of view. Until now they have received only very little attention, although they are observable with binoculars in the case of bright satellites. The whole phenomenon lasts for about half a minute. The entire process occurs in reverse as the satellite emerges from the Earth's shadow; then, however, observations become very difficult because of the uncertainty of the position in the sky.

14.4. Time Services

14.4.1. Time Signals

Every satellite observation must be accurately timed, with an accuracy which is otherwise required only for star occultations by the Moon. How important the necessary time service becomes is shown by the fact that observers in the Soviet Union have been supplied with standard time during their observing periods by telephone, in spite of the high costs. European observers can always use regular hourly time signals, with which they can check a synchronous clock or a stop watch. It is much more convenient and reliable, however, to make use of continuous second pulses.

Continuous time service is ideally provided by the transmitters of the National Bureau of Standards: WWV in Washington (e.g., on 10, 15, 20, and 25 MHz) and WWVH in Hawaii (e.g., on 10 and 15 MHz). They give continuous second signals and in addition every 5 minutes, spoken hours and full minutes (EST) or corresponding Morse signals (UT). They are switched off only between the minutes 45 and 49, when they can be replaced by a stop watch or another transmitter (for instance, CHU in Canada on 7.34 and 14.67 MHz); see page 94).

14.4.2. Time Registration

It is not necessary here to give details of chronograph design. But it may be mentioned that the recording can be shared among various observers by arranging for each to use a different height-of-recording impulse. Again, one can separate transits through the north and south by opposite directions of the pen deflection. More possibilities still are offered by the oscillograph; those familiar with its operation will have plenty of ideas. Tape recorders are very popular in amateur circles, since these do not require any special expenditure and permit simultaneous time registration and recording of the whole observation.

The tape recorder should be run at its greatest possible speed. When played back the machine is run at its slowest speed, and we obtain in this way a kind of "acoustic slow motion." The tape will be repeatedly measured with a stop watch to increase the accuracy. Any missing impulses can be bridged easily.

It is possible to move the tape, when playing back, very slowly by hand and to mark the impulses heard in the loudspeaker directly on the magnetic tape with a wax crayon. It is of course necessary to perform this marking always at the same relative position. The tape can then be measured exactly as a chronograph band and can later be cleaned.

Arrangements should be made for simultaneous recording of the time signals (WWV, seconds pendulum, or metronome), handmade impulses of the observers, and oral remarks. Special possibilities are offered by a twin-track recorder. In any case, the observers should use push-button switches since spoken time signals are not precise enough. The tape recorder should be kept at a uniform temperature and should be started early to guarantee uniform running.

A reliable watch with a fast-moving second hand can furnish a time record by using an auxiliary camera, which is set up with full aperture in a dark room. Using an electrically monitored electronic flash one can photograph the watch dial and perhaps also additional data about day, station, etc. Each negative should be restricted to at the most two observing photos and one check photo (full minute). The photograph of a swinging second pendulum would permit a rather fine subdivision of the second if calibrated for amplitude.

14.4.3. Time Markings on the Photographic Trace

In order to insert time markings on the satellite trace itself, we can either interrupt or deflect this trace at certain moments. This deflection can be done either in a rough manner mechanically by knocking the stand, or optically by inserting a plane glass disc in front of the entrance aperture by a synchronized relay. A disadvantage of the deflection method frequently noted is the diffuseness of the markings. Interruptions of the trace can be measured easily in the microscope and are therefore usually more suitable.

If the interruptions are carried out with the help of a shutter we must make allowance for its inertia. It is safer to direct a photographically ineffective infrared light ray from outside the field of view onto the shutter in such a way that it excites a photocell inside the camera, and records the amplified current by pen or oscillograph. Such an effort, however, will not be within reach of everybody.

The simplest procedure is to cover the objective by, for example, the slide of a plate holder at a safe distance. To be more easily distinguished these times of covering can be made unequally long. A large number of measurable points can be provided by uniformly chopping up the trace, as is often done in meteor observation.

The technique of chopping described below can with some practice easily guarantee an accuracy of a tenth of a second. The observer flaps the slide of the plate holder up and down in front of the objective lens, keeping to the rhythm of the WWV transmitter or of another acoustical signal. For better distinction he makes the downward interruption shorter (at every odd second) and the upward interruption a little longer (at every even second). Each tenth second he performs the upward motion outside the field of view, so that a trace section of double length results. The movements must start well in time so that the synchronization during the critical transit phase is a perfect one. In order to save all writing during the operation, if there is one exposure per minute, every time we put a plate holder away we place a special card on it which indicates the particular minute. A set of sixty cards is easily prepared beforehand.

14.5. Radio Observations

14.5.1. Satellite Transmitters and the Ionosphere

It is extremely valuable to observe satellites in the range of the meter and decimeter waves, since we thus obtain for the first time reliable information about the electrical condition of the high atmosphere. This information is not available from ground-based observations of the limiting frequency, nor from radio astronomy or short-lived rockets. This applies particularly to the reception of the pure carrier wave. This is not the place to deal with the decoding of the measuring data transmitted by the satellite; we restrict ourselves to some remarks on the astronomical use of radio observations.

Optical refraction can nearly always be neglected in the case of satellite observations, but the refraction of radio waves is very marked. It shows itself in the lengthening of the apparent diurnal arc (i.e., an extension of the time an object is observable). There are also anomalies in the form of irregular refraction, double refraction, and polarization, which show themselves with simple receiving equipment as variations of the incident field strength. It is for this reason that the state of the ionosphere is of great importance for the evaluation of satellite reception, and vice versa.

Numerous ionospheric survey stations are engaged in a continuous measurement of limiting frequencies, quantities that give a measure of the electron density of the high atmosphere. They are also the basis for the calculation of the refraction coefficients:

$$n = \sqrt{1 - \left(\frac{\nu_0}{\nu}\right)^2}.$$

Here ν_0 is the limiting frequency and ν the transmitter frequency in which we are interested. The phase velocity is then: $c' = c/n$, where c is the velocity of light in vacuum. The optical and the radio signals thus do not arrive at exactly the same time. The lengthening of the radio paths because of the ionospheric refraction also contributes to this effect. The actual path length is

$$E' = E\frac{H}{\displaystyle\int\frac{dH}{n(H)}}.$$

Here we have again called E the geometric distance of the satellite. In practice we replace the integral by a summation over several ionospheric layers.

14.5.2. The Receiver

Radio observations are not restricted to the short twilight periods and are also independent of the weather. Within a short time, therefore, extensive material becomes available. It is interesting to note that during the very first months of the original Sputnik observations the Soviet observers collected more than ten times as many radio observations as optical data, although the objects were very bright and the radio transmitters were only working for short periods.

However, the amateur observer must in general restrict himself to "active" satellites that are within his range. The location of "silent" satellites according to the radar principle is the privilege of a few radio stations since its technical demands exceed even those required for the transmission and reception of Moon echoes. Large interferometers, as used in professional radio astronomy, are also beyond the reach of the amateur astronomer.

Successful reception of satellite signals requires an expert in radio techniques who is equipped with a first-class receiver. Thus this field has essentially become the realm of well-organized amateur radio enthusiasts around the world. The interested amateur astronomer will be well advised to contact his local amateur radio club to receive information about the design and operation of special receivers and converters, frequency measuring devices, and aerials. Those who have neither the time nor the skill for the construction of a homemade station can buy commercial receivers of the required quality, but these will certainly not be inexpensive and are obtainable only through specialist dealers. The electronics industry has often lent valuable support to the aims of amateur astronomers with advice and active help.

14.5.3. General Measurements

As soon as we have verified that the satellite is "transmitting" and that its first signals have been recorded on the tape of our recorder, the first technical problem arises: the registration of the field strength of reception. We usually employ a millivoltmeter at an intermediate-frequency stage. Its deflections are noted in relative S units or (though absolute calibration is not essential) microvolts, together with the time of the observation. The principal maximum of the resulting curve indicates the time of the nearest "electrical approach" of the satellite. The disturbing other maxima can be repressed by changing the characteristics of the aerial.

The next step is the continuous "homing" of the satellite azimuth. For this purpose the amateur astronomer will only rarely possess the necessary equipment. He will be more successful in studying the frequency shifts, which are discussed in the following section. However, there are still other possibilities for the amateur astronomer. When the satellite, toward the end of its life, traverses somewhat denser layers of the atmosphere it produces a trail of ionized gases. This phenomenon is well known from the study of meteors. Sometimes such ionization regions show themselves by the reflection of short radio waves which can then be received beyond their normal range. It was thus possible to observe the disintegration of the first Sputnik into several fragments in the course of a few successive days.

14.5.4. The Doppler Effect

The astronomer is familiar with the Doppler effect, which makes it possible to determine the radial velocity of stars and spiral nebulae by the displacement of spectral lines. It is represented by a shift in the frequency

$$\mathrm{d}\nu = \nu \cdot v/c'.$$

Here ν is the emitted frequency, v the relative radial velocity between the source of the radiation and the observer (taken with its correct algebraical sign), and c' the phase velocity.

If the satellite passes directly above the observing site, the effect is more pronounced than if it passes at some distance from it, since it is only the radial component of the velocity that counts. The total frequency change is

$$\Delta\nu = 2\nu \frac{v_{max}}{c'}$$

and can amount at a velocity of 8 km per second and a satellite frequency of 20 MHz to about 1 kHz. It is therefore necessary in the course of the transit to continually tune the receiver, which must have very good frequency resolution. If we follow the frequency shift accurately, say by mixing with a standard frequency and recording the difference signal, we obtain pairs of values (ν, t), where t is the time; the result is a symmetrical S-shape curve. Each measured value can therefore be used twice: once before and once after the instant of nearest approach of the satellite.

At this moment of nearest approach the relative velocity changes its algebraic sign; this is the point of inflexion of the curve. The tangent at this point is steeper, the more rapidly the maximum of the receiving frequency changes. This depends on the orbital velocity and the nearest approach of the satellite. By the cooperation of several stations it is possible to devise a special method of orbital determination solely from the time of nearest approach and the gradient of the tangent.

14.6. Elementary Calculation of the Orbit

14.6.1. Circular Orbital Velocity

In the case of a satellite near the Earth a first good approximation to the orbital shape is represented by a circle. If we neglect air resistance and consider the Earth to be a perfect sphere, we obtain for the orbital velocity K of a satellite moving in a circular orbit the expression

$$K = K_0 \sqrt{\frac{R}{R+H}}.$$

Here R is the radius of the Earth (about 6370 km), H again is the altitude of the orbit, and K_0 the theoretical circular velocity for $H = 0$, that is, 7.9 km/sec. The period U (in seconds) is then

$$U = 2\pi \cdot \frac{R+H}{K}.$$

Written in the form

$$U = \frac{2\pi}{K_0} \cdot \frac{(R+H)^{3/2}}{\sqrt{R}}$$

we recognize KEPLER's Third Law. If we insert the numerical values we obtain $U = 0.01 \cdot (R+H)^{3/2}$. The transformation $H = [(100\,U)^{2/3} - R]$ furnishes, with the help of the period, an approximate value for the mean altitude of the orbit. By the term period we always mean the time between two successive transits of the satellite above the same latitude circle and in the same direction. Here the rotation of the earth is irrelevant.

14.6.2. Inclination of the Orbit

The orbital inclination i of an artificial satellite is the angle at which the satellite intersects the equator when it moves from the Southern to the Northern Hemisphere; see Fig. 14–7. If the satellite moves exactly from the west to the east we have $i = 0$. In this case the orbit remains exactly above the equator. In all other cases it goes just as far north as south, because the orbital plane must always pass through the center of the Earth.

If at a given place (geographic latitude φ and longitude λ, here always reckoned to the east of Greenwich) the orbit has an inclination i' to the latitude circle, we obtain the orbital inclination from the formula

$$\cos i = \cos i' \cos \varphi.$$

Correspondingly we obtain

$$\cos i' = \cos i \cdot \sec \varphi,$$

if we know the inclination i and the azimuth A of the satellite's orbit. We have (counting astronomically from the south through the east) $A = 90° + i'$.

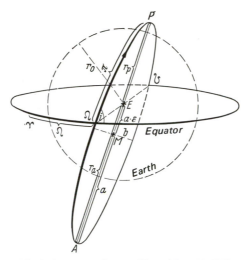

Fig. 14–7 The orbital elements of a satellite. After H. WÖRNER: *Die Sterne* **34**, 59 (1958). The notations are the same as for planetary orbits, but refer to the equator and not to the ecliptic.

14.6.3. Longitude of the Nodes

The point at which the north movement of the satellite intersects the equator is the ascending node, simply called "the node" in what follows. If we assume that its position in space is fixed (the cases where this is not so will be dealt with below) then its geographic longitude changes continuously because of the rotation of the Earth. We calculate the longitude of the node at the last equator transit of a satellite that at a given moment moves to the north at the northern latitude φ.

We now obtain the pathlength W covered by the satellite from

$$\sin W = \sin \varphi \cdot \operatorname{cosec} i.$$

We then have for the longitude difference L between observing site and node

$$L = L' - \frac{Z}{240}.$$

If L' is the difference in longitude for a fixed Earth, we have

$$\tan L' = \cos i \cdot \tan W.$$

The second term on the right-hand side makes allowance for the rotation of the Earth if the time $Z = W \cdot U/360$ seconds has elapsed since the transit through the equator; here U is also expressed in seconds.

14.6.4. Movement of the Nodes

The equatorial bulge of the Earth causes a precession of the orbital plane, which for a satellite is much more rapid than for the faraway Moon. Its value depends on the altitude of the orbit and the inclination and is approximately

$$p = 10 \cdot (\cos i) \cdot \left(\frac{R}{R+H} \right)^{7/2}$$

degrees per day. This is the amount by which the node moves daily along the equator toward the west. The actual values of p are of great scientific interest since they enable us to calculate the flattening of the Earth with an accuracy never achieved before.

In order to follow the geographic longitude of the node over a longer period, we must also take into account that in the course of one year of 365.25 days the Earth itself makes a full revolution of 360° around the Sun. Thus the node moves westward by an additional amount of 360/365.25 = 0.986 degree per day.

14.6.5. Geographic Ephemeris

We now possess all the data necessary to calculate, for a given time T, the geographic coordinates φ and λ of a satellite; we need to know the time T_0 and the geographic longitude λ_0 of a transit of the node, the inclination i, and the period U. We ignore the fact that the satellite does not move with a uniform velocity because its orbit is in reality not a circle but an ellipse. Furthermore, without additional data we cannot make allowance for the changes of the orbital elements that are due to air resistance and other external influences. Of course, it is just the evaluation of these effects that makes the reduction of precision observations so fruitful. Here we are only interested in obtaining a general picture of the observational possibilities, as expressed by a geographic ephemeris.

If we write $T - T_0 = Z$, we obtain the geographic latitude from

$$\sin \varphi = \sin i \cdot \sin \frac{360}{U} \cdot Z.$$

Here U and Z are to be expressed in the same units of time. The determination of the geographic position of the satellite is carried out in two steps:

$$\tan L' = \cos i \cdot \tan \frac{360}{U} \cdot Z$$

and

$$\lambda = \lambda_0 + L' - 0.25 Z_m - (p + 0.986) Z_d.$$

Here, Z is expressed as Z_m in minutes and as Z_d in days.

If Z is considerably larger than the orbital period U, we start the calculation with the value of the last node. We form Z/U and insert that part of Z in the formula for λ which represents an integer multiple of U. Here $L' = 0$. We then enter the complete system of our formulae with the remainder of Z, using our just calculated λ in place of the original λ_0.

It is useful to make a template that represents the first half of the orbit and ends at the equator; this part of the curve shows the times after the transit of the node as a scale, with each time marking shown below the corresponding geographic latitude. The template can be calculated from i and U, or can be made in accordance with the published timetables which list the places above which the satellite passes; it can then be used also for the determination of orbital inclination and longitude of the node, as well as for the transit of the node and the period. Because of the hemispherical symmetry of the orbit one can also use values from the other half of the orbit. The template can then be set on a suitable map of the world by simply moving it along the equator to any particular value T_0.

14.7. Elliptical Orbits

14.7.1. Eccentricity and Altitude

Every satellite that moves in an elliptical orbit continuously changes its distance from the center of the Earth, and thus its altitude varies between a smallest and a largest value, which correspond to the points of the perigee and apogee. The more these limiting values of H differ, the larger the orbital eccentricity e. If we ignore the fact that the Earth is not a perfect sphere and that the observer is generally not at sea level, we can write

$$e = \frac{H_A - H_P}{2R + H_A + H_P}.$$

Here H_A and H_P denote the altitudes of apogee and perigee, respectively, while R is again the radius of the Earth.

14.7.2. Motion of the Perigee

The position of the perigee of the orbit is expressed by an angle called the "argument for the perigee," the angle subtended at the center of the Earth between the satellite and the node. If we know this angle we can, with the help of some other quantities and some mathematics, calculate the altitude for every point of the orbit. These values would be constant for every given geographic latitude if the position of the perigee did not change. In fact, however, the perigee changes by

$$D = 5 \left(\frac{R}{R+H} \right)^{7/2} \cdot (5 \cos^2 i - 1)$$

degrees per day. Furthermore, we have to use for H the mean value

$$H = \tfrac{1}{2}(H_A + H_P).$$

The cosine term has a special consequence, since for $i = 63°26'$ we have $D = 0$. If the orbital inclination is larger than this, the motion of the perigee becomes retrograde. In the case of the Sputniks, with i about $65°$, the values of D were already very small. As with the precession of the nodes, the perigee motion also yields important data concerning the figure of the Earth.

14.7.3. Semimajor Axis

The size of an elliptical orbit is determined by the semimajor axis a. This is evident if we consider the formulae for the apogee and perigee distances, given respectively:

$$R_A = a(1+e) \quad \text{and} \quad R_P = a(1-e).$$

In connection with the relationships

$$R_A = R + H_A \quad \text{and} \quad R_P = R + H_P,$$

where R is the radius of the Earth, we use these formulae to calculate the limiting values for the height of the orbit as a function of a and e.

We now introduce Kepler's Third Law giving the ratio between the orbital period and the semimajor axis and write it in the form

$$a^3 \approx 36.3 \cdot 10^6 U^2.$$

Here, U is expressed in minutes and a in km.

14.7.4. Influence of Air Resistance and Gravitational Variations

In celestial mechanics, the semimajor axis represents a measure of the orbital energy. If this energy decreases under the influence of air resistance, then the semimajor axis and the period also decrease. The satellite approaches the Earth, and its motion becomes faster. This apparently paradoxical phenomenon, whereby a body moves faster through braking, is valid until the orbit is transformed to free fall in ballistic conditions. Furthermore, as confirmed by experience, the theory of motion in a resistant medium shows that with decreasing semimajor axis the eccentricity, too, decreases. The orbit approaches a more and more circular shape. The distance of the perigee remains almost constant for a long time, while the distance of the apogee decreases rapidly.

Exact calculations of these processes, on the basis of all available data, have already led to important insights into the density of the high atmosphere. The first result is that this density is surprisingly large. Various other changes with height and time have been found, which disclose that even in the highest layers of the atmosphere a kind of "weather" still exists.

The Earth is not a perfectly uniform sphere, nor even a perfectly regular flattened sphere. Its geoid (mean sea-level surface) has a complex and irregular shape, and this variation in gravitational field causes observable changes in satellite orbits, which can be used with high precision to calculate the Earth's true shape.[2]

For all these purposes, a continuous survey of all artificial Earth satellites is necessary. In the United States this survey (under the code name "Moonwatch") is organized by the Smithsonian Institution.[3] In Great Britain the Royal Observatory, Edinburgh, coordinates the observing program, while the main center for orbital calculations is the Royal Aircraft Establishment, Farnborough.[4]

In both these continuing surveys a worldwide network of amateur observers have played and are playing an important part in making the many thousands of observations necessary. With their help, extremely valuable calculations have been made of the properties of the upper atmosphere, and of the figure of the Earth. This is an impressive illustration of the scientific importance of reliable observations made by the amateur astronomer.

[2] KING-HELE, D. G.: *Observing Earth Satellites*. London: Macmillan, 1966.

GAPOSCHKIN, E. M., and LAMBERT, K.: Smithsonian Astrophysical Observatory Special Report 315 (1970).

KING-HELE, D. G.: Heavenly Harmony and Earthly Harmonies, R.A.E. Technical Memo. Space 174, R.A.E. Farnborough, December 1971—to be published in *Vistas in Astronomy*, A. BEER, ed., Vol. 16. Oxford: Pergamon Press, 1973.

[3] NELSON HAYES, E.: *Trackers of the Skies*. Cambridge, Mass.: H. A. Doyle, 1968.

Smithsonian Institution Bulletin Series, M.T.B.1., April 1967, *et seq.*, from Smithsonian Astrophysical Observatory Moonwatch Headquarters, 60 Garden Street, Cambridge, Mass.

[4] Notes on Satellite Tracking, B. McINNES, Royal Observatory, Blackford Hill, Edinburgh, EH9 3HJ.

15 | Observation of the Planets

W. Sandner

15.1. The Observation of Planetary Surfaces

This field has long been particularly favored by amateur astronomers, and several observers have achieved significant success. Names such as DENNING (1848–1931), R. M. BAUM, McEWEN in English speaking countries; SCHRÖTER (1745–1816), FAUTH (1867–1941), LÖBERING in German regions; and DU MARTHERAY (1892–1955) in France will always remain closely interwoven with the history of planetary research. In spite of this, however, in view of the great optical achievements and the multitude of observation and measuring methods at the disposal of today's scientific institutes, it is justified to ask whether planetary observations by the amateur can still be a promising undertaking. Apart from the purely personal value of every observation, of whatever kind, the work of amateur astronomers in this field still has a special, if restricted, value, as long as the observer concerns himself only with fields appropriate to his possibilities, say regular surveys, and provided he possesses the necessary personal and instrumental equipment.

Lunar observations provide an excellent exercise (see Chapter 11), since lunar details are absolutely constant and subject only to different illumination according to the Sun's position; thus, with the help of maps or good photographs, it can be immediately tested whether the observer has seen and drawn things correctly. Also simultaneous observations by two amateur astronomers with the same instruments, but otherwise completely independent, are of value and important educationally from the point of accuracy. Beyond that, they give the observer a personal judgment of his own observations.

It is essential from the very beginning not only to describe what has been seen but to fix it at once in a drawing. It is an old experience that the intention to draw greatly promotes the observing as such, since the draftsman is forced to look at the object more closely to be able to put it on paper. He therefore sees more, and in more detail, than the mere observer. Special drawing ability is not required, since it is essentially only the drawing of contours with which we are concerned; the rest comes quickly with practice.

It is no use trying to observe everything; rather one should restrict the work to some very special objects. These, however, should be followed up at every possible opportunity. If our observations are to be of real value they must be continued with perseverance over years and decades. This applies in particular to surveys—one of the most important personal assumptions for our activity.

It is regrettable that, except in rare cases, most amateur observers use only the first half of the night for the observation. Of course, in view of professional obligations this is quite understandable; nevertheless observations should be extended beyond midnight. First, the air is generally better then, and second, the observer, if he has slept for a few hours in the evening, is then fresher than after a day's work. In particular, observations of Mercury and Venus in the morning hours are absolutely necessary for scientific reasons.

15.1.1. Instrumentation

Whether the observer uses a refractor or a reflector will depend on his financial circumstances as well as on the local site. A large focal length is most advisable. It is for this reason that we prefer a Cassegrain arrangement to a Newtonian one for our reflector, although we do not deny the usefulness of a Newtonian mirror. Also for refractors we shall find a small aperture ratio useful and in no case should this be larger than 1:15, if possible 1:18 or smaller. An E-objective 1:20 possesses an image quality which is about equal to that of an Apochromat. A long focal length brings about a certain clumsiness, but this can be avoided through the use of a SCHAER-type instrument (see Sec. 2.2.4.3). The relatively large detail on the planets Mars and Jupiter, which observers in the second half of the Seventeenth Century, such as HEVELIUS and CASSINI, saw with simple (nonachromatic) objective lenses, can only be explained by the extremely long focal length of their telescopes (air tubes). PFANNENSCHMIDT suggested years ago that it would be a very rewarding task for amateur astronomers to imitate these attempts with single lenses of smallest aperture ratio (1:30 and less).

Uncomfortable posture, which is the rule during observation with an ordinary reflector, reduces accuracy particularly when drawing special surface detail. The use of a coudé reflector is of advantage, and also a Newton refractor [see Fig. 2–12(b)], which has the additional advantage of the shortening of the length to about one-half the total length. Also a zenith prism can be useful.

As for the eyepiece, the monocentric eyepieces of the STEINHEIL type (see page 71) have proved most suitable. That they have only a small field of view is unimportant here.

Concerning the magnification we should accept the rule that for an instrument of normal focal length (refractor 1:15, reflector 1:10), the normal magnification should be numerically equal to the diameter of the objective expressed in millimeters. For long-focus instruments one can go to $1\frac{1}{2}$ times this quantity, but beyond that only in exceptional cases. The successful planetary observer M. DU MARTHERAY of Geneva has coined the phrase "*l'observation des planètes est un art délicat.*" He summarizes his experiences in a diagram [Orion, *Mitteilungen der Schweizerischen Astronomischen Gesellschaft*, No. 28, p. 109 (1950)] from which Table 15.1 was derived.

Table 15.1. Choice of Magnification for Planetary Observations

Objective	Magnification							
	Mercury	Venus	Mars	Jupiter	Jupiter's satellites	Saturn	Uranus	Neptune
75 mm	150	150	175	150	150	175	175	—
135 mm	200	225	275	200	375	250	300	300
250 mm	300	300	325	275	400	350	350	350
300 mm	350	350	350	300	500	375	450	500

The table might appear somewhat optimistic, but we have to consider that DU MARTHERAY possessed exceptionally good eyesight and was working with an excellent instrument under favorable atmospheric conditions.

Somewhat different results have been reported by H. WICHMANN [*Sterne und Weltraum* **4**, 175 (1965)] separated according to different atmospheric conditions. WICHMANN plots a curve, reproduced in Fig. 15–1, from which he derived Table 15.2. This table, too, demonstrates that smaller instruments are less sensitive to poor atmospheric conditions than the larger ones. It is for this reason that for objectives of large diameter (10 inches and larger) the use of a diaphragm is sometimes to be recommended (for fair or poor seeing); in this way FAUTH was sometimes able to improve considerably the quality of the images produced by his 385-mm refractor.

Special auxiliary instruments—micrometers, photometers, spectroscopes—are generally beyond the scope of the amateur astronomer's planetary equipment. A small micrometer, however, since it is practically indispensable

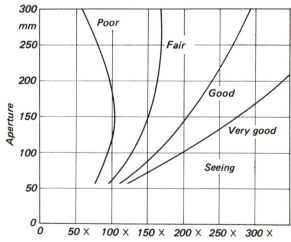

Fig. 15–1 Conditions for planetary observation.

Table 15.2. Magnification and Seeing

Objective diameter	Applicable magnification			
	Seeing poor	Seeing fair	Seeing good	Seeing very good
100 mm	90	125	160	190
150 mm	100	150	200	275
200 mm	110	170	240	340

for certain purposes, should be included in the equipment, if possible (see Sec. 15.5).

Observations with color filters are very useful. These, however, must fulfill high demands: they must be exactly plane, with a completely colored glass mass, and their exact absorption curve must be accurately known. The filters of SCHOTT and GENOSSEN are very suitable and reasonably priced.

G. D. ROTH recommends the following filters: RG2 (strong red), OG5 (orange), VG6 (green), GB12 (blue); the author of this article uses: RG5 (red), GG11 (yellow), VG8 (green), BG23 (blue). For the observation of Venus with its excessively bright light, a neutral filter can be of advantage [see also G. D. ROTH: *Handbook for Planet Observers*. London: Faber and Faber (1970)]. Better still is a wedge of neutral glass, which permits a continuous variation of the brightness of the image.

15.1.2. Observing Conditions

The amateur astronomer will be essentially interested in physical observations visually or photographically, since the recording of occultations of stars by planets is still beyond his instrumental reach.

It is essential to fully appreciate the apparent diameter of the planets. From such consideration follow the magnifications required to give to the small discs of the planets in the telescope the same apparent size as the Full Moon when viewed with the naked eye. This is shown in Table 15.3.

Table 15.3. Apparent Planetary Diameters

Planet	Apparent diameter	Magnification
Mercury	$4\rlap{.}{''}8$–$13\rlap{.}{''}3$	$280 \times$ in elongation ($6\rlap{.}{''}5$)
Venus	10 –64	$70 \times$ in elongation ($25''$)
Mars	4 –25	$70 \times$ in opposition
Jupiter	31 –48	$40 \times$ in opposition
Saturn	15 –21	$100 \times$ in opposition
Uranus	3 –4	$500 \times$ in opposition
Neptune	2.5	$750 \times$ in opposition

It is evident that the figures in the table can give only a rough indication of the situation, because the faintness of the object greatly reduces the image quality in the telescope. It is best for Jupiter, followed by Mars and Saturn, but because of the atmospheric conditions it is poor for Venus and still poorer for Mercury, not to mention the faint, distant planets Uranus and Neptune. Furthermore observation is unfavorably influenced by seeing and other conditions (e.g., uncomfortable posture of the observer or coldness of the winter night). From this it follows that on Mars an object must have at least the size of a small county, say some 60 km diameter, to be at all recognizable as a kind of undefined small spot in the telescope of a well-equipped amateur astronomer; for Jupiter the area must be, say, of the size of Australia, and for Saturn of that of Africa. It is best to be aware of these conditions.

Furthermore, it is important to have correct insight into the coordinate systems on the observed planets. The position of north and south in the reversing telescope is obvious. Concerning the terms "east" and "west" on a planetary disc, however, opinions differ and it would be most important to achieve a uniform designation for the future. If one transfers the terrestrial directions to the telescope image, we have east on the right and west on the left. It would be more correct to start with the notations "east limb" and "west limb" of the planetary disc and then to mark the limb where the Sun rises (in the telescope on the right) as the west limb, since an observer positioned at this part of the planetary surface would see the Sun in front of him in the east on the opposite horizon; he himself is therefore in the west. Accordingly, we should call the limb of the planets, visible on the left in the telescope (the sunset limb), which now disappears in the darkness of the night, the east limb of the disc. This notation corresponds also to the terms "oriens" (east) and "occidens" (west) used, for instance, on the maps of Mars by SCHIAPARELLI, ANTONIADI, and others, and to the fact that when we look to the south we have east on the left and west on the right. In this respect confusion in the notation can easily lead to error, for instance, in describing certain effects of haziness on the limb of Mars.

For physical observation of the planets it is important to note the phase angle, which can be taken from the Ephemerides, giving the measure of the unilluminated part of the planetary disc. This phase angle is the angle Sun–planet–Earth, or in other words, the angle subtended by the rays from the planet to the Sun and to the Earth (Fig. 15–2). It can be calculated from

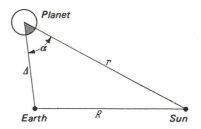

Fig. 15–2   Phase angle. After G. MÜLLER.

the distances Sun–Earth (R), Sun–planet (r), and Earth–planet (Δ), according to the formula

$$\cos \alpha = \frac{\Delta^2 + r^2 - R^2}{2r\Delta}.$$

The phase angle $0°$ therefore corresponds to full illumination (Full Moon), the phase angle $180°$ to invisibility (New Moon). For the planets within the Earth's orbit, Mercury and Venus, it can assume all values between $0°$ and $180°$. For the outer planets except Mars, it is always so small that it can be neglected. In the case of Mars it reaches a maximum value of $46°$, when nearly one-eighth of the disc remains invisible.

15.1.3. Some Special Hints

Without, as a rule, being aware of it, the experienced observer takes into account some physiological effects that may under certain circumstances strongly influence the observations. It is evident of course that only a completely dark-adapted eye can achieve good results. Beyond this, however, to avoid fatigue of the eye, it is recommended to take the eye away from the telescope frequently and look into the darkness for 5 to 10 seconds. This is of special importance for discs with little contrast, such as Venus and Uranus.

A further fact to be considered is that we see objects of vertical extent differently from those of horizontal extent. This is of particular importance in the case of Jupiter with its dark belts running parallel to the equator. It is therefore useful to view the planet after the completion of the drawing, in a position rotated by $90°$ (using a prism or an eyepiece turnable by $90°$), and then to complete the drawing accordingly.

In the case of Mars, whose surface features as a whole are invariable, it can be a hindrance that the experienced amateur astronomer soon knows the whole map by heart, for if, as frequently happens, he recognizes a conspicuous object at first glance, such as Syrtis major, M. Acidalium, or Solis Lacus, the entire map of that particular region of Mars comes to mind. It is then difficult for him to decide for certain whether he actually sees the object or only believes he sees it, since he knows it must be there. To avoid this illusion and possible resulting errors, it has been found useful to observe one of the oppositions through a reversing prism. In this way the familiar picture is reversed and the details are in a totally different arrangement.

Observation with both eyes facilitates the discovery of very fine details. Some experienced observers therefore recommend the use of a binocular attached to the eyepiece collar. C. RECLA (Verona) achieved very good results with a 5-inch double refractor.

A thorough discussion of errors and disturbances arising in planetary observations has recently been published by W. W. SPANGENBERG [*Astronomische Nachrichten* **281**, 6 (1952); and *Die Sterne* **41**, 94 (1965)].

It is evident that when all possible sources of error have been avoided only really established details should be incorporated in the drawings. To facilitate proper orientation of the entries, some printed "template" will be useful here which clearly indicates the axes of the coordinate system. Exact orientation of the image is indispensable. The horizontal axis should correspond to the east–west direction. To find this line, one simply stops the driving clock and allows the planet to move along the horizontal wire of the cross wire at the beginning of the observation. Jupiter requires a specially drawn "grid" because of its strong ellipticity; the same applies to Saturn because of its ellipticity and the changing appearance of its ring. For all the other planets the same form can be used, with different diameters. Such templates are, e.g., obtainable from the section for lunar and planetary observation of the "Vereinigung der Sternfreunde" (see page 550), and from other astronomical organizations. There are also within the "British Astronomical Association" special sections for the study of the various planets. Information concerning the purchase of templates, such as for Jupiter and Mars drawings, can be obtained from the BAA's Jupiter Section, which is directed by W. E. Fox, 40 Windsor Road, Newark, Notts, England. Reference is made to the corresponding work in the United States, the "ALPO," Harvard College Observatory, Cambridge, Massachusetts.

As far as Mars and Jupiter are concerned it is advisable to calculate a reticule to facilitate the rapid reading-off of the coordinates of any surface feature. Such grids can also be bought and enlarged on film to the appropriate size for measurement on the drawings.

As a matter of principle each amateur astronomer should, as far as possible, undertake the analysis of his observation himself; we return to this point below. The required Ephemerides can be extracted from the Yearbooks (see page 548).

It is very helpful for the astronomer to join a special group of planetary observers, so that he will always be up to date regarding unforeseen phenomena and the latest results of research, and thus able to hand over to the particular experts those of his observations that he cannot analyze himself.

15.2. The Observation of the Various Planets

15.2.1. Mercury

Mercury, the large planet nearest to the Sun, deserves its reputation of being the most difficult to observe. The old fable of Copernicus, who is supposed to have regretted on his deathbed never having seen Mercury at all, certainly has all signs of improbability but is nevertheless not to be dismissed altogether and is always put forward as a proof of the difficulties in the observations of the planet. In any case I would like to strongly recommend its observation. It may serve as an encouragement that such a demanding observer as Fauth once wrote that even a 4-inch telescope can achieve remarkable results in the study of Mercury's surface features.

Mercury, moving inside the Earth's orbit, is never in opposition to the Sun but in the best case can occupy elongations of about 28°. In the tropics, where the ecliptic descends steeply below the horizon, observing conditions are relatively the best; they deteriorate rapidly the more we move away from the equator north or south, and in mean latitudes we see Mercury only in the twilight shortly after sunset in the west (in its eastern elongation) or shortly before sunrise in the east (western elongation) but always in the haze of the horizon. Day observations are discussed below. Observations from high mountains have a chance of success only if it is possible to observe Mercury as close as 2° from the Sun, as demonstrated by the achievements of French astronomers on the Pic du Midi in the Pyrenees, at 2900 m, with the application of the most exacting precautions. Because of the position of the ecliptic horizon, and the inclination of Mercury's orbit toward the ecliptic, the best observing conditions in the Northern Hemisphere of the Earth prevail during evening visibility in spring and morning visibility in autumn. Mercury shows phases similar to the Moon, in the evening (eastern elongation) with a narrowing crescent, and in the morning (western elongation) with increasing crescent.

In the tropics, where the ecliptic descends steeply, as mentioned above, and where the twilight is therefore short, Mercury can be easily observed. However, at average latitudes the inclination of the ecliptic toward the horizon is small, smaller still at higher latitudes, and even at the time of its greatest elongation the planet reaches only a moderate altitude above the horizon. Observing conditions are made still more unfavorable by the fact that Mercury's orbit deviates strongly from a circle. In the morning sky at the beginning of April, Mercury can reach an elongation of as much as 27°50′ but at the end of September one of only 17°52′. In the evenings, conditions are reversed, the largest possible apparent distance from the Sun occurring in autumn, the smallest distance (in the largest elongation) in the northern spring. Nevertheless for observers in North America and Europe better evening visibility occurs in spring and the better morning visibility in autumn. In spring, the Sun moves north in its orbit in the ecliptic. In the evening visibility Mercury precedes the Sun and is thus farther north, reaching a higher altitude above the horizon for an observer in the Northern Hemisphere. The opposite occurs in autumn when the Sun wanders south, and in the evening visibility Mercury is more in the south than the Sun and does not reach high altitudes.

At lower latitudes Mercury becomes an easily observable object, since at maximum it reaches the brightness of Sirius. Nevertheless, it never becomes a conspicuous object at average latitudes, being too near the horizon. Its apparent brightness varies according to its position between $-1^m\!.6$ and $+1^m\!.7$. Its color is decidedly yellowish; this is best seen when it is near Venus so that both planets can be seen in the same field of view. On such an occasion in January 1956 the Mexican planetary observer F. J. ESCALANTE (1881–1972) found Mercury "definitely yellowish," in contrast to the "white

Venus." Measurements of the Russian astronomer L. N. RADLOWA in 1949 with a visual colorimeter showed its color more reddish than expected.

It is extremely difficult to distinguish details on the little disc; this is caused not so much by the small size of the image but by the unfavorable visibility conditions. Since the days of SCHIAPARELLI it has become customary to observe it telescopically in daylight, and in this way ANTONIADI achieved an important observational series. Of course the disc is then so faint that the amateur astronomer with his restricted means cannot make out many details, as these require objectives of more than 30 cm aperture. It is essential to catch the right moment during twilight observation. Since this cannot be predicted, one will often have to wait a relatively long time at the telescope, and most of the waiting will be in vain. The American planetary observer W. H. HAAS writes, "the most important factor in achieving success is to observe the planet at a time when the contrasts are strongest; surface detail is most conspicuous when the image is neither too faint as in the daylight sky, nor too bright as in the evening or morning twilight" [*Popular Astronomy* **55**, No. 3 (1947)]. Similar remarks apply to Venus. If one of the short moments of steady air arrives (often only of a few seconds' duration) then details on Mercury appear clear and in good contrast, similar to those of Mars; but these moments are extraordinarily rare. Nevertheless, the English planetary observer PATRICK MOORE assures us that he has seen the Solitudo Criofori with a 3-inch refractor—a remarkable achievement, indeed.

Experience shows that color filters of average density make an essential contribution to the steadiness and contrast of the image during twilight. In the author's experience, blue filters are particularly advantageous.

In spite of these difficulties, years of careful observation have led to *maps* of Mercury. Up to the middle of the last century the basis of these was a rotational period proposed by SCHIAPARELLI in 1889. However, according to radio-astronomical researches in 1965, the rotational period of Mercury was determined as being only 59 days (which could easily be reconciled with the old optical observations), and a revision of these maps became inevitable. The latest map of Mercury, by CRUIKSHANK and CHAPMAN (see Fig. 15–3), uses this 59-day period. Since maps of Mercury are not so well known as the numerous maps of Mars, we attach a list:

Year	Author	Reference
1889	G. V. SCHIAPARELLI	Sulla rotazione di Mercurio. *Astronomische Nachrichtene* **123**, No. 2944, Kiel.
1896	P. LOWELL	New Observations of the Planet Mercury. *Memoirs of the American Academy of Arts and Sciences* **12**.
1920	R. JARRY-DESLOGES	Observations des surfaces planétaires.
1934	E. M. ANTONIADI	La Planète Mercure et la rotation des satellites. Paris: Gauthier-Villars.
1936	H. MCEWEN	The Markings of Mercury. *J. Brit. Astron. Assoc.* **46**, 10.

1947	W. H. HAAS	Ein zehnjähriges Stadium des Planeten Merkur und seiner Atmosphäre. *Popular Astronomy* **55**, No. 3. (German version by E. PFANNENSCHMIDT).
1948	L. RUDAUX	*L'Astronomie*, Paris.
1953	Pic du Midi	*L'Astronomie* (Bulletin de la Société Astronomique de France), p. 65, February 1953.
1960	G. WEGNER	*Strolling Astronomer* **14**, 11–12.
1964	W. W. SPANGENBERG	*Die Sterne* **40**, 26.
1967	P. CRUIKSHANK and C. R. CHAPMAN	*Sky and Telescope* **34**, No. 1.

For designation of the features on Mercury astronomers use the names introduced by E. M. ANTONIADI [La Planète Mercure et la rotation des satellites. Paris: Gauthier-Villars (1934)].

Some observers, such as ANTONIADI with the 80-cm reflector of the Meudon Observatory, believe that they have seen certain changes in surface features on Mercury, partly of a periodical, partly of a permanent nature. In any case such observations, which after all might be illusions, are certainly beyond the scope even of a well-equipped amateur astronomer. On the other hand the latter can easily follow the changes in Mercury's appearance due to the pronounced libration of the planet; the map by WEGNER of 1960 is the first one to take the libration into account.

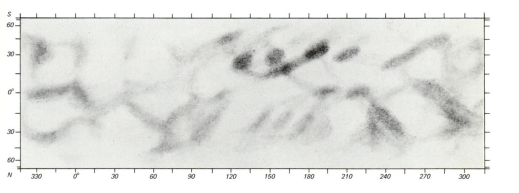

Fig. 15–3 The new composite map of Mercury, based on 130 visual observations, with south at the top. A rotation period of exactly two-thirds the orbital period was assumed. The map projection is a cylindrical orthographic equal area. Positions are accurate to no better than 10°, and the rather small effect of libration in latitude was ignored. (Dr. C. R. CHAPMAN and Dr. D. P. CRUIKSHANK, 1967, with kind permission of *Sky and Telescope*.)

In 1974, the spacecraft "Mariner 10" obtained most informative close-ups of Mercury, revealing a Moon-like, richly-cratered surface, with details far beyond anything shown by any terrestrial telescope.

In contrast to Venus the determination of the exact time of an observed dichotomy (phase angle 90°) is of little value for Mercury. Here only measurements with a filar micrometer are of value, while the derivation of phase drawings using graphical methods, so successful for Venus, usually fails for Mercury because of the unfavorable observing conditions. We also note that because of the orbital eccentricity the dichotomy does not coincide with the greatest elongation.

We do not wish to conclude the chapter on Mercury without a discussion of the transits of this planet in front of the solar disc. Although the scientific value of these transits is small, no serious observer would like to miss them, if only because of their rarity. They take place when the lower conjunction Mercury–Sun occurs near the nodes, which at present is around May 7 and November 9. Since at the time of the November transits Mercury is nearer the Sun and also the apparent solar diameter is larger than in May, the limits for the occurrence of a Mercury transit lie between May 4 and May 10 and between November 4 and November 14; therefore November transits are more frequent than May transits. On the average, 13 transits occur every century, following each other at time intervals of at least 3 and at the most 13 years. The following transits between 1974 and 2000 A.D. are to be expected: 13 November 1986, 6 November 1993, and 15 November 1999.

The duration of a Mercury transit depends, of course, on the length of the chord traversed by Mercury across the disc. It can reach $8\frac{1}{2}$ hours if Mercury crosses the center of the solar disc, and it can be very short indeed if the transit is close to a tangent. Both extreme cases will be realized in the 20th century: in 1973 the transit took 8 hours, in 1999 it will be nearly grazing. The duration of the immersion and emersions also depends on these conditions and amounts to about 3 to 5 minutes.

The observer is disappointed when he sees how minute Mercury's disc appears on the bright Sun, although the planet has its largest possible apparent diameter (in May somewhat larger than in November). The apparent diameter of Mercury on 9 May 1970 was 12″0; on 10 November 1973 it was 9″9. The observation itself can be made directly or in projection. The times of contact can also be determined with the help of a prominence spectroscope or coronograph (Mercury in front of the chromosphere); see "Observaciones del paso de Mercurio del 14 de Noviembre de 1953." *Boletin astronómico del Observatorio de Madrid* **4**, No. 7.

The following features of the phenomenon should be noted:

1. The times of contact. First and second contact: outer and inner contact when entering the solar disc; third and fourth contact: inner and outer contact when leaving the disc.

2. Possible visibility of an aureole around the black planetary disc.

3. Optical phenomena, in particular that of the black "drop" because of the difficulties it introduces in the measurement of the second and third contact.

An attempt to determine the diameter of the planet from the duration of the passage of Mercury's disc can be made; this was first done by HEVELIUS at Danzig in 1661. These diameter values are, however, rather uncertain for optical-physiological reasons.

In comparison with the very black little disc, the sunspots, and the umbrae, appear only very faintly violet or matte gray. Good photographs, with accurately recorded times, are therefore suitable for a photometric comparison between sunspot and disc of Mercury. In 1970 we had the rare case that the planetary disc crossed exactly through the center of a sunspot; it remained distinctly visible in front of the spot.

It is important to note that the appearance of the small absolutely black planetary disc in front of the large bright solar disc causes numerous optical illusions, for instance the observation of a bright spot inside Mercury's disc. Great caution is therefore recommended.

15.2.2. Venus

The next planet, Venus, is much easier to observe than Mercury. Although Venus moves around the Sun within the Earth's orbit so that it never comes into opposition and always remains a morning or evening star, it can reach an elongation of 47° from the Sun and is therefore observable for a long time after sunset or before sunrise.

Between one upper conjunction and the next there is a time interval of 584 days (synodic period). Venus goes through the same phases as does Mercury: from the lower conjunction via the western elongation (in the morning sky, increasing phase) and the upper conjunction (Venus in full light) and the eastern elongation (in the evening sky, decreasing phase) back to the lower conjunction (Venus equivalent to New Moon). The interval between the largest eastern and the largest western elongation (via the lower conjunction) is on the average 144 days; between the largest western and the largest eastern elongation (via the upper conjunction), 440 days, on the average.

Besides the Sun and Moon, Venus is the brightest celestial body, reaching at maximum (about 35 days before and after the lower conjunction) the magnitude $-4^{m}3$. In each elongation it remains accessible to photometric observation for about 220 days, that is, from 60 days after the upper conjunction to 12 days before the lower conjunction, and 12 days after the lower to 60 days before the upper conjunction, respectively. The brightness during this epoch changes as given in Table 15.4.

Because of its brightness Venus can even be seen with the naked eye for several weeks in daylight during each of its elongations, provided one has knowledge of its accurate position, stands in the shadow of a house, and has really adapted one's eyes to "infinity."

Table 15.4.

| Days before or after | | | | |
Upper conjunction	Lower	Distance from Sun	Phase angle	Brightness
60	232	15°4	21°6	$-3^{m}06$
100	192	25.3	36.3	-3.15
140	152	34.6	51.7	-3.29
180	112	42.3	68.6	-3.48
220	72	46.3	89.3	-3.76
240	52	44.0	103.2	-3.95
250	42	42.1	112.0	-4.07
260	32	37.4	122.9	-4.09
270	22	29.7	136.7	-3.92
280	12	18.2	154.4	-3.56

Using a telescope, it is of advantage to observe Venus in the daytime. An equatorially mounted instrument will be very useful; an azimuthally mounted one, the axis of which is perfectly vertical while the tube is connected to a simple pendulum quadrant, will also serve well. The altitude of Venus above the horizon can be determined from its coordinates, for instance, graphically and without calculation with sufficient accuracy, using a nomogram shown in Fig. 15–4, constructed by GRAMATZKI. It is of course necessary to have the telescope focused on the previous night, using some star image.

Another simple method for finding Venus or another bright star in the daytime with an azimuthally mounted telescope is the following. One selects from a star catalog or a star atlas a fixed star that has the same declination as our required object (e.g., Venus) but is so far away in right ascension that it can be set on comfortably on the previous night. The difference in declination should be as small as possible, and in any case less than half the diameter of the field of view. On the previous night one sets the telescope on the selected star, the right ascension of which we call α'_1, notes the time t_1, and clamps both axes. If the right ascension of our object is α, then at

$$t = t_1 + (\alpha_1 - \alpha)$$

(expressed in Sidereal Time), it will cross through the field of view. Again it is essential to focus very accurately, otherwise the faint disc of the star would disappear in the bright daylight sky.

For observations in twilight a faint neutral filter is useful or, better still, colored filters, which give astonishingly good results for surface observations on Venus. However, the experiences of different observers vary considerably. We now turn to the various items of possible Venus observations.

Bright and dark surface areas. All detail on Venus, if real, is extremely faint and in all cases very diffuse and difficult to draw; after all, the telescope

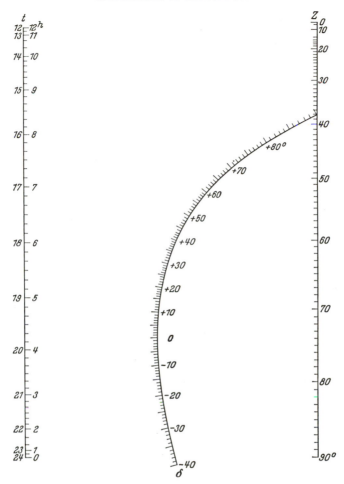

Fig. 15-4 Nomogram by GRAMATZKI; t = hour angle, δ = declination, Z = zenith distance.

does not show us the solid planetary surface but only a cloud cover which envelops the whole planet. A good comparison of the situation might be with that of a closed cover of stratus clouds, sunlit, and observed from an airplane flying at high altitude.

Certainly, the occasional brightenings at the ends of the crescent of Venus (which were sometimes rashly called the "polar caps") appear to be real. However, they do not resemble in any way the well-known polar caps on Mars. Whether, as is sometimes mentioned, a dark border around these brightenings is actually real or only a contrast phenomenon remains an open question.

However, the dark bands close to the terminator and running parallel to the latter, which are frequently reported, must be relegated to the realm of optical illusion. On the other hand, gray stripelike shadings, which run about perpendicular to the line connecting the end points of the crescent, could be real; they agree well with photographs by KUIPER.

Continued observations in this field are needed urgently, for which the use of colored filters is strongly recommended. Needless to say, the rigid critical faculties of the observers are essential. The choice of the instrument plays a decisive role, because it is a well-known experience that "detail" that appears conspicuous in small telescopes disappears in average and large instruments. It is desirable to extend the observation over many hours, at western elongation, say, from the rising of Venus up into the hours of noon or of the early afternoon. Even if such series do not yield any new results, they help to clarify how far things that have been "seen" can be considered as real features or only as illumination effects.

Under these circumstances it is not surprising that the rotational period and the position of the axis of Venus remained for a long time rather uncertain quantities. The literature quotes values for the period between 20 hours and 225 days (W. SANDNER: On the rotation of the planet Venus. *Vega* No. 10, 1953, and No. 30, 1955).

As for the diameter of Venus various of the quoted data disagree among themselves. We would like to mention three different methods of observation: (1) direct micrometer measurements at the telescope [see H. PAUSCHER: Drei Messreihen zur Bestimmung des Venusdurchmessers. *Astronomische Nachrichten* **285**, 174 (1960)]; (2) from Venus transits; and (3) from star occultations (see Sec. 15.3). The differences are probably due to the influence of the Venus atmosphere.

Deformation of the terminator. Evidently such deformations have occasionally been simulated by contrast effects, but as a whole we may consider them as realities. Every Venus observer with some years experience has recorded such phenomena. Whenever they are seen their exact shape and position should be recorded in a drawing (with care to avoid exaggerated deformations), with the duration of visibility and possible faintness.

Here we should also mention the relatively frequent observations of an unevenness of the end points of the crescent. Sometimes one of the horns is pointed, while the other appears blunted; sometimes one of the horns "overlaps." One must beware of illusions caused by the air spectrum.

Phase and dichotomy. The exact recording of the phase is made difficult by contrast effects (too faint an image in the daylight sky, too bright an image in the twilight). Twilight phenomena in the Venus atmosphere itself also play a role (the terminator appears slightly diffuse in contrast to the sharp planetary limb). The exact determination of the dichotomy, that is, of the time when the disc is exactly half illuminated, is of interest because of systematic differences between calculated and observed times, the so-called Schroeter effect [see E. SCHOENBERG and W. SANDNER: Die Dichotomie der Venusscheibe. *Annales d'Astrophysique* **22**, 839 (1960); and B. POLESNY: Die Beobachtung der Dichotomie des Planeten Venus, *Die Sterne* **42**, 165

(1966)]. The above-mentioned difference amounts to several days and is in the same sense for all observers; it may perhaps be different for different filters. It is improbable that it is due to a physiological effect and may rather be the consequence of an atmosphere of Venus.

Since it is practically impossible to determine the exact time of the dichotomy at the eyepiece (that is, to estimate whether the dichotomy has already taken place, is just beginning, or is already over), it is advisable to use a graphical method. Observations should start as early as possible, some weeks before the expected dichotomy, and should be continued as long as possible. On every clear evening or morning the phase should be drawn with utmost accuracy into templates, combining every 10 to 12 estimates to calculate a mean value. After conclusion of the whole series all drawings are measured again together with the thickness of the crescent, expressed in fractions of the Venus diameter as measured from horn to horn. These values are then plotted in a right-angle coordinate system, with the observing dates in Julian days as abscissae and the crescent diameters as ordinates. These points are then connected by a curve; this curve intersects the value 0.5 at the time of dichotomy.

Determinations of dichotomy photographically or with the help of a micrometer are also very useful (see Sec. 15.5).

Overlap of the horns and the ashen light. At the time of small crescents (near the lower conjunction) an overlapping of the horns occurs beyond half of the circle (see page 391). Values up to 25° have been measured with certainty, larger ones are probable. Here, too, an effect of the Venus atmosphere shows itself.

Many observers state that at the times of a very narrow crescent they have frequently observed the unilluminated part of Venus, shining in a faint light, similar to the earthshine on the Moon. The cause of this phenomenon is not yet fully known, and even the suspicion of an optical illusion has been mentioned. It is also not possible to say with certainty whether the unilluminated part is brighter or darker than the sky background. The color is given differently by different observers, usually as brownish, or brownish red, occasionally as purple-gray, faint gray, or even as dark violet. It is suspicious that this phenomenon has frequently been seen by inexperienced observers "at first glance" while experienced ones could see nothing. The late Berlin astronomer H. J. GRAMATZKI lists the following tasks for the Venus observer. Figure 15–5 shows the notations used below.

1. Estimation of the width d in relation to the planetary diameter D.

2. Do the end points a and g of the horns and the center M of the planet lie on one line or do they overlap? Determination of the overlap angles φ_N and φ_S through a measurement on the drawing, or only of the whole overlap angle φ.

3. Which of the "horn triangles" *abc* and *gef* appears to be the brighter one?

4. Is there a color difference between *abc* and *gef*?

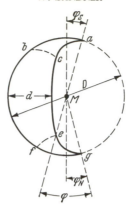

Fig. 15–5   The observation of Venus (data according to GRAMATZKI).

As for transits of Venus across the Sun we need not say very much. These phenomena are certainly very interesting, but no Venus transit will take place this century. The previous ones were on 9 December 1874 and on 6 December 1882. The two next ones will occur on 8 June 2004 and on 6 June 2012.

15.2.3. Mars

Mars is the first planet that circles the Sun outside the Earth's orbit, hence it can come into opposition. These oppositions are on the average two years and fifty days apart (the synodic period of 780 days). Because of the larger eccentricity of the Mars orbit they are of a very different character as far as a terrestrial observer is concerned, since Mars reaches an apparent diameter of only 13″ for the January–February aphelion oppositions, while perihelion oppositions in the late summer show diameters of 25″.5. In addition, observing conditions are modified by the fact that during the perihelion oppositions, Mars is in the most southerly part of the Zodiac, so that for a European observer at average northern latitude it hardly rises above the hazy layers of the horizon. (See Fig. 15–6.)

> The inclination of the rotational axis of Mars against the plane of its orbit means that at the time of the spring opposition the north pole is facing us, while during the perihelion oppositions in the late summer it is the south pole of the planet. Since it has unfortunately become the habit only to observe the "favorable" perihelion oppositions, the southern hemisphere of Mars has become much better known than its northern hemisphere. Serious amateur observers, therefore, should attempt the "unfavorable" aphelion oppositions, provided the necessary instrumental equipment is at their disposal.

As a general rule we might say that the observations should begin when the apparent diameter of Mars exceeds the value of 10″, and that they should be continued until it becomes smaller than this limit. For instance, during the perihelion opposition of 1956 this observing period lasted from 16 May to 18 December (about 7 months) but during the aphelion opposition of 1948 only from 27 December 1947 to 10 April 1948 ($3\frac{1}{2}$ months). Amateur astronomers whose telescopes have a diameter of more than 30 cm are able to extend the observing period; such observations are, of course, also of importance in view of the seasonal changes on Mars.

Unlike the more distant outer planets the phase angle of Mars must not be neglected. It can amount to 46°, and the phase angle effect is quite conspicuous. Its omission from a drawing proves with no further evidence either that the drawing was not carefully done or that the observer lacks the ability to grasp essential features.

Daylight observations of Mars have been carried out occasionally, and it is possible to distinguish some surface details at this time, but otherwise such observations have little importance.

The use of color filters is often of advantage. Generally an orange filter will suffice, but occasionally better results are obtained with red filters. In general, blue filters are unsuitable for surface observations. The Russian astronomer BARABASCHEW remarks that cloud formations are an essential

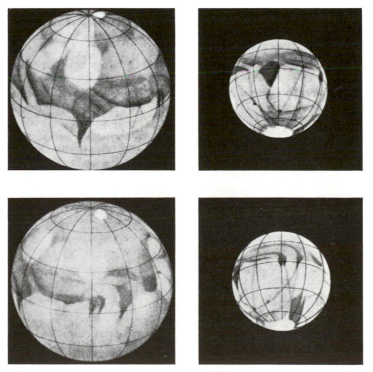

Fig. 15–6 Mars: orientation of axis and the relative sizes at perihelion (on the left) and aphelion (on the right).

object for the amateur astronomer; they are best in red light, only faintly visible through green filters, and cannot be seen at all with a blue filter. W. D. HEINTZ recommends the use of thick cobalt glass for blue, and an RG 1 filter of 2 mm thickness for red (in order to minimize scattered light); yellow and green filters have no advantage.

It is strongly suggested that printed templates be used for the drawings. The Mars Section of the British Astronomical Association recommends such templates of 2″ diameter. The author of this article uses printed forms adapted to 21, 27, and 39 mm for the diameter of Mars.

When the observer starts to draw surface details, he first of all enters the most conspicuous feature into a not-too-small template. This will generally be the polar cap [see Fig. 15–7(a–c)]. Starting with the cap has the additional advantage that it at once provides an approximate orientation. It is recommended, however, at least for stronger magnification, to use a micrometer wire for the determination of the north–south direction, since an eccentric position of the polar cap and the phase sometimes considerably falsify the real situation. Then one enters the contours of the dark features with a sharp soft pencil and makes this moment the basis for the calculation of the central meridian. Normally, strongly brightened areas, such as clouds, are surrounded by broken lines. Some drawings show a dark border region around the bright polar cap. This is no illusion. That this border feature is real is also confirmed (as particularly emphasized by C. FEDTKE) by the fact that other very bright regions, such as Hellas, never show such a dark border. When the outlines of the drawing are finished, the shading is added. In general, detail on Mars is hard and definite; however, one should also pay attention to the shading in the dark clouds which in many cases shows seasonal and sometimes daily variations.

In order to establish such brightness variations with certainty and to be able to follow them by drawing light curves, the observer is advised to estimate the color intensity of the various regions according to an empirical scale. According to G. DE VAUCOULEURS, whose empirical scale should be used, the bright polar caps are given the value 0, and the dark sky background the value 10; accordingly extended bright areas ("deserts") near the center of the disc have the value 2 and the darkest parts of the Martian surface the value 6 or perhaps 7. Experience shows that the estimation according to the scale is not more difficult and not less accurate than the well-known brightness estimates of stars by the ARGELANDER method. [See G. DE VAUCOULEURS: Physique de la planète Mars, Paris (1951); see also Orion, *Mitteilungen der Schweizerischen Astronomischen Gesellschaft* **7**, 139 (1962).]

We now turn to the determination of positions of details on Mars in arcographic longitude. Here transit observations are used, which at the same time provide a check on the rotation, that is, permit the determination of the time of transit through the central meridian. In 1954, W. D. HEINTZ published a list of selected objects near the equator:

1. Aryn, the bright tongue between the prongs of the Gabel Bay (Sinus Meridiani), which according to definition is the zero meridian.
2. Aurorae Sinus, western end point.
3. Solis Lacus, center.
4. Titanum Sinus, the north end of Mare Sirenum.
5. Trivium Charontis, center.
6. Cyclopum Sinus, northwest end of Mare Cimmerium.

(a) 27 June 1956, CM = 318°

(b) 1 August 1956, CM = 338°

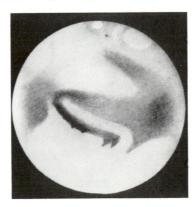

(c) 5 September 1956, CM = 343°

Fig. 15–7 (a–c)   Mars: drawings by E. ANTONINI for which a 6-inch refractor was used. From *Mitteilungen für Planetenbeobachter* **9**, 50 (1956).

7. Great Syrte, north point.

8. Portus Sigeus, on the northern side of Sinus Sabaeus.

HEINTZ remarked: "Frequent observation of a few localities is much more valuable than isolated uncertain notes about many points. Transit observations should be restricted to the period of about ten days before and after opposition (which is not identical with the day of greatest approach), since otherwise the phase, in addition to the nonuniform brightness of both borders, is rather disturbing and adds uncheckable errors to the observation."

The color of the bright object ranges from bright red to ochre. Observers with good color vision emphasize again and again the conspicuous presence of color in the dark areas. Different observers are of different opinion but besides a general gray the presence of greenish to bluish tones appears established [see SCHMIDT, I.: Sind die "grünen" Gebiete auf dem Mars wirklich grün? *Weltraumfahrt under Raketentechnik* **11**, 9 (1960)]. To make these facts the basis of a whole theory as has sometimes been suggested ("astro-botany") appears very daring. For such observation the use of a mirror or a good apochromat is preferred to an ordinary reflector.

To be able to note occasional irregularities (protrusions and indents) as well as light phenomena at the Mars limb usually requires very powerful optics and strong magnification, which are rarely at the disposal of the amateur astronomer.

The observation of cloud formation in the Martian atmosphere is an important field of work for the adequately equipped observer. Here, of course, an accurate knowledge of the background in all its fine detail is necessary, therefore only observers who have the experience of several complete Mars oppositions should attempt this task. Such cloud formations contrast in radiance with their surroundings and can even reach a certain similarity with the polar cap (as, for instance, a long-lived cloud formation that appeared in 1952 above the feature "Chryse"). Frequently they are only yellow and difficult to distinguish from the general background; or they can be seen only as a haziness effect or as faint brightenings, sometimes indirectly through the obscuring of some conspicuous surface detail. Such haziness effects occur above certain regions of Mars every day in the afternoon hours, sometimes over long periods. The continuous observation of compact cloud formations, which are relatively rare, permits the determination of the prevailing direction and velocity of the wind, and a comparative analysis of all available observations makes it possible to obtain pictures of the total weather situation on Mars (as has been shown by HEINTZ). We also recall the extensive haziness (sandstorm?) that affected nearly the whole southern hemisphere of Mars in the period from the end of August to the middle of September 1956 and at times even covered the polar cap [see HEINTZ, W. D.: Die Marsannäherung 1956. *Die Sterne* **35**, 129 (1959)].

Even more important was the great "sand storm" which raged on Mars between October and December 1971 – just at the time when the spacecraft "Mariner 9" reached the planet to turn into an orbit to circle it.

The so-called "violet layer" of the Martian atmosphere prevents the observation of the surface in blue light. Sometimes, however, this cover breaks, as, for instance, in August 1956 shortly before the large "dust storm," and the detail of the surface can then be recognized even in a blue filter. All this lies within the range of the amateur's tools, and since these are problems of continuous survey we have here a valuable field of activity.

During the most recent observing periods it has been demonstrated that photography of the Martian surface in black and white and even in color is certainly promising.

An interesting photometric experiment, which every amateur astronomer who masters ARGELANDER's method can carry out visually with an ordinary pair of binoculars, has been described by E. PFANNENSCHMIDT in *Mitteilungen für Planetenbeobachter* [No. 14 (1948)]: "A comparison of the light curve of Mars with a map of the planet demonstrates that the light variation (the amplitude of which is about 18%) is caused by the spot formation on the surface. The greatest brightness, therefore, exactly coincides with the time at which the region of the surface which is richest in bright spots is visible on the disc; the brightness minimum coincides with the visibility of the dark spots. The light variation is plotted in a system with the coordinates' rotational phase and brightness, respectively. According to GUTHNICK and LAU the maximum lies between the areographic longitude 117° to 120°, and the minimum between 300° and 6°. Comparison stars are best to be chosen amongst the red stars."

In the analysis of our observations first of all we calculate the central meridian for every single drawing. This can be done with the data published in the astronomical Yearbooks and with a table of the hourly changes of the central meridian given on page 513. The construction of the grid, however, must be repeated every time because of the inclination of the axis of Mars; the necessary formulae have been developed by K. GRAFF [see *In Publikationen der Sternwarte Bergedorf* 2, No. 7 (1924)]. When the positions of as many conspicuous points as possible have been determined these can be combined in a general map of Mars in Mercator projection.

Such a construction has been dealt with in detail by H. OBERNDORFER [*Sterne und Weltraum* 8, 66 (1968)]. He draws a rectangle 217 mm long and 100 mm high; its horizontal midline is the equator, the vertical midline the zero meridian. Starting at the equator parallel lines are drawn at distances of 6, 7, 8, 9, 10, and 11 mm. These correspond to southern latitudes in the upper hemisphere and to northern values in the lower hemisphere (10, 20, 30, 40, 50, and 60°). Then parallel lines are drawn on the right and left to the vertical zero meridian at distances of 6 mm; these correspond to the longitudes 10, 20, 30, . . . 180° (from the zero meridian to the right) and to 350, 340, 330, . . . 180° (from the zero meridian to the left). Then all the details observed at the eyepiece are transferred onto these drawings dealing with a belt of only 10° (areographic) width to the right and left of the central meridian at each session. Thus an observer who draws Mars at the same hour

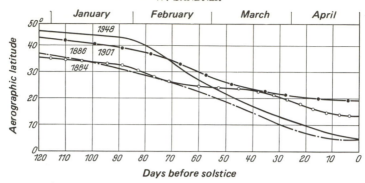

Fig. 15–8 Melting of the north polar cap of Mars at different oppositions. After DE MOTTONI.

once every night will need about 30 days to account for everything visible on the planet. Naturally, if he observes more than one time each night, the necessary time is reduced. If unfavorable weather conditions occur, of course, even a very industrious observer might miss certain meridians during an opposition period.

The measurement of the drawings can also lead to a determination of the boundaries of the polar cap. It will be found that the center of the cap does not always coincide with the rotational pole. This has to be taken into account if one wishes to plot the "melting" of the cap (Fig. 15–8) and to derive the speed with which the spot recedes in spring and autumn. Such measurements, extended uniformly over many oppositions, can inform us whether the speed of "melting" perhaps depends on the state of solar activity.

For such investigation it is necessary to know when the various seasons begin on Mars, as given in Table 15.5. The heliocentric longitudes are to be taken from the Ephemerides.

A great many maps of Mars are available. Unfortunately this has brought about a certain confusion in the nomenclature of single features and canals; a new internationally recognized list, together with the corresponding positions, is given in the table on page 512.

Table 15.5. The Martian Seasons

| Heliocentric longitude | Season beginning | |
	Northern hemisphere	Southern hemisphere
88°	Spring	Autumn
178°	Summer	Winter
268°	Autumn	Spring
358°	Winter	Summer

The close-up photographs obtained by the spacecraft "Mariner 9" made it possible to map accurately about 90% of the surface of Mars. A multitude of craters and rills, and apparently also volcanoes, were recorded. Actually, these maps do not agree in all details with those shown by telescopic observations from the Earth (i.e., with the so-called "Albedo Maps" and their brightness-darkness shadings).

More than for other planets, work on Mars requires a thorough knowledge of the older literature, including that of the last century, since some features and other peculiarities that were neglected for decades have again become topical. We recall here the so-called "Kerb" phenomenon of 1924 and 1956, which is described by F. KIMBERGER in Das "Kerbphänomen" auf Mars [*Mitteilungen für Planetenbeobachter* **11**, 1 (1958)].

Mars has two satellites which are among the smallest bodies of the solar system. Although their magnitude is about 12^m, they are so close to the bright Mars that they can be discovered only with large telescopes. The inner one of the two is one of the most difficult objects to observe in the whole solar system; under favorable conditions, the outer one can be observed with telescopes of 300 mm aperture, provided an occulting disc in the focus of the telescope is used to exclude the light of the planetary disc.

15.2.4. The Minor Planets*

The asteroids are usually very faint objects. Only two of them (Vesta and Ceres) exceed 7^m at the time of opposition, that is, they are 70 times brighter than 10^m. The oppositions are repeated every 1.2 to 1.4 years. Since the orbits of the minor planets are rather eccentric and inclined toward the Earth's orbit, the opposition distances, and thus the apparent magnitudes, can vary greatly; some minor planets even move far outside the Zodiacal belt.

The minor planets are distinguished by a number and a name, for instance, (6) Hebe. In so far as they have not been sufficiently observed for a definite orbital determination, they carry a preliminary notation according to the date of discovery, for instance, 1939 EA.

Even for the largest of them, one having a diameter of a few hundred kilometers, the apparent diameter remains below $1''$; most of them are 10 or 100 times smaller. Therefore no surface details can be seen, and the interest of the observer is concentrated on the exact photographic determinations of position (which after all are of great importance for celestial mechanics and on photometric surveys).[1]

Comparison stars of suitable brightness, whose positions can be compared on the plate with the positions of the minor planets, are usually found in the relevant catalogs (AGKs or the Carte du Ciel; see also the literature cited on page 547). Complete theories of the orbits of the first four asteroids discovered, Ceres, Pallas, Juno, and Vesta, are available. The Ephemerides for observation are given in the Yearbooks. (See also the literature noted on

---

* This section was prepared by W. D. HEINTZ and G. D. ROTH.
[1] See also ROTH, G. D.: *The System of Minor Planets*. London: Faber and Faber, 1962.

page 548.) For the other objects, elements and approximate Ephemerides are given in the annual publication, *Efemeridy Malych Planet*, of the Theoretical Astronomical Institute in Leningrad.

The magnitudes of the asteroids and data about their possible light variations are rather unreliable. Only precision photometric work can help here (see Sec. 19.9).

There are various observational tasks accessible to the amateur astronomer. The current literature reports the systematic work of some of them.[2] As to visual photometry even good field glasses or small telescopes of two- and three-inch aperture will suffice. The serious amateur astronomer will also find useful photographic work available to him; those asteroids with a suspected rotational light variation, usually of very short period, can be investigated to advantage only by photographic photometry. For this purpose very sensitive emulsions of 24/10 DIN and faster should be used. Experience shows that they will still be perfectly pointlike for an exposure time of about a quarter of an hour if the camera is guided properly with the help of a neighboring star. In this way it is possible to examine objects down to 10^m. A long-focus instrument ($F = 100$ to 1000 mm) is definitely preferable to a miniature camera. Of course, no observer can do without the help of good star maps and Ephemerides; see also the references on page 540.

Careful photometric surveys of suspected objects can lead to insight into the rotational light variation of the brighter minor planets. This requires frequent magnitude estimates, visually or photographically, during an evening, say once every hour. If the observer possesses a device to displace his plateholder at the telescope (e.g., a screw of about 0.5 mm thread) he can make several exposures next to each other on the same plate. The image of a minor planet is particularly suitable if the plateholder is turned to such a position angle that the displacement of the plate after every exposure is approximately perpendicular to the direction of motion of the planet.[3] The rows of star images are then parallel on the plate, while the one caused by the planet is somewhat inclined. These photographs should cover several hours of the evening, with intervals of about 10 minutes between exposures. The exposure time itself will be about 15 minutes for the brighter objects. The brightness is then interpolated with reference to the surrounding stars, similar to visual step estimates. This is carried out in the simplest way by direct comparison of two images and one calibration scale photographed simultaneously. Comparison fields with stars of known brightness are, for instance, the North Polar Sequence and the Pleiades; see page 534.

Simple color measurements on minor planets are also of great promise. Here we are aiming at the determination of a color index. The knowledge of a

---

[2] See, for example, M. Beyer's work on the light variations and the position of the rotational axis of the asteroid 433 Eros during the opposition 1951–1952 in *Astronomische Nachrichten* **281**, 121; also W. Malsch's observations of minor planets in 1958, *Astronomische Nachrichten* **285**, 90.

[3] Leutenegger, E.: Photographie von Planetoiden. Orion, *Mitteilungen der Schweizerischen Astronomischen Gesellschaft* **5**, 496 (1958).

sufficiently large number of such indices makes it possible to derive the albedo of some minor planets. The principle of the observation is that visual or photographic magnitudes are determined through insertion of light filters of known transparency, such as the SCHOTT filters.

15.2.5. Jupiter

Jupiter promises more success than any other planet. Its large diameter, the numerous details and rapid variability of its surface, and its rapid rotation, which permits the derivation of many positions within a single night, account for the fact that Jupiter is the most favored object for the amateur astronomer. Indeed, a large portion of our knowledge of this planet was obtained by amateurs.[4]

A detailed study of Jupiter is due to B. M. PEEK: *The Planet Jupiter*, 2nd ed., London: Faber and Faber (1958). The Jupiter Section of the British Astronomical Association is directed by W. E. Fox, 40 Windsor Road, Newark, Notts. See also the ALPO work (Sec. 15.1.3).[5]

Jupiter's visual brightness in opposition is -2^m39, which makes it, except for Venus, the brightest planet. The brightness variations dependent on the phase are inconspicuous and those due to the varying distance of the planet are very small indeed. The oppositions follow each other at intervals of 13 months.

In addition to its large diameter and wealth of detail, observations are facilitated by two other circumstances: (1) The phase angle is at the most only 12° and therefore prevents the observation of only a very small crescent, whose thickness is at the most 1/100th of the diameter and can be ignored altogether. (2) The position of the axis is very favorable since its largest change amounts to only 3°5, so that the parallels for all observations are practically straight lines, using medium or even large telescopes.

On the other hand, the flattening of 1:16.3 must always be taken into account. The use of printed templates cannot be avoided; those with a diameter of 67:62.5 mm have proved of particular advantage, and have also been used by FAUTH and GRAFF. The amateur astronomer should check carefully the axis ratio of any template he uses. For further reduction he will find it very useful to place a transparent reticule with degrees on his drawing.

The central meridian must be calculated separately for each drawing. We know that the equatorial zone of Jupiter rotates faster (9^h50^m) than the medium latitudes (9^h55^m), in which most of the spots occur. One therefore distinguishes in the calculation of the central meridians a System I (equatorial zone, hourly rotational angle 36°6) and System II (medium latitudes, hourly change of the central meridian 36°3). The central meridians for 0^h UT of every day are given

[4] See, for example, MÄDLOW, E.: *Astronomische Nachrichten* **280**, 161 (1950).

[5] The amateur astronomer will always be well served by the studies of newly published scientific researches and observational models; here we mention, for instance, G. P. KUIPER's Lunar and planetary laboratory studies of Jupiter. *Sky and Telescope* **43**, 4–8, 75–81 (1972).

in the Almanacs; simplified calculation can be carried out with tables of the changes of the central meridian, which are given in the Appendix on page 513. Certain details shown by the telescope on the planetary disc do not belong to the actual surface but are atmospheric structures; they show considerable proper motion relative to the actual rotational system. It appears that the solid planetary body has a rotational period of $9^h54^m30^s$, which was recently termed System III in the Ephemerides of the radio astronomers.

Small telescopes already show two or more bands which cross Jupiter's disc parallel to its equator. Between those bands are areas called "zones"; the International Nomenclature is shown in Fig. 15–9. The brightest zone is marked *EZ*.

Among the spots the so-called "Great Red Spot" is the most remarkable; it can be traced on old drawings at least as far back as the 1830's. However, even much earlier drawings by HOOKE in 1664 and by CASSINI, the first of about 1670, indicate the presence of this spot, which can be followed up to 1713. At times it was very conspicuous and of a decidedly red color, at other times it was only indistinct and without definite coloring. According to LÖBERING, it lies deeper than the cloudlike formations that we observe on Jupiter, and it is red only in periods when there are no outside disturbances.

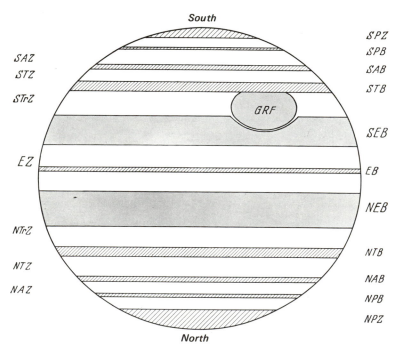

Fig. 15–9 Nomenclature of the bands and zones on Jupiter (as seen in the reversing telescope). Abbreviations: N = North, S = South, B = band (dark), Z = zone (bright), E = Equator, Tr = tropical, T = temperate zone, A = arctic zone, P = polar, GRS = Great Red Spot. After W. BÜDELER.

Details of its changes in position (proper motion) in longitude will be discussed below.

The nomenclature of the other spots is not as uniform as that for the bands and zones. It would be useful if some uniform classification could be adapted, and we recommend the one proposed by E. MÄDLOW,[6] who distinguishes and illustrates with drawings the following features (apart from the Great Red Spot and the classical Veil):

1. *Gray spots.* Usually short-lived objects in the zones, partly sharp and of regular boundaries, partly indistinct.

2. *White spots.* Round or oval white structures with boundaries, usually sharp, appearing within the dark bands as well as on the more diffuse background of the bright zones.

3. *Rifts.* Broad lines, about 1–2 degrees wide, usually sharp. They intersect a band from one zone to the next, as a rule, under an angle of 45–60 degrees.

4. *Indents.* Dark or light indents of the bands.

5. *Kerbs.* Sharp semicircular bays in the smooth border of a band.

6. *Bridges.* Wide or narrow dark bands crossing the bright zones from one band to the next, which they enter in usually rather conspicuous bays.

7. *Garlands.* Dark lines that start from a bay of a band, extending for a few latitude degrees into a zone, to follow afterward this zone, either ending in it or returning to the band, where they enclose a bright region of the zone like an island.

8. *Rods.* Longish, narrow, dark objects usually of gray or red color, which are mainly within a band and are characterized by their narrowness.

9. *"Granat" spots.* Very small round and bright-red spots, no larger than the shadow of a Jupiter satellite.

10. *Bars.* Distinctly bounded dark spots of characteristic forms in the bands, considerably larger than the rods or the "granat" spots.

11. *Dark concentrations in the bands.* These exhibit more or less diffuse outlines.

12. *Knots.* Knotlike thickenings of the narrow dark bands.

It is important to point out that it is of greater value to observe throughout one or two nights for 10 hours consecutively (and to make a drawing every 50 minutes, determining as many positions as possible through meridian transits) than to produce one drawing each night for 20 nights. The magnification should not be less than 150. The following fields lend themselves to the observer.

Complete Drawnigs of Jupiter's Disc. Here we should note that because of the rapid rotation the time spent on such a drawing should be at the most 10 minutes and, if possible, only 5 minutes. It is advisable to have ready beforehand preliminary drawings of these bands in their rough outlines; to achieve this, and thus the best possible accuracy of the positions in latitude, one may use very large magnifications at the basis, or micrometric measurements of the bands. Then the most conspicuous objects are drawn in, in

---

[6] MÄDLOW, E.: *Himmelswelt* **55**, No. 7/8 (1948).

longitude and latitude, proceeding, because of the rapid rotation, from left to right. Finally the drawing is completed in its fine details. The time attached to the drawing, which determines the central meridian, is the time when the rough details are drawn. Observers with good color vision should make color notes.

Determinations of Positions. Great emphasis should be placed on exact position determinations in longitude. This can be done in a rough manner by measuring a drawing with the help of a transparent cover sheet. According to LÖBERING it is much more accurate to enter on a preliminary drawing several positions before and after the transit through the meridian, adding exact times, and to measure those to arrive at an average value. Most accurate is to estimate the exact moment at which a spot crosses the central meridian; one will be surprised how accurately this can be done. An experienced observer can determine this moment to the nearest minute, corresponding to an accuracy of $0°6$ in Jupiter longitude. If the objects are very extended, like the Great Red Spot, this method can be used only for the center of the object since otherwise (e.g., preceding or following border) the asymmetry of the image introduces errors in the estimate that cannot be checked. In one single night it is possible to obtain easily two or three dozen positions.

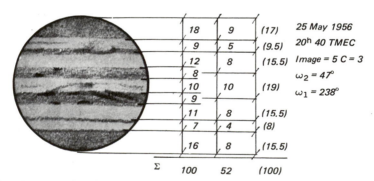

Fig. 15–10 Estimating the widths of Jupiter's stripes and bands. After CORTESI.

To get a reliable basis for the drawings of Jupiter's surface it is useful to start by estimating the relative width of the zones and bands; according to S. CORTESI,[7] the width of a well-defined band (usually the *EZ* band or the bordering bands) is best, using a scale up to 10. This band is then called 10 and other widths are interpolated. Several measurements are combined to a mean, and the final values are converted into percentages of Jupiter's diameter (N–S); see Fig. 15–10.

Experienced observers with sufficiently powerful optics (apertures of 15 cm and more) will select particularly interesting structures and follow their changes. Figure 15–11 gives an example, showing a group of detailed drawings in the neighborhood of the Great Red Spot, obtained in 1950 with a

---

[7] CORTESI, S.: *Orion* 7, 137 (1962).

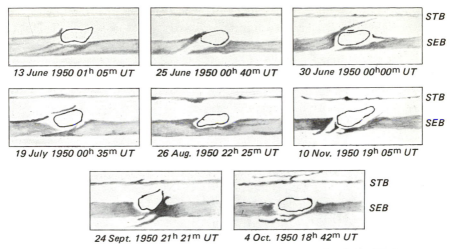

STB
SEB

STB
SEB

STB
SEB

13 June 1950 01ʰ 05ᵐ UT   25 June 1950 00ʰ 40ᵐ UT   30 June 1950 00ʰ00ᵐ UT

19 July 1950 00ʰ 35ᵐ UT   26 Aug. 1950 22ʰ 25ᵐ UT   10 Nov. 1950 19ʰ 05ᵐ UT

24 Sept. 1950 21ʰ 21ᵐ UT   4 Oct. 1950 18ʰ 42ᵐ UT

Fig. 15–11   Jupiter's Great Red Spot in successive months. After W. SANDNER.

refractor of 20 cm aperture; they show very clearly the flow of cloud masses around this Spot.

Intensity Estimates of Details. Little work has been done in this promising field. A photoelectric procedure, used by W. VOIGT and FAUTH in the latter's observatory in Grünwald, goes technically beyond the means of most amateur astronomers. GRAMATZKI achieved results similar to the photoelectric procedure by photometry of photographs. Most promising, however, because it does not require any instruments and is sufficiently accurate, is the method of estimating intensity according to a memory scale; this method has been developed by FAUTH, who used it through several decades.

G. DE VAUCOULEURS developed a scale in 10 steps (see page 366 in the section on Mars). Applied to Jupiter this scale, according to S. CORTESI, yields on the average the following values:

0.5 = greatest brightness of very bright zones.
  1 = normal brightness of the zones.
  3 = most frequent average brightness of the polar zones.
  6 = average brightness of the large equatorial bands (NEB, SEB) during normal activity.
7–8 = small, very dark condensations, which sometimes occur inside the bands.
  9 = satellite shadows on the planetary disc.

These estimates, which do not give any trouble and do not require any extra time, can easily be carried out side by side with the other Jupiter observations.

Great care must be devoted to the interesting task of analyzing the results. The drawings of the various views of Jupiter's surface can be combined in a complete chart of Jupiter in MERCATOR projection, as this has been done on

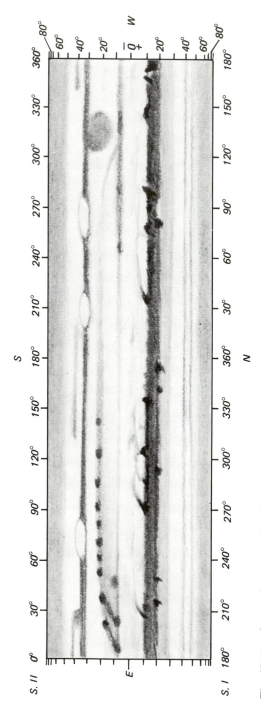

Fig. 15–12 A complete map of Jupiter. Drawn from 30 April to 3 May 1958. After S. CORTESI: *Mitteilungen für Planetenbeobachter* **11**, 27 (1958).

many previous occasions. Because of the relative displacements of the rotational systems I and II, not more than two or at the most three successive nights should be combined (see Fig. 15–12).

To design a complete map of Jupiter, S. WILLIAMS[8] used a somewhat different method. He does not use the single drawings but prepares for every observing night a special form sheet with a reticule for MERCATOR projection, in which he does not designate the meridians with the corresponding degrees in longitude, but rather with the times at which these coincided on the particular night with the central meridian. In this kind of "template" those objects of Jupiter's surface that are free of distortion at or near the central meridian are entered about every 20 minutes. The longitude degrees are calculated subsequently and entered. This method has the advantage that it yields immediately a complete map without the bulk of many detailed drawings, and that it gives ample time to the observer to carry out his determinations of positions and estimates of brightness.

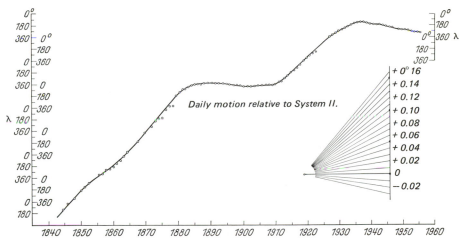

Fig. 15–13 The motion of Jupiter's Great Red Spot in the period 1840 to 1950. After W. LÖBERING.

The longitude determination of single spots can be combined into curves permitting an insight into the streaming conditions and their variations, which may be partly caused by changes in the level of the particular spots. Particularly interesting longitude determinations of the Great Red Spot have been analyzed by LÖBERING[9] and have yielded interesting results (Fig. 15–13). Finally, intensity estimates can be tabulated and averaged in annual "Relative

[8] PFANNENSCHMIDT, E.: *Mitteilungen für Planetenbeobachter*, September/October 1950.
[9] LÖBERING, W.: The motion of the Great Red Spot on Jupiter. *Astronomische Nachrichten* 284, No. 2.

Numbers," which can, for instance, be compared with the solar activity.[10]

The Satellites. While Jupiter's satellites V to XII are accessible only to very large telescopes, and even then in part only by photography, the four "old" moons deserve special treatment. To follow the relative position with a small measuring micrometer certainly has little research value, but it represents a good exercise and every young observer should do so for some time. The changing pattern of Jupiter's satellites offers a multitude of phenomena, the prediction of which can be extracted from the Almanacs. Here we mention:

1. Transits of the satellites in front of the disc (ingress and egress).
2. Transits of the satellite shadows over the disc (ingress and egress).
3. Occultation (disappearance of the satellites behind the planetary disc and their reappearance.
4. Eclipse (disappearance of the moons in the shadow of Jupiter and their reappearance).

The occultations, listed as item 3, show the satellites as small discs even in average instruments.

It has been shown by P. AHNERT[11] that the observation of eclipses (item 4) is of considerable value even if only small instruments are used. As for the satellites I and II we can only observe the entry into the shadow before the opposition, and after the opposition only the reappearance. It is essential to fix accurately the times of the beginning and of the end of the eclipse. For this purpose it is indispensable that these observations always be carried out by the same observer with the same instrument under conditions that are as much alike as possible; for instance, the same number of observations should be made before as after the opposition. Observations in Germany have been collected and analyzed by the German Academy of Sciences at the Observatory Sonneberg.

AHNERT[12] makes use of observations of the beginning and the end of the satellites' transits in front of the planetary disc (item 1 above) and of the occultations by the planet (item 3), but remarks that these cannot be fixed with the same certainty as the beginning and end of eclipses.

Mutual occultations of Jupiter's satellites or occultations of one satellite by the shadow of another are rare.[13] Twice in the course of a Jupiter year, that is, about every six years (when Jupiter's right ascension is about 8 or 20 hours), such phenomena are more frequent, and their observation should not be neglected as it is excellent exercise in efficient astronomical observa-

[10] MÄDLOW, E.: The periodicity of surface structures on Jupiter. *Sternenwelt* **3**, No. 5/6 (1951).

[11] AHNERT, P.: The system of the Jupiter satellites I–III. *Die Sterne* **39**, No. 9/10 (1963).

[12] AHNERT, P.: Observations of Jupiter's satellites in 1964–65 and in 1965–66, *Die Sterne* **42**, No. 9/12 (1966).

[13] Details are given in FAUTH, P.: *Jupiter Observations Between 1910 and 1938–39*, Part 2. Berlin: G. Schönfeld, 1940.

tion. The phenomena should be followed with large magnifications, as well as with lower magnifications, since then both satellites, when observed photometrically, will behave as an Algol variable.

Table 15.6 gives an idea of size and brightness of the satellites I to IV.

Table 15.6. Size and Brightness of Satellites I–IV

Satellite	Diameter	Brightness
I	$1''05$	$5^{m}43$
II	$0''87$	$5^{m}57$
III	$1''52$	$5^{m}07$
IV	$1''43$	$6^{m}12$

The largest of them, Jupiter III (Ganymede), appears in the 30-cm telescope about the same as Mars in a 2-inch telescope. Anyone with an instrument of this kind will sometimes be able to study shadow effects; with lesser optics, however, this will be impossible.[14]

15.2.6. Saturn

There are so many similarities in so many respects between Saturn and Jupiter that what has been said about the latter can largely be applied to Saturn. Its oppositions follow each other after 1 year and 13 days. Its apparent total brightness very much depends on the position of the rings. Wide open rings correspond to 0^{m}, and the most unfavorable case of a disappeared ring to only $1''5$. The largest possible phase angle is $6°$.

The surface of the planet itself shows bands similar to those on Jupiter but much fainter and more diffuse. The nomenclature is the same in both cases. Well-defined spots are rare, so that the rotational period of Saturn can only be approximated (for *EZ* and mean latitudes $10^{h}14^{m}$ to $10^{h}16^{m}$). It is for this reason that until recently no physical Ephemerides had been calculated for Saturn corresponding to those obtained for Jupiter and Mars. Finally in 1966, the WILHELM FOERSTER Observatory in Berlin for the first time calculated central meridians on the basis of a rotational period of $10^{h}14^{m}$.

Saturn shows well-defined bright spots only on very rare occasions: the last time was in August 1933, before that in 1903 and 1876. These remain for a few weeks until they become diffuse and fade away. Whenever such features are observed it is important to follow their structural changes as carefully as possible. Since the rotational periods of these objects differ considerably from those of other parts of Saturn's surface, determinations

---

[14] For drawings and charts of Jupiter's satellites, see, for instance, LYOT, B.: L'aspect des planètes au Pic du Midi. *L'Astronomie* **67**, No. 1 (1953); and SANDNER, W.: *Satellites of the Solar System*, London: Faber & Faber, 1965. New maps have been published in 1973 by the Soviet astronomer Chodak.

of passage through the central meridian are always valuable. Dark objects (spots) also appear occasionally to obey some independent laws of rotation; this was demonstrated, for example, by a dark zone in the northeastern band in 1949–50.[15]

In order to estimate the intensity distribution on Saturn and its rings, Fauth uses the same scale as in the case of Jupiter.

Because of the large flattening of Saturn itself (1/10.4), and owing to the changing appearance of the rings, it is necessary to use preprinted templates for the drawings.

The appearance of Saturn's rings changes according to the relative position of the Earth and Saturn; twice in the course of a Saturn year the rings appear to us wide open, and twice we look at them edgeways on. The latter situation takes place when Saturn is in Leo (172°) or in Pisces (352°); the widest opening when the planet is at 82° and 262° ecliptical longitude.

In the case of disappearing rings it is essential to note possible thickenings or knots and other irregularities of the ring to the east and west of the planet.

At the time of the open ring, only a 3-inch telescope is needed to show the so-called Cassini Division, a dark line separating the outer ring A from the inner ring B. Besides this "Cassini line" we also find a distinct "Encke line" inside ring A. All other divisions reported by various observers are still uncertain.

Within ring B we find a faint ring C, which is diffuse at its inner border. Whether there exists a further ring outside C, only occasionally visible and indicating a very delicate structure, is uncertain. The dimensions of the various rings are the following:

Ring A, outer diameter	39″97
Ring A, inner diameter	34″60
Ring B, outer diameter	33″67
Ring B, inner diameter	25″93
Ring C, inner diameter	20″43
Saturn sphere, equatorial diameter	17″24
Saturn sphere, polar diameter	15″49

The various rings differ essentially by their brightness, and from time to time their intensity should be measured. Drawings should show the shape of the shadow on the rings as well, and the shadow of the rings on the sphere. Certain anomalies which occur occasionally can probably be explained mainly by physiological causes.

When the rings are narrow, the brightest of the ten Saturn satellites must result in the same phenomena (occultations, etc.) that we know from the Jupiter system. However, only the brightest of them, Titan, produces transits of its

---

[15] Roth, G. D.: Aussergewöhnliche Rotation eines Dunkelobjektes im NEB auf Saturn. *Astron. Nachr.* **281**, 89 (1953).

small disc and of its shadow in front of its Saturn sphere; already modest optical means (20-cm aperture), when the disc has a diameter of $0''.6$, show this phenomenon. With more powerful optics (diameter 30 cm) it might be possible to observe the satellites Rhea, Thetis, and Dione. Predictions are given in the *Yearbook of the British Astronomical Association* and (in German) in the annual publication *Der Sternhimmel* of the Schweizerische Astronomische Gesellschaft.

Important, too, are brightness estimates of Saturn's satellites, particularly of Japetus, whose light variation exceeds two magnitudes. Its brightness maximum coincides with the western and its minimum with the eastern elongation, indicating the existence of synchronous rotation. For small and medium telescopes only the six brightest satellites are accessible, whose magnitudes are as follows.

Satellite	Brightness
II Enceladus	$11^{m}.7$
III Tethys	$10^{m}.3$
IV Dione	$10^{m}.5$
V Rhea	$9^{m}.7$
VI Titan	$8^{m}.3$
VIII Japetus	variable $10^{m}-12^{m}$

15.2.7. Uranus, Neptune, and Pluto

The size and brightness of these planets are given in the brief accompanying table. To most amateur astronomers only Uranus and Neptune among the distant planets offer the possibility of brightness estimates; in the case of Pluto we shall be content if we merely see it once as a small point of light.

Planet	Brightness	Diameter
Uranus	$5^{m}.58$	$3''-4''$
Neptune	$7^{m}.75$	$2''.5$
Pluto	$14^{m}.75$	smaller than $0''.3$

Uranus can be distinguished at once from its neighboring stars by its distinct greenish color. Telescopes of 10-inch aperture and greater occasionally show some delicate shading; this was compiled for the first time on a chart by KIMBERGER.[16] Since Uranus' axis of rotation nearly coincides with its orbital plane, there are times when we look at its pole and others when we view its equator. The latter was the case in 1923 and 1965, while the dates of polar viewing are 1902, 1944, and 1986.

Only the largest telescopes are capable of revealing faint bands on Neptune, which otherwise, because of faintness and diffuse outline of the bands, cannot be distinguished from an ordinary star.

[16] KIMBERGER, F.: Die Oberfläche des Planeten Uranus, *Die Sterne* **41**, No. 7/8 (1965).

Brightness estimates of the total light of Uranus and Neptune with Argelander's method (see page 446) can be valuable. It appears that there exists some relation with solar activity, which is similar for all outer planets. In the case of Uranus brightness variations can inform us about inclination and the rotational period. Fedtke, as early as 1938,[17] drew attention to a peculiar behavior of its color index; the amateur astronomer will find interesting possibilities here.

15.3. Star Occultations by the Planets

The rarity of star occultations by planets is due to the small apparent diameter of the planetary discs, so that if an occultation is to occur the difference in geocentric latitude between the occulting planet and that of the star must be at the most only a few seconds of arc. These are also discussed on page 322.

On 7 July 1959 a very rare phenomenon took place: Venus made a transit over Regulus, an event which had never been observed since the invention of the telescope. Nevertheless there are certain reports mentioned by Bode, dating from the pretelescopic era, referring to two earlier events of this kind which are supposed to have taken place on 16 September 1574 and 25 September 1598.

Occultations of faint stars are of course more frequent, but still very rare. There were, for instance, two in 1958: once when Mars occulted the $7^{m}_{.}3$ star BD + 15°450 on 25 August, and again the $8^{m}_{.}9$ star BD + 16°624 on 26 October 1958. A list of all observed occultations of stars by planets has been published by W. Sandner in an article in *Die Sterne*, Vol. **49**, No. 2, pp. 242–246, 1973.

The duration of an occultation depends on (1) the apparent diameter of the planet, (2) the actual amount of the apparent motion of the planet, and (3) the chord along which the occultation takes place.

During the observation special attention should be paid to (1) the times of contact (ingress and egress), (2) position angles at these times, and (3) the effect of a possible planetary atmosphere.

If there exists a noticeable planetary atmosphere of sufficient height and density, instead of a sudden disappearance and reappearance, we might expect a gradual decrease and increase of brightness. It is particularly so in the case of Venus, which should attract the observer's special attention. For instance, when on 26 July 1910 the star η Geminorum was occulted by Venus, the observations revealed a gradual subsequent increase of its brightness over some 1.5 seconds. Similarly, when Venus occulted Regulus on 7 July 1959, Güttler[18] noted a duration of the ingress phase of 1.8 ± 0.5 seconds. He notes: "Perusal of the records obtained indicates that some

---

[17] Fedtke, C.: *Beobachtungs-Zirkular der Astronomischen Nachrichten* **20**, 33 (1938).
[18] Güttler, A.: Die Beobachtungen der Regulus-Bedeckung durch Venus am 7 Juli 1959. *Mitteilungen für Planetbeobachter* **12**, 37 (1959).

observers were unprepared for such a short duration of the ingress, a fact which could have affected the accuracy of the record. It therefore appears advisable on future occasions to precede the occultation observation by a rehearsal, using suitable signals of some kind to achieve a reliable determination of such short time intervals. Furthermore, it would be very valuable, even if fairly difficult, to measure the limiting magnitude at the time of the observation and on the same part of the sphere. The artificial star of Graffs' photometer could be very useful for this purpose."

15.4. The Photography of Planetary Surfaces*

Amateur astronomers are rather new to this field. It was mainly the pioneering work of GRAMATZKI that has changed the situation. The progressive perfection of miniature photography has also opened up new possibilities. By now, amateurs on both sides of the Atlantic have demonstrated that even relatively modest optical equipment can yield astonishing results, not very far behind those achieved with the large telescopes. (See Figs. 15–14 and 15–15).

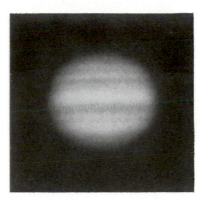

Fig. 15–14 Photograph of Jupiter taken with a KUTTNER reflector, aperture 110 mm; 8 January 1966, 21^h10^m UT; plate Ilford FP3; exposure 1 second; focal length 13 m in eyepiece projection (PETER HÜCKEL).

15.4.1. Conditions

The photography of planetary surfaces is governed essentially by the same principles as that of the Moon. We can thus refer back to what has been said in that section, only the small diameter of the objects and their relative faintness require some amendments. Thus, for instance, in a telescope of 4000 mm focal length the disc of Jupiter in opposition (with an angular diameter of 48") has a diameter of 0.93 mm; and that of Mars in the most favorable case (24") a disc of 0.47 mm, which decreases to only 0.33 mm in

* This section was prepared by F. KIMBERGER (Fürth) and G. NEMEC (Munich).

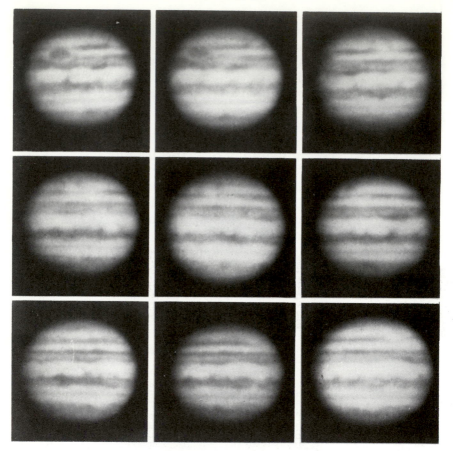

Fig. 15-15 Jupiter photographs taken with a reflector 200/4000 mm at an equivalent focal length of 36 m, exposure time 6 sec, plate 14/10 DIN (G. NEMEC, Munich).

mean opposition (17″). We thus require long-focus objectives, and consequently (except for Venus) rather extended exposure times. It is therefore of paramount importance to have a very accurately constructed telescope mount, vibration-free drive, and perfect guiding; on occasions when the seeing is bad, photography should be given up altogether.

Both refractor and reflector are equally suitable, and the larger the objective the better the image. In the case of a refractor an E-objective is sufficient if a yellow filter is being used.

In his essay on planetary photography G. NEMEC[19] remarks, "To the old saying that no chain is stronger than its weakest link may be added that it is useless to strengthen further the strong links of the chain. This refers

[19] NEMEC, G.: Planetenfotografie. *Mitteilungen der Volkssternwarte Köln* **10**, No. 1 (1966).

not only to the design of instruments, but is of general validity. It is important both to recognize the source of the error, and also its deteriorating effect on the image quality. For instance, it is of little value for our purposes to study the more or less damaging refractions on secondary mirrors, if these effects are smaller than those of scintillation. It is also just as meaningless to discuss the advantages of absolutely achromatic reflectors, if the light rays reflected from the mirror reach the film only after having passed through a series of more or less chromatic filters, BARLOW lenses, and eyepieces. We shall hope in vain for good seeing (an air quality 1) if by some clumsy handling of the instrumental outfit we produce air turbulence which never permits the planetary disc to come to rest."

15.4.2. Procedure

Apart from telescopes of very long focal length, as described in the following, focal pictures of planetary discs are out of the question, except when we want to take those of, say, Jupiter's satellites together with their planet. Excellent results were achieved relatively early with horizontal mirrors of long focal length, set up without a tube. The light from the planet falls onto a plane mirror, which is rotated by a clock drive; it then falls on a parabolic mirror, which is fixed, and from which it is reflected into the observing hut. The late H. STREBEL used, at his private Observatory in Herrsching/Ammersee, three different systems of mirrors of 33 cm to 940 cm, 35 cm to 3100 cm, and 60 cm to 1400 cm. The focal image of Jupiter in the 33-cm mirror has a diameter of 2 mm, in the 35-cm mirror one of 6 mm; the negatives permit a strong magnification. However, the drive has to fulfill the highest requirements. STREBEL's reproductions[20] demonstrate the high performance of this method. B. SCHMIDT in Mittweida achieved outstanding results with a mirror of 30-cm aperture and 31-m focal length.

Detailed data for the design and the use of such an instrument have recently been given by A. MÜLLER in Meilen. Using a KUTTNER reflector and Cassegrain systems of very long focal length we may, too, expect good results.

A method that can be used in nearly all cases is to take the photograph in the extended focal image of the telescope with the help of a Barlow lens or eyepiece. Both methods give equally good results. More recently short-focus miniature film optics have also been used, e.g., by P. HÜCKEL in Weilheim. The amateur astronomer will frequently possess a suitable eyepiece and can therefore avoid the purchase of special projection optics. Any type of eyepiece is suitable, as long as it provides sharp definition. Experimental trials will help to select the most suitable projection eyepiece. The focal lengths should not be too large; rather, at least for the first tests, we should stay near the lower limit of magnification and only then go on to increase it. The size of the projected image is, of course, also dependent on the distance from the eyepiece to the film. Thus by varying this distance the magnification

[20] STREBEL, H.: *Die Sterne* No. 5/6, Plate 5 (1928).

can be changed accordingly; but here, too, it is advisable to find the optimum magnification in every case by experiments, in which the quality of guiding and seeing conditions should be taken into account. Furthermore, it is important to prevent vignetting by the tube supports, and ensure that there are no shiny reflecting spots where screws enter the instrument; this can be achieved by the attachment of some velvet paper covers. Of course, all the projection systems must be well cleaned (first roughly and then by applying a badger-hair brush). Color filters must be built into the beam in such a way that any unavoidable fine dirt does not interfere with the projection: they should therefore not be too near to the focal plane.

Cameras with reflex systems are preferable. For practical reasons the use of miniature camera size is advisable—this is cheaper and offers a rich selection of emulsions. Of course, the camera must be rigidly connected with the telescope. It is therefore useful to have a number of connecting tubes in readiness, to be able to make variations easily for different photographs.

Even with a fine ground glass it is difficult to get the planetary surfaces in focus. Under all circumstances a magnifying lens should be used that does not affect the image quality by the grain of the ground glass. It is easier to get the Moon in focus, and with the help of a fiducial mark this focus can then always be reproduced. However, one should allow for temperature changes of the instrument. It is best to start photographic experiments on the much easier object of the Moon before one proceeds to the planets.

It is, of course, useful to take photographs in series on one and the same plate, with small variations of the setting in right ascension or declination, since this saves photographic material. However, moonlight near the photographed region, or a bright twilight sky, produces a veil if too many photographs are taken. And sometimes solarization effects occur on the previous images. If the secondary optics is not well corrected one is restricted to the center of the negative which also limits the extent of the series.

15.4.3. Photographic Material

The sensitivity of the emulsion should not be too great (14/10 to 17/10); otherwise the size of the grain would be obtrusive. Because of the predominance of red-yellow tones panchromatic emulsions are preferable. For some special work, however, we will have to use different material.

Until recently it was thought that sensitivities up to 17/10 DIN led to the best results. It was assumed that the negative would be much magnified afterwards. Today, however, first-class mounting and most accurate guiding of amateur instruments are in the foreground. Therefore, under satisfactory atmospheric conditions, another approach is advisable: The technique of projection of the negative onto high-sensitivity film and only limited enlargement later has certain advantages, particularly in view of the small effects by grain. Experiments of this kind by various amateur astronomers have very recently led to encouraging results. This method is recommended for further tests.

Of course, we should start with black and white photography. However, as the color film technique is perfected, the application of color emulsions will be useful. There have already been several promising attempts by amateurs, e.g., H. BERNHARD in Munich. Loss of focus by increased emulsion thickness is to be feared less than diminished contrast. Unfortunately color films show very little of certain celestial objects, particularly if one uses relatively moderate optical means. Planetary pictures in color have at present more of an aesthetic value.

The position is very different if color filters are used for black and white photography. Observatories possess extensive experience in this field. Amateur astronomers should devote more attention to it, and not only for the purpose of increasing contrast. Unfortunately color filters (e.g., by SCHOTT & GEN.) demand an excessive prolongation of the exposure times, so that we must omit their use in general. G. NEMEC in Munich recommends yellow, orange, and bright-red filters, which lead to highly contrasting planetary details. At an observatory particularly assigned for amateur work and popular demonstrations in Munich, it was recently possible to obtain very good ultraviolet photographs of Venus; R. PRINZ used a UG2 filter by SCHOTT & GEN. of Mainz, designed for the wavelengths 3600 Å in connection with blue-sensitive negative material.

15.4.4. Work in the Darkroom

It is natural for the amateur astronomer concerned with photography to develop and further process his delicate pictures of planetary surfaces. The negatives should be developed with fine-grain developer, avoiding those that smooth out contrast.

The copying of the negatives, too, requires a considerable amount of practice. They must be exposed in such a way that the fine details of the photograph are not affected by veil, while at the same time the grain should not be enhanced. Exposure and developing must, therefore, be adapted with great care. To achieve sufficient contrast hard paper with a glossy surface is recommended.

Unfortunately, small negatives of planetary surfaces do not permit much enlargement. KIMBERGER remarked on this point: "There are different ways in which this situation can be improved. Very elegant, but equally difficult, is the method of superpositioning several negatives. This assumes that several negatives of equal quality are available. Here the grain is partly eliminated by a method which can be hardly faulted from the photographic point of view. Another method is usable if only one negative exists. Then the grain might be so to speak suppressed if during the process of enlargement the enlarger is brought a little out of focus. This leads to only slightly diffuse brightness values with diffused grain structure. By subsequent hard copying or reproduction of these pictures one can achieve strong contrast. Unfortunately such photographs do not represent the astronomical reality with respect to the range of contrast. Errors can easily be introduced in this way.

But if used with sufficient caution such photographs can still provide certain topographical information which cannot otherwise be achieved photographically. This method was also formerly used by GRAMATZKI." KIMBERGER recommended it strongly; with it he obtained rather detailed pictures of Mars.[21]

Special methods to be used in darkroom work have been described by NEMEC. He writes,[22] "If the picture requires subsequent linear magnification of at least ten times linear, the planetary discs become rather grainy, even on 14/10 DIN film, since optimum contrasts can be achieved only on extra-hard bromide paper. Furthermore, especially in the case of Jupiter, the pictures show such a strong limb darkening that one-third of the planetary diameter disappears in the sky background. In order to reduce the graininess, a composite process is applied, projecting the negative onto paper fixed in the enlarging frame, on which the outlines of the planet as well as a few outstanding details are sketched in with a pencil. Then the extra-hard bromide paper is exposed for one-third of the predetermined total exposure time, then removed, while another negative is placed on the sketch and exposed for another third of this time. After the third negative the procedure is complete. This means an underexposure for each negative, which prevents a strong appearance of the silver grain; for the next negative the position of the grain is a different one with the same effect. The results are nearly grainless pictures with planetary details of rich contrast."

Furthermore, NEMEC removed the limb darkening on the planets nearly completely by his ring-diaphragm technique which he describes as follows: "An iris diaphragm is placed between the enlarging lens and the photographic paper in such a way that it permits the projection of the planetary negative to go through. It is somewhat tilted so that it is imaged in a slightly oval shape, corresponding to the planetary flattening. During about one-third of the exposure time the diaphragm is continually closed down so that the border of the planet is made less conspicuous. The remaining exposure time is used without the diaphragm and leads to the exact covering of the sky background and the remaining coverage of the planet.

"The optimum effect is achieved by a combination of both methods whereby a picture is produced by six single exposures. The results are fine-grained and uniformly covered large planetary images without limb darkening, as are demonstrated by professionally obtained photographs."

The working methods of KIMBERGER and NEMEC described above show that the last word on the application and usefulness of darkroom procedures has not yet been spoken. Here, too, lies a wide experimental field for the amateur astronomer and hobby handyman, who will go to any end to produce the very best pictures possible.[23]

[21] KIMBERGER, F.: Visuelle Beobachtungen und photographische Arbeiten während der Marsopposition 1956. *Die Sterne* **34**, No. 9/10 (1958).

[22] NEMEC, G.: *Sterne und Weltraum* **5**, 94 (1966).

[23] Further important discussions on the subject were given by ALEJANDRO L. DE LA BARRA: Cálculos Básicos de Fotografía Astrónomica. El Universo (Mexico), No. 62, 1963.

15.5. Micrometer Measurements and Photometry of the Planetary Surfaces

Very few amateur astronomers will have the instrumental means to carry out successful micrometric or even photometric measurements of the planets. Since, however, in our technological age there are many technically trained amateur astronomers who have achieved remarkable results in the design and application of complicated physical instruments we will briefly mention some of the possibilities.

As for the micrometer measurements, we first consider the simplest form of micrometer that can be attached to an azimuthal telescope, namely, the ring micrometer. This, however, is unsuitable for planetary measurements. Young observers may use it to get practice in the measurement of the relative position of the Jupiter satellites. It might also be used for mutual occultations of such satellites in order to follow their approach.

The best but also the most expensive form of such an instrument suitable for our purposes is the filar-micrometer, which, however, also demands good clockwork (see page 94).

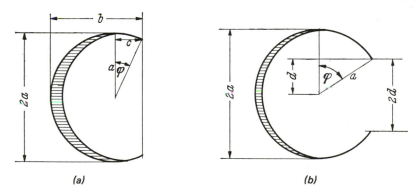

(a)   (b)

Fig. 15–16 Venus: observations by N. RICHTER.

Which features should be measured by the planetary observer? Let us first of all consider Venus, which offers two possibilities: (1) the determination of the moment of dichotomy by measuring the diameter of the crescent from day to day, and (2) the measurement of the overlapping of the horns (overlapping angle near the lower conjunction) and of its changes with decreasing distance from the Sun.

The measurement of this angle φ is discussed by N. RICHTER.[24] If the angle is small it should be measured according to Fig. 15–16(a); if the angle is large we measure, according to Fig. 15–16(b), the quantities $2a$ and b,

[24] RICHTER, N.: Das Übergreifen der Sichelspitzen beim Planeten Venus. *Die Sterne* **17**, No. 7 (1937).

or $2a$ and $2d$, respectively. Since $c = b - a$ the angle can be found from

$$\sin \varphi = \frac{b-a}{a} \quad \text{and} \quad \cos \varphi = \frac{d}{a},$$

respectively. It is possible to make the corresponding measurements for Mercury.

In the case of Mars we have the following three possibilities.

1. The measurement of the position of conspicuous features of the planetary surface.

2. The determination of the position and extent of compact cloud formation.

3. Measurements of the polar cap from the point of view of (a) determining the position of its center (which does not always coincide with the pole of rotation); (b) measuring the diameter of the polar cap; and (c) measuring the widths of a possible dark border (due to melting water). These measurements are the most significant.

Finally, we deal with Jupiter and Saturn. It is important to measure the latitude of the single zones and bands as a basis for accurate drawings (which should be repeated at regular time intervals). Furthermore, the latitude on Jupiter of certain distinct spots and of the diameter of the Great Red Spot should be determined.

If Saturn were to produce another conspicuous spot, as in 1933, it will be important to obtain accurate measurements of its diameter in longitude and latitude, and its position in latitude in Saturn coordinates with the purpose of determining the period of rotation.

The difficulties involved in photometry of planetary surfaces are much greater, but neither are these insuperable, as was demonstrated some decades ago by GRAMATZKI. He designed a surface photometer, the little mirror of which had the diameter of 0.03 mm, that is, its surface was only about 1% the size of Mars in its mean opposition, as measured with his telescope. The source of light was a small lamp with ground glass, and the variation of the brightness of the mirror was achieved by using the principle of the cat's eye photometer. Of course, the brightness of the mirror must be calibrated with the help of bright stars. This instrument has given interesting results,[25] obtained from measurements of the discs of Jupiter and Mars, based on 10 to 16 measuring points on the planetary surface; see also Sec. 19.8.

Such observational series are of real scientific importance and it should be possible to obtain them with reflectors of 30-cm aperture, provided the telescope is most carefully set up and possesses a perfectly functioning drive.

---

[25] GRAMATZKI, H. I.: Zur Krisis der Amateur-Planetenforschung. *Die Sterne* **28**, No. 5/6 (1952).

16 | The Observation of Comets

A. Güttler
Revised by N. B. Richter

16.1. Introductory Remarks

Although comets are common in the neighborhood of our planet it is rare that one of them approaches the Sun close enough to develop impressively. On such an occasion, however, we may observe in rapid succession a multitude of interesting phenomena, and the amateur astronomer will therefore be well advised to prepare for the observation of such an infrequent and transient visitor to our sky.

Reports on observations of comets, particularly about the discovery or the rediscovery of periodic comets, are published in the circulars distributed by the International Astronomical Union's Central Bureau for Astronomical Telegrams. Particularly urgent reports are announced by a special telegram service. The circulars give orbital elements derived from the early observations, and these elements are improved as more observations become available (see page 514). Ephemerides are also supplied. These give values of the equatorial coordinates at intervals of ten days, and usually the distances r and Δ of the comet from the Sun and the Earth (in astronomical units) as well. Frequently the expected apparent magnitude of the cometary head is also included, calculated according to the semiempirical formula

$$m = m_0 + 5 \log \Delta + 2.5\, n' \log r. \qquad (1)$$

The index n is unfortunately rather uncertain, and it is intended to include the influence of the solar radiation of the comet. The reduced brightness m_0 (corresponding to heliocentric and geocentric distances of 1 A.U.) is then adjusted to the available observations.

Accounts of comets can be found in numerous journals and astronomical Yearbooks, particularly in the *Quarterly Journal of the Royal Astronomical Society* and the publications of the Astronomical Society of the Pacific. References to the observations of each comet are provided in *Astronomy and Astrophysics Abstracts*. Predictions concerning periodic comets are given in great detail in the *Handbook of the British Astronomical Association* as well as in the IAU Circulars. Besides the well-known astronomical journals (see page 548), current reports on cometary observations by amateur astronomers are given in *The Strolling Astronomer*, which is published by the Association of Lunar and Planetary Observers (USA). This working group of amateur astronomers has a special section for cometary observations (see also page 549). The British Astronomical Association also has such a section.

The calculation of an ephemeris from the specified parabolic, elliptical, or circular orbital elements of a comet is explained in Sec. 8.5.

New comets are given the names of their discoverers, and they retain these names whenever they return. Up to three names are sometimes given as discoverers (for examples, see page 514). In addition, each comet is given a provisional designation, consisting of the year and a current letter *a*, *b*, *c*, and so on, in order of discovery. After a short while this designation is replaced by the year in which the comet came closest to the Sun (i.e., to perihelion) and Roman numerals I, II, III, and so on, in order of perihelion passage. These final designations can be found in the IAU Circulars.

16.2. Visual Observations

Before discussing astrometric and photometric observations, we will first provide some data on instruments for making visual observations, in particular, for comet hunting.

16.2.1. Instruments

When a comet is farther than some 1.5 to 2 A.U. from the Sun, it usually does not show much of a tail and appears in the telescope under magnification as a diffuse light spot. For the visibility of a given comet it is only the surface brightness of the image that counts. It is therefore advisable to use for the visual discovery of a comet an instrument of large aperture-to-focal-length ratio, which has the additional advantage of a large field of view. Special comet seekers are available.

At the Skalnaté Pleso Observatory in Czechoslovakia, where several comets were discovered between 1945 and 1959 as the result of systematic searches, large binoculars, of 10-cm aperture and 25× magnification, were used. These instruments have great light-gathering power and a large field of view. The Japanese comet hunter TSUTOMU SEKI, who was a codiscoverer of the comet IKEYA-SEKI 1965 VIII, used a homemade 3-inch telescope with 17× magnification. Except for the faintest comets, any amateur astronomers' telescope, refractor or reflector, is suitable for the discovery of a comet—provided one uses a low magnification. To make good use of the whole field of view, it is therefore recommended that one use a correspondingly corrected eyepiece, such as one of the MITTENZWEY type (see Sec. 2.2.6.2). Large binoculars are also very useful for comet hunting if they are equipped with a firm support.

It is worthwhile to go to larger magnifications only when the comet is nearer the Sun and showing further development. To find it easily after the change of eyepiece, it is necessary to check the setting accurately in the finder.

16.2.2. Locating a Comet

Locating a large comet, of course, offers no difficulties. For a fainter comet one must use the ephemeris to plot its path on a star map that goes down to stars of at least sixth magnitude (see page 546). To do so it is necessary to correct the cometary positions to the equinox of the chart by allowing for precession, discussed in Sec. 6.4.1.2. It is then easy to interpolate between the positions of the comet for 0^h Universal Time (or Ephemeris Time) for the time of observation. If the telescope does not have an equatorial

Fig. 16–1 Homemade comet finder: two C-objectives with $f = 500$ mm and $f = 80$ mm, with two wide-angle eyepieces of $f = 20$ mm. After P. DARNELL, Copenhagen.

mounting with reliable setting circles, the observer can orient it using nearby stars, which he can identify with the help of obvious geometrical configurations. When looking at a star map immediately before the observation any bright illumination should be avoided in order not to disturb the dark-adaptation of the eye. If it has been necessary to extrapolate the ephemeris, the predicted position may be considerably in error, by even as much as 1°. Furthermore, one must always be aware of confusion with permanent celestial objects of similar appearance.

16.2.3. Discovery

Those who will try their luck at the discovery of new comets must be prepared to expend a very great amount of time and patience. They should scrutinize the sky carefully with a powerful instrument of small magnification, such as is described above. The best chance of success lies in the western sky after sunset and the eastern sky before sunrise, because that is where the comets near the Sun are, that is, those that are at their brightest. About twice as many comets are discovered in the morning sky as in the evening sky, and there is a conspicuous accumulation of comets near the ecliptic. Something else to consider is that they can be found more easily in regions less densely populated with stars where there are also fewer objects that can easily be confused with comets, such as star clusters, galaxies, and nebulae. However, in other parts of the sky, particularly in the constellations Coma Berenices and Virgo, concentrations of extragalactic nebulae that look very much like comets make searches very difficult.

Still other possibilities for confusion exist which one must try to avoid. For instance, "ghosts," secondary images of bright stars, may be produced by reflection on the lenses. These can be identified at once, however, by jiggling the telescope slightly. A further source of false cometary reports is the cloud searchlights used by meterologists at airports for the measurement of the heights of clouds. With a suitable distribution of clouds, these searchlights can simulate to the naked eye the most beautiful of comets, but one that will not participate in the daily rotation of the celestial sphere. If it is certain that all these possible traps have been eliminated, one should look for possible motion of the suspected celestial object; in the course of half an hour this motion will usually be unmistakable and will rule out the possibility of one of the above-mentioned permanent celestial objects. Of course, there is still the possibility that the comet has already been discovered by somebody else. There is little likelihood of picking up by mistake a minor planet, because of a comet's diffuse appearance.

After the discovery of a new comet it is necessary to establish

1. Its physical appearance (see Sec. 16.5.1)
2. Its estimated brightness (see Sec. 16.6.1)
3. Its exact or approximate position (see Sec. 16.4)
4. The time of observation (in Universal Time)
5. The meteorological conditions

After the discovery the observer should at once get in touch by telephone or telegraph with either a large observatory, the director of the Comet Section of the British Astronomical Association, or the IAU Central Bureau for Astronomical Telegrams directly.

16.3. Visual Determination of Position

Approximate values for the equatorial coordinates of a comet may be obtained much as for any other celestial object. Estimate the position relative to a conspicuous geometric configuration of stars, then look up the configuration on a good star map. The best procedure is to make a sketch, adding the comet last because of its motion. The time (to the nearest 0.5 minute) should be recorded at once. If a telescope with small magnification or a finder is being used, one should also draw the boundary of the field of view and the east–west direction, which with an equatorial mounting can be immediately found by operating the fine motions or switching off the drive. The position of the comet may then be transferred onto a star map and the coordinates determined (for the equinox of the map). If the reference stars are favorably distributed, the result is often surprisingly satisfactory; in any case the error will probably not exceed a very few minutes of arc.

If there is only one known star in the field of view, we can use cross hairs in the focal plane. Such cross hairs can easily be made with thin wires mounted on a small cardboard disc and fitted into the positive, that is, not Huyghenian, eyepiece.[1] See also Sec. 2.2.6.2. During the observation one estimates the distances x and y of comet and star from the meridian wire and the perpendicular wire in fractions of the diameter d of the field of view. From this one finds, with d in minutes of arc,

$$\alpha_{\text{\textsection}} = \alpha_* + \frac{d}{15'} \frac{x_{\text{\textsection}} - x_*}{\cos \delta_*}, \tag{2}$$

$$\delta_{\text{\textsection}} = \delta_* + d(y_{\text{\textsection}} - y_*). \tag{3}$$

Putting one of the objects at the center of the cross hairs will simplify the situation.

Visual measurement of the position of a comet is usually obtained with a position-wire micrometer. However, since the amateur astronomer rarely has one of these at his disposal he might use another very simple but efficient device, the so-called cross-staff micrometer.[2] Such a device can be made very easily, since the precision required for such a measurement is provided entirely by the daily motion of the celestial sphere. The comet (or any other celestial object) is referred to a star that crosses the field of view shortly before or afterward. The star's position may be found later from catalogs that give accurate astrometric data for stars down to ninth magnitude.

The procedure is illustrated in Fig. 16–2, which shows the field of view

---

[1] Rapp, K.: Anleitung für Sternfreunde zum Selbstaufzeichnen der Bahn des Kometen Cunningham 1940 c. *Die Sterne* 21, 112 (1941). In the examples given there, the term $\cos \delta_*$ in the subsequent Eq. (2) has been erroneously omitted.

[2] A description of all varieties of this micrometer can be found in Becker, E.: Mikrometer und Mikrometermessungen, in *Valentiners Handwörterbuch der Astronomie*, Vol. 3, Part 1, p. 64. Breslau: Verlag Eduard Trewendt, 1899.

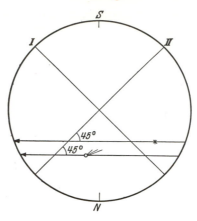

Fig. 16–2   Field of view of a telescope with a cross-wire eyepiece micrometer.

of a fixed telescope with a cross-wire eyepiece. The cross wire may be rotated in such a way that the first wire runs from northeast to southwest and the second northwest to southeast. The equatorial coordinates of an immobile object on the sphere are represented by α and δ, those of the reference star, by α^* and δ^*. If we now measure the differences in right ascension in minutes of time and the differences in declination in minutes of arc we find for the times t_1, t_2 and t_1^*, t_2^* of the transits of the object and of the reference star across both wires, in minutes,

$$\Delta t_{1,2} = t_{1,2} - t_{1,2}^* = \alpha - \alpha^* \pm \frac{\delta - \delta^*}{15 \cos \dfrac{\delta + \delta^*}{2}},$$

from which follows

$$\Delta \alpha = \alpha - \alpha^* = \frac{\Delta t_1 + \Delta t_2}{2}. \tag{4}$$

Transits to the north or south of the center of the cross wire are denoted with a plus or a minus sign, respectively. Furthermore,

$$\Delta \delta = \delta - \delta^*,$$

$$\bar{\delta} = \frac{\delta + \delta^*}{2},$$

$$\gamma = 15 \frac{\Delta t_1 - \Delta t_2}{2},$$

$$\Delta \delta = \pm \gamma \cos \bar{\delta}.$$

$$\tag{5}$$

Since we know δ^* and the diameter of the field of view (see Sec. 2.1.2.4), the position of the object can be calculated accurately enough.

One can also use the following equation, which can be derived in a similar manner, and which contains on the right-hand side the first two terms of a TAYLOR series:

$$\Delta\delta = \gamma \cos \delta^* [\pm 1 - 0.0022 (t_2 - t_1) \sin \delta^*]. \tag{5a}$$

The second term will rarely exceed 1 % and therefore can usually be neglected. The same applies to refraction (see Sec. 5.3.1) if the reference star is not near the horizon and far away from the object.

With moving celestial bodies such as comets and planets their daily motions μ_α and μ_δ on the celestial sphere have to be taken into account. (One should not confuse these usual notations for the daily changes in right ascension and declination with the daily motion of the sphere.) If an ephemeris is available, these quantities can be extracted from it with an accuracy sufficient for this purpose. If one calculates the values α' and δ' using the above equation from the observations, the latter must be corrected for the daily motion, in minutes of time and minutes of arc respectively, and with a minus sign for northern or plus sign for southern transits:

$$\alpha = \alpha' \mp \frac{\mu_\delta}{15} \frac{t_2 - t_1}{2 \cos \bar{\delta}},$$

$$\delta = \delta' \cdot \mu_\alpha \frac{t_2 - t_1}{2 \cos \bar{\delta}}. \tag{6}$$

These corrected coordinates correspond to the time $t = (t_1 + t_2)/2$; before they are used further, they must be reduced to the beginning of the year of observation because of precession. For the equinox of the values α^* and δ^* (Catalog), see Sec. 6.4.

If possible, one should repeat this procedure with additional reference stars, but not over too long a time, and form the corresponding means.

On the other hand, it is possible to obtain the daily motions from two observations sufficiently far apart in time. Here the differences in time are expressed in days.

$$\mu_\alpha = \frac{\alpha_{II} - \alpha_I}{t_{II} - t_I} \quad \text{and} \quad \mu_\delta = \frac{\delta_{II} - \delta_I}{t_{II} - t_I}. \tag{7}$$

If we do not have an ephemeris we can use two such observations by replacing the α and δ in Eqs. (5) by α' and δ'. Approximate values for μ_α and μ_δ are obtained if one calculates the corrected coordinates according to Eqs. (6) and then from Eqs. (7) the improved daily motions.

Since unilluminated cross wires cannot be seen against the dark sky background and since illumination might cause faint comets to become invisible, the observer should use instead a pair of thick perpendicular cross wires or small strips, whereby, in order to increase the usable field of view, the point of crossing is displaced toward the northern or southern boundary. The disappearance or reappearance of the object behind this cross staff can then be clearly observed.

The 45° orientation of the cross can be very easily achieved by marking the cross eyepiece and pasting a strip of millimeter paper around the eyepiece tube. One first sets one of the two cross staffs in the east–west direction, by observing the daily motion of the star on the sphere, and then turns the eyepiece until the marking has moved by a quarter of the circumference of the tube. This setting can then be marked once and for all by a second line on the eyepiece tube. It is advisable to practice this method on stars.

16.4. Photographic Determination of Position

The most common method nowadays for the determination of the positions of comets and minor planets is by photography, making use of very small images of the object, as far as possible undistorted by orbital motions. The exposure time should be as short as possible. Although an astrograph is generally used for this purpose, a refractor adapted for photographic work, or even a plate camera, is certainly adequate. A miniature camera is unsuitable, however, because of its small film scale.

If the cometary head or the minor planet is bright enough, one simply exposes the film as for an ordinary celestial photograph by following the daily motion of the celestial sphere. The exposure time should be only long enough to record a sufficient number of stars, at least four, well distributed around the comet, and of apparent photovisual magnitude of not brighter than seventh magnitude. One should avoid very fast plates as well as rapid development because of the coarser plate grain.

It is convenient to make three or four exposures on the same plate, with the camera or plate holder moved each time by a small amount in, say, declination. This protects from plate flaws, and furthermore reduces the observational error by making use of all these images. Because of the shortness of the exposure time, automatic guiding is adequate.

It is important to record the meteorological conditions, the time of the beginning of each exposure (to the nearest 0.5 minute), as well as the exposure times, which for safety may be varied a little, say, 1.0, 0.5, or 2.0 minutes.

The plate may conveniently be reduced by TURNER's method (see Sec. 8.4). The rectangular coordinates x and y of the comet and the comparison stars should ideally be obtained with a measuring machine. These are usually available only at large observatories. Without such a machine, it is useful to make an enlarged reproduction of the plate on photographic paper (again, if possible, as a negative, i.e., black stars on a white background). By placing transparent millimeter paper on the photograph one may easily read off all the coordinates.

It will sometimes be necessary to make observations near the horizon. If the zenith distance is more than 85°, it is very important in such a case to make allowance for differential refraction; the rapid increase in refraction causes TURNER's method to give poor results.

At the rediscovery of a returning periodic comet the astrograph is made to follow not the daily motion of the star but the expected motion of the comet derived from the Ephemerides. This method is always used when the moving object is very faint, provided its orbit is well known. One sets the telescope on a star using a position-wire micrometer, which during the exposure is continuously moved in the direction opposite to the calculated motion of the object.

16.5. Observations of Cometary Structure

We shall now describe the physical observation of comets. This differs from the astrometric determination of the positions and motions of cometary heads, dealing instead with cometary structure, brightness, spectra, and polarization of their light.

16.5.1. Visual Observation

Comets that are far away from the Sun and appear only as faint diffuse patches of light can often be observed simply by averted vision (see Sec. 3.2), and even under strong magnification they exhibit very little structure. It is desirable to record the diameter of the head, called the coma, and to note any deviations from circular shape, particularly the position angle of the tail, if one exists. One should also look to see if there is a brighter condensation in the coma. If a drawing is made, the scale should be referred to the diameter of the field of view. The orientation and time, to the nearest minute, should be indicated. Figures 16–3 and 16–4 show the manner in which such drawings should be laid out.

Generally, much more detailed observations are possible when the comet is closer to the Sun. If the head is viewed under strong magnification, say, 400 times or more, surprisingly rich structure may be seen. The following details should be recorded:

1. Central condensation or "nucleus": size, appearance, and number of discrete condensations.
2. Coma, the cloud of gas and dust that envelops the nucleus: size, outline, and structure.
3. Streamers and jets originating from the nucleus.
4. Dark spaces and bright clouds.
5. Envelopes of luminous material clearly separated from and on the sunward side of the coma, sweeping around toward the tail.

At this magnification, however, notes about color, even in the most favorable cases, will be restricted to the central condensation. Of course, it can't hurt to try, with the naked eye or with field glasses of low magnification, to determine the color of a cometary head.

Magnification of this strength or greater shows little of the tail, for the surface brightness is too low. It is better observed with field glasses or simply

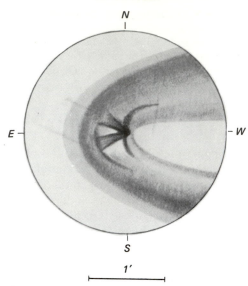

Fig. 16–3   Head of Halley's comet on 9 May 1910 with nucleus, two rays, two ray fans, with a bright cloudy stripe at the end of one of these, and a dark space behind the envelope surrounding the nucleus and the coma. (Negative drawing by R. INNES, Johannesburg, Union Observatory, Circular No. 4, 1910).

with the naked eye, making a drawing of its size and orientation by using reference stars which are later identified and used to determine the length of the tail. A comet may even have more than one tail. According to BREDEKHIN we may distinguish the following types:

Type I: Narrow, straight tails that may also be associated with a fan of tail rays.
Type II: Wide, curved, fan-shaped tails.
Type III: Strongly curved tails.

Structure within the tails, as well as knots of material appearing in the tails and occasionally completely separated from them, are not visually observable.

When publishing drawings or noting interesting details, it is important to include data concerning the instrument—its aperture, focal length, and magnification—and to state the meteorological conditions, particularly the presence and phase of the Moon, since this can very much affect the features observed.

Observations should be continued each night for as long as possible, and any changes in the comet's physical appearance should be studied carefully. Attention should be given both to variations in shape and to position angle.

16.5.2. Photographic Investigation

Photographic work, as far as it can be carried out by an amateur astronomer, is directed first of all to the investigation of tail structure. Since considerable changes might take place even within a few minutes, the observer who wants more than just a pretty picture should restrict himself to relatively short exposure times. It may be advisable to take in succession a series of pictures with different exposure times, say, 3, 1, and 9 minutes, and to repeat this sequence as often as possible. Should there be clouds near the comet it is better to interrupt the exposure than to continue and obtain plates of dubious value.

Of course, it is necessary to use powerful instruments for this work, and to get a complete picture of a large comet, a large field of view is necessary. A miniature camera with a powerful lens is perfectly suitable. It can easily be attached to the telescope, which then serves as a guide for keeping the camera on the head of the comet during exposure. If the exposure times are shorter than a minute there is no need for any guiding. In spite of the large light-gathering power the lens should not be faster than f/2.5 (composite

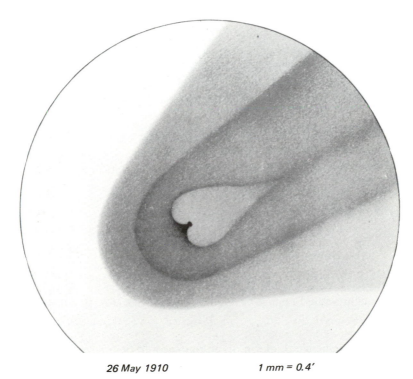

26 May 1910   1 mm = 0.4′

Fig. 16-4 Head of HALLEY's comet on 26 May 1910 with nucleus, ray fans, heart-shaped dark area, and an envelope surrounding the coma; south is on top. (Negative drawing by K. GRAFF, *Mitteilungen der Hamburger Sternwarte*, No. 12, 1913).

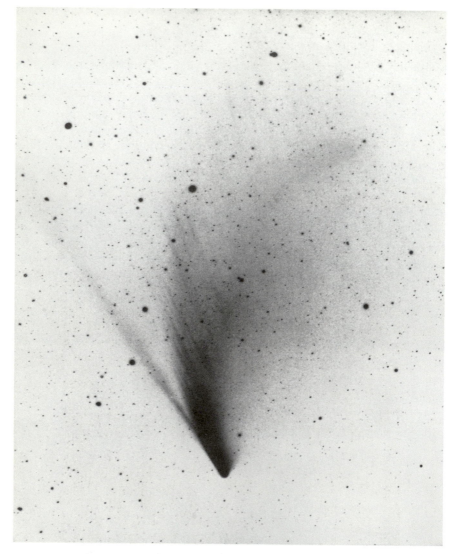

Fig. 16–5 Comet Mrkos (1957 d), photographed on 13 August 1957, 12ʰ UT, by A. McClure in Los Angeles, using a 10-cm refractor ($f = 50$ cm), plate Kodak 103 a–E, and exposure 13 min [*Sky and Telescope* **16**, 568 (1957)].

lenses are recommended), and a diaphragm should be used to correct errors near the edge of the field.

The most suitable film is panchromatic film of medium or fine graininess so that enlargements will be possible. If there is a Moon or noticeable twilight, a red filter, such as the 2-mm Schott RG1, would be useful, and it is then necessary to expose at least four times longer. Unfortunately, the longer exposure will diminish the value of the photographs for the study of tail structure.

A record should be made of the meteorological conditions, the instrumental data, the type of film, and method of development, and the time of the beginning of the exposure and its duration to 0.1 minute.

Successful photographs may reveal the raylike structure of the cometary tail (see Fig. 16–5). Particularly interesting are such features as tail clouds, the outward motion of which can be studied during the sequence of exposures. In this way it is possible to derive the velocity of the material. Frequently screwlike distortions may be observed.

Of particular scientific value are monochromatic photographs in the light of characteristic emission lines. An interference filter can be placed in front of the camera objective (not in front of the photographic plate or the film). These filters should have very narrow passwidths (50 to 100 Å half-width) and be centered on the wavelengths 3883, 4050, 4216, 4380, 4737, 5165, and 5635 Å, respectively. (See Fig. 16–6.)

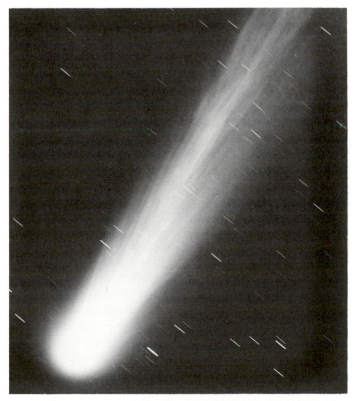

Fig. 16–6 Comet BENNETT (1969 i) photographed in the light of the CN emission band at 3883 Å with the 2-m Schmidt telescope of the KARL SCHWARZSCHILD Observatory in Tautenburg-Jena, German Democratic Republic, by Dr. F. BÖRNGEN on 12 April 1970 at 2.36–3.16 CET.

16.6. Photometry

The diffuse and tenuous nature of comets causes particular difficulties for their photometry, and this, together with the rarity of conspicuous comets, has retarded the development of this branch of cometary observation. To a large extent these difficulties have been overcome by photoelectric methods, but these are hardly accessible to the amateur observer. The methods to be described below differ considerably in principle, according to which part of the comet is to be observed.

The aim of these investigations is not only the exploration of the local brightness distribution within the comet, particularly within its tail, but also the study of the brightness development with time and with the varying distance from the Sun. It is possible to represent the total magnitude of the comet's head as a function of time in the form of a light curve. More significant, however, is another diagram, in which according to Eq. (1) on page 393 ($m - 5 \cdot \log \Delta$) is plotted against $2.5 \cdot \log r$ (the geocentric and heliocentric distances Δ and r being taken from the Ephemerides). The result will by no means always be a straight line, as one might expect. One will often have to be satisfied if portions of this curve can be represented by straight lines, from which one can read off the "reduced magnitude m_0" as ordinate, and the exponent n as gradient. Frequently the brightness shows variations or even outbursts.

16.6.1. Photometry of the Cometary Head

If the head of the comet is clearly visible to the naked eye, but still appears essentially starlike, it is possible to estimate its apparent visual magnitude by comparison with stars. The procedure is the same as in the case of variable stars (see page 445), and here too it is necessary to correct for atmospheric extinction if the comet and stars differ considerably in altitude or are near the horizon. The atmospheric extinction must be redetermined every time, except for exceptionally good nights, in the same way as has been described for the photometry of the eclipsed Moon, which is discussed in Chapter 12.

For telescopic comets visual comparison with stars is very difficult because of their faintness and diffuseness. Sometimes the observer can obtain diffuse images of the stars by putting them out of focus while always keeping the comet itself in good focus, but this method is not completely satisfactory. As with conventional photometry, if possible, one should choose for comparison stars only those whose color is similar to that of the comet and that are at a similar altitude. Because there are systematic differences from instrument to instrument, if the development of the comet requires a change of instrument (or even just a change of eyepiece), extended series of overlapping observations should be carried out with both instruments in order to be able to refer one set of observations to the other.

The determination of the total brightness of the cometary head can be carried out photographically as described on page 400 for the determination of positions. Of course, it is important that the image of the cometary head on the plate be essentially starlike and that the motion of the comet be such that sufficiently long exposure times do not produce elongated images. Comparison with neighboring stars, or perhaps with the polar sequence, is practically the same as in the case of photographic photometry of variable stars and minor planets (page 454).

16.6.2. Photometry of the Tail

With the means available to the amateur astronomer, visual photometry of the cometary tail is very difficult, even if one restricts oneself to local brightness differences in the most conspicuous part of the tail of bright comets. This is because of the very small surface brightness. Recent experience has shown that most successful tail photometry is done photoelectrically.

It is also possible to use photographic methods, although this time one should use longer exposures, so as to smooth out the complex and rapidly varying details. The plates should be calibrated in the standard way and then measured with a microphotometer. But one can produce a photographic enlargement on mat paper and then draw lines of equal surface brightness, the so-called isophotes. One thus obtains outlines of the cometary tail in the form of a contour map. It is advisable to use very soft photographic film for both the original and the enlargement.

16.7. Spectrum and Polarization

16.7.1. Spectral Photography

If we place an objective prism in front of the objective (see Fig. 16–7), the image of the comet on the plate is drawn out into a small spectrum. To prevent the tail spectrum from being superimposed onto that of the head so that its refracting edge is approximately parallel to the tail, the prism has to be oriented. In this way we obtain a sequence of monochromatic cometary images side by side. Because of diffraction in the prism it is necessary to incline the optical axis of the camera to that of the telescope by an appreciable angle; this can easily be determined in the daytime by observation of a terrestrial object on the ground glass of the camera. The exposure time must, of course, be much longer than for a comparable direct photograph, perhaps

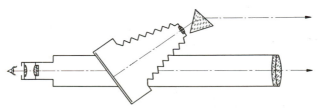

Fig. 16–7 Photographic camera with objective prism mounted on the telescope.

even as much as an hour or more. A successful exposure will at least show clearly the molecular bands in the cometary head, particularly the extremely intense band of cyanogen (CN) at 3883 Å, as well as the fainter bands of the C_2 molecule at 4737, 5165, and 5635 Å. A sensitive panchromatic emulsion is recommended.

Using an objective of 10.5-cm focal length and an *f*-ratio of 4.5 yields, for a 30° flint glass prism set at minimum deviation, stellar spectra nearly 6 mm long.

16.7.2. Polarization

If one places a polarization filter (such as is used in photography to suppress reflections) in front of the eyepiece during the telescopic observation of a bright comet and rotates it, a variation in the brightness of the cometary head can be observed. Such investigations have not been conducted very often, but it has always been found that the light of the comet is most polarized in a plane perpendicular to the one defined by the comet, the observer, and the Sun. The electric polarization vector is therefore most intense perpendicular to this scattering plane, which is also plausible from a theoretical point of view.

This vector can be checked for every reasonably bright comet by placing the filter, oriented in the direction of the electric vector, into a temporary tube that carries a graduation on a cardboard disc. An index on the eyepiece tube then permits one to read off the position angle of the filter. This is compared with the position angle p of the scattering plane, which is calculated from the following equations:

$$\Delta\alpha = \alpha_\odot - \alpha_\psi,$$

$$\tan \eta = \cos \Delta\alpha \cdot \cotan \delta_\odot,$$

$$\tan p = \frac{\tan \Delta\alpha \cdot \sin \eta}{\cos (\eta + \delta_\psi)}.$$

Here α_ψ and δ_ψ are the equatorial coordinates of the cometary head extracted from the Ephemerides; those of the Sun, which can readily be taken from the Astronomical Ephemeris, are $\alpha_\odot$ and $\delta_\odot$.

17 | Meteors and Fireballs

F. Schmeidler

17.1. General Notes About Meteors

With the term meteor we characterize all phenomena that occur when a body from space enters the Earth's atmosphere; the body or its fragments are called meteoroids or, if they happen to land on the Earth's surface, meteorites.

The observation of meteors is a field that offers the amateur astronomer many possibilities for scientifically useful work. The recent extensive photographic investigations by professional astronomers, as well as the use of radio methods, mean that this statement requires qualification, but visual meteor observations are still of great value.

We distinguish between ordinary shooting stars and large fireballs, often also called bolides. The latter are sometimes bright enough to turn night into day and may be accompanied by noise. The border line between meteors and fireballs is not sharply defined; generally speaking, we may use the term fireball if it is as bright as or brighter than Venus (magnitude -4).

It was only about the year 1800 that it was recognized that meteors are cosmic bodies that penetrate our atmosphere and then burn up. Astronomers are interested in the problem of their origin and their physical structure. The first question is mainly answered by the determination of the orbit in space, the second by studies of the luminous processes involved. Only in the case of large fireballs are we sometimes lucky enough to be able to examine fragments of meteorites in the laboratory.

A single observer can obtain only a part of the necessary observations; to answer all the questions, one generally has to consider the results of at least two observers. This chapter is intended to show the single amateur astronomer how to reduce his observations. Those who wish to contribute on a larger scale can send their results to the nearest large observatory or to the Meteor Section of the British Astronomical Association. Of course, only really conspicuous meteors should be reported. Observations of any trails that may remain after the meteor has disappeared are also of interest.

17.2. Methods of Observation

Meteors can be observed visually (with or without a telescope), photo-graphically, or, since a quarter of a century ago, with radio-astronomical devices. The amateur astronomer will as a rule be able to participate only in visual observations, but we shall mention at least the other two modes of observation as well.

17.2.1. Visual Observations

A single observer can record only the apparent orbit of a meteor. If two observers at a suitable distance (most favorably 80 to 100 km) record the same meteor, its height and thus its orbit in space can be determined from the parallactic displacement; the two observers must pay attention to the same details. It is therefore sufficient to mention here only the most essential points:

1. Data about the apparent orbit. The point at which the meteor was first seen must always be given, and it is desirable to determine three points of the orbit, including the beginning point if observed and the end point. All data must be written down at once to avoid errors of memory.

The best procedure is to refer these points to stars, for instance, "half-way between α and β Ursae majoris, or one quarter of the distance between α Cygni and α Lyrae." If it is impossible to refer the orbit to the stars, as in the case of bright fireballs observed in daytime, the path can be referred to the Sun or perhaps to the Moon. There is little point in relating the data to terrestrial objects, although estimates of the apparent height and the azimuth are also used even if less certain.

2. Time and place of the observation.

3. Duration of the phenomena. Considerable errors are commonly made in this determination, and only extensive experience can yield good estimates. Poor data about the duration are worthless. With a bit of practice it is possible to count seconds quite accurately.

4. Light phenomena. The maximum brightness of the meteor can be estimated only by experienced observers. It is advisable to state the brightness not in magnitudes, but, because of the effect of extinction, by comparison with stars at the same altitude; for instance, "same brightness as β Aquarii and somewhat fainter than δ Aquilae, but appreciably brighter than β Aquilae." If a luminous train remains, one should note its changing position and intensity. One might also remark the colors observed.

5. Acoustical phenomena. Since fireballs are sometimes accompanied by thunderous noises, one should take the opportunity to estimate the time lapse between light and sound.

6. Statements about the weather, particularly the cloud situation, are very desirable.

The observer may, of course, pay attention only to a selected few of these questions, in accordance with his special aims.

Meteor observation requires great patience. It is necessary to watch a particular region of the sky for a considerable time and to record all meteors observed within its boundaries. There is not much point in attempting to survey the whole sky at the same time. Binoculars will enable fainter meteors to be detected, but they involve a severely restricted field of view. A trained observer will be able to determine the point of disappearance of a meteor with an accuracy of a few degrees. It may be useful, although possibly less accurate, to mark on a star chart the points of beginning and end, and thence the apparent orbit as a whole.

17.2.2. Photographic Observations

The problems involved in the photographic observation of meteors, except in a few details, are essentially the same as in visual work. Accurate guiding is important. An ordinary photographic camera will rarely yield useful results because of the small scale.

The principal drawback of photographic observation is that only the brightest meteors and those whose apparent motion is slow can be recorded. The problems with the not-so-bright meteors have largely been overcome by the so-called Super-Schmidt camera,[1] but problems with speed are so far unavoidable. An important advantage of photographic observation is the possibility of measuring the velocities of meteors. If the trail of the meteor can be interrupted at regular intervals, for example, by a rotating sector in front of the objective or the plate, the velocity is readily determined by measuring the distances between the breaks (typically every 0.1 second).

If photographic exposures are made simultaneously at two different observing sites, the cameras must be directed so that their lines of sight intersect at the average height of meteors, which is 80 to 100 km. The most appropriate distance between the sites will be something like 50 km.

In summary, photographic observation of meteors can be recommended to the amateur astronomer only if he has both suitable optical equipment and devices for measuring his plates.

17.2.3. Radio Astronomy Observations

The observation of meteors by radio-astronomical instruments is based on the fact that along the path of a meteor the air is more strongly ionized than elsewhere and that this air reflects radar beams. Systematic radio observations were started before the Second World War, and since 1945 tremendous progress has been made in this field.[2]

Radar observations also make it possible to determine two quantities that cannot be derived from a single visual or photographic observation, namely the direction from which the meteorite has come—the so-called radiant—and the velocity. The radiant is determined by the fact that a radar

---

[1] The design is by J. G. BAKER, and details have been given in an article by WHIPPLE in *Sky and Telescope* **8**, No. 4 (1949).

[2] See LOVELL, A. C. B.: *Meteor Astronomy*. Oxford: Clarendon Press, 1954.

echo can be achieved only if the beam intersects the meteor path approximately at right angles. The velocity is obtained from small variations in the amplitude of the oscillations of the radar beam depending on the velocity. Details are given in the book *Meteor Astronomy* by LOVELL (see Footnote 2).

17.3. Special Observational Problems

17.3.1. Meteors

We distinguish between sporadic meteors and members of meteor streams. The physical phenomena are the same in both cases, and it is thought that the most important process is, not friction, but ionization of air molecules by impact. Meteors appear at heights of 100 to 150 km and generally disappear at heights of 50 to 100 km. Most meteors have masses considerably smaller than 1 gram and are no more than grains of dust.

One of the simplest observations is that of the meteor frequency, defined as the number of meteors per hour. The meteor frequency is actually much larger than might be expected, and on a clear night an observer without optical aid can see on the average 8 or 10 meteors. There are large variations depending on the time of both the day and the year. Most meteors appear when the point towards which the Earth is moving is high above the horizon. The reason for this is that meteors observed near this point—the "apex" of the Earth's motion—are approaching the Earth and thus have high relative velocities. Since the apex lies in the ecliptic and is 90° west of the Sun, most meteors appear in the early morning hours. Furthermore, most meteors appear in the autumn, because the apex then has its highest altitude. The above statements depend on the assumption that meteoroids move in orbits inclined at all angles to the Earth's orbit, although it is now clear that there are many more in direct orbits, i.e., moving about the Sun roughly in the same direction as the Earth.

There have in the past been considerable differences of opinion concerning the velocities of meteors. From the meteor's relative velocity and direction one can derive the heliocentric velocity, and if this is larger than 42 km per second, the meteor is traveling on a hyperbolic orbit (otherwise the orbit is elliptical). Meteors on hyperbolic orbits would have originated outside the solar system. It is now clear, particularly from recent photographic observations and radar data, that meteors do not have hyperbolic orbits. Some of the old visual observations suggested the existence of hyperbolic meteors, but the velocities derived were evidently strongly affected by systematic errors.

At certain times of the year meteors are particularly frequent and these meteors all come from the same radiant. It then follows that the meteors must be traveling on parallel orbits in space, that is, that they are members of a stream. Perhaps the best known of the meteor streams is the Perseid stream, which appears from the beginning to the middle of August. For several of the streams it has been established that their orbits coincide with the orbits of comets, which implies that meteoroids are remnants of cometary matter. Table 17.1 summarizes the data about the most important meteor streams.

Of these streams, the Draconids (or Giacobinids) and the Leonids are called periodic streams, i.e., meteors are frequent only when the parent comet is near the Earth. There were great Draconid displays in 1933 and 1946, and somewhat lesser displays in 1926, 1952, and 1953. The greatest displays of Leonids occur every 33 years, and there are records of them going back more than a millenium. The only spectacular display during the present century has been in 1966.

Table 17.1. Meteor Streams

Stream	Time	Radiant		Frequency	Comet
		α	δ		
Quadrantids	2.1 – 4.1	230°	+50°	Great	No comet known
Lyrids	19.4 –23.4	273	+31	Small	Comet 1861 I
η-Aquarids	28.4 –16.5	340	0	Average	Comet HALLEY
δ-Aquarids	22.7 –10.8	344	–15	Average	No comet known
Perseids	27.7 –17.8	40	+55	Great	Comet 1862 III
Draconids	10.10	267	+56	Variable	Comet GIACOBINI-ZINNER
Orionids	15.10–25.10	94	+14	Small	Comet HALLEY
Taurids	26.10–22.11	54	+15	Small	Comet ENCKE
Leonids	15.11–17.11	151	+22	Variable	Comet TEMPEL-TUTTLE
Geminids	6.12–16.12	113	+32	Great	No comet known
Ursids	21.12–23.12	217	+76	Small	Comet TUTTLE

The richest of the regular, annual showers is now the Quadrantids, which for a short while yield a frequency of more than 100 meteors per hour. The Perseids and Geminids are also very rich showers and extend for longer periods.

Meteor observers should keep in mind the following points.

1. One should observe both sporadic and stream meteors. The richness of a stream can change radically from one year to another.

2. Each of the major streams should be observed each year.

3. Observations of meteor frequency are more valuable if the brightnesses of the meteors are recorded.

4. Observations at zenith distances greater than 70° should be avoided in moonlight.

5. Observational material from the Southern Hemisphere is rather scarce, and amateur astronomers observing meteors there can render a great service, even without the use of special instruments.

Radio observations have also proved the existence of daytime streams, which are meteor swarms that are inaccessible to conventional observing methods. Most of these streams, which actually seem to be related to each other, appear in May and June.

17.3.2. Fireballs

A large meteoroid may penetrate deep enough into the Earth's atmosphere for its outer surface to be heated by friction. The meteoroid may break up into several pieces. If the velocity is sufficiently braked by friction, the fireball fades out and the meteoritic fragments fall to Earth.

Generally meteorites are made of either stone or iron. The latter consist of almost pure iron, but stony meteorites may also contain some iron.

Meteorite fragments often weigh several kilograms and typically exhibit crustlike surfaces, as can be seen in Fig. 17–1. Identification of a particular

Fig. 17–1 Iron meteor found at Cabin Creek, Arkansas. After BERWERTH: *Ann. d. naturhistorischen Hofmuseums*, Vienna, 1913.

object as a meteorite can be made only by an expert. A characteristic feature of many, but not all, metallic meteorites is the so-called Widmannstetten figures, which appear if the surface is treated with acid. More information on the subject can be found in such books as—

F. G. WATSON: *Between the Planets*, 188 pp., revised ed., pp. 75–179. Cambridge, Mass.: Harvard University Press, and Oxford: Oxford University Press, 1956.

B. MASON: *Meteorites*, 274 pp. New York–London: John Wiley & Sons, 1962.

C. B. MOORE (Ed.): *Researches on Meteorites*, 227 pp. New York–London: John Wiley & Sons, 1962.

B. M. MIDDLEHURST and C. P. KUIPER: *The Solar System*, Vol. V, pp. 161–526. Chicago: University of Chicago Press, 1963.

B. MASON (Ed.): *Handbook of Elemental Abundances in Meteorites*, Vol. 1 of the Series on Extraterrestrial Chemistry, 555 pp. New York–London–Paris: Gordon and Breach, 1971.

17.4. Determination of Meteor Orbits

The original orbit of a meteoroid in space is a conic section (almost always an ellipse) with the Sun at a focus. When the meteoroid is near the Earth the orbit can be regarded as a hyperbola, with the center of the Earth at a focus. The gravitational attraction of the Earth causes the meteor to fall more steeply than it otherwise would, that is, the radiant is moved toward the zenith.

Since the meteoroid is observed on only a very small section of the hyperbolic orbit, for practical purposes the orbit can be considered as a straight line, which projects into an arc of a great circle on the celestial sphere. The determination of the heliocentric orbit involves the reconstruction of the orbit in space from observations of these small arcs.

17.4.1. The Orbit in the Atmosphere

First of all we have to determine the geometrical parameters of the motion within the Earth's atmosphere. The necessary equations, summarized below without derivation,[3] use these notations:

ρ = radius of the Earth

λ_1, φ = geographic longitude and latitude of the observer

h_1, A = apparent altitude and azimuth of the end point of the orbit

α_1, δ = right ascension and declination of the end point

α', δ' = right ascension and declination of any other point of the orbit

z = linear height of the end point above the ground

A_0, D_0 = right ascension and declination of the radiant

We must calculate for each observing station the auxiliary quantities J, N, K, and i according to the formulae

$$\sin J \sin (\lambda - N) = \sin A \sin \varphi \qquad \tan i \sin (\alpha' - K) = \tan \delta'$$

$$\sin J \cos (\lambda - N) = \cos A$$

$$\cos J = \sin A \cos \varphi \tag{1}$$

$$\tan i \cos (\alpha' - K) = \frac{\tan \delta - \tan \delta' \cos (\alpha - \alpha')}{\sin (\alpha - \alpha')}.$$

Then the unknowns x, y, X, and Y follow from the equations

$$x \sin N \sin J - y \cos N \sin J + \cos J = 0$$

$$X \sin K \sin i - Y \cos K \sin i + \cos i = 0. \tag{2}$$

If there are two observing sites, the unknowns can be fully determined by

[3] Detailed mathematical treatments are provided by these books: BAUSCHINGER, J.: *Die Bahnbestimmung der Himmelskörper*, 2nd ed., p. 587. Leipzig: ENGELMANN, 1906, and STRACKE, G.: *Bahnbestimmung der Planeten und Kometen*. Berlin: Springer, 1929.

two equations of the form (2); if there are n observing sites, the most probable values of the unknowns can be calculated according to the method of least squares. The procedure then continues with the formulae

$$\cotan \chi \cos \eta = x \qquad \cotan D_0 \cos A_0 = X$$

$$\cotan \chi \sin \eta = y \qquad \cotan D_0 \sin A_0 = Y$$

$$\sin^2 \frac{s}{2} = \sin^2 \frac{\varphi - \chi}{2} + \cos \varphi \cos \chi \sin^2 \frac{\lambda - \eta}{2} \qquad (3)$$

$$z = 2\rho \sin \frac{s}{2} \frac{\sin \left(h + \frac{s}{2} \right)}{\cos (h+s)}.$$

This furnishes the coordinates A_0 and D_0 of the radiant and the height z above the ground of the end point of the orbit. The length of the path follows with the help of three auxiliary quantities σ, τ, and d from the formulae

$$\cos \sigma = \sin \delta \sin \delta' + \cos \delta \cos \delta' \cos (\alpha - \alpha')$$

$$\cos \tau = -\sin D_0 \sin \delta - \cos D_0 \cos \delta \cos (A_0 - \alpha)$$

$$(\rho + z)^2 = \rho^2 + d^2 + 2\rho d \sin h \qquad (4)$$

$$l = \frac{d \sin \sigma}{\sin (\sigma + \tau)}.$$

Finally, we obtain the velocity in the orbit as the quotient of the length and duration of the path.

In many cases it is possible to determine the radiant with sufficient accuracy by graphical methods. One marks on a star map the apparent paths as seen from two observing sites: the point at which their extensions (extended backwards) intersect in the radiant.

17.4.2. The Orbit in Space

The observed motion of the meteor in the atmosphere refers to the orbit on which the meteor traveled under the additional influence of the Earth's attraction. To relate this to the true orbit in space, one must allow for the zenith attraction of the radiant and the increase in linear velocity. The additional displacement of the radiant due to diurnal aberration can nearly always be neglected.

Here, too, the relevant formulae can only be summarized without proofs. A detailed derivation is given in the book by J. G. PORTER, *Comets and Meteor Streams*, p. 81, London, 1952. We use the following definitions:

V = velocity of the meteor in its original orbit
V_1 = original velocity of the orbit relative to the Earth
V_E = velocity of the Earth in its orbit
V_2 = observed velocity of the meteor in the atmosphere

Celestial mechanics proves that

$$V_1^2 = V_2^2 - 125,$$

where V_1 and V_2 are expressed in km sec^{-1}. This furnishes V_1 and using the following formula we find the zenith attraction of the radiant:

$$\tan \tfrac{1}{2}\Delta\zeta = \frac{V_2 - V_1}{V_2 + V_1} \tan \tfrac{1}{2}\zeta.$$

This is the amount by which the observed zenith distance ζ of the radiant must be increased.

The radiant corrected in this way is still only an apparent radiant since it corresponds to the direction of motion of the meteor relative to the Earth, but not to the direction of the meteor's absolute motion in space. As to the determination of the true radiant we refer to the above-mentioned book by PORTER. The original heliocentric velocity V of the meteor can be found with the simple formula

$$V^2 = V_1^2 + V_E^2 - 2V_1 V_E \cos \beta \cos (\lambda - \lambda_A),$$

in which λ and β are the ecliptical coordinates of the apparent radiant corrected for zenith attraction and λ_A is the ecliptical longitude of the apex. The latter is related to the ecliptic longitude $\lambda_\odot$ of the Sun by the expression

$$\lambda_A = \lambda_\odot - 90° + 57\rlap{.}'6 \sin (\lambda_\odot - 102\rlap{.}°2).$$

The heliocentric velocity of the meteor found in this way furnishes the semimajor axis of its orbit. By means of the formula

$$V^2 = k^2 \left(\frac{2}{r} - \frac{1}{a} \right),$$

where r is the distance from the Earth to the Sun, which to great accuracy can be considered as constant. If r and a are given in astronomical units and $k^2 = 1$, we obtain V in units of the mean velocity of the Earth V_E, which is equal to 29.8 km sec^{-1}. As mentioned on page 412 this same formula furnishes the velocity $\sqrt{2} \cdot V_E = 42$ km sec^{-1} in the parabolic limiting case ($a = \infty$).

18 / Noctilucent Clouds, Aurorae, Zodiacal Light

W. Sandner

18.1. Introduction

In this chapter we shall be concerned with the group of luminous phenomena that take place partly in the high and highest layers of the Earth's atmosphere—noctilucent clouds, luminous bands, and aurorae—and partly in interplanetary space—the Zodiacal Light. Actually, the study of these phenomena belongs to a large extent in the field of geophysics rather than astronomy, but usually the observing astronomer includes them in his nighttime work because according to our present knowledge, they are probably, or at least possibly, caused by extraterrestrial events. Because these are all faint, extended luminous phenomena and brightenings of the night sky and their observation reveals many traits in common, for technical reasons it is advisable to deal with them together.

This group of phenomena is of particular importance to the observing amateur astronomer, because they provide him with a type of work that promises results without elaborate technical tools. Furthermore, the professional astronomer will be grateful for the collaboration of the amateur, since this field more than many others requires a permanent and continuous survey of the sky. In the case of luminous clouds, bands, and aurorae, even the statement that nothing was visible, i.e., the recording of negative results, is of value.

In all cases the observer must have an eye that is capable of distinguishing even the smallest differences in brightness and the finest extended brightenings —and not every eye is suited. Only aurorae can sometimes reach considerable brightness. The observations are made even more difficult by the fact that in most cases we are dealing with extremely elusive objects, which become visible only to the completely rested and dark-adapted eye and only where no other light, from the Moon or bright planets or artificial sources, interferes. In or around towns or cities such observations become impossible, and as modern lighting extends more and more, even to the flat countryside and the high mountains, the opportunities become continuously rarer. This means that the observer must look for sites far away from any traffic lanes, somewhere in the mountains or on the high seas, to achieve successful results.

Apart from the disturbing light from the Moon, the possibilities are further narrowed by the fact that these phenomena are restricted totally or partially to certain latitude zones.

18.2. The Zodiacal Light

At average northern latitude the Zodiacal Light can best be observed on moonless evenings in February and March in the west, or in October in the eastern morning sky. In spring, when the last remnants of the twilight have disappeared in the west, and when the Sun has reached a depression of 17 to 18°, we can find the pyramid of light steeply rising, and a little inclined to the south, above the point where the Sun has set. The basis of this pyramid is about 30°; according to J. SCHMIDT (Athens), the full development of the Zodiacal-Light pyramid is already reached when the Sun is 15°9 below the horizon. Objective photometric measurements of the Zodiacal Light show that the pyramid is in reality much wider than one would estimate visually. Its lowest part is obscured by the haze band of the horizon which reaches up to 10 or 15°. The pyramid of light is not sharply defined, but rather its brightness decreases from the innermost brightest core (which in its lowest part, nearest the Sun, lies immediately above the haze band of the horizon) to both sides, and toward the apex of the cone. The brightest parts of the Zodiacal Light are, at least in the tropics and in the experience of the author, certainly brighter than the brightest clouds of the Milky Way in Sagittarius. Some observers have claimed rapid pulsations of the brightness of the Zodiacal Light, but such reports appear to be based on illusions caused by physiological effects. On the other hand, some seasonal variations may well be real. The Swiss observer F. SCHMIDT points out that the eastern Zodiacal Light (on the morning sky in autumn) seems to be larger but fainter than the western Zodiacal Light. He remarks that this is caused by the varying transparency of the Earth's atmosphere in the course of the day. The color of the Zodiacal Light is usually given as yellowish; SCHMIDT even speaks of a reddish color of the core. The author is in agreement with PLASSMANN, calling the color "silvery." On color photographs, taken from balloons in the stratosphere at altitudes of 30 km, the Zodiacal Light appears somewhat yellowish. Judging from objective photometric measurements, the color of the Zodiacal Light does not differ essentially from that of the Sun.

The axis of the Zodiacal Light coincides approximately—though not exactly—with the ecliptic, as indicated by the name of the phenomenon; this causes its changing visibility with the various seasons. As can be easily shown with a celestial globe or a rotatable star map, in spring at every average northern latitude, the ecliptic rises in the western sky steeply above the horizon after sunset; in autumn, the corresponding situation occurs before sunrise—the various phases in reverse order. During the other seasons the ecliptic is but little inclined to the horizon, and the delicate shimmer of the Zodiacal Light disappears in the haze. In average southern latitudes, the observing conditions are, of course, principally the same, except that now the best visibility occurs in February and March in the eastern morning sky, and in October in the western evening sky. In equatorial regions the Zodiacal Light is equally visible during all seasons.

Unfortunately, observations in average latitudes are always strongly affected by the haze of the layers near the horizon. Consequently, the brightness distribution with reference to the light axis, which here is always somewhat inclined, is always asymmetrical. The more steeply the ecliptic rises above the horizon, the more pronounced becomes the pyramid of light. Really successful observations (which essentially means reliable photometric data) are therefore possible only in low latitudes. There, in the tropics, the ecliptic descends steeply below the horizon at all seasons, and therefore the twilight is considerably shorter and the pyramid of the Zodiacal Light rises nearly perpendicularly above the horizon, so that it very soon leaves the haze of the horizon regions. Since vacations to the tropics are nowadays a real possibility for many amateur astronomers, those who are interested in these phenomena should arrange their journey for a time when observing conditions are favorable, including an absence of moonlight. Particularly

Fig. 18-1 The western Zodiacal Light photographed from Mt. Chacaltaya in the Bolivian Andes on 2 August 1958 by Dr. M. F. INGHAM, Oxford.

promising, of course, are observations from high mountains in these low latitudes; indeed, the best results of photometric measurements were obtained in the Andes of South America. (See Fig. 18–1.)

Although observations from a balloon are outside the possibilities of the amateur astronomer, in 1962 E.P.NEY obtained some beautiful color photographs from an unmanned stratosphere balloon at a height of 30 km using a camera of the aperture ratio 1:1.

At a distance of 180° from the Sun, that is, just opposite to it, the ecliptic shows another faint, usually somewhat oval-shaped brightening, the "Gegenschein," which is considerably fainter than the main light. Much fainter still is a "light bridge," which connects the main evening light via the Gegenschein to the morning light. We could therefore speak of a Zodiacal band; it crosses the whole sky along the circle of the Zodiac, is brightest in the immediate neighborhood of the Sun, and shows a faint brightening at a point opposite to it. Many observers, even experienced ones, have never seen the Gegenschein; still fewer have witnessed the light bridge. There is no doubt that the latter is one of the most difficult phenomena to observe in the whole sky. (See Fig. 18–2.)

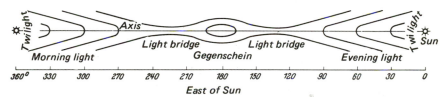

Fig. 18–2   The structure of the Zodiacal Light. After NEWCOMB-ENGELMANN.

Nature itself poses many difficulties the observer must try to overcome. Bright planets, particularly Venus, can make proper observations absolutely impossible. Also chains of faint stars can simulate a false boundary to the Zodiacal Light. The well-known phenomenon whereby the upper point of the cone seems to remain for a considerable time near the Pleiades (above the western sky in spring) also belongs to the group of illusions: this conspicuous star concentration attracts the observer's eye. One should, therefore, try to identify what is mere illusion and, in particular, to avoid all biased opinion.

HEIS, for instance, who for some 30 years during the last century observed the Zodiacal Light in the town of Münster in Westphalia, West Germany, and who had a very sensitive eye, excluded disturbing light by using a cardboard cylinder, blackened inside, with both a diameter and a length of 30 cm. SCHMIDT, however, was against the use of such a device, since it strongly affects the general aspect of this very extended, faintly luminous structure. HEIS also remarks that the light of the Moon, if it is only three or four days old, is not necessarily disturbing.

The exploration of the Zodiacal Light is essentially a photometric problem. The observer will therefore try to determine the boundaries and the brightness variation by estimating isophotes. Because there are no definite boundaries, but only continuous areas of transition, this is a difficult task and much affected by arbitrary judgment. As far back as the 1850's, Rev. GEORGE JONES distinguished four degrees of brightness, every one of which overlaps with the preceding one, and C. HOFFMEISTER provided additional isophotes when he began his observations some 60 years ago. It is not easy to estimate the exact position of the light axis to be able to enter it on the star map. It is therefore advisable to determine the position of the apex of the Zodiacal-Light cone every evening and to enter these points on the map, and then connect them when the observational series has been concluded.

Really useful results will not be achieved by pure brightness estimates, but rather by exact photometric measurements. HOFFMEISTER, therefore, on several tropical expeditions, used a surface photometer with which he measured sections across the cone of light. This was successful even on his particular observing site—the bottom of a rocking vessel. Similar photometric and also polarimetric measurements were carried out during the 1950's by R. DUMONT and his collaborator on the Pico de Tejde on Tenerife.

According to this work, the central plane of the Zodiacal Light does not coincide exactly with the plane of the ecliptic, nor, as had been suggested, with that of the solar equator, but is inclined to it by a small amount. HOFFMEISTER's conclusion of 1932 was that the surface of symmetry of the inner Zodiacal cloud is not a plane and that its position is determined by the orbit of the inner planets.

The problem as to the nature of the Zodiacal Light can be considered as solved. For a long time two fundamentally different groups of opinion were opposed to each other. One thought that the Zodiacal Light was of terrestrial origin, the other considered it a cosmic, interplanetary phenomenon. The main representative of the terrestrial hypothesis during the last few decades was F. SCHMID. Today the question has been decided in favor of the cosmic theory, mainly owing to the early work of VON SEELIGER, the later detailed photometric investigations of HOFFMEISTER, and the recent work of BLACKWELL. This now-accepted explanation is in principle also the oldest, because in 1684 FATIO DE DUILLIER is supposed to have put forward the hypothesis that the Sun is surrounded by a swarm of diffracting particles that extends beyond the Earth's orbit. It could then be supposed that Zodiacal Light is caused by the scattering of sunlight by interplanetary matter, both free electrons and dust particles. As early as 1934 GROTRIAN put forth the theory that there exists a transition from the corona to the Zodiacal Light; but it was only through the thorough investigations of BLACKWELL[1] that the still existing gap between the measurements of the corona on the one hand and those of the Zodiacal Light on the other has been closed, so that today a

[1] BLACKWELL, D. E.: *Endeavour* **19**, 14 (1960); see also INGHAM, M. F.: *Scientific American* **226**, 78 (1972).

continuous transition from the F-corona to the main body of the Zodiacal Light has been established with certainty (Fig. 18–3). According to this view the F-corona and the Zodiacal Light are fundamentally the same phenomenon observed from different directions, that is, at different distances from the Sun. F. L. WHIPPLE in his recent work does not exclude an additional geocentric ring of interplanetary dust at a distance of more than 20 Earth radii, although this appears rather improbable to this author.

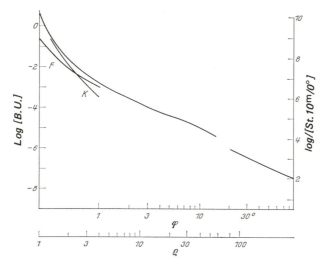

Fig. 18–3 Brightness of the solar corona and the Zodiacal Light. F = F-corona; K = K-corona. In B.U. (BAUMBACH-unit = 10^{-6} of the surface brightness at the Sun's center). St. $10^m/\square$ = stars of apparent magnitude 10^m per square degree. φ = elongation; unit of ρ is the solar radius. From H. ELSÄSSER: "Materie im interstellaren Raum. *Orion* **86**, 183 (1964).

Essential for our understanding of the nature of the particles that cause the phenomenon of the Zodiacal Light are measurements of the polarization. These have been carried out since the beginning of the 1960's by R. DUMONT on Tenerife. Unfortunately, however, such measurements are beyond the possibilities of the amateur astronomer.

Photographing the Zodiacal Light requires a fast camera, if possible 1:1.8 or a minimum of 1:3.5, with a large field of view. The exposure time is at least 10 minutes, and the camera must follow the diurnal rotation of the sky. Color pictures can also be obtained. Photography of the Gegenschein requires considerably longer exposure times, at least 30 minutes. The photometric evaluation of the photographs is made rather difficult because of the large extension of the phenomenon.

An experienced and well-equipped astronomical photographer can obtain spectra of the Zodiacal Light. On his observing station in Windhock in southwest Africa, HOFFMEISTER successfully used a spectrograph, which he described as follows: ". . . the spectrograph was built in the workshop of the observatory at Sonneberg. It consists of slit, collimator lens, prism, photographic objective, and plate holder, that is, it is of the simplest design. To make possible a successful measurement of spectral lines, I applied a step slit . . . . With the help of a slide with corresponding openings it is possible to cover the middle part of the slit and to give free access to its upper and lower part to take the comparison spectra. . . . The area of the sky that is covered in by the spectrum has a radius of $5°5$. The width of the slit was 0.105 mm up to October 6, and 0.07 mm from October 7 on. . . . A neon-argon-lamp provided good comparison spectra. . . ."[2]

18.3. Observation of Aurorae

The observation of aurorae in the north and in the south was one of the most important tasks of the International Geophysical Year 1957–58, and also of the Year of the Quiet Sun 1964–65. It is a field in which the collaboration of numerous amateur astronomers is important, since we are dealing here with very short-lived phenomena, and even a negative finding can be of value.

The prospects of seeing an aurora at an average northern latitude are, for an attentive observer, not so small as one might think. As shown in Fig. 18–4, the zone of greatest frequency of the northern lights is not exactly around the pole, but runs along the northern coast of Eurasia, south of Iceland and Greenland, across northern Labrador, Canada, and Alaska, and through the Bering Straits to complete the circle again at the northern coast of Asia. Thus the "Northern Light Pole" lies between the geographic and the magnetic pole on the islands north of Canada. Because of the eccentric position of the zone, the northern lights reach generally farther southwest of the Atlantic than on the European side. From the zone of greatest frequency, the number of aurorae decreases toward the pole as well as toward low latitudes. Over a long period, the zone of greatest frequency will average more than 100 auroral nights per year. Scotland will have 30, northern Germany 3, and southern Germany one per year, and in southern Italy, only one aurora will occur every 10 years. But even near the equator, one can see an aurora on very rare occasions. Witness the great aurora of 21 January 1938 (to which corresponded a simultaneously visible southern light) observed on board the vessel *Meteor* in the Atlantic in the tropical zone at only 20° northern latitude. The aurora of 25 September 1909 reached even farther south—to Batavia at 6° southern latitude, and Singapore at 1° northern latitude.

As can be seen in Fig. 18–4, the prospect of observing an aurora in the same geographic latitude in North America is greater than in Europe; this corresponds to the difference in the position of the magnetic poles with reference to the rotational poles. For instance, we can expect some ten

[2] HOFFMEISTER, C.: *Z. Astrophys.* **19**, 116 (1939), Mitt. d. Sternwarte Sonneberg, No. 34.

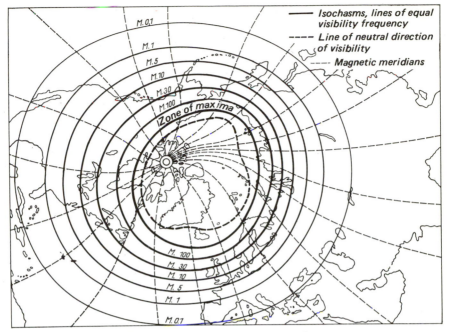

Fig. 18-4   The zone of greatest aurora frequency.

northern aurorae per year in Chicago (at $+42°$ geographic latitude, corresponding to Rome on the European side) and one aurora night per year at the latitude of New Orleans (geographic latitude $+30°$, the latitude of Cairo); and even on the Tropic of Cancer in Mexico an aurora occurs on the average of every ten years, that is, with the same frequency as in southern Italy at $40°$ geographic latitude.

In the Southern Hemisphere, the chances of seeing a southern aurora are much smaller, simply because the zone of greatest frequency runs nearly exclusively over regions with little sea traffic. The magnetic pole is at $-67°$ (1965) geographic latitude in Antarctica, that is, close to New Zealand. Extensive observational series have been obtained at the Carter Observatory in Wellington, New Zealand, which is situated at $-41°$ geographic latitude. On the South American side observing possibilities are much less favorable.

The chance of observing an aurora varies with several periods and is therefore different at different times. Whether a diurnal variation with a maximum at between 20 and 22 hours local time in central Europe is real, or whether it is only simulated by observing conditions, cannot be decided with certainty. On the other hand, a period of 26 to 28 days is very conspicuous and corresponds to the rotational period of the Sun. Also an annual double wave with maxima at the time of the equinoxes and minima at the time of the solstices can certainly be distinguished in average latitudes; in higher latitudes it is masked by the annual variation of the brightness of the

sky at daytime. It runs parallel to the variation of the magnetic disturbances
and is caused by the changing position of the Earth's axis relative to the Sun.
Most pronounced is the 11-year frequency variation, which follows the
behavior of the solar activity. Still longer periods probably exist, which are
superimposed on the 11-year period just as for sunspots.

The brightness of aurorae ranges from very faint brightenings of the sky,
which cannot with any certainty be identified as aurorae, up to the intensity
of the most conspicuous Milky Way clouds, and in high latitudes even up to
the brightness of the Full Moon, when it becomes bright enough to read
large print! As for the colors, red clouds usually dominate at low latitudes,
while elsewhere yellow-green, in particular, and also bluish and silvery tones
have been recorded. For aurorae that are sunlit and therefore situated outside
the shadow of the Earth, the usual yellow-green is less prominent, due to
nitrogen bands, than the blue-violet.

In describing the various shapes we use the international classification
listed below, which is taken from the publication of the "Seewetteramt
Hamburg," Some examples of aurora types are illustrated in Fig. 18–5.

Forms without structure

G = diffuse ground type.

HA = homogeneous quiet arcs, usually appearing near the horizon and
as a rule showing a sharp lower boundary and diffuse upper part.

HB = homogeneous bands. These do not show the regular shape of the
arcs, but indicate more or less lively motions.

PA = pulsating arcs; the intensity of an arc, or part of it, changing
periodically.

PS = diffuse luminous surfaces, which appear or disappear frequently at
the same spot of the sky in periods of several seconds.

S = diffuse surface type.

Forms with ray structure

C = corona; the rays converge toward the geomagnetic zenith. Very
frequently only half the corona is distinct. The corona can also be formed by
bands or draperies in the vicinity of the geomagnetic zenith.

D = drapery.

R = rays appearing in single structures or in bundles.

RA = arcs with ray structure, a quiet homogeneous arc frequently
transforming after a little while into an arc with ray structure. These rays can
reach different heights.

RB = band with ray structure, the band consisting of short rays. If the
rays get longer, we have type D.

Flaming aurorae

F = large waves of light rolling up from below toward the zenith.

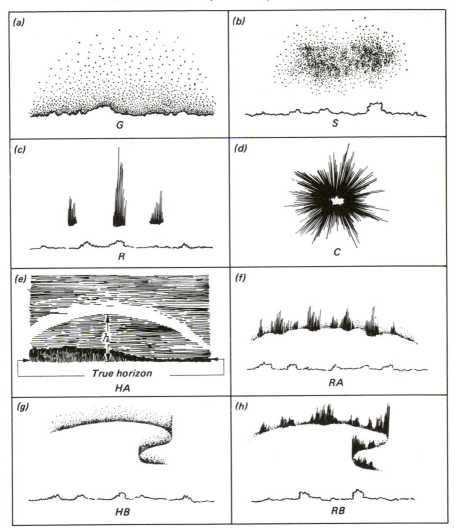

Fig. 18–5 Examples of various types of aurorae. The meanings of the symbols are listed in the text.

This scheme can be further refined by the designations A, R, and S in the following manner:

QA = quiet arc
RR = rapidly moving rays
DS = diffuse surfaces

An important aid in the classification of aurorae is the *International Atlas of Aurorae* (University Press, Edinburgh), published in 1963 on behalf of the

International Union of Geodesy and Geophysics. On 4 color plates and in 52 black-and-white illustrations, this atlas illustrates all important types of aurorae.

Some observers recommend the use of filters to exclude the general light of the night sky and possible disturbances through light from terrestrial sources or from the Moon. Interference filters for the wavelength 5577Å, corresponding to one of the most intense emissions in the auroral spectrum, are preferred for this purpose.

To determine the altitude of the phenomenon above the horizon a simple pendulum quadrant is very useful. The determination of the azimuth can be carried out with a compass or better still using fixed points on the ground, the azimuth of which is measured on the next day or derived from a map. For the announcement of the observation the following outline of information may be used.[3]

(a) Each night should be characterized by a double date (ordinary date and weekday).

(b) Time in Universal Time.

(c) Observing site in geographic longitude and latitude, with two free columns for geomagnetic coordinates.

(d) Observing conditions according to the following code:
L = aurora present (without any doubt)
O = aurora not present (a definite finding, provided the weather conditions are good
? = aurora probable (but not certain because of haze, etc.)
X = aurora not recognizable because of fog, etc.

(e) Description of the shape of the aurora according to Fig. 18–5.

(f) Azimuth (A) and altitude (H) in degrees, greatest height marked by arrow.

(g) Color.

(h) Brightness, described according to the following scale:
1 = faint (as bright as the Milky Way)
2 = average (as bright as cirrus clouds illuminated by the Moon)
3 = bright (as bright as cumulus clouds illuminated by the Moon)
4 = brilliant
Under normal conditions in central Europe only grades 1 and 2 will normally be observable.

It is desirable to supplement the descriptions by sketches. They should be sent immediately for scientific evaluation to one of the astronomical societies mentioned in the Bibliography on page 549.

The classic country for aurora research is Norway. It was here that C. Störmer in Oslo developed his methods of photographic altitude determination through the use of simultaneous photographs from several stations that were interconnected by telephone. L. Harang, in charge of the Aurora Observatory in Tromsö, has also made important contributions. In Germany, contributions were made before World War I at the special Meteorological-

[3] *Der Seewart* **18**, 6 (1957).

Geophysical-Observatory on Spitzbergen, where series of photographs of aurorae were obtained. In 1939, W. BAUER produced, on behalf of the Allgemeine Elektrizitäts Gesellschaft, a special aurora film.[4]

STÖRMER's work shows that aurorae have their greatest frequency at heights of 100 to 115 km above the Earth's surface. Sometimes they come down to 70 km, and the upper limit is in general 300 km although occasionally heights of up to 1000 km have been measured with certainty. Although it has sometimes been reported that aurorae have apparently reached down almost to the surface of the Earth (in one example, they appeared to be in front of nearby mountains), all these observations should probably be rejected as illusions—quite apart from relevant theoretical objections (see Fig. 18–6).

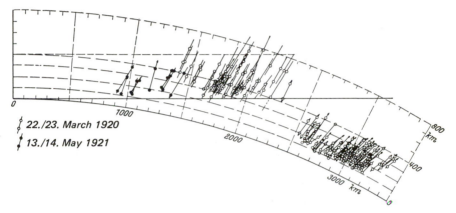

22./23. March 1920

13./14. May 1921

Fig. 18–6 The position of aurora rays relative to the Earth's shadow. After C. STÖRMER.

In the polar regions one frequently meets with the opinion that bright aurorae are often accompanied by a crackling noise resembling that of grass or cornfields burning. Here we must emphasize that the organized network of scientifically experienced observers have never, over a period of several decades, recorded any such aurora noises. Nevertheless, HARANG mentions some cases in which even with very critical inspection certain testimonials for aurora noises can hardly be refuted.

The close connection of aurorae with solar eruptions and the disturbances of the Earth's magnetism is common knowledge. These disturbances are caused by the arrival in the Earth's atmosphere of electric particles originating in the Sun. It is interesting to observe and record these magnetic disturbances in their connection with the aurorae. This is quite practicable with the means at the disposal of the amateur astronomer, as demonstrated by C. A. PFEIFFER in his construction of a homemade one-thread magnetograph.[5]

[4] Archivfilm No. B407, Reichsstelle für den Unterrichtsfilm, Berlin (1939).
[5] PFEIFFER, C. A., and RÜGEMER, H.: *Sternenwelt* **2**, 138 (1950).

In 1890, the Göttingen astronomer Brendel, jointly with the geographer Baschin, succeeded in photographing an aurora at Bossekop in northern Norway. Today, auroral photography has become a promising field for the amateur astronomer. Because of the rapid variability we require powerful optics—1:1.8, if possible—to be able to achieve this short exposure time. The brightest aurora visible in, say, the M.1 zone shown in Fig. 18–4, can be photographed with highly sensitive plates with exposure times of only 10 to 20 seconds, but usually it will not be possible to go below one minute. General figures, of course, cannot be given, owing to the large differences in the brightness of the objects.

Color photography of aurorae, too, can be a rewarding undertaking for the experienced photographer. With an aperture of 1:1.8 we can usually obtain pictures of bright aurorae within one to three minutes.

It is also possible for two amateur astronomers to work at different sites, 20 to 200 km apart, to achieve a photographic determination of altitudes. Bright stars photographed at the same time serve as reference points for the measurement.

For educational purposes an aurora film was produced by W. Bauer (AEG Forschungsinstitut) before World War II. This film also presents the "Terella" experiments of E. Brüche.

Spectrograms of the aurorae are out of the question for the amateur astronomer. The photographs obtained by Harang in Tromsö were obtained with powerful spectrographs of small or medium dispersion, the exposure times ranging from a few minutes to some hundreds of hours, according to the light-gathering power of the spectrograph and the required "depth" of the spectrum.

18.4. Noctilucent Clouds

The noctilucent clouds (NLC) are a rather rare phenomenon, occurring sometimes less than once a year, whose visibility, moreover, is restricted to certain latitude regions and seasons. An experienced observer, B. Albers in Hamburg, describes them as follows: "On some extremely clear evenings in the summer months, more exactly between June and August, when the bright twilight on the northern sky never disappears in the course of the whole night (midnight–twilight), the NLC appear—one or two hours after sunset, very low above the northern horizon, as longish banks or flat bands and fine, tubelike patterns, frequently crossed by long, irregular wave trains. They can be seen shimmering and shining in a delicate silver-white, or sometimes more bluish, light, with a faint golden yellow tone near the horizon, sometimes so to speak phosphorescent and, particularly in the delicately luminous veils, of a kind of ethereal delicacy."[6]

From this description, then, we may gather that these are rather inconspicuous formations, somewhat similar to cirrus clouds, which in accord-

[6] Albers, B.: Leuchtende Nachtwolken, Briefe aus München, No. 24 (1958).

ance with their appearance are sometimes also called "silver clouds." They are transparent to such a degree that it is possible to observe some bright stars through them. In no case, however, should they be confused with the "Moon's mother-of-pearl clouds," phenomena of considerable brightness in the night sky, which are situated at very different altitudes, about 30 km or so.

The familiar name "noctilucent clouds" is not strictly correct, because we are dealing here not with actual luminous structures, but only with objects made visible by reflected sunlight.

The one condition necessary for their visibility is particularly transparent air with good seeing (since we are concerned with a phenomenon close to the horizon). Most frequently such conditions exist after the arrival of large, very dry and dust-free Arctic air masses. The occurrence of the NLC is restricted to regions between 45 and 60°, both in northern and southern latitudes, and to the summer months from June to the beginning of August, with a pronounced maximum at the end of June and the beginning of July, although in the Soviet Union some NLC have been reported as early as the beginning of March and up to the end of October. The monthly distribution according to W. W. SPANGENBERG is given in Table 18.1.[7]

Table 18.1. Monthly Frequency of the NLC in the Period 1932–1941

Month	No. of cases	Percentage
April	3	7
May	0	0
June	8	19
July	25	59
August	5	12
September	1	3

Most of the observations were obtained between 55 and 60° North, and only once is mention made of an observation at $+71°$. At latitudes below $\pm 45°$ the NLC have never been observed. On 15 June 1963, however, a cloud was observed at Tucson, Arizona, 32° North, that had similarity with a noctilucent cloud and whose distance was determined as 71 km; however, it appears that this was an artificial cloud caused by the start of a rocket.[8]

There are some observations from the Southern Hemisphere, particularly from Chile, but they are much rarer than those from the north, since observing conditions in the south are much less favorable, the latitude zone between $-45°$ and $-60°$ being nearly wholly covered by water. The sparse southern observations of the noctilucent clouds show a frequency maximum for the

[7] SPANGENBERG, W. W.: Über die Leuchtenden Nachtwolken der Jahre 1932–1941, Wetter und Klima, Heft 1–2, 1949.
[8] Steward Obs. Contr. No. 46, Tucson, Arizona, 1963.

Southern Hemisphere during the months December–January, that is, just what we would expect.

Still unexplained, and therefore to be checked again, is a result obtained by British observers, according to which in the late summer noctilucent clouds seem to appear more frequently nearer the pole.

For the observer of upper-atmosphere phenomena it is of interest to know at what negative solar altitudes luminous night clouds have been observed. Observations at Sonneberg, latitude $+50°.4$, gave values between $-10°.6$ and $-16°.1$; theoretically, observations should be possible even at $-6°$ under favorable conditions.

It has been impossible to find a periodicity in the NLC occurrence. Usually they appear after periods of anomalous twilight phenomena, as they have been frequently observed during the years 1963 to 1965 at many places. Possibly these are connected with the outbreak of the Mount Agung volcano on Bali on 17 March 1963, which produced vast quantities of dust thrown up into the high atmosphere. These may also be responsible for the extreme darkness of the lunar eclipses during these years. Furthermore, years with relatively frequent NLC are followed by longer periods in which these are completely lacking. They were mentioned for the first time in literature in 1648 by P. EMANUEL MAIGNON, but it was only in 1885 that they were described in detail by O. JESSE, at a time when a period of greater frequency was setting in.

Whoever is lucky enough to see noctilucent clouds should at once transmit his observation to the nearest meteorological station or to an observatory. Even the simple information that noctilucent clouds have been seen can be of value, provided that exact time and place are reported. If the observer can do more, he will add a detailed description of their appearance, changes with time, position in the sky in altitude and azimuth, and their color. Since noctilucent clouds move about with considerable speed, an attempt should be made to determine as accurately as possible the direction of their motion. Of particular importance are measurements, or at least estimates, of the height above the horizon of their upper boundary; a simple pendulum quadrant will be extremely useful, since experience shows that altitudes above the horizon are in most cases considerably overestimated, particularly by observers of limited experience.

According to W. SCHRÖDER[9] the noctilucent clouds can be classified within the following abbreviated scheme:

Type I:	Veils, of a tattered structure, whitish
Type II:	Bands, with sharp boundaries
Type III:	Waves, occurring in subgroups a and b
Type IIIa:	Parallel short bands
Type IIIb:	Bands showing wave structure
Type IV	Vortices

[9] SCHRÖDER, W.: Über die Leuchtende Nachtwolken, *Zeitschr. f. Meteorologie*, Vol. **21**, 11–12, pp. 340ff. Berlin: Akademie-Verlag, 1970.

W. W. SPANGENBERG derived from his records during the years 1932 to 1941 the data given in Table 18.2 on motions of NLC (see Footnote 7 for source reference). Thus we see that in more than 70% of all cases the clouds came from eastern directions, a fact that is in good agreement with investigations concerning the motions of meteor tails. According to STÖRMER the motion is directed from the sunlit atmosphere toward the direction of the Earth's shadow. It was JESSE who first recognized the large horizontal velocities that can occur. STÖRMER determined an average velocity of 44 m per sec

Table 18.2. Direction of Motion of the NLC in the Period 1932–1941

Direction arrived from	Percent	Total percentage
SE	5	
E	29	72
NE	38	
N	5	
NW	4	
W	9	13
SW	0	
S	0	
Doubtful	10	

in 1932, and in 1934 81 to 83 m per sec. SPANGENBERG's values for 1932–35 are all below 100 m per sec. All previous observations had originally led to considerably higher values. However, SÜRING mentions velocities up to 300 m per sec. PATON, in 1964, combined all available measurements and arrived at velocities of about 50 m per sec. We should mention that rocket measurements have yielded data of wind velocities at the height of 180 km of the order of 40 to 45 m per sec. These figures alone show the great importance of all such observations.[10]

Even the simplest observation discloses the wavelike structure of the luminous night clouds; it is therefore important to record the distances between the various wave trains. Exact measurements of the distances between the wave crests gave values of the order of 9 km.

Measurement of the apparent altitude, obtained visually and photographically, can lead to the determination of the true (absolute) altitude above the Earth's surface. JESSE found from his 1885–91 observations values of 80 to 89 km, with an average of 82 km: this is exactly the mean altitude of the Mesopause of the atmosphere. In the 1930's STÖRMER extended his aurora-observation service to these rare clouds and obtained several series of parallactic photographs in south and central Norway: here heights of

[10] HOFFMEISTER, C., *Z. Meteorologie* **1**, Nos. 2–3 (1946).

74 to 92 km (with a mean of 81.4 km) were derived in 1932, and values of 78 to 85 km (mean 82.2 km) in 1933–34, in good agreement with the earlier visual results of JESSE. The experienced amateur astronomer can make a valuable contribution to this problem.[11]

Except for the above-mentioned observations by STÖRMER, we did not possess any systematic series until 1957. At that time a special observing program was established within the framework of the International Geophysical Year, and in 1962 systematic observation began in the United States; in 1964, in the Southern Hemisphere.

Everything indicates that the NLC represent structures of fine dust particles which, floating at great height, are still lit by sunlight when for the observer on the ground the Sun has long since sunk below the horizon. It is not yet clear why the dust should give preference to the latitude zones between 45 and 60°, on both sides of the equator. It is possible that the magnetic field of the Earth may play a role, but it is also possible that the system of circulations in the upper atmosphere is responsible for this. It is also as yet unclear as to whether these dust masses are thrown up into these great heights by volcanic eruptions on the Earth, or whether they have perhaps penetrated from outside, that is, have come in from interplanetary space. The large frequency in the 1880's following the eruption of the volcano Krakatau in 1883 was evidence for the first suggestion; indeed corresponding volcanic eruptions are on record for all periods of conspicuous luminous night clouds, outbursts in which large quantities of dust and stone material were produced, for example, in 1902 in Martinique and in 1912 in Alaska. On the other hand, however, NLC also appeared after the fall of the famous Tunguska meteorite on 30 June 1908 and after the transit of the Earth through the tail of HALLEY's comet in May 1910. Also frequent phenomena in the middle of June 1935 could be related to the comet 1930d, whose orbit had just crossed that of the Earth. Probably any of these phenomena could play a role in producing the NLC. In August 1962, a group of Swedish-American investigators sent up four rockets from Northern Sweden into the upper atmosphere to collect the responsible particles from the altitude region of the NLC and bring them down to Earth. Laboratory examination showed that the collected material was meteoric dust covered with a thin layer of ice.

Photographic records of the NLC, obtainable with relatively modest means, should be attempted whenever possible. The use of highly sensitive films and apertures of 1:1.8 or 1:2 in 15 seconds' exposure time will yield useful records. Because of the rapid variability the exposure must not be extended too long. To be able to study the changes in the clouds and in their direction of motion, it is always most desirable to obtain whole series of

---

[11] A good survey of today's situation in the exploration of the noctilucent clouds is given by G. DIETZE, in the "International Symposium on the Study of Noctilucent Clouds as Indicators of Processes in the Upper Atmosphere," in *Die Sterne* **42**, 163 (1966). See also B. HAURWITZ (National Center for Atmospheric Research, Boulder): Leuchtende Nachtwolken, *Sterne und Weltraum* **11**, 180 (1972).

photographs.[12] (See Fig. 18–7.) Color photographs, of course, would be even better.

Even motion pictures of noctilucent clouds, obtained on 8- or 16-mm film, are within the reach of the amateur astronomer. Movie cameras with apertures 1:2 are normally used. Making single exposures of, say, 15 seconds, and using highly sensitive films should yield good results. The normal projection speed is 16 frames per second, and one scene should last for 8 to 10 seconds. We therefore need 160 successive exposures of 15 seconds each, requiring a total of some 40 minutes. The projection would then last 10 seconds (acceleration 1:240). Such a film should bring out impressively the relevant changes in shape and position of the luminous night clouds. For the first time in the 1950's some Soviet observers produced several such impressive films illustrating the motions of luminous night clouds.

Fig. 18–7 Noctilucent clouds, photographed by G. ARCHENHOLD on 16 June 1935, 22^h12^m to 22^h13^m CET, at Horwich, Bolton, Lancashire, England.

18.5. Luminous Bands

Certain phenomena known as "luminous bands" are most probably of extraterrestrial origin. They appear as extremely delicate structures without a sharp boundary, and only those who are fortunate to work under really good conditions will ever have an opportunity to see them. Many persistent observers have never yet seen these luminous bands, which are particularly difficult to distinguish because of the lack of any characteristic form. Their systematic survey started only in 1920 through the work of HOFFMEISTER at the Sonneberg Observatory. Sometimes they appear as a kind of light spot. Frequently their structure shows stripes with a distinct point of convergence near the horizon, and sometimes they are arranged as a kind of network.

[12] MEIER, L.: Mitt. d. Sternwarte Sonneberg, No. 47, 1959.

The presence of luminous bands is often recorded simply as "bright night sky." They are certainly quite unlike aurorae. (See Fig. 18–8.)

Luminous bands differ distinctly from luminous night clouds in both appearance and distribution in time and place. For instance, they are not

Fig. 18–8 Photographs of luminous bands. After C. HOFFMEISTER.

confined to certain latitudes—Hoffmeister saw them not only at Sonneberg, but also regularly at Windhoek in southwest Africa, and occasionally in the Gulf of Mexico—and the annual behavior of their frequency is the same in both hemispheres. As a rule, intensity and frequency of the luminous bands increase during the course of the night.

According to Hoffmeister, these phenomena are caused by cosmic dust entering the atmosphere. The annual variation of the frequency is explained by him as follows:[13]

> The annual curve has two typical properties: a long, deep minimum, covering the months March to September, and a maximum from October to the middle of February; and very irregular behavior with high peaks at the time of the maximum. We should also mention the rise in July and August, which, however, does not produce any really large values. In the second half of September the curve descends for a short while nearly to zero. In 1951 I drew attention to the fact that this situation can be explained by the superimposition of two effects. These are, first, incoming meteoritic material from interplanetary space and second, the condition of the upper atmosphere controlled by the Sun. This condition apparently weakens the luminous effect in the case of intense ultraviolet radiation, that is, through the strong increase of the electron concentration, and would cause weakening of the density in the summer and strengthening in winter. This annual variation makes a connection with meteor swarms seem quite likely, both in single cases and in the general annual frequency distribution. However, the various meteor swarms behave differently, for example, contrary to the general rule, the Lyrids do not produce any luminous bands, while the Leonids produce relatively few meteors but many bright bands.

The average height of the luminous bands is about 125 km, derived from photographic-trigonometric measurements; the actual values show a scatter, and the luminous bands are considerably higher than the luminous night clouds.

It is important for the inexperienced observer to avoid a confusion with the light bridge or the Gegenschein. In central Europe luminous bands will usually be found on the northern sky.

The observers should attempt a determination of the bands' direction of motion and velocity since from this information the direction and speed of the air streams in these great heights can be derived. The occasional occurrence of whole systems of luminous bands with different points of convergence and different direction of motion (as has been established by Hoffmeister) makes the presence of several layers probable.

> In this way, Hoffmeister postulates the existence of the high atmospheric Passat System (Ultra-Passat). He also finds that at heights around 120 km, a wind blows from the southwest with great regularity at an average velocity of

[13] Hoffmeister, C.: Interplanetare Materie und Verstärktes Nachthimmelleuchten, *Z. Astrophys.* **49**, 233–242 (1960); Mitt. d. Sternwarte Sonneberg No. 51.

50 m per sec. In winter there are additional winds from the northwest to the northeast, whose velocity is somewhat greater than in summer. In 21% of the cases the values are greater than 100 m per sec; the highest value found up to now is even 267 m per sec.

The photography of the luminous bands requires very powerful optics. Using the most sensitive emulsions, we might require an exposure time of about 5 minutes.

18.6. The Blue Sun

A very rare and unpredictable phenomenon is the "Blue Sun" or the "Blue Moon." Although not of an astronomical nature, it will certainly be immediately noticed by an observing astronomer and should therefore be mentioned briefly. It consists of a curious, quite unreal blue to blue-violet coloring of the Sun, the Moon, and other stars and planets. This coloring lasts at least a few hours, sometimes a few days. Particularly well observed in its details was the phenomenon of September 1950. Giant forest fires in northwest Canada on 22 and 23 September produced enormous fire clouds, which were driven by the wind to the east, reached Newfoundland on 25 September, crossed the Atlantic on 26 September (watched by planes), and reached the European coast the morning of 27 September. In northern Germany they appeared as very high, brownish clouds, not unlike altro-stratus. They remained visible all over Europe from northern Scandinavia down to Portugal until the first days of October, producing the above-mentioned phenomenon of the Blue Sun. Again, every observer of such an event should report it to the nearest observatory or weather station immediately, with exact data as to time and place.

19 | The Photometry of Stars and Planets

W. Jahn

19.1. On the Measurement of the Sensitivity of the Eye

Photometric work requires a comparison scale. Daily life teaches us to compare equal with equal. It is essential to note that the measurements of events affecting our senses can be based on scales and measuring methods that are independent of our sensations. Thus, light intensity is measured by chemical effects on a photographic plate, or by the electric current from a photocell. If, on the other hand, our light sensations themselves are used to measure the light intensity that causes them, we have to take into account that our memory for light sensations is only a very restricted one. Our eye can easily be misled as to the intensity and kind of the stimulus. Only when the stimulus L is changed by a minimum value ΔL does the eye register another sensation E. The ratio $\Delta L/L$ is well known within an average range of sensation (the so-called FECHNER range). Starting from an initial sensation E_0, whether or not E increases in large or small steps with increasing L depends on the sensitivity of a particular eye. This building-up of a sensation E, starting from E_0 and triggered off by many $\Delta L/L$ quantities, leads to the following result:

$$\text{Constant} \times (E - E_0) = \log (L/L_0).$$

In these "equations" we compare totally different things. The last-mentioned equation states simply that the intensity of the sensation depends on the logarithm of the intensity of the stimulus (WEBER-FECHNER Law).

19.2. The Photometric Capabilities of the Eye

We refer the reader to Sec. 3.2, but summarize in what follows the characteristic photometric properties of the eye.

The amount of radiation that the eye can receive without being blinded lies between 1.3×10^{-9} HK/m^2 and 5 HK/m^2, i.e., within a range of 24 magnitude classes. The limiting magnitudes are not constant; HK indicates HEFNER candles and m stands for meters.

The eye is capable of arranging light sensations according to their intensity. In the same circumstances equal sensations are triggered off by equal intensities.

The eye is capable of recognizing equal intensities in equal circumstances, i.e., it can construct a memory scale and interpolate into this scale other light intensities.

The eye can recognize small differences in brightness. These are more or less fixed for the same observer and change only very little. Equality can be established within an error of about ± 0.1 magnitude class.

The eye is capable of distinguishing a multiple of the smallest recognizable magnitude difference (up to four times its size).

Finally, the eye can compare brightness differences among three sources of light and distinguish them quantitatively.

19.3. Sources of Error in the Estimation and Measurement of Brightness

19.3.1. *Influence of the Errors of Instrument and Eye*

The size, shape, and light distribution of the image that the observer sees in the telescope depend on the form and quality of the objective lens or mirror. If, because of vignetting at the center, much of the light is redistributed into the diffraction rings, we perceive this more strongly for bright than for faint stars. Placing a diaphragm at the free aperture changes the light distribution in the diffraction disc. The brightness of the telescopic image cannot therefore be determined accurately from the equations given on page 18, but only by comparison with other images of known brightness.

The same effect comes about if objective or mirror is not sufficiently corrected for spherical aberration. The images of the stars are then surrounded by rings whose appearance depends on brightness, color, and the aperture of the telescope.

If the star images are surrounded by a violet halo, which is always the case for a two-component achromat, then the brightness of this halo changes with the color of the star. The images of red stars contain a much larger proportion of the incident light, and thus red stars gain in brightness in comparison with stars of other colors.

Furthermore, the effect of colored objectives is important. Greenish objectives, e.g., FRAUNHOFER achromats, change the color of a celestial object.

A photometer should exhibit the comparison light source in the same manner as the object to be measured. This can be achieved only partially. A very sharply imaged star with its brightness and position fluctuations can always be distinguished from an artificial star. A slight defocusing of the natural star can diminish these differences.

Also important are losses in the sharpness of the image caused by aberrations in the eye, particularly astigmatism. If the diameter of the light bundle received by the eye is large, i.e., if the magnification is small, the observation of fine details suffers.

19.3.2. Influence of the Earth's Atmosphere, Position and Brightness of the Object, and Comparison Stars

In the Earth's atmosphere, light suffers a certain amount of extinction. On its way to the observer, the light of a star at the zenith passes straight through the whole height H of the atmosphere. This is the shortest path.

If the star is by z degrees away from the zenith, then the path s of the light through atmospheric layers of height H is increased from H to

$$s = H \sec z.$$

At the zenith distance $z = 60°$ the path is twice as long as at the zenith distance $z = 0°$. The intensity of the observed brightness is then

$$i_{60} = i_0 \cdot p^H,$$

if the transparency of the air $(1 - p)$ is constant. Because of the complex structure of the atmospheric layers it is not always possible to calculate the "extinction" of the light from this simple relation, and it is necessary to use empirically determined, or calculated, but observationally confirmed, values; see, for example, Table 2 of the Appendix. The tabulated values give the decrease of the light, expressed in magnitude classes, as a function of the observed zenith distance. These "extinction" values have to be subtracted from the observed brightness, and the outcome is the magnitude "reduced to $z = 0°$."

Brightness measurements are likely to be in error when the observation stretches over a considerable period of time during which the star changes its zenith distance, unless we use the correct mean value of the zenith distances. Because of the rapid increase of s with z the best way is to reduce every single observation to $z = 0°$. Furthermore, we have to take into account whether the stars are at such different zenith distances that their differential extinction must be allowed for. We can avoid errors if we arrange the observations near the zenith or symmetrical to the meridian. Another source of error is such an extremely high or anomalous transparency of the air that our tabulated values are invalid.

Another and an unavoidable source of error is air turbulence. This is very rarely so small that the stars do not scintillate during one second. If the air is slightly opaque or hazy, the stars seem to keep very still; however, in this case only micrometric, not photometric, measurements will be reliable enough. Altogether we should avoid photometric observations at zenith distances beyond $70°$.

We now turn to some other sources of error which apply mainly to brightness estimates and visual measurements.

1. Error in zenith distance. Estimates and measurements frequently show an increasing error with decreasing zenith distance. This error is caused by the tiring of the observer in a tense body position and can therefore be reduced simply by a comfortable seat or couch.

2. Error in position angle. Nearly all observers estimate the brightness of a star in the field of view at the bottom right a little brighter than a star on the top left. Such an error can be determined from a very extended series of observations and allowed for afterwards. It can be avoided by the use of prisms, which reverse right and left of an image, or by observing the star at the same place on the retina. Never look at two stars at the same time!

3. The interval error. In the case of step estimates, the smaller steps are affected by the smallest errors. The errors of the larger steps are multiples of this. This error is countered by assigning "weights" to the observations, as outlined on page 448.

4. The distance error. The larger the distance between objects being compared, the less reliable the observation. A comparison star not included with the variable star in the field of view should be reached without difficulties. This source of error should be taken into account when the comparison stars are selected.

5. The neighborhood error. A star is estimated as too faint if it is in the neighborhood of a brighter star. Magnitude estimates in the Milky Way are more difficult than in fields that are poor in stars. This error can sometimes be reduced through a choice of stronger magnification.

19.4. The Preparation of a Brightness Estimate

19.4.1. Choice of the Observing Instrument

A variable star, which we intend to measure with the step method, must appear in the telescope equal in brightness to stars in the range 0^m5 to 3^m5 when seen with the naked eye (FECHNER range); see page 439. In Table 19.1 we find the required aperture ratios of the observing instruments (field glasses, refractor, reflector).

If the variable star appears to be too bright, the objective lens will have to be diaphragmed down. If its brightness is the same as that given in the very first column, then the star is at the limit of the FECHNER region; then, too, we shall use a diaphragm. Any change of telescope requires a new determination of the step scale (see Sec. 19.5.2.3).

Table 19.1. Required Apertures for Brightness Estimates of Variable Stars

Brightness of the variable in magnitude classes	Aperture	
0^m5–3^m5	7 mm	(eye)
2.5–5.5	18	(field glasses)
4.5–7.5	44	(2-inch telescope)
6.5–9.5	110	(4-inch telescope)
8.5–11.5	260	(10-inch telescope)

Table 19.2. Variable Stars Observable with the Naked Eye

Name	Right ascension	Declination	Brightness		Period
			Maximum	Minimum	
γ Cassiopeiae	0^h51^m	$+60°2$	1^m6	2^m9	Irregular
o Ceti (Mira)	2 14	$-$ 3.4	3.4	9.2	331 days
β Persei (Algol)	3 2	$+40.6$	2.4	3.5	2.867
ζ Aurigae	4 55	$+40.9$	4.9	5.6	972
ε Aurigae	4 55	$+43.7$	3.3	4.1	9900
α Orionis	5 50	$+$ 7.4	0.5	1.1	2070
β Lyrae	18 46	$+33.2$	3.4	4.1	12.908
η Aquilae	19 47	$+$ 0.7	3.8	4.5	7.176
μ Cephei	21 40	$+58.3$	3.7	4.7	Irregular
δ Cephei	22 25	$+57.9$	3.6	4.2	5.366

NOTE: The brightness of Aurigae is photographable. The data given for Mira Ceti are mean values.

From CAMPBELL and JACCHIA: *The Story of Variable Stars.*

19.4.2. The Choice of the Variable Star

Table 19.2 gives the selection of stars observable with the naked eye.

Table 19.3 lists the most extensive and useful catalogs, available in any observatory library. For the classification of variable stars, see Table 21 in the Appendix.

19.4.3. Information Service for Observers

To facilitate the selection of comparison stars we mention only a few of the photometric catalogs (see page 547–548). In addition, the following associations publish instructions and results:

1. American Association of Variable Star Observers, (AAVSO), 187 Concord Avenue, Cambridge, Massachusetts 02138.

2. Berliner Arbeitsgemeinschaft für veränderliche Sterne, Wilhelm-Foerster-Sternwarte, Berlin-Schöneberg, Papestrasse 2, Germany.

19.4.4. Finding the Variables in the Sky

The observer should draw a little chart showing the variable star and the neighboring comparison stars together with the necessary identifications (Fig. 19–1). In the case where we use an inverting telescope we shall mark this chart in such a way that the right ascensions increase from left to right and the declinations from the top to the bottom. The size of the field of view and the eyepiece used will be indicated on the chart (see page 11). If the observer uses field glasses he should select a star that is visible with the naked eye and that has the same right ascension as the variable, then move

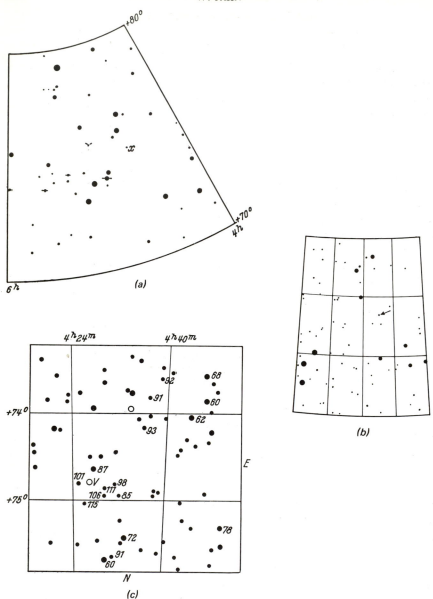

Fig. 19–1   Maps for the variable star X Camelopardalis. (a) From A. Bečvář: Atlas Coeli, 1950.0. (b) From Bonner Durchmusterung 1855.0. (c) After the publication by the American Association for Variable Star Observers. The magnitudes are marked in units of $0^m\!.1$. The position of (c) is turned by 180° with respect to (a) and (b).

Table 19.3. Catalog of Variable Stars

Title	Author	Place of publication
Obshii Katalog peremennykh zvezd—OKPZ (catalog)	B. V. KUKARKIN, et al.	Moscow: Akademija Nauk USSR
Katalog zvezd zapodozrennykh y peremennosti—KZP (catalog)	B. V. KUKARKIN, et al.	Moscow: Astr. Soviet Akad. Nauk USSR
Peremennye Zvezdy (bulletin)		Moscow: Astr. Soviet Akad. Nauk
Rocznik Astr. Observ. Krakowskiego (yearbook)		Krakow; Polska Akad. Nauk
Preliminary Catalog of Supernovae (up to 1967)	M. KARPOWICZ, K. RUDNICKI	Warsaw: Warsaw Univ. Press
Katalog und Ephemeriden veränderlicher Sterne (up to 1943)	H. SCHNELLER	Kl. Veröff. Berlin-Babelsberg
Katalog und Ephemeriden veränderlicher Sterne (up to 1923)	Vierteljahrsschrift der Astr. Ges.	Leipzig: Poeschel u. Trepte
Astronomische Jahresberichte (annual)	K. HEINEMANN, u.a.	W. de Gruyter
Atlas Stellarum Variabilium (with catalog)	J. G. HAGEN	Berlin: L. Dames
Peremennie Swesdie (journal)	Astron. Bulletin of the Soviet Acad. of Sci. USSR	Moscow

his field glasses, starting from the star, in the direction of increasing or decreasing declination, until he has found the variable. If an azimuthally mounted instrument is used, the observer should first look directly with the naked eye at a bright star that has the same altitude or the same azimuth as the variable, then point his instrument to the star and by changing azimuth or altitude reach the variable star. If an equatorially mounted instrument is used, he should set the coordinates on the circles, or otherwise set it on a star visible to the naked eye and having the same declination, then change the instrument only in the direction of right ascension until he reaches the variable.

If the comparison stars do not appear simultaneously with the variable star in the field of view, then the finding of the comparison stars and the setting of the instrument on them must be thoroughly practiced until this process succeeds very easily.

19.5. Brightness Estimates

19.5.1. Direct Magnitude Estimates

Under this heading we mean a rough, usually improvised estimate derived from naked eye comparisons with other celestial objects, such as satellites, meteors, or comets. Such an estimate is achieved (1) by indicating stars of equal brightness and equal color, (2) by giving a series of brighter or fainter stars of the same color, or (3) by forming a series of stars arranged according to brightness, into which the estimated object can be interpolated. The accuracy may be about half a magnitude.

If we use an instrument, for instance, field glasses or a comet finder, then the mobility of the observer's eyes is restricted. The comparison stars must be observed through the telescope and fixed within a memory scale, which must frequently be checked and corrected. Years of practice in the use of a memory scale lead to estimates of an accuracy of 0.1 magnitude. Reliable visual magnitudes are given on page 547.

19.5.2. The Step Methods of Herschel and Argelander

19.5.2.1. Herschel's Steps

The estimation measures previously described can be modified, according to W. HERSCHEL,[1] by separating the single terms of a magnitude sequence by strokes, the length of which characterizes the brightness difference, e.g.,

$$\alpha \text{——} \gamma\beta - \iota\theta \text{——} \varepsilon\eta\zeta \text{———} \nu\pi\tau \qquad \text{Aurigae.}$$

Instead of the actual length of the strokes, HERSCHEL also used numerous short strokes and gave, for instance, the brightness of 17 Cygni as: 8 — 17, 21 — — — 17.8. In other words, star 8 was a little brighter than star 17, star 21 much brighter than 17, and afterwards star 17 equally as bright as star 8. This method uses the smallest constant differences of brightness sensation. It is the first step method, and it was by its use that HERSCHEL surveyed the stars in Flamsteed's Star Catalog. Later on, ARGELANDER took over this method, introducing four steps (denoting them 1, 2, 3, 4).

19.5.2.2. Argelander's Steps

This method is used extensively for brightness estimates of variable stars. The star is compared with several comparison stars that are a little brighter or fainter, but whose brightness must be known extremely accurately. The brightness differences are expressed in "steps." The estimated steps are arranged in a step scale and converted into a magnitude scale. ARGELANDER distinguishes:

Step 0: "If both stars appear to me of equal brightness, or if I feel inclined to estimate first one then the other to be the brighter star, then I call them equally bright and indicate this by putting their symbols side by side, whereby it is irrelevant which symbol comes first. If, therefore, I compare the stars *a* and *b*, then I write either *ab* or *ba*. Better still, the notation *a0b* or *b0a*."

Step 1: "If at first glance both stars appear to be equally bright, but on

---

[1] Royal Society and Royal Astronomical Society: *The Scientific Papers of Sir William Herschel*, Vol. 1, p. 530, London, 1912.

closer examination and frequent moving from a to b and b to a, a appears nearly always (with very rare exceptions) just a little brighter, then I call a brighter than b by one step, and denote this by $a1b$. If, however, b is the brighter, I write $b1a$, so that the brighter star is always before, the fainter one after the number."

Step 2: "If the one star always and without any doubt appears brighter than the other, I call this difference two steps, and denote it by $a2b$, if a, and by $b2a$, if b is the brighter star."

Steps 3 and 4. "If there is a difference that is immediately apparent to the eye we assign to it three steps and call it $a3b$ or $b3a$. Finally $a4b$ denotes a still more conspicuous difference in favor of a."

In addition to these whole steps ARGELANDER frequently also used half steps, for instance, $a2.5b$. Those observers who restrict themselves to whole steps will generally be unable to exceed four or at the most five steps.

Concerning the connection of a variable with its comparison stars, ARGELANDER states: "Every observation should contain as many as possible of the comparison stars, without involving too great a number of steps, and always at least one brighter and one fainter comparison star, even when the variable appears exactly or very nearly equal to one of them." The result of such an estimate is written as follows: amV indicates that the variable star is fainter than the comparison star a by m steps, while Vnb means that the variable is brighter than b by n steps.

19.5.2.3. The Formation of a Step Scale[2]

Let us assume that a variable star has been observed on three evenings and has been compared on each of them with four comparison stars a, b, c, and d. The number of steps between the variable and the other stars may be called m, n, p, and q. If a, b, c, and d are arranged in order of their brightness we may, e.g., obtain

First evening		Second evening		Third evening	
	am_1V		am_2V		am_3V
	Vn_1b		bn_2V		bn_3V
	Vp_1c		Vp_2c		cp_3V
	Vq_1d		Vq_2d		Vq_3d

From these estimates we form the mean values M of the step numbers from a to b, from b to c, and from c to d.

The mean M_1 of the step numbers from a to b on three evenings, then, is

$$M_1 = \frac{m_1 + n_1 + m_2 - n_2 + m_3 - n_3}{3}.$$

[2] See also BLASBERG, H. J.: Der Sternfreund als Beobachter veränderlicher Sterne. *Die Sterne* 33, 55 (1957).

The mean value M_2 of the step numbers from b to c on 3 evenings is

$$M_2 = \frac{-n_1+p_1+n_2+p_2+n_3-p_3}{3}.$$

The mean value M_3 of the step numbers from c to d on 3 evenings is

$$M_3 = \frac{-p_1+q_1-p_2+q_2+p_3+q_3}{3}.$$

The sums $(m+n)$, $(n+p)$, and $(p+q)$ can be directly estimated for checking the calculations.

Using these mean values we construct the step scale, starting from the brightest comparison star down to the faintest. If a is the brightest comparison star we assign to it the step 0, and obtain the step scale:

$$a = 0 \text{ steps}$$
$$b = M_1 \text{ steps}$$
$$c = (M_1+M_2) \text{ steps}$$
$$d = (M_1+M_2+M_3) \text{ steps}$$

Within this step scale the variable star, if its brightness has changed, occupies different positions on different evenings, for example:

Step value S_1 on the first evening:

$$S_1 = \frac{0+m_1+M_1-n_1+M_1+M_2-p_1+M_1+M_2+M_3-q_1}{4}.$$

Step value S_2 on the second evening:

$$S_2 = \frac{0+m_2+M_1+n_2+M_1+M_2-p_2+M_1+M_2+M_3-q_2}{4}.$$

Step value S_3 on the third evening:

$$S_3 = \frac{0+m_3+M_1+n_3+M_1+M_2+p_3+M_1+M_2+M_3-q_3}{4}.$$

It is advisable to assess different weights to the step number by which the variable differs from 0, M_1, (M_1+M_2), and $(M_1+M_2+M_3)$, i.e., to give the difference only the weight $\frac{1}{2}$ if this difference is larger than 3. If, for instance, the differences between V and (M_1+M_2) and between V and $(M_1+M_2+M_3)$ should only receive the weight $\frac{1}{2}$ we have

$$S_1 = \frac{2(0+m_1)+2(M_1-n_1)+(M_1+M_2-p_1)+(M_1+M_2+M_3-q_1)}{6},$$

$$S_2 = \frac{2(0+m_2)+2(M_1+n_2)+(M_1+M_2-p_2)+(M_1+M_2+M_3-q_2)}{6},$$

$$S_3 = \frac{2(0+m_3)+2(M_1+n_3)+(M_1+M_2+p_3)+(M_1+M_2+M_3-q_3)}{6}.$$

With this step method we are able to establish the whole light curve of a variable star, therefore also the length of its period and the positions and depths of its maxima and minima.

19.5.2.4. The Formation of the Magnitude Scale

The step scale will proceed in large or small steps according to the skill of the observer. If we subdivide the abscissa in a rectangular coordinate system in as many divisions as a step scale contains whole steps, and also subdivide the ordinate according to whole magnitude classes for the variable, then enter the photometric brightness (catalog brightness) of the comparison stars to the step values 0, M_1, $(M_1 + M_2)$, and $(M_1 + M_2 + M_3)$, we obtain a sequence of points which in the ideal case lie on a straight line. In this case, every step is just as large as a certain uniform magnitude difference, say, 0.08 magnitude class. The smaller the gradient of this straight line, the more sensitive is the eye and the ability of the observer to make the estimates.

Normally the points do not lie on a straight line, and we then draw a curve in such a way that the distances of the points from the curve in the direction of the ordinates are as small as possible. We can do this successfully with a careful drawing using only the eye and a flexible ruler. The smaller the curvature of the resulting curve the more uniform the steps of the observer in the brightness interval surveyed. Using this curve we are now able to establish the step values of the variable star in magnitude classes. The latter ones are used to draw the final light curve (see Fig. 19-2). Again, we draw the curve through the sequence of points in such a way that the points are as close as possible above or below the curve.

19.5.2.5. Allowance for Observational Errors

On page 441 we dealt with a number of errors that can be established in nearly all observations, as investigations of long series have shown. Their existence becomes evident if we form the ratio

$$\frac{\text{Magnitude difference between 2 comparison stars}}{\text{Estimated step numbers between them}}$$

for the single evenings, weeks, and years, observations before and after midnight, large and small zenith distances, brighter and fainter comparison stars, nearby or more distant comparison stars, and so on. These comparisons show which observations can be used with full weight and which will only have partial weight for the determination of the M and S values. Such an assessment of weights is carried out on page 448. The observer experienced in numerical methods will be able to calculate corrections.

19.5.2.6. Other Methods of Step Estimates

In POGSON's method we estimate the brightness difference between two stars not in HERSCHEL-ARGELANDER steps but in units of 0.1 magnitude classes.

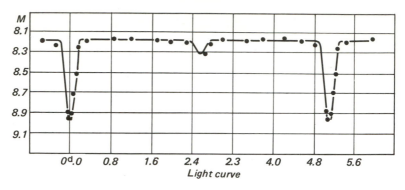

Light curve

Star	BD	Size	Δ α	Δ δ
f	+ 65° 1774	7^m.9	− 2^m 50^s	− 40.0
a	+ 66 1535	8.3	+ 7 34	0.7
b	+ 66 1517	8.7	− 1 55	− 14.3
y	+ 66 7528	9.0	+ 2 49	+ 0.8

Comparison stars

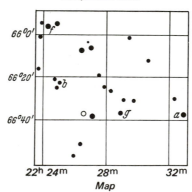

Map

Fig. 19–2 The variable star XZ Cephei. After H. SCHNELLER, *Veröff. Univ. Sternwarte Berlin-Babelsberg*, Vol. 8, No. 6, p. 45.

Using stars of accurately known and small magnitude differences, the observer will be able to form a memory scale of the corresponding brightness differences. Here the estimation of steps furnishes at once the required brightness in magnitude classes. However, in its accuracy potential, this method lags far behind the HERSCHEL-ARGELANDER step method.

Without going into detail we just mention the interpolation method by E. C. PICKERING and the methods by PEREPELKIN, NIJLAND, SCHÖNEFELD, and HAGEN.[3]

---

[3] See *Handbuch der Astrophysik*, Vols. 1 and 2. Berlin: Springer, 1931.

19.6. Brightness Measurements

19.6.1. Visual Measurements

Visual measurements with photometers are all based on estimates of equal brightness. We make use of the excellent capability of our eye to distinguish, within a certain brightness range, very small differences between pointlike and extended objects (see page 440). The smallest brightness difference which can be recognized is about 0.05 to 0.15 magnitude classes. For smaller differences both objects appear equally bright to the eye. The accuracy of a single estimate is therefore about 0.1 magnitude. The same, or approximately the same, brightness of the observed objects is produced in the relevant photometers produced by weakening an artificial light source by a neutral wedge to such a degree that the source becomes equal to the object (GRAFF's photometer) or by changing the surface brightness of the extrafocal image in the telescope by displacing the eyepiece distance in such a way that it appears equally bright to an artificial luminous surface (GRAMATZKI's surface photometer, see Sec. 2.2.10.2). Thus, surface photometers permit the comparison of extrafocal star images among themselves, or of extended objects (nebulae, cometary heads) with extrafocal star images.

The artificial light source must have a constant brightness. This is achieved by operating the small electric bulb of GRAFF's photometer with an accumulator, taking a moderate current from the accumulator which is on floating charge from the rectifier. Active luminous material, such as mesothorium, changes its luminosity by only very low percentage during a year.[4]

The observed object must always be compared with stars of the same color, always using the same catalog. If we use a color filter we must do so during all the observations of our series of the same object, and furthermore we must determine the wedge constant of the filter. The measured brightness is valid only with reference to the particular catalog that has furnished us with the comparison stars.

If we deal with extended objects we measure either the total brightness or the brightness per unit area (surface brightness, intensity of illumination). If the eye sees the image in the eyepiece subtending a small apparent angle, e.g., on an absolutely dark sky an angle smaller than 20 minutes of arc and on a bright sky an angle smaller than 1 minute of arc, it considers the extended object as pointlike. (See page 18.) In this case we do not require a special surface photometer.

If, for instance, the object in the sky has a diameter of 1′, we can only magnify it to 20′ on a dark sky, that is, magnification 20×, if we wish to use a point photometer. It must be tested whether with such a small magnification the exit pupil is also really smaller than the eye pupil (see page 10).

If the size of the object we want to measure exceeds the indicated limits

[4] See RICHTER, N.: Radioaktive Kristallphosphore als photometrische Standardlichtquellen. *Die Sterne* **29**, 196 (1953).

we must use a surface photometer. A GRAFF photometer in its modified version can be changed into a surface photometer (see Fig. 2–29).

The brightness of each object is increased by the sky brightness, which on the average can amount to 4 magnitude classes per square degree. The brightness of a star S, whose image reaches—through diffraction, imaging errors, air unsteadiness, and inaccurate focusing—a diameter of 5″ and covers an area of 1.5×10^{-6} square degrees, thus appears brighter by the equivalent of a star F of the magnitude

$$m_F = 4 + 2.5 \log \frac{1}{1.5 \cdot 10^{-6}} = 16.5.$$

The total brightness m (star image plus sky background) is then

$$m = m_S - 2.5 \log \left(1 + \frac{i_F}{i_S}\right),$$

where
$$\frac{i_F}{i_S} = 2.512^{-m_F + m_S}.$$

A star of magnitude 12^m, on a sky background of 4^m per square degree, is brighter by 0.027 magnitude classes. Such an error is smaller than the accuracy of a visual photometric measurement and can therefore be neglected.

The sky background—both for focal and extrafocal settings—is imaged in the same surface brightness, as long as the eye diaphragm remains unchanged. We can state, "The luminosity density of the sky background is proportional to the undiaphragmed surface area of the exit pupil." It is therefore possible to vary the observed brightness of the sky background by a suitable choice of the eye diaphragm.[5]

19.6.2. Measurements with Photoelectric Photometers

The most accurate brightness measurements can be achieved with photoelectric photometers (see Sec. 2.2.10.4). The deflection of the measuring instrument, which measures the photocurrent, is within a wide range proportional to the intensity of radiation received. The size of the image of the observed object as produced by the objective lens is irrelevant. For a star as well as for a nebula the total incident radiation is recorded.

The light spot on the layer of the photocell should not be too small. For this purpose we displace the cell a little along the optic axis. If we wish to record a circular area of w seconds of arc diameter, then the diaphragm must have the diameter y' calculated from Eq. (24) in Chap. 2 from the focal length f of the objective.

The normal unsteadiness of the air is irrelevant for work with a photoelectric photometer. The color sensitivity, however, varies considerably with

[5] See *Handbuch der Astrophysik*, Vol. 2, p. 570. Berlin: Springer, 1931.

the alkali metal hydrides used. The color sensitivity of a cesium cell, with the sensitivity maximum at 0.000550 mm wavelengths, is nearest to that of the eye. Radiation measurements with the photoelectric photometer must be connected with a catalog of star magnitudes in the same way as every other measurement of radiation or brightness. If the photoelectric cell has the same color sensitivity as the receiver with which the magnitudes of the catalog stars are measured, then the latter can be used for brightness comparison without any consideration of color. Otherwise we should only use stars of the same color, that is, spectral type.

In photoelectric photometric work we must measure the comparison stars alternatively with the neighboring star-free sky background. The readings for the latter must be subtracted from those obtained during the photometric measurement of our object (star, nebula, cometary head, etc.).

We only just mention the important problems involved in the guiding of the star image within the opening of the photometer, in the removal and reduction of the dark current, and finally also in the checking of the fatigue of the photoelectric cell.

Photoelectric measurements should if possible be referred to the specially determined star magnitudes of H. L. JOHNSON (see Table 19 in the Appendix). This succeeds best if we adapt the whole instrumental equipment to that used by JOHNSON. JOHNSON's magnitudes are called the UBV system (U = ultraviolet, B = blue, V = visual). Table 19.4 indicates the instrumentation required for this magnitude system. A compilation of other magnitude systems and their technical realization for the stellar photometry in certain narrow spectral regions can be found in LANDOLT-BÖRNSTEIN's tables (see the Bibliography, page 537).

Table 19.4. Technical Realization of the UBV System of H. L. JOHNSON[6]

Symbol	Cell	Filter	Peak sensitivity in nm	Instrument
U	1 P 21 (SbCs cathode)	Corning 9863 or Schottglas UG 2 (2 mm)	350	Reflector (Al-surface)
B	1 P 21 (SbCs cathode)	Corning 5030 with Schottglas GG 13 (2 mm) or Schottglas BG 12 (1 mm) with Schottglas GG 13 (2 mm)	435	Reflector (Al- or Ag-surface)
V	1 P 21 (SbCs cathode)	Corning 3384 or Schottglas GG 11 (2 mm)	555	Reflector (Al- or Ag-surface)

[6] JOHNSON, H. L.: A photometric system. *Ann. d'Astrophysique* **18**, 292 (1955); see also Table 19.5.

19.6.3. *Measurements from Photographic Plates*

Photographic instruments and photographic plates are discussed in Sec. 2.2 and Sec. 3.3.2. Every celestial photograph is a document, which, even decades later, can be checked and reevaluated. The latter process may concern the position and the brightness of some stars, the number of stars, and the density and distribution of interstellar matter. The most valuable property of the photographic plate, however, is its ability to store the received radiation, and by this accumulative process to make stars visible whose images have a very small surface brightness only, just above the sky background. When the amateur astronomer makes use of these advantages of the photographic plate he must count on the fact that the plate as such is frequently not very suitable to immediate precise statements. The professional astronomer uses rather complicated methods and instruments for the evaluation of photographic plates, in particular for the most difficult work—photometry. The amateur astronomer will not attempt to imitate photographic photometry on a scale which is accessible only to large observatories. Rather, in this field, too, he will apply a method that is adapted to his particular possibilities and effective at the same time. Again the only method of this kind is that of the step estimates of Herschel and Argelander, discussed earlier in this chapter, as the best observing method of the amateur astronomer in visual photometry.

Photographic observations are performed in exactly the same manner as visual work: the unknown brightness of a variable star is referred to known magnitude values of comparison stars. The latter must be photographed under the same conditions as the variable. Frequently it will be impossible to find on the plate a sufficient number of comparison stars whose brightness and color closely approach those of the variable. Furthermore, the comparison stars should also be at nearly the same distance from the optical axis as the variable. In this case the imaging errors act in the same way on the variable as on the comparison stars and both can be compared with each other immediately (see page 40).

Astronomers have established star fields which have been measured photometrically with great accuracy. We mention the International Polar Sequence (see also Pleiades–Sequence p. 534, end of the Appendix), Kapteyn's Selected Areas, and the Harvard Standard Regions. Furthermore, there exist accurate catalogs of the stars in the clusters Pleiades and Praesepe, so that these, too, can be used as comparison stars. Of course we shall always select a star field that is nearest to the variable. The star field containing the comparison stars must be photographed on the same plate as the variable star, and room left for it on the plate. The exposures of variable and comparison stars must follow each other immediately. In order to avoid imaging errors that depend on position, we equip the plate holder with an arrangement for displacing it in such a way that each half of the plate is symmetrical to the optical axis during exposure. Then the variable star will be photographed in the same way as the comparison stars close to the center of the field of view. But even without such a displacement a suitable choice of the star to which we point the telescope axis during the exposure, i.e., our guiding star, will make it possible to place the variable and the comparison stars at equal distances from the center. The brightness of the variable is determined by the HERSCHEL-ARGELANDER step method. We must make sure that the brightness of the comparison stars be corrected for differential extinction with respect to the variable (with the help of Table 2 in the Appendix).

Each photographic photometer (telescope plus filter plus plate) has a certain color sensitivity. Reddish stars on blue plates produce fewer blackenings than white stars of the same total brightness. The gradation of the blackening produced by stars of different brightness is therefore different for every photographic combination. If a star of variable brightness and color, or several stars of different brightness and color, are to be interpolated into the magnitude scale of a star catalog, then our equipment must have the same color sensitivity as the one on which the magnitudes of the catalog were based. If that is the case then all catalog stars of all colors and spectral types can be used for comparison purposes. The sensitivity of an equipment need not to be investigated separately for every wavelength. The color sensitivity of two arrangements is considered equal if the center of gravity of the energy of the effective radiation lies at the same wavelength, the so-called isophotal wavelength. The determination of the isophotal wavelength has been dealt with by such authors as BECKER.[7] Table 19.5 lists a number of plate-filter combinations whose sensitivity maximum has been investigated. Measurements should be in a well-defined narrow color region, and in internationally agreed narrow limits.

In place of the visual and photographic magnitudes of the "International Photometric Systems" (IP_v and IP_g) the International Astronomical Union (IAU) has recommended the extensive use of the system initiated by JOHNSON[8] and by JOHNSON and MORGAN (the UBV System). This system is characterized by the combination of plates and filters listed in Table 19.5, according to JOHNSON or GÜSSOW. Table 19.5 also gives W. BECKER's RGU system.

Not all images appear equally sharp on the photographic plate. If they are sharp on the optic axis, that is, at the center of the plate and closest to the focus, then the sharpness decreases toward the edge of the plate (see page 40). If we photograph a sequence of stars of different brightness repeatedly, so that they are imaged at different distances from the center of the plate, we find that somewhat different stellar magnitudes correspond to equal blackenings. These differences can be established by careful comparison of the blackenings, and zone and field corrections can be applied with the

---

[7] BECKER, W.: *Sterne und Sternsysteme*, 2nd ed., p. 6. Dresden and Leipzig: Theodor Steinkopff, 1950.

[8] JOHNSON, H. L.: A photometric system. *Ann. d'Astrophysique* **18**, 292 (1955), and JOHNSON, H. L. and MORGAN, W. W.: Fundamental stellar photometry. *Astrophys. J.* **117**, 313 (1953).

Table 19.5. Combination of Photographic Plates and Filters for Brightness Measurements in Narrow Color Regions

Symbol	Plate	Filter (Schott)	Sensitivity maximum in nm	Source*
⌐U	Kodak 103a O	UG 2 (2 mm)	350	Johnson
	Agfa V 8301	UG 11 (2 mm)	356	Güssow
	Agfa-Spectral Blue Rapid	UG 1 (2 mm)	357	Jaschek
U ⌐	Blue-Plate	UG 2 (2 mm)	373	Becker
	Blue-Plate	BG 3 (1 mm) + GG 13 (4 mm)	424	Becker
	Agfa V 8301	GG 13 (2 mm)	433	Güssow
⌐B	Kodak 103a O	GG 13 (2 mm)	435	Johnson
	Agfa Astro	BG 5 (2 mm)	446	Jaschek
	Agfa-Spectral Blue Rapid	VG 4 (2 mm)	476	Jaschek
G ⌐	Blue-Plate	GG 5 (2 mm)	481	Becker
	Agfa Astro	GG 14 (2 mm)	505	Jaschek
	Agfa Isochrom	GG 14 (2 mm)	546	Jaschek
	Raman panchrom.	BG 23 (1,2 mm) + GG 14 (3 mm)	552	Güssow
└V	Kodak 103a D	GG 11 (2 mm)	555	Johnson
	Orthochrom. Plate	GG 11 (2 mm)		Becker
	Agfa Mikro	OG 2 (2 mm)	575	Jaschek
	Gevaert ultra panchro 8000	OG 2 (2 mm)	602	Jaschek
	Gevaert ultra panchro 8000	RG 2 (2 mm)	634	Jaschek
R ⌐	Isopan Plate	RG 1 (2 mm)	638	Becker
	Agfa-Spectral Red Rapid	RG 5 (2 mm)	660	Jaschek
	Infrared Plate	RG 2 (2 mm)	715	Becker

* Becker, W.: Die vier Standard-Spektralbereiche der astronomischen Integralphotometrie und die Helligkeit der Polsequenz in ihnen. *Veröffent. der Univ.-Sternwarte Göttingen*, No. 80, 1946.

Güssow, K.: Zur photographischen Photometrie im U, B, V-System. *Die Sterne* **32**, 30 (1956).

Jaschek, W.: Die Prüfung von photographischen Astroobjectiven. *Photographische Korrespondez* **87**, 11 (1951).

Johnson, H. L.: A photometric system. *Ann. d'Astrophysique* **18**, 292 (1955). See also Table 19.4.

use of a correction curve. These corrections usually increase uniformly from the plate center to the edge of the plate. If, in addition, a dependence of the correction on the direction, i.e., position angle, is found, then either the plane of the plate is not perpendicular to the optic axis, or the objective or the mirror of the camera is affected by zone errors. In these cases the correction becomes rather complicated.

The comparison of the star images on a plate can be suitably performed with the help of an auxiliary scale. This should include a dense sequence of images that belong to exactly defined star magnitudes. It is produced on a special plate by carefully arranged exposure times using a natural or artificial pointlike source of light of constant brightness, or alternatively with a sequence of stars of known brightness. If t_1 and t_2 are two exposure times then the images of the same source of light differ in magnitudes by

$$m_2 - m_1 = 2.5p \, (\log t_1 - \log t_2).$$

The factor p can be taken as 0.85. This scale is calibrated by comparison with the images of the particular catalog, which will serve for the determination of the brightness of the variable star or of the minor planet, etc. For every new plate the scale has to be freshly calibrated.

19.7. The Reduction of Photometric Observations of Variable Stars

19.7.1. The Drawing of the Light Curve

The time of observation is noted to the nearest minute (year, month, day of the Gregorian calendar, and hour and minute in UT). The clock correction should be determined before and after the observation (see page 94), but not applied during the observation. Later on, during the reduction of the observation, it is added, expressed in Universal Time. Then the year, month, and day of the Gregorian calendar are converted to the Julian Period. This time scale started with the astronomical date -4712 January 1, Mean Greenwich Noon, so that every successive day starts with 12 hours UT; for instance, on 1960 January 0 ($=1959$ December 31 in the Gregorian calendar), 12 hours Standard Time, we have the Julian Date 2436934.000. The time of the observation is expressed in fractions of a day of the Julian Period, e.g., 1960 January 1, 19 hours MEZ (18 hours UT) = 2436935.250. A table for the conversion of our calendar date into Julian Date and an additional table for the conversion of the time of the day into fractions of a day are found in the Appendix, Tables 7 and 8.

The observing time corresponding to the time when the light of the variable star has reached the Earth is the so-called "geocentric time." If, for instance, the star is in the ecliptic, then the Earth during its revolution around the Sun is alternately nearer and farther away, by up to one astronomical unit. Correspondingly the light of the variable star reaches us by 8.317 minutes, on the average, earlier or later. This difference is of importance for the observation of short-periodic variables and eclipsing binaries. In order to have uniform time data we reduce the observing time to the center of the Sun; this gives the "heliocentric time." The calculation involves the geocentric longitude L of the Sun and the latitude β and longitude λ of the star according to the relation:

$$\text{Heliocentric Time} = \text{Geocentric Time} + 8\overset{m}{.}317 \cdot \cos\beta \cdot \sin(L + 270° - \lambda). \tag{7}$$

The single brightness estimates and measurements of the variable are affected by accidental errors. These can be eliminated by the formation of mean values, if a sufficiently large number of observations during a light period are available. The mean values should be formed over a small section of the period only, say ten days, if the period covers several months. These sections can overlap when means are calculated. The whole operation can also be carried out graphically, by combining each point with the two following ones to give a triangle, whose center of gravity is plotted according to an eye estimate. This yields a somewhat smoothed curve. This process can be repeated, and we obtain a sequence of points which permits us to recognize the general run of the light curve, which then can easily be drawn in.

The determination of the light maximum and of the light minimum can, however, only be derived from curves that are free from the accidental observational errors that we discussed above. We draw through the curve a number of straight lines parallel to the time axis (abscissa). The curve divides these lines into separate segments. The line that joins the midpoints of these segments intersects the curve rather accurately at the maximum of minimum.

The length of the period of the light variation is obtained with rather good approximation if the observations cover several periods. We choose two distinct minima or maxima, at the beginning and the end of the observing interval, and divide the time between them by the number of periods. The result is the mean period.

A practical method for the determination of the light curve consists in the subdivision of the whole sequence of single observational points covering several periods into parts of one period. If, as is usually the case, the minima are more pronounced than the maxima, this division will be taken approximately at the maxima. At these places we shall cut the graph of the observational values and then superimpose the parts by parallel displacement until

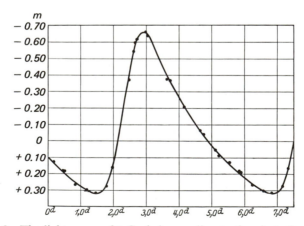

Fig. 19–3 The light curve of δ Cephei according to the photoelectric measurements of P. GUTHNICK. The period is 5.366 days. From *Astronomische Nachrichten* **208**, 171 (1919).

we obtain a coincidence of the minima or maxima. The parts of the curve are drawn on transparent paper and superimposed in such a way that they differ from each other as little as possible. We can then draw a mean light curve by inspection. In Figs. 19–3 to 19–10 we reproduce characteristic light curves of different variables. See also Table 21 in the Appendix.

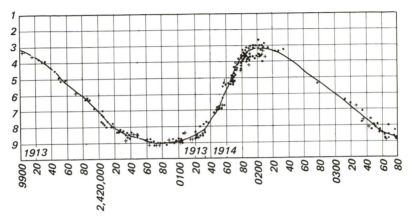

Fig. 19–4 Light curve of *o* Ceti. From *Handbuch der Astrophysik* Vol. 6, p. 100. Berlin: Springer, 1931.

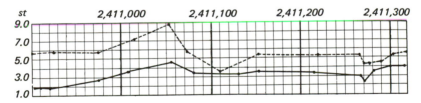

Fig. 19–5 Light curves of α Cassiopeiae (upper curve on the graph) and μ Cephei (lower curve). After PLASSMANN: *Handbuch der Astrophysik*, Vol. 6, p. 167. Berlin: Springer, 1931.

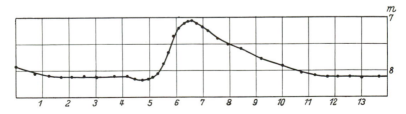

Fig. 19–6 Light curve of RR Lyrae from photographic-photometric measurements by HERTZSPRUNG. From *Handbuch der Astrophysik*, Vol. 6, p. 188. Berlin: Springer, 1931.

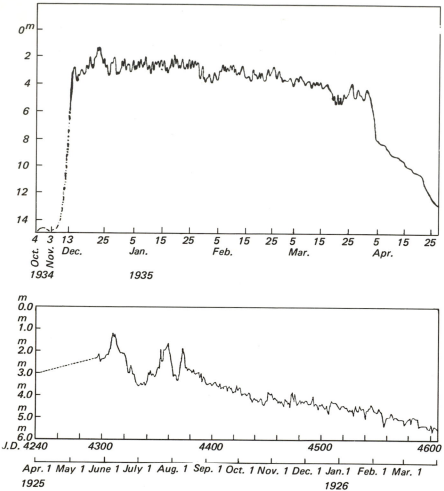

Fig. 19–7 Light curves of novae. Above, Nova Herculis 1934, after A. BEER: The Nebular Spectrum of Nova Herculis, *Monthly Notices Roy. Astron. Soc.* **96**, 238 (1936). Below, Nova Pictoris 1925, after H. SPENCER JONES: *Handbuch der Astrophysik*, Vol. 7, p. 675. Berlin: Springer, 1936.

19.8. On the Photometry of the Major Planets

The purpose of the photometry of the surfaces of the major planets (as for the Moon's surface) is to determine and compare the brightness of small surfaces. These brightnesses lie between 10^{-4} and 2 stilb (see Table 19.6). Their average can be calculated from

$$B = \frac{4.10^{-0.4m}}{zw^2},$$

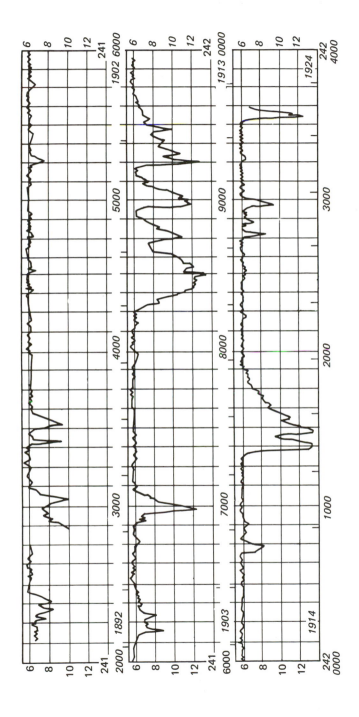

Fig. 19-8 Light curve of R Coronae Borealis from 1892–1923. From *Handbuch der Astrophysik*. Vol. 6. p. 72. Berlin: Springer, 1931.

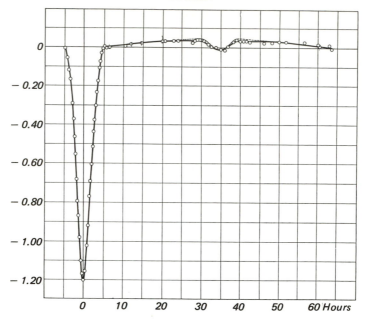

Fig. 19–9 Light curve of the eclipsing binary β Persei (Algol). From *Handbuch der Astrophysik*, Vol. 6, p. 428. Berlin: Springer, 1931.

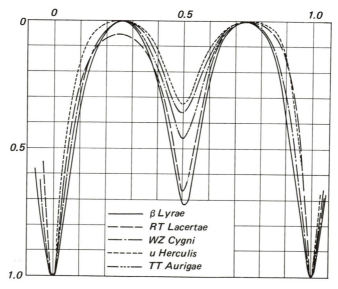

Fig. 19–10 Five light curves which show the typical characteristics of the β Lyrae stars: β Lyrae, RT Lacertae, WZ Cygni, u Herculis, and TT Aurigae (*Astronomische Nachrichten* **211**, 357 (1920).)

Table 19.6. Brightness B of the Major Planets and the Moon

Planet	B in stilb
Mercury	0.6
Venus	2.0
Mars	0.2
Jupiter	0.070
Saturn	0.028
Uranus	0.0054
Neptune	0.00037
Moon (1–2 days)	0.03
Moon (First Quarter)	0.1
Moon (Last Quarter)	0.1
Moon (Full)	0.6

where z is the fraction of the illuminated disc, and w the radius of the planet in seconds of arc.[9]

Our eye possesses the unique ability to distinguish very small differences in the illumination of bright areas, and is therefore unsurpassed as an instrument for carrying out the photometry and drawing of small regions of the planetary discs. Figure 19–11 shows how large at least the brightness

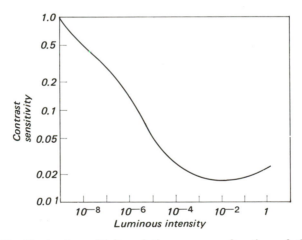

Fig. 19–11 Contrast sensitivity of the eye as a function of the luminous sensitivity B_A. Intensities of the order of 10^{-3} to 10^{-1} permit the measurement of contrasts as small as 0.018.

contrast must be to be noticed by the eye. The smallest contrast that is still noticed is caused by energy densities that correspond to those on the planetary discs.

[9] See GIFFEN, C. H.: Foundations of visual planetary astronomy. *The Strolling Astronomer* **17**, 59–72 (1963).

These densities noted by the eye can be made larger or smaller by changing the magnification of the telescope. The larger the magnification, the smaller the light flux produced by an illuminated surface on our retina. It is therefore advisable to move the light flux into the region of greatest contrast sensitivity by a suitable choice of magnification.

"There does not exist a relation that can be arrived at by purely geometrical consideration involving aperture, focal length, and illumination, for the visual observation of surface brightnesses at different instruments, or with different magnifications."[10] If, however, we assume with GIFFEN that the illumination B_A of the image on the retina of the eye is in the same relation to the illumination B' of the light bundle that enters the eye at normal magnification as the square of its diameter (d'^2) to the square of the diameter (p^2) of the pupil, then we are in agreement with the empirical fact that in the case of normal magnification the observed surface brightnesses are largest.

$$\text{(a)} \quad B_A = B' \frac{d'^2}{p^2} \quad \text{when } p \geq d',$$

$$\text{(b)} \quad B_A = B' \quad \text{when } p < d'.$$

According to GIFFEN this relation can be further explored in connection with Fig. 19–10. Using the expressions for the magnification V and for the normal magnification V_n [see Eq. (3) in Chap. 2 and surrounding text],

$$V = \frac{d}{d'}, \qquad V_n = \frac{d}{p}.$$

We obtain from (a)

$$B_A = B' \cdot \left(\frac{d}{V}\right)^2 \cdot \left(\frac{V_n}{d}\right)^2.$$

Here (V/d) is the variable magnification per cm of the free objective aperture, and (V_n/D) the corresponding normal magnification. If the latter is fixed, V_n, then the magnification V to be chosen determines the surface luminosity B_A, and thus the contrast sensitivity c. The B_A values at which the smallest contrasts can still be noted are read off from Fig. 19–11. The values are in the region of $B_A = 0.1, 0.01$, and 0.001. The corresponding values of contrast are $0.018, 0.017$, and 0.018. Figure 19–12 shows the relation between B_A, B', V/d, and (V_n/d).

With the help of Fig. 19–11 we are able to determine, for known B' values, those magnifications per cm of free objective aperture at which we are still able to note the small contrasts $c = 0.018$ and 0.017. The B' values can be derived from the B values of Table 19–6, if the transparency of the atmosphere and the instrument, including filters, is known. Assuming for the total

[10] See *Handbuch der Experimentalphysik*, Volume "Astrophysik," p. 107. Leipzig: Akademische Verlagsgesellschaft, 1937.

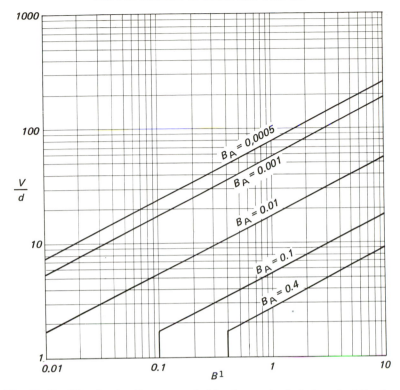

Fig. 19–12 The density B_A as recorded by the eye is plotted against the density B' which applies to normal magnification and fixed aperture $p = 0.578$ cm $(1/p^2 = 3.00)$ of the eye pupil.

transparency the values 0.6, 0.8, and 1.0, the optimum magnifications (V/d) have been calculated in Table 19.7.

We must also take into account the specific ability of the eye to resolve details on the bright planetary discs. It is necessary to choose a suitable

Table 19.7. Optimum Magnification per cm of the Objective Aperture for Optimum Values of Contrast $B_A = 0.1$, 0.01, and 0.001

Planet	Total transparency								
	0.6			0.8			1.0		
Mercury	3.3	10.2	33	3.8	12.0	38	4.2	13.4	42
Venus	6.0	18.98	60	6.9	21.92	69	7.8	24.5	78
Mars	1.9	6.0	19	2.2	6.93	21.9	2.45	7.75	24.5
Jupiter	1.1	3.5	11	1.29	4.10	13	1.45	4.58	14
Saturn	—	2.2	7.1	—	2.6	8.2	—	2.9	9.2

magnification, so that the light bundle that enters the eye is so narrow that the resolving power becomes as large as possible. In what follows, we shall refer to the calculation suggested by GIFFEN given at the beginning of this section.

We stated on page 15 that two equally large diffraction discs can be separated if the distance of their centers is equal to their radius. In this case we obtain for the linear distance within the focal plane of the objective (see page 12), for ideal image quality,

$$\frac{b}{2} = 1.22 \cdot \frac{\lambda f}{d}.$$

If we assume for the wavelength of the light the value $\lambda = 550$ nm, and express the angular distance w which corresponds to the linear distance $\frac{1}{2}b$ in seconds of arc, we have

$$w = 13''4/d \text{ (in cm)} \qquad \text{or} \qquad 5''3/d \text{ (in inches)}. \qquad (8)$$

In this way, we have also assigned to the smallest still resolvable angular distance w a certain objective aperture d; w is proportional to $1/d$. This applies to complete steadiness of the air. If the diffraction discs are enlarged or blown up by air turbulence, the smallest separable angular distance is also enlarged. To this larger angular distance w^* corresponds according to Eq. (8) a hypothetical d^* in analogy to the observed w^* in the case of perfectly steady air. In this case we can consider d^* as a measure of the resolving power. If w is the effect of the air unsteadiness, so that

$$w^* = w + \Delta w,$$

then evidently we have

$$\frac{d^*}{d} = \frac{w}{w + \Delta w}.$$

The value of Δw can be measured or estimated (see page 23), and in this case d^* is a measure of the instrumental resolving power. Its value depends for a given instrument only on $\Delta w/w$. The ratio d^*/d is at the most equal to 1.

If we judge two diffraction discs by visual observation, the resolving power of our eye becomes important. Just as for the telescope, this parameter depends on the diameter of the entrance pupil. For light of the wavelength $\lambda = 550$ nm, the eye is able to separate in the best case

$$w_A = \frac{2''31}{p} \qquad (p \text{ in mm}).$$

This value is reached for $p < 3$ mm, because beyond this value the optical aberrations of the eye become considerable, and w_A is not proportional to $1/p$. It is therefore possible to assign to every value of w_A that has been enlarged by Δw_A a certain hypothetical p^*.

Such a relationship is represented in Fig. 19–13. The value of p for the observer's eye at the telescope has to be replaced by the exit pupil of the telescope:

$$p = \frac{d}{V}.$$

According to the choice of magnification, we can change d/V, and thus make the light bundle entering the eye more or less narrow. In any case, p must be chosen such that the decrease of the resolving power from d to d^*, as caused by the influence of Δw, is not further decreased by too large a p value. We must have

$$\frac{p^*}{p} \geqq \frac{w}{w + \Delta w}.$$

Equality indicates the minimum magnification to be applied: $V = d/p$. Table 19.8 lists for given values $(w/w + \Delta w)$ the values V/d that have been obtained in connection with Fig. 19–13.

MACKAL[11] investigated the visual brightness estimates, the problems of

Table 19.8. The Minimum Magnification V/d per cm of the Objective Aperture as a Function of Different Magnifications of the Diffraction Discs

$\dfrac{w}{w + \Delta w}$	$\dfrac{V}{d}$	$\dfrac{w}{w + \Delta w}$	$\dfrac{V}{d}$
0.30	1.7	0.70	4.4
0.35	2.0	0.75	4.9
0.40	2.3	0.80	5.4
0.45	2.6	0.85	6.1
0.50	2.9	0.90	7.0
0.55	3.2	0.95	8.2
0.60	3.6	1.00	10.0
0.65	4.0		

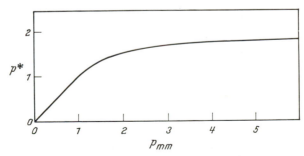

Fig. 19–13 The resolving power of the eye. The effective diameter p^* of the pupil is plotted against its real diameter p.

[11] MACKAL, P. K.: Rationale and procedures for a Jovian rank and color report. *The Strolling Astronomer* **21**, 114–116 (1969).

the standard 10-step intensity scale, in particular, made by observers of Jupiter. He proposes an equivalent procedure which uses comparison series. On the other hand, CHAPMAN[12] has demonstrated how such really valuable information can be obtained from brightness estimates.

19.9. The Photometry of the Minor Planets

The largest of the minor planets is (1) Ceres; it has a diameter of 740 km and its distance from the Earth, at opposition, is 1.77 A.U., where 1 A.U. = 149,600,000 km. According to Eq. (8) in Chap. 2, Ceres will appear, as seen from the Earth, with angular diameter

$$w = \frac{740 \cdot 57°296}{1.77 \cdot 1.496} \cdot 10^{-8} = 0.6 \text{ second of arc.}$$

According to Eq. (14) in Chap. 2, the telescope should have an aperture ratio of $13''8/0''6 = 23$ cm in order for Ceres to appear as a small disc. As a rule the minor planets (except for very near objects, for example, Eros, at times) are seen with ordinary telescopes as simple points and can therefore be photometered like any other pointlike light source. The photographic plate, too, does not reach the necessary resolving power when photographing a minor planet, as we can see from Eq. (24) in Chap. 2.

An asteroid, i.e., a small planet, discloses its nature by its motion. Since the mean distances of the asteroids are about 3 A.U., their periods according to KEPLER's third law are about 5 years and the mean daily heliocentric motion, therefore, about 0.2 degree. One sidereal day later we observe that the distance of the minor planet from a star at the same latitude has changed by 0°4. If we are guiding our camera, $f = 100$ cm, by setting it on a neighboring star and not on the minor planet, then the latter will draw in one hour a little stroke y' [according to Eq. (24) in Chap. 2], namely,

$$y' = 100 \text{ cm} \cdot \frac{0°4}{24 \cdot 57°296} = 0.03 \text{ cm.}$$

The resolving power of the plate (0.03 mm), is therefore exceeded with an exposure time of 6 minutes, or for $f = 10$ cm in 1 hour. We can thus comfortably photograph most of the minor planets as pointlike images.

Since the distance of the minor planets from the Earth and from the Sun varies continuously, we have to allow for certain corrections to be applied to the photometric observations. For these we require the quantities listed below, the first of which can be taken from a Yearbook, the others from minor planet Ephemerides (see also Sec. 15.2.4). All are expressed in astronomical units (A.U.).

---

[12] CHAPMAN, C. R.: Intensity estimates and a seeing-scale, *The Strolling Astronomer* **21**, 116–117 (1969).

R = the distance from the Sun to the Earth

Δ = the distance from the Earth to the minor planet

r = the distance from the Sun to the minor planet

r_0 = the semimajor axis of the planetary orbit

The light time. Light travels the distance of 1 A.U. in 8.317 minutes and therefore the distance from the minor planet to the Earth in $8.317 \times \Delta/R$. To obtain the time at which the minor planet possessed the observed brightness, we must subtract from the observing time the amount $8.317 \times \Delta/R$.

Reduction to a standard distance. The observed magnitudes of the minor planets depend on Δ and r. To obtain comparable values, convert the directly observed magnitudes m_b into the brightness the minor planets would have at a distance $\Delta = 1$ and $r = 1$. We therefore obtain this reduced magnitude g as

$$g = m_b - 5 \,(\log r + \log \Delta). \tag{9}$$

Occasionally, observed magnitudes m_b are reduced to the mean distance at opposition, i.e., to $(r-1)$. This gives the brightness m_0:

$$m_0 = m_b - 5 \,(\log r + \log \Delta) - \log r_0(r_0 - 1). \tag{10}$$

Correction for phase angle. The phase angle α is, in the triangle Sun–minor planet–Earth, the angle at the planet. It can be calculated from tan $p/2$, according to the formula on page 216, which for small phase angles leads to more accurate results than the formula for cos α, given on page 353, which suffices for larger phase angles.

If now the reduced magnitudes, corrected for light time and distance, are plotted against the phase angle α, we usually note increasing brightness with decreasing phase angle. By extrapolating the brightness curve beyond the observed value, we can estimate the brightness for the phase angle $\alpha = 0°$. Empirically or by the method of least squares it is then often possible to represent the brightness m_0 as a function of the phase angle α as follows:

$$m_0(0) + \alpha\varphi = m_0(\alpha). \tag{11}$$

The constant φ is the so-called "phase coefficient." We quote a few values which may serve as indicators for attempts to represent the observed magnitudes, according to Eq. (11).[13]

(1)	Ceres	0.0698	(5)	Astraea	0.030
(2)	Pallas	0.0380	(16)	Psyche	0.030
(3)	Juno	0.0908	(39)	Laetitia	0.030
(4)	Vesta	0.0386			

Detailed observational series of certain planets show that the brightness

---

[13] See HAUPT, H., and AUZINGER, H.: Helligkeitsbeobachtungen von kleinen Planeten, *Mitteilungen der Universitäts-Sternwarte, Wien* **5**, 119 (1952).

depends not only on the phase angle, but also on another phenomenon. One assumes that the minor planets are irregular fragments of a planet that rotates and in doing so reflects different amounts of light in the direction of the Earth. Such an additional rotational light variation will also be expected for those planets where it has not yet been observed. It has been found for the following seven planets:[14]

(433) Eros	$0^{d}212$	(129) Antigone	$0^{d}10$
(15) Eunomia	$0^{d}127$	(345) Tercidina	$0^{d}37$
(39) Laetitia	$0^{d}092$	(944) Hidalgo	—
(44) Nysa	$0^{d}132$		

The minor planets are also discussed in Sec. 15.2.4, and the observer may be interested in the Minor Planets Circular published by the Cincinnati Observatory, Observatory Place, Cincinnati, Ohio.

19.10. Notes on the Measurement of Colors

Colorimetry is performed in astronomy by brightness measurements referring to two different color regions, i.e., wavelength ranges. If we measure, for instance, the brightness of a star in the blue, measurements in the red will lead to different magnitudes. A white star has more blue radiation, a reddish star more red. The ratio, expressed as a magnitude difference between the two color regions, is a measure for the color and is called the "color index," i.e.,

$$\text{Color index} = (\text{shortwave magnitude}) - (\text{longwave magnitude}) \qquad (12)$$

The brightness of a star in a certain color region is obtained by the introduction of a filter. Only colored glasses that give sharp boundaries for a given wavelength region are suitable as filters (see Fig. 2–13). The limitation of the color region can reach great accuracy by an appropriate adaptation of filter and radiation receiver (see page 42).

The difference in brightness shown by a star in two color regions depends not only on the color as such but also on the filters used. It is essential therefore to indicate as accurately as possible the wavelength that is most effective in the particular color region. A more convenient, and as far as the amateur astronomer is concerned the only advisable, method is to refer the star to be measured to a definite color system as in a catalog (see page 456). Here, we may mention as an example, the color measurements of GIALANELLA on the Monte Mario near Rome. Using a 155/2240 mm refractor he first measured the light behind a blue filter, then through a red filter, using a wedge photometer. The stars were selected from a catalog giving their color

[14] See GRAFF, K.: Die physische Beschaffenheit des Planetensystems. *Handbuch der Astrophysik*, Vol. 5, p. 413. Berlin: Springer, 1931.

index. The readings of the wedge, K (blue) and K (yellow), were noted and for each star the ratio formed:

$$Q = \frac{(\text{color index})}{K\,(\text{yellow}) - K\,(\text{blue})}.$$

Q changes somewhat with the color of the star. Afterwards he measured the brightness of a star with the same photometer and the same filters under the same observing conditions. He then obtained a difference [K (blue) − K yellow)], which when multiplied by the above Q values, gave the required color index.

Color index catalogs for the determination of the above Q values can be found in LANDOLT-BÖRNSTEIN: Zahlenwerte und Funktionen: New Series, Group VI, Vol. 1, pp. 347ff. Heidelberg: Springer, 1965.

The photographic determination of the color index is carried out in exactly the same manner as the photographic determination of brightness in two exactly known and sharply defined color regions.

20 / Double Stars

W. D. Heintz

20.1. The Visual Double Stars

Double and multiple stars figure prominently in the sky gazers' lists of worthwhile objects, and also serve, in various degrees of difficulty, as tests for telescope optics (see Table 22 in the Appendix). As a regular working program for the amateur observer, however, they appear to offer little scope. This is by no means due to unusually high requirements of instrumentation or theoretical knowledge but simply that this kind of work is a long-term project demanding continuous attendance. The present chapter will begin with a brief, general outline, restricted to the visual double stars, and will discuss the measuring techniques in subsequent sections.[1]

The components of a binary star move about their common center of mass in ellipses similar to each other, according to Newton's laws. Since the center of mass cannot be observed directly, the relative orbital motion—likewise elliptical—of one component with respect to the other is followed. The great majority of pairs with medium or large separations, i.e., those accessible to small telescopes, show only very slow orbital motions corresponding to revolution periods above a thousand or ten thousand years. Yet distinct changes of position can frequently be found by comparing measurements made a few decades apart. In some cases where the periods are less than a century (ξ Sco, 70 Oph) the motion is noticed within a few years.

Besides these real (physical) binaries, there are many instances of so-called optical pairs consisting of independent stars, the apparent proximity of which is merely a matter of perspective. The only rigorous proof of the physical nature of a pair is the existence of a curved (orbital) relative motion, as the motions of optical components are rectilinear. Numerous pairs have not shown any appreciable relative motion so far, so that the curvature criterion is undecided, yet the presence of a proper motion common to the components permits us to conclude their physical relationship. The cataloged double stars by now number more than 50,000, the majority of which in all probability are physical, although this can be individually proved for only part of them.

---

[1] HEINTZ, W. D.: *Doppelsterne*. München: Wilhelm Goldmann Verlag, 1971; and HEINTZ, W. D., ed.: Proceedings of the Fifth IAU Colloquium (Nice, 1969). *Astrophysics and Space Science* **11**, No. 1 (1971).

Optical objects are least probable among the close and the bright pairs, and very rare, for instance, among components brighter than eighth magnitude at separations under 5″. The proper motion of most of the brighter stars is large enough to reveal a rectilinear displacement relative to an optical companion within a century. At separations over 30″ the optical objects predominate although many physical pairs with common proper motions occur. Once a pair is found to be optical it is usually dropped from observing programs.

Triple and multiple systems are rather frequent but their components are very rarely at approximately equal separations from each other. Usually there are a close pair and a distant third body (or another close pair) at a distance at least ten times the separation of the inner pair. This arrangement of distances is vital for the stability of the system. Sometimes the presence of invisible companions can be deduced when the visible ones deviate from Keplerian motion.

Double and multiple systems are very frequent in the space around the Sun. In fact, the real single stars (nonmembers of binary systems, like the Sun) form a minority of, at best, 20%. A relationship between galactic clusters, stellar associations, and multiple and double stars may be understood in evolutionary terms. The division of the binaries into the classes of visual, spectroscopic, and photometric (eclipsing) pairs is merely dictated by the different techniques of observation. These classes cover different (and partly overlapping) ranges of true separations and periods, yet there are no indications for basic genetic differences between the close and the wide pairs. All orders of periods and orbital dimensions can be found, from the closest photometric pairs revolving with their surfaces almost in contact, up to the widest objects of the Proxima Centauri kind: these are often so widely separated that their attraction is almost submerged in the general interstellar gravity field, and only their common proper motion and parallax show that they just manage to keep together. The orbital motions of binary stars offer virtually the only way to determine stellar masses and, in conjunction with this, a fairly reliable method of distance determination, the so-called dynamical parallaxes.

With regard to color and brightness differences, too, the double stars show a great variety. There are pairs of equal brightness and color, pairs with small magnitude differences where the companion (the fainter component) is frequently a little redder than the primary star, and also some combinations of a red giant star with a white companion. At large magnitude differences the color and the spectrum of the companion usually cannot be ascertained, owing to the interfering glare of the primary. Since the eye tends to strongly exaggerate color contrasts, visual color observations of double-star components are entertaining though of no scientific value. Some variable stars are also members of binary systems.

The position of the companion relative to the primary on the celestial sphere is determined by two quantities, as shown in Fig. 20–1. The micrometer gives the polar coordinates, viz, the separation ρ in seconds of arc, and

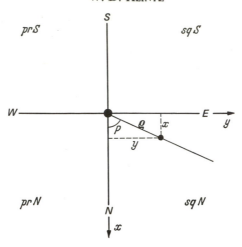

Fig. 20–1 Double stars: relative coordinates.

the position angle p, reckoned in degrees from north through east, south and west (counterclockwise) from 0 to 360°. Photographic plates yield the rectangular coordinates x and y, both in seconds of arc. The conversion is

Declination difference $\qquad \Delta\delta = x = \rho \cos p,$

Right ascension difference $\quad 15 \cos\delta \cdot \Delta\alpha = y = \rho \sin p,$

or, inversely, $\qquad \tan p = y/x, \qquad \rho^2 = x^2 + y^2.$

(The factor 15 cos δ converts seconds of time into arc seconds.)

The N, E, S, and W directions divide the field of view into four quadrants (designated as stated in Sec. 1.3). The zero point of the position angles, i.e. the North direction, is subject to a slow precessional displacement. Hence all position angles are changing, at the rate of $+0°.557 \sin \alpha \sec \delta$ per century.

Apart from the "seeing" conditions, the keenness of vision, and so on, the difficulty of observing a certain double star with a given telescope may be graded according to three parameters:

1. The separation ρ. The well-known DAWES formula states that pairs can just be resolved (if bright enough, and composed of equal components) at a separation of 4″.6 divided by the telescope aperture in inches (or 11″.7/ aperture in cm). In closer systems, the star discs appear to touch, finally merging into an elongated image the noncircular shape of which can be seen, under favorable conditions, down to about half the DAWES resolution limit. The DAWES formula was derived from observations with some small to medium-sized refractors, and it is a good approximation in this range while, at larger apertures, the observed resolving power is slightly poorer than predicted.

2. The magnitude difference Δm. The resolving power can be exploited fully only for near-equal components. A one-magnitude difference of brightness already makes for more difficult measuring. At larger differences, the risk of serious errors (both accidental and systematic) increases.

3. The component magnitudes m_A and m_B. If the stars are too faint for convenient direct-vision observation the attainable accuracy of the measurements will drop sharply. There is a great difference between a *visible* and a *measurable* light source. Therefore, the usual rules (e.g., that a four-inch telescope reaches to 12th-magnitude stars) are worthless in this respect. The brightness limit for safe measurements greatly depends on atmospheric conditions and on the sensitivity of the eye. As a guideline, it may be said that $10\overset{m}{.}5$ stars require a 10-in. telescope, and 12th-magnitude stars at least a 20-in. instrument, though the measuring accuracy begins to drop already two magnitudes earlier.

From the total light m_t and the magnitude difference Δm of a system, the component magnitudes can be computed and vice versa. The accompanying table shows the relation between Δm and the quantity $m_A - m_t$ by which the primary is fainter than the combined light.

Δm	$m_A - m_t$	Δm	$m_A - m_t$	Δm	$m_A - m_t$
$0\overset{m}{.}0$	$0\overset{m}{.}75$	$1\overset{m}{.}0$	$0\overset{m}{.}36$	$2\overset{m}{.}0$	$0\overset{m}{.}16$
0.2	0.66	1.2	0.31	3.0	0.07
0.4	0.57	1.4	0.26	4.0	0.03
0.6	0.49	1.6	0.22	5.0	0.01
0.8	0.42	1.8	0.19	6.0	0.00

For example: The components of 70 Oph are $4\overset{m}{.}19$ and $5\overset{m}{.}98$. Then $\Delta m = 1.79$ yields $m_A - m_t = 0.19$ and hence the combined brightness $m_t = 4\overset{m}{.}00$. Data for Castor are: $m_t = 1.54$ and $\Delta m = 0.96$, hence $m_A - m_t = 0.37$. Then the primary is $1\overset{m}{.}91$, and the companion $2\overset{m}{.}87$. Evidently, equal components are fainter by $\frac{3}{4}$ magnitude each than their combined light. If Δm exceeds 3^m the companion contributes little to the total light.

Visual double stars are referred to by a discoverer's code (key letters plus numbers). For stars listed in the ADS (AITKEN's *General Catalogue of Double Stars*, containing all systems known before 1927, and north of $-30°$ of declination), the ADS number also serves as a reference. Since the code usually gives some indication of the difficulty of pairs, the most frequently occurring codes are mentioned below.

The oldest systematic catalog, containing about 3000 objects, and the earliest good measurements are due to W. STRUVE (Σ). Pairs previously observed by W. HERSCHEL (H) and even earlier are also listed by their Σ-

number, if any. Most of the objects accessible to small telescopes are Σ-pairs. In the southern sky, they are paralleled by J. Herschel's discoveries (h).

Subsequent discoverers proceeded to more difficult pairs, i.e., to closer and fainter objects or those with large magnitude differences. These are O. Struve (OΣ), Burnham (β), Hussey (Hu), Hough (Ho), Aitken (A), Kuiper (Kpr), and for the southern sky Innes (I), See (λ), van den Bos (B), Rossiter (Rst), and Finsen (φ). Most stars in the A and later lists require more than 10 inches of aperture; most of the φ objects are interferometric discoveries.

The search for visual binaries has attained a high degree of completeness for the brighter stars although, of course, some new pairs are added once in a while. Most of the more recent discoveries concern pairs fainter than 9^m and 10^m, viz., wider objects by Jonckheere (J), Espin (Es), Milburn (Mlb), close pairs by Couteau (C), Muller (Mlr), and others.

Within a system, components are distinguished according to brightness by the letters A = primary, B = companion, and C, D, etc., for additional components in multiple systems.

20.2. Micrometers and Visual Measurements

The exploration of visual double stars, in particular the measurement of the relative positions of the components with the aim of determining the orbits, is one of the few astronomical tasks that still relies for the most part on the oldest technique, visual observation. It has always been largely the work of quite a small number of observers, including some outstanding amateur astronomers, who continued it for years and decades. The accidental errors of measuring are, in comparison with the quantities to be determined, significant enough that some practice is needed until worthwhile results can be obtained. Moreover, even observers of long standing may have systematic errors, and it is only by extensive series of data, covering long time intervals, that these errors can be determined and eliminated.

The instrumentation required consists of a long-focus telescope with high-quality optics, a well-adjusted equatorial mounting with drive and slow motion, in conjunction with a position micrometer, or with a photographic tailpiece plus a measuring machine. The reductions are very simple, and the work is not connected with certain seasons of celestial regions since there are double stars all over the sky.

Not all the numerous double stars can be constantly surveyed, and for many faint pairs the gaps in the observations extend over decades. Therefore, a permanent and considerable need for good measurements is felt. To be more specific:

(a) Close and, as a rule, fast-moving pairs (separations of 0″5 and less) require frequent observation (visually or interferometrically) at larger telescopes.

(b) In the separation range around 1″ to 5″ the brighter pairs (combined magnitude brighter than 8."5) have almost always been amply observed, at least those in the northern sky. Only very precise measurements will compete.

(c) If the components are brighter than 9^m or 10^m, and at a separation of 3" or more, they are best observed by photography. Nearly all these objects move very slowly so that a small, well-spaced number of good measurements suffices.

(d) The data on pairs fainter than about 9."5 frequently are for the most part still quite incomplete. This is a rewarding field of work, containing many thousands of "neglected" pairs; but it needs at least 15 or 20 inches of aperture to be tackled.

It is seen that small instruments will not find much of a worthwhile field of activity. Nevertheless, a review of a versatile and educationally valuable type of equipment, the micrometer, should not be omitted from this chapter.

All micrometers in use are screw micrometers. Old devices without movable parts (such as cross-staff and ring micrometers), operated by estimating transit time differences, are useless for the small separations in double stars. They are far too inaccurate even for large separations, and have long been superseded by photography.

Complete micrometers appear not to be commercially available at present, yet their construction is not excessively difficult. Screws meeting the high precision requirement have become available in recent years at moderate cost.

The filar micrometer is the most widely used. Its results seem, on average, not to suffer from significant systematic errors though individual observers may show such errors.

In this instrument, two metal frames are placed in the focal plane, each frame carrying a wire (*a* and *b* in Fig. 20–2) and perhaps some spare wires. The screw shifts one frame relative to the other at right angles to the optical axis. The connection of the moving frame with the screw should be free from backlash, and the wires *a* and *b* lie as nearly as possible in the focal plane—yet

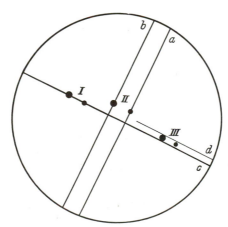

Fig. 20–2 Measurement of a binary star with the filar micrometer.

without touching when shifted past each other—because they have to appear equally sharply focused in a short-focus eyepiece. For the protection of the wires, the double frame is placed in a box that can be rotated about the optical axis, the amount of rotation in degrees (position angle) being read off a divided circle. (A circle not less than 8 inches in diameter, with full- and half-degree marks, is recommended for safe readings to $0°1$.) The microm- eter box carries the eyepiece on a separate, sliding front plate. For purposes of adjustment, it is recommended that the "fixed" frame also be movable over a certain range, although this motion should be tight and perhaps clamped during observation.

The center of the field of view should be close to the optical axis so that stars do not move across the field while the micrometer is rotated. The screw head carries a drum to read the position of the moving wire in revolutions, hundredths, and (estimated) thousandths thereof. Check the adjustment of the polar axis as well as the centering of the optics to avoid systematic errors; the stellar image and its diffraction rings should look exactly circular.

An adjustable light source is required. Some observers use field illumina- tion (preferably by red light); others use directly illuminated wires, the brightness being adjusted so that the wire is sharply visible in the field yet dark in front of a star. The light bulb should be far enough from the microm- eter that no heat can reach the frames.

Because of their low scatter of light, spider-web wires of homogeneous thickness, if available, are preferred to quartz or similar material. Stretching and fastening the wires requires some patience and skill, but a carefully pre- pared set of wires can last for years. A wire that is losing its tautness or collecting dust particles is removed before it starts knocking out other wires. Keep the wires protected from wind and dust from both sides as far as possible, and cover the micrometer box when not in use. Finely engraved, thin glass plates could be used in place of wires.

The magnifying power chosen is usually the highest permitted by the telescope and the seeing, except that faint pairs may profit from a lesser power since the image deterioration under high power reduces their visibility. Changing the eyepiece may endanger the wires, which are usually quite close to it. Therefore, the eyepieces should have a close-fitting thread, and should be changed with both hands, one pressing the eyepiece against the frame to prevent tilt, the other screwing it out or in. Opinion of observers is divided as to whether or not the reduction of aperture by a diaphragm yields a worth- while image improvement for brighter stars.

The procedure for measuring a double star is as follows. A fixed wire is set on the line joining the stars so that it bisects both images (position I in Fig. 20–2). To check the accuracy of the angle setting, the wire is placed alternately just above and below the stars by applying a light pressure to the tube. Once the direction of the wire appears accurate the circle is read. This process is repeated a few times. The separation is measured with the wires perpendicular to the line joining the stars (position II). One component is placed under the fixed wire, and the moving wire set to bisect the other star; then the setting is repeated with the wires interchanged. The drum is read each time, the difference between the readings corresponding to double the separation. (This method is more accurate than single separations, and free from unsafe coincidence settings.) Again the measurement is repeated. Between the settings, the wires are always offset (in angle or separation) from the measuring position in order to improve the independence of the settings, and to reduce the bias. Any observation may consist of four or five angles and three double separations. The trick taught by practice is to make the settings just in the right number and at the right speed so that no accuracy is lost nor the eye becomes tired.

A wire parallel to the screw (c in Fig. 20–2), i.e., perpendicular to a, is useful for measuring the position angles; otherwise the wire a has to be used and the micrometer rotated by 90° for the separations. Some observers feel that the angle is more safely determined in a narrow space between a double wire (c, d, and position III) since the appearance of stars (especially unequal components) under wires may be accompanied by optical illusions.

The line joining the eyes is kept parallel, or else perpendicular, to that joining the stars, but at no other angle, or considerable errors may result. The parallel (..) or perpendicular (:) position is recorded in the observing book, together with remarks referring to the seeing, the magnification (if different ones are used), the approximate hour angle, and the quadrant of the companion. To check on possible gross reading errors a rough estimate of the separation may be added. Experienced observers frequently add estimates of the component magnitudes or of Δm. These results are useful particularly for faint and unequal pairs the photometric data of which are still uncertain. Color estimates are of no avail, as was stated earlier.

Two quantities are needed to reduce the observations, viz, the zero point of the circle and the screw value. Errors of the circle or of the screw, if present, have to be allowed for.

1. Mean of the double separations multiplied by half the screw value = separation in seconds of arc.

2. Mean of angles minus zero point of circle = position angle (change by 180° if necessary to conform with the quadrant recorded).

3. Convert the observing time into decimal fractions (thousandths) of the year, using a table found in any astronomical almanac.

The screw value (the angle corresponding to one screw revolution in the focal plane) can be determined by repeatedly measuring the Sidereal Time interval of transit of a circumpolar star (δ about 75°) near the meridian over a set number of revolutions, using a stop watch or the eye-and-ear method (see Sec. 13.4.3). The time is converted into seconds of arc by the factor 15 cos δ, with the apparent (instantaneous) declination being used for δ. Another method is to measure some widely separated pairs of stars with well-known positions, for instance, in the Pleiades. (In earlier times, the recommended standard sequence was the so-called "Perseus arc" in the cluster h Per, consisting of the stars BD + 56°498, + 56°543, and eight intermediate ones. Yet this sequence has not been remeasured of late.) Since the screw value found is used once and for all it should be precisely known, say to better than 1 part in 3000. Therefore, the measurements are repeated until the probable error is small enough. The result can be compared with the formula: Screw value = 206,264″8 multiplied by screw pitch, divided by focal length of telescope (pitch and focal length being expressed in the same unit).

A screw may have periodic and progressive errors (i.e., repeating or accumulating, respectively, on successive revolutions). Their accurate determination requires auxiliary tools. A rough check may be afforded by a terrestrial object (scale) measured on various parts of the screw. Modern high-precision screws should be expected to be free from significant periodic errors, and a table of progressive errors should be supplied by the manufacturer.

The zero point of the circle is determined each observing night. A star, near the equator and the meridian, trails along that wire used for the position angle settings. By repetition, the parallelism of the wire and the star's motion throughout the field of view is determined to 0°1. This fixes the east–west direction; subtract 90° to obtain the zero point (north reading). The zero-point observation should be made at the beginning and the end of the observing session, and perhaps also during the session. The observer will soon find out for what length of time he may rely on a constant zero point.

Another item to be examined is the division of the circle, since division errors, if present, affect the observations in a systematic fashion. Frequently the errors result from a slight eccentricity of the circle. Opposite the pointer used for circle readings, a second pointer is attached, and the observer checks whether the difference of readings at both pointers remains constant in all parts of the circle. The procedure is then repeated with smaller angles between the pointers, say, 90° and 60°.

The observation of a double star is repeated on several nights during the same observing season. The amount of random errors can be assessed from the comparison of the results, and a single night's observation carries little weight. In order to check on systematic errors, comparisons are made with extended series of observations obtained at larger telescopes, particularly with photographic measurements, and also with the Ephemerides from reliable orbits. Certain standard stars have been proposed by P. MULLER to check on errors of separation (see Table 22 in the Appendix). The use of a prism reversing the field of view has sometimes been suggested for eliminating possible systematic errors in the position angles. In this case, half the readings are taken with the reversing prism, half without it.

The double-image micrometers are designed to set stellar images against each other rather than against a wire. Among some applications of double refraction tried, the MULLER-type birefringent prism deserves special mention (Fig. 20–3): The crystal axes of the two cemented parts of the prism are perpendicular to each other, and at 45° to the optical axis. The dispersion between the ordinary and the extraordinary rays is proportional to the

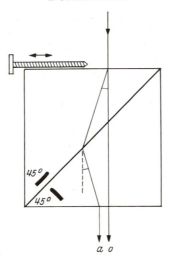

Fig. 20–3 The light path in the MULLER prism.

lateral displacement of the prism by the micrometer screw, and the screw value becomes small enough to render periodic errors insignificant.

The total of four star images representing a double star in the field is arranged on a straight line at equal distances, or in a square, since the regularity of either pattern can be safely rated (Fig. 20–4): Let A—B be one image of the pair, and a—b or a′—b′ the other. The angles are read in one of the indicated positions, and the shift a—a′ = b—b′ corresponds, respectively, to twice or four times the separation. The observer soon learns to decide which pattern he prefers. Settings of image coincidence are too inaccurate and are avoided. Apart from this, the measuring procedure corresponds to that by wires.

The division into two images causes a loss of brightness by 0.75 magnitude, which is further increased by some light loss in the prism, yet largely compensated for by the absence of illumination. It appears that the double-

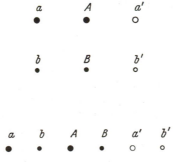

Fig. 20–4 Double-star measurements with the double-image micrometer (two methods).

image instrument permits a faster gain of practice than the filar one, and is also less susceptible to personal systematic errors. It is distinctly advantageous on bright, extended objects (planetary diameters), and is less affected by an imperfect telescope drive. For these reasons, this versatile micrometer can be particularly recommended to amateur observers, although it is more expensive.

A rotatable polaroid plate (with a reading scale) enables one to vary the relative brightness of the images. This feature is advantageous for micrometric measurements and, moreover, renders the instrument a useful comparison photometer for close double stars.

A wire in the field serves to determine the screw value (transit time difference of the images of a circumpolar star) and the zero point. It is not used for the measurements themselves.

The principle of comparing the double star with an adjustable pair of artificial star images reflected into the field of view is rarely used. The comparison-image micrometer by DAVIDSON[2] might be mentioned. It consists of a point light source (pinhole with collimating lenses), a WOLLASTON prism splitting the image, two crossed NICOL prisms, and mirrors reflecting the image into the focal plane. The position angle is read on a circle and pointer connected with the WOLLASTON while the brightness is adjusted by rotation of the NICOLS. This device has more light power than the MULLER prism yet has a serious disadvantage: The artificial image never looks exactly like a real, scintillating star. This fact may cause systematic errors in the measurements of close pairs and is particularly hazardous for visual photometry. The device is technically difficult, and specialized for use on double stars, that is, not as versatile as the previously discussed types. It is best used on faint pairs.

The micrometer suggested by GRAMATZKI also belongs to the double-refraction kind, having, in addition, a grating plate which attaches diffraction rays to the stellar images. The prism is fixed, thus the distance between the images is constant, and the measurement of separations effected by angle readings, the companion being placed on the ray of the primary. This theoretical advantage is more than offset by several drawbacks, viz., the very limited useful range of separations, a serious loss of light, and the risk of bad errors in close pairs due to image distortion. No results obtained by this design are known.

Interferometer observations are very difficult and require large instruments. They are beyond the scope of this book, and a brief comment may suffice. Monochromatic light λ, entering through two slits, combines to form an interference pattern. In the case of a double star, the pattern disappears or reaches minimum visibility in a certain position, namely with the line joining the slits paralleling the position angle, and the slit separation being $D = \lambda/2\rho$, by which formula the star separation ρ is determined. An eyepiece interferometer has been constructed and applied by FINSEN.[3]

[2] DAVIDSON, C. R.: *Monthly Notices R.A.S.* **98**, 176 (1938).
[3] FINSEN, W. S., *Astron. J.* **69**, 319 (1964).

Let us add a comment on the faint, wide pairs ($m_t > 9^m$, in the separation range of about $3''$ to $30''$). As shown by the Index Catalogue (IDS), many data in this area are still uncertain or missing. An experienced, assiduous observer can do a good deal of useful work with an instrument powerful enough to render 12th and 13th magnitude stars measurable. Desiderata are magnitudes (photovisually measured, or estimated on a reliable scale checked with the North Polar Sequence), and careful identifications using the Bonner Durchmusterung and the Astrographic Catalogue. For approximate relative positions (to degrees, and tenths of seconds, in order to pick out the cases of significant relative motions) the micrometer might be dispensable, as a scale in the field combined with a small circle will suffice.

20.3. Photographic Observations

As compared with visual work, photographic measurements are restricted to wider pairs but are more accurate, the greater accuracy owing mainly to the multiple exposures obtained on one plate in a short time.

The separation ρ (in radians) equals the distance between the focal images of the stars, divided by the focal length. Converted into seconds, 1 millimeter on the plate corresponds to $206''.265$ divided by the focal length in meters. Modern photographic plates successfully combine the requirements of high speed and fine grain so that well-exposed images can be measured at separations under 0.15 mm, sometimes down to 0.10 mm. Even so, the photographic use of long-focus instruments with the usual focal ratios is limited to about 15 times the DAWES resolution. The use of magnifying optics, as in planetary photography, is promising some extension of the range.

Certain photographic effects (neighbor effects) appear in closely adjacent photographic images, spoiling the measurement. This range of separations is avoided, and left to the visual observer.

The plate is strictly perpendicular to the optical axis. Exposure times giving images of just the right size and density for good measuring are found by experience. The optimum exposure times greatly depend on the seeing. Emulsion shifts are insignificant at the small separations concerned; they may be troublesome in other kinds of astrometry, and may then call for precautions in the processing of plates. The series of exposures of a double star on one plate is completed by a star trail showing the orientation of the plate. The image scale can again be checked by measuring some very wide pairs. Beware of coma!

The color equation, that is, a relative displacement of components of different color, and a possibly disturbing atmospheric spectrum are eliminated by filters. The magnitude equation (inaccuracy and systematic errors in the measurements of unequally bright components) is removed by use of a suitable grating, the first-order images of the primary then being measured relative to the central image of the companion. If Δm is very large, special diaphragms can be used (also in visual work), for instance, a regular hexagon producing a six-rayed image, so that, at a suitable orientation of the diaphragm, the faint companion is seen in the dark space between the rays even when close to the primary star.

Photoastrometric work with long-focus instruments cannot be discussed here in detail. It aims to determine gravicentric motions and mass ratios of binary stars by measuring a star relative to a suitably selected reference frame of distant background stars over a long interval of time. The main requirement is a measuring machine of the highest precision.

Differential refraction affects all measurements of wide pairs, and has to be taken into account in precise photographic positions particularly, as well as in screw-value determinations. The effect is to lift the lower of two stars more than the upper one, thus decreasing the separation. For a pair with a difference of 1 minute of arc in altitude, the differential refraction amounts to $0''.04$ or $0''.07$, approximately, at the respective altitudes of $40°$ or $30°$. Special refraction tables (e.g., by DE BALL, Leipzig, 1907) are provided for detailed computations.

The plates are measured twice in each coordinate, in direct and reverse position. It is better still for a second observer to duplicate the measurements. Plate measurements require a bit of practice, and personal systematic errors can easily occur.

Electronic techniques have been introduced of late, particularly photography by electronic cameras of the LALLEMAND design (similar to television tubes). The equipment affords a high accuracy, but despite some improvements, it is still expensive and very delicate, and the reductions are so laborious as to require a computer. There is also photoelectric scanning of the focal image (dispensing with photography altogether), which, however, has not yet succeeded in matching the visual and photographic accuracy.

20.4. Orbital Elements and Ephemerides

The orbit of a double star is described by seven quantities called orbital elements defined by analogy to the elements of planetary orbits (see Sec. 8.5). The motion, in general, is not at right angles to the line of sight, and the apparent (projected) orbit therefore differs from the true one, the amount of foreshortening depending on the inclination.

> a, the semiaxis major, in seconds of arc, and
> e, the numerical eccentricity, fix the size and the shape of the orbit.
> The time element of the motion is given by
> P, the period of revolution, in years (or the mean motion $\mu = 360°/P$, in degrees per year), and
> T, the instant of passage through the periastron, the point of smallest true distance between the components. The time of passage through any other point of the orbit then is given by KEPLER's Second Law, stating that the line joining the components sweeps equal areas in equal times.

Finally, the relation between the true and the apparent orbits involves three angular elements:

i is the inclination between the plane of the true orbit and the plane of projection (the latter tangential to the celestial sphere), and is counted from 0 to 180°. Direct motion (position angles increasing) is indicated by i being less than 90°, otherwise the motion is retrograde.

Ω, the node, is the position angle of the line in which the true and the apparent orbits intersect. The two nodes differ by 180°, and the one smaller than 180° is usually quoted as an orbital element. The node is affected by precession, as are all position angles (see page 474).

ω, the angle between the node and the periastron in the orbit, is counted in the direction of motion.

In place of these angles, the orbit can also be expressed in rectangular coordinates, using the so-called THIELE-INNES elements A, B, F, G. They are transformed from the classical elements a, ω, Ω, and i by the equations

$$A = a(\cos \omega \cos \Omega - \sin \omega \sin \Omega \cos i),$$
$$B = a(\cos \omega \sin \Omega + \sin \omega \cos \Omega \cos i),$$
$$F = a(-\sin \omega \cos \Omega - \cos \omega \sin \Omega \cos i),$$
$$G = a(-\sin \omega \sin \Omega + \cos \omega \cos \Omega \cos i).$$

Up to the present, the orbits of about 600 visual systems have been computed.[4] Many orbits are very accurately known, many others are still quite uncertain. Most of the calculations refer to close pairs. The orbit computation for a pair on the basis of all measurements to date is not difficult mathematically, but lengthy, and it requires skill and care in the interpretation of the measurements to obtain a good result. Wherever both the amount of orbital motion and the number of measurements obtained to date suffice to define the orbit (approximately, at least), one or more orbits have actually been calculated. Thus, new computations cannot be encouraged as there is very little need for them. In fact, most of the work published in recent years by computer-happy people consists of redundant duplications failing to improve upon previous orbits.

It may be desired to compute positions (Ephemerides) from an orbit for the purpose of comparison with the measurements.[5] The formulae needed to obtain, for any time t, the polar coordinates p and ρ or the rectangular ones x and y from the elements, are given below. The computation proceeds via an auxiliary quantity E to the coordinates in the true orbit (polar v and r, or rectangular X and Y), which are then projected into the corresponding positions of the apparent orbit.

$$\mu(t-T) = M = E - e \sin E$$

$$\tan v/2 = \sqrt{(1+e)/(1-e)} \tan E/2 \qquad X = \cos E - e$$
$$r = a(1-e^2)/(1+e \cos v) \qquad Y = \sqrt{1-e^2} \sin E$$
$$\tan (p-\Omega) = \tan (v+\omega) \cos i \qquad x = AX + FY$$
$$\rho = r \cos (v+\omega)/\cos (p-\Omega) \qquad y = BX + GY.$$

[4] FINSEN, W. S., and WORLEY, C. E.: Third Catalogue of Orbits of Visual Binary Stars, *Rep. Obs. Circular* 129, Johannesburg, 1970.

[5] MULLER, P., and MEYER, C.: Troisième Catalogue d'éphémérides d'étoiles doubles. Observatoire de Paris, 1969.

The first formula, called KEPLER's equation and containing the Second Law, is solved by iterative methods (see Sec. 8.5), with the quantity $e \sin E$ converted into degrees by the factor $180/\pi = 57.296$. Tables giving v, or X and Y, directly with the argument M are available,[6] and permit one to dispense with the computation of E.

[6] *Publ. Allegheny Observatory*, Vol. 2, No. 17 (Tables of v); *Astron. Papers of the American Ephemeris*, Vol. 19, Part 1, Washington, 1964 (Tables of X and Y).

21 | The Milky Way and the Galaxies

R. Kühn

Revised and completed by *F. Schmeidler*

21.1. Introduction

There are fields in observational astronomy that are less suited to scientific work by the amateur observer. The following section describes one of them, for only rarely will the amateur astronomer be able to take an active part in the attack on the unsolved problems of galactic research. Should he therefore not concern himself in any way with the relevant objects? Not at all. It is the mere observation of the objects of the Milky Way and of other galaxies that brings to every amateur astronomer a wealth of interesting and, I must say, really wonderful experiences and no observer, *regardless of the instrumental means at his disposal*, should omit to make such observations.

Nothing can replace the experience gained by the observer in actually seeing for himself the objects he wishes to learn about. The knowledge that everybody should have about the structure of the Universe can be extended and consolidated only in this way. This chapter, therefore, is addressed to those who wish to encourage the pursuit of and teach others their own knowledge of astronomy, and who have access to an observatory for amateur astronomers. As things are, the understanding that most people have about the structure of the Universe does not extend beyond the boundary of our own solar system. Indeed, most people know more about the planets than they do about all the remaining Universe. Because the programs of many amateur observatories, as well as many amateurs who wish to show the sky to their friends, give preference to observations of the Moon and planets, the wrong idea is conveyed that these objects are the "whole Universe." The misconception is difficult to remedy. To me it is important to try to transmit to as many people as possible the correct ideas about the structure of our Universe. Teaching alone, then, offers the amateur astronomer a wide field of most beneficial activity.

Above all, we should not forget that the main purpose of our observations is, in the last analysis, the delightful pleasure in the study of the object itself. And this joy will be found in abundance by all who dedicate their observations to the study of the Milky Way and of the distant galaxies. Those who wish to carry out work of value, however, should certainly discuss their plans, programs, and methods with the astronomers of a large observatory.

21.2. The Instruments

All the objects whose observation concerns us here have one thing in common: They are relatively faint and many of them are rather extended. Therefore, from the very beginning we demand of our instruments that they fulfill two requirements: They must have as large an aperture as possible and they must have a field of view as large as practicable. High magnifications are rarely required.

21.2.1. Visual Instruments

Although the human eye is not an instrument in the usual meaning of the word, it must be emphasized that many observations can be carried out even with the unaided eye. The prime requirement, however, is that observations should be made immediately after the eye has been completely rested in the dark. Furthermore, the observer should be as far away as possible from all sources of artificial light.

The instrument most suitable for the observation of many extended galactic and extragalactic objects are powerful binoculars. Among the many commercial types we shall select those that fit best our particular aim. Their optical qualities are usually characterized by two numbers, the first of which is the linear magnification, and the second the aperture of the objective lens, expressed in mm. A 7×50 binocular therefore has 7 times magnification and an objective diameter of 50 mm. We prefer an instrument with an objective as large as possible, preferably 50 or 60 mm. Unfortunately, the price increases with the size of the lens. Which magnification is the most favorable cannot be stated in general terms since it depends on the special interest of the observer. Nevertheless, I believe that in most cases an instrument with $10 \times$ magnification might represent the optimum. Even binoculars cannot be used for serious celestial observations without a tripod. An instrument of $10 \times$ magnification can be used without a tripod for terrestrial observations, provided the observer supports his elbows; in the case of astronomical observations, however, this is generally not possible. A suitable stand can be homemade without too much effort. It is particularly favorable if from the very beginning one can combine a comfortable observing chair with a suitable setup of the instrument.

A particular advantage of binoculars for the observation of faint objects is that both eyes can be used. This increases the subjective reception of the light of the objects. We therefore discourage the use of monocular field glasses.

If we have a normal small refractor at our disposal, it is most essential that we obtain a good eyepiece of small magnification. Unfortunately many beginners are inclined to apply high magnifications, and so spoil for themselves the pleasure of the observation. We could say that for most of our objects the best magnification would be about equal to the "normal" magnification, that is, that the exit pupil should be equal to the pupil of the eye. If we assume the pupil of the eye in the dark to be 5 mm (the value of 6 mm will only rarely be reached) then, for an instrument of 108 mm aperture (i.e., about a 4-inch telescope), this magnification will be 108/5, or 22 times. The corresponding focal length of the eyepiece should then be $\frac{1}{22}$ of the focal length of our objective lens.

Some galactic and extragalactic nebulae even stand three times the normal magnification, and for planetary nebulae such a magnification should really be used. But in general it will always be better to use too low than too high a magnification.

The same applies to the reflecting telescope. Here, too, magnifications that are rather close to the normal magnification as defined above are preferable. Some instruments require modification; for instance, if the secondary mirror is so small that it fills with light only the central part of the field of view taken in by the eyepiece, it may be necessary to change to a larger secondary mirror.

Particularly if we wish to show an inexperienced layman a faint object in the telescope, a very small magnification should be used, even one below the normal magnification. Often the untrained person will only then be able to see faint objects at all. In most cases a small magnification will enable the experienced observer to distinguish all details and in general to enjoy observing much more than with a larger magnification. Small magnifications also facilitate the finding of celestial objects.

We might mention here that larger binoculars, such as were used during the war by the air force and the navy, can render excellent service for the observation of faint objects. These instruments are frequently obtainable at a relatively low price, and although their optical quality is sometimes not of the very highest, for faint extended objects this does not matter too much.

21.2.2. Finding Faint Celestial Objects

Setting on faint objects can be carried out in two different ways, either with graduated circles on the instrument and a sidereal clock, or by the use of a good star map. Which method is followed essentially depends on the personal taste of the observer. The former procedure has the advantage that, if correctly applied, it functions, so to speak, automatically. The second method requires some skill on the part of the observer, but no equatorial mounting or circles are necessary. The study of the star map, moreover, has the additional advantage of familiarizing the observer with the objects of the sky.

21.2.2.1. Setting with Graduated Circles

A good catalog will provide the coordinates of the required objects. Those of them that have to be searched for with larger magnifications, e.g.. planetary nebulae, absolutely require converting the coordinates to the particular equinox of the year of observation. The declination can be set immediately on the declination circle. The appropriate hour angle is found by the equation

Hour angle of an object = Sidereal Time – the right ascension.

If the resulting values are larger than 12 hours we know that the object is east of the meridian. In this case we subtract the hour angle from 24 hours and set the new hour angle toward the east. If the equation gives negative values, the object is also east of the meridian. The Sidereal Time is best read from a sidereal clock, the performance of which is constantly checked by time signals. For this purpose, it is quite sufficient to regulate a good pocket or wrist watch in such a way that it keeps Sidereal Time. If no sidereal clock is available it is possible to set on a star whose right ascension is known and to read off the hour angle directly on the telescope. The above equation furnishes the Sidereal Time at the time of the observation.

21.2.2.2. Setting with a Star Map

The star map method requires first of all finding out whether the desired object is contained in a good star atlas. If not, we transfer its position onto the map with the help of the coordinates given in a catalog. If one does not wish to use the Star Atlas at the telescope it will be necessary to make a sketch of the neighborhood of the object, in a handy size, which will be usable on all future occasions. This little map, as shown in Fig. 21–1, will contain, starting out from a bright star, all objects that are visible in the finder of the telescope. The required object will be marked by a cross, and this map will make it

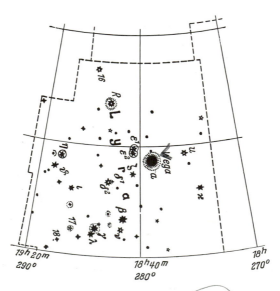

Fig. 21–1 Finding chart for the Ring Nebula in Lyra. After SCHURIG-GOETZ: Tabulae Caelestes, 7th ed.

possible to set accurately on the object in the finder. The main telescope is then used first with an eyepiece of very low magnification and large field of view, and then if suitable at a higher magnification. The map should be oriented according to the appearance of the field in the instrument. Although this method may appear somewhat complicated, it nevertheless leads rapidly and safely to a good result.

21.2.3. The Photographic Instruments

The observation of the objects of the Milky Way and of galaxies has always been an outstanding domain of celestial photography. The amateur astronomer, too, is making more and more use of photographic methods, because this is the very field where relatively modest tools lead to good results, provided, of course, that the observer has patience enough for the necessary long exposure times, which sometimes cover several hours.

21.2.3.1. Photographs with Commercial Cameras

The familiar and very popular miniature cameras are excellently suited to our purpose. It is not necessary to use the most expensive cameras, but certain optical requirements should be fulfilled. The light-gathering power of the objective lens determines the exposure time. It is not absolutely necessary that the objective have an aperture ratio of 1:2 although this would save time and effort. It should certainly not be smaller than 1:3.5. In any case we always use it at full, or nearly full, aperture. Since short exposures are out of the question, we can do without a special shutter; we set the shutter on "B" and then use a cable release that can be clamped in the open position.

The way in which the camera is attached to the telescope is very important. Because of the great stability necessary, the camera can be fixed on the end of the declination axis, in place of the counterweight, only if the mounting of the telescope is very massive and rigid. It will be of advantage in most cases to attach the camera directly to the instrument near to the objective. See Fig. 21–2.

Although the demands on the accuracy of the guiding are not too great for our focal lengths (between 35 and 80 mm), the equatorial mounting must be perfectly adjusted. It is just for these celestial photographs, which cover a large area, that the danger from an inaccurate mounting is great. Frequent correction in declination distorts the images of the stars: the guiding star will be imaged as a point, but stars near the edge of the plate will describe small circular arcs around the guiding star. The guiding itself is most comfortably done with a good driving clock, although this is by no means absolutely necessary. A perfect slow motion in right ascension makes it possible, too, to obtain excellent photographs. After some practice the observer will find that the operation of the slow motion is no more troublesome than the constant checking of the driving clock.

A powerful miniature camera is particularly useful for photography of the star clouds and the dark nebulae of the Milky Way, and always when it is

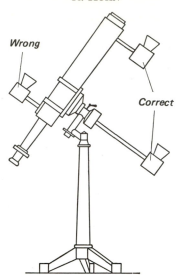

Fig. 21-2 The mounting of a miniature camera on a telescope. After R. BRANDT: Das Fernrohr des Sternfreundes, Stuttgart: Frankh, 1958.

important to cover a large area of sky. The films used should be of fairly high speed (21/10 to 25/10 DIN), but not of the highest sensitivity, since these have too coarse a grain.

21.2.3.2. Photographs with a Small Astrograph

If it is our intention to photograph single objects, for instance, a certain spiral nebula, our miniature camera will be unsuitable because of its small image scale. Here we need instruments of focal lengths of some 20 cm up to 100 cm. In addition to astrophic lenses, amateur-made SCHMIDT cameras, and also the newer design of the MAKSUTOV camera (in which the SCHMIDT corrector plate is replaced by a meniscus lens with spherical surfaces), have now become possible. In principle, the photographic procedure here is the same as for a short focal length, except that the demands on the accuracy of guiding, adjustment, and stability of the mounting increase with the focal length of the camera. Just as a high magnification should be avoided in visual observation, here we have to avoid very long focal lengths. An objective of 50 cm focal length and an aperture ratio of about 1:5 will in this respect meet most of the needs of the amateur astronomer. Instead of films we use plates which, if possible, we develop ourselves.

If the telescope is a refractor it is not suitable as an imaging system for the photography of faint extended objects because of the usually great focal length. On the other hand, reflecting telescopes of the Newtonian and Cassegrain types, with an aperture ratio of between 1:4 and 1:6, can be highly recommended for this kind of photography.

21.3. The Most Important Objects

A list of suitable objects can be found in many publications.[1] The instrumental equipment, the geographic position of the observing site, and also personal preferences will play an important role in the selection. In what follows, we shall only deal with the most important types of objects; a catalog will be found on page 532.

21.3.1. The Milky Way

The term "Milky Way" is used with two meanings. We use it to denote either the phenomenon of the faintly shimmering band of light which crosses the sky, or to mean our own star system, although nowadays it is more usually called simply "the Galaxy." Even a few observations that we can make with the naked eye will reveal to us the relation between these two concepts.

On a clear, moonless night in summer or autumn the band of the Milky Way rises nearly perpendicularly from the south and passes through our zenith to the north. The Milky Way is not uniformly bright but is condensed into several brighter or fainter clouds. Superposed on this patchy structure we can also find without much difficulty that the brightest parts of the Milky Way lie in the constellation Sagittarius, which in summer evenings can be found near the southern horizon.

This brightening of the Milky Way in Sagittarius is particularly obvious if we are observing from a site in the far south, say, in the Mediterranean, and if our horizon is clear and free of haze. Correspondingly, if not so distinctly, we can observe that the Milky Way in winter is much less impressive than in the summer. Thus, observation with only the naked eye has led us to two conclusions: The Milky Way surrounds us as a closed circular band, and the brightest region of it lies in the direction of Sagittarius.

Now if we take our binoculars or a telescope of small magnification, we find without much difficulty that the brightness of the Milky Way clouds as seen with the naked eye indicates the number of faint stars that without magnification are indistinguishable as single points. Knowledge of all these facts are steps toward the proper understanding of the structure of our whole star system, a structure into which studies during just the last few decades have given us a new and tremendous insight.

Our star system has the shape of a flat disc. The inner portion of this disc is populated with stars that are more tightly packed the nearer they are to the center of the system. Our Sun with its planets is near the central plane of the system but not at its center. Of course, we now see immediately how the

---

[1] For instance, in VEHRENBERG, H.: Mein Messier-Buch, Düsseldorf, 1966; NORTON, A. P.: Star Atlas; BEČVÁŘ, A.: Atlas Coeli and Atlas Eclipticalis; see pp. 546–547 and 548–549 in the Bibliography. See also the series: A Messier-Album, *Sky and Telescope*, e.g., **37**, 25 (1965).

phenomenon of the Milky Way occurs: If we look in the direction of the central plane of the system, we notice a much greater number of stars in the sky than if we look in the perpendicular direction where our line of sight rapidly reaches the border regions of the system. In a certain direction, however, namely toward the center of the system, the star clouds are particularly dense, as we have just observed. Indeed, the center of our star system lies, as seen from the Solar System, in the direction of the constellation Sagittarius.

These very simple observations and inferences should be made by every beginner; they will give him a much livelier and more impressive understanding of the interrelation of cosmic objects than any book alone could ever demonstrate to him.

Finally, we mention that one of the most impressive and beautiful observations is provided by binoculars or a more powerful refractor with a large field of view, e.g., a comet finder, if we wander with it over the whole course of the Milky Way. In this way we immediately encounter numerous beautiful objects and gain, in particular, a just impression of the widespread distribution of the dark clouds. The same applies to a photographic exploration made with a powerful miniature camera.

21.3.2. Star Clusters

Star clusters are assemblies of stars. Some of them contain only a few members, so that they are recognized only as a small increase in the star density above the general stellar field of their surroundings. Others, however, represent small star systems of their own and contain sometimes hundreds of thousands of stars.

Star clusters fall into two groups, the open star clusters and the globular clusters.

There are a few hundred open clusters in our star system, all more or less near the Milky Way. The brightest of these can be seen with the naked eye, as, for instance, the Pleiades, the Hyades, and Praesepe, "the Beehive." For the observation of open clusters, the magnification must be rather small. Some objects, such as Praesepe, can be seen only with a low magnification and a very large field of view. Others permit more normal magnifications.

For globular clusters the situation is different. Here it is sometimes necessary to use the most powerful magnification of the instrument. It is advisable, however, to start with a small magnification. These star clusters then appear as nebulous and circular small clouds, which can be resolved into stars only in their outermost parts. If we then proceed to higher magnifications, more and more parts of a cluster are resolved into single stars; the center of the cluster, however, even in the largest instrument, always remains a nebulous, diffuse object. In photographic work it is advisable to use cameras whose length is not too short; in this case the emulsion of the film should have a rather fine grain.

The value of all these observations is primarily a didactic one. Research work in this field is difficult and requires expensive instruments, as well as contact with an observatory. The same applies to the observation of nebulae, planetary nebulae, dark clouds, and extragalactic nebulae.

The globular clusters populate a region of space that reaches far beyond the disc of our star system. We know about one hundred of these objects over the sky; some of them contain more than one hundred thousand stars.

Table 21.1. Data on a Sampling of Open Star Clusters

Name	Messier No.	Brightness	Diameter	Constellation
Pleiades	M 45	1^m	$100'$	Taurus
Praesepe	M 44	4	95	Cancer
h Persei	—	4	33	Perseus
χ Persei	—	5	36	Perseus
—	M 41	5	30	Canis major
—	M 50	6	10	Monoceros

A few open clusters suitable for observation by amateurs are listed in Table 21.1. The magnitudes given represent the total brightness. Because of the diffuse appearance of the clusters the limiting magnitude of the visibility for a certain instrument (and of course also for the naked eye) is by one or two magnitude classes less favorable than for the stars. A similar listing for some globular clusters is given in Table 21.2.

Table 21.2. Data on a Sampling of Globular Star Clusters

Constellation	Messier No.	Brightness	Diameter
Hercules	M 13	6^m	$10'$
Canes Venatici	M 3	6	10
Sagittarius	M 22	6	17
Scorpius	M 4	6	14
ω Centauri	—	4	30

21.3.3. Galactic Nebulae

Galactic nebulae are gas or dust clouds, condensations in a substrate of interstellar matter extended through the central plane of our Galaxy, and are situated within our own stellar system. Three distinct types exist. There are genuine gaseous nebulae which radiate light that is excited by the radiation of neighboring hot stars, pure dust nebulae that just reflect the light from stars, and nebular clouds in which both gas and dust are present.

The faintness of these objects is generally the greatest obstacle to their observation. It is for this reason that the observer should obey the following fundamental rules:

1. Observations should only take place in complete darkness, i.e., when there is no disturbance from artificial illumination, twilight, or the light of the Moon.
2. The eye must be rested and completely adapted to darkness.
3. The observer should use, as far as possible, indirect vision.
4. The instrument should have large aperture, and the magnification should provide an exit pupil of at least 5 mm diameter.
5. Beginner's work in this field should concentrate on easy, i.e., bright objects.
6. Observations should be restricted to objects that are sufficiently high above the horizon.
7. If some visitors or laymen wish to see something, demonstrate to them only the brightest objects of this kind to avoid their disappointment.

Those who have some experience in the observation of these faint objects should also try to make drawings of such nebulae. Although nowadays such drawings have little scientific value, they provide excellent training in telescopic viewing.

Three bright galactic nebulae, observable with the equipment of the amateur astronomer, are listed in Table 21.3. Although rather faint, the Crab Nebula is interesting as the probable remnant of a supernova explosion about a thousand years ago.

Table 21.3. Some Bright Galactic Nebulae

Name	Messier No.	Total brightness	Diameter	Constellation
Orion Nebula	M 42	2^m	30′	Orion
"Lagoon" Nebula NGC 6523	M 8	6	10	Sagittarius
"Crab" Nebula	M 1	8	4	Taurus

21.3.4. The Dark Nebulae

Where there is dustlike interstellar matter, which is not illuminated by stars, it appears, if it is sufficiently dense, as a dark cloud. These clouds absorb the light of the stars or of bright nebular clouds behind them. They appear to us as rather empty star fields and dark holes in the sky. The larger ones can be observed with the naked eye, if we follow the course of the Milky Way. Observing within bright star clouds with a telescope we find again and again such dark regions that come about through absorption of light by these clouds of dustlike material. For the telescopic observation of these objects we require powerful binoculars or a comet finder.

The main obstacle to the observation of the dark clouds is that no list of these objects that are suitable for visual observation exists. The best procedure therefore is for the observer, with a suitable instrument, to wander about over the surface of the Milky Way. Particularly rewarding are the constellations that make up the Milky Way between Cassiopeia and Sagittarius. Finally, we may mention that some of these dark clouds are in front of not a star field but a bright galactic nebula. These can be seen with particular ease. An example is the bay in the large Orion Nebula, which is caused by a dark cloud.

21.3.5. The Planetary Nebulae

The designation of these objects is misleading. They have nothing to do with either the planets or the planetary system. They owe their name only to their appearance (particularly in small instruments), which resembles in size the faint, but not sharply defined, disc of a planet.

Planetary nebulae are gaseous clouds that surround a faint central star in a more or less spherical form. They have probably originated in a nova-like outburst of the star. The central star is usually very hot, and by its powerful radiation excites the light of the nebula. In some cases the central star is surrounded not by a cloud in the form of a uniform sphere, but rather by a relatively thin spherical shell of gas; the nebula then appears to us as a ring surrounding the central star. The most beautiful example of this kind, perhaps the most beautiful planetary nebula of all, is the Ring Nebula (M 57) in the Constellation Lyra.

The observer of planetary nebulae must take into account that these objects are usually very small, and that with a low magnification they can hardly be distinguished from stars. He will therefore, if possible, use a magnification that is near the upper limit of the performance of the instrument. This of course makes the finding of these objects more difficult, but careful observing using the methods described earlier will lead to good results. As an indication we may mention that most planetary nebulae demand a magnification of at least 100×. Photographic observation of planetary nebulae is possible only with advanced equipment.

The best-known and brightest planetary nebula is the Ring Nebula in Lyra, mentioned above. Most other objects of this type are rather faint and require fairly large optical equipment for a satisfactory view.

21.3.6. Galaxies

Objects outside our own Galaxy are called "extragalactic." Galaxies, which are really great star systems like our own, are scattered throughout the whole Universe. Sometimes, however, because they appear in the amateur's telescope as faint, nebulous patches, they are called extragalactic nebulae.

The nearest galaxies, which are about 180,000 light years distant and are the two companion galaxies to our own Milky Way system, are known as the Magellanic Clouds. They are readily visible to the naked eye, but only from southern latitudes. The next closest of the large galaxies is the spiral

Fig. 21-3 A section of the northwestern part of the cluster of galaxies in the constellations Virgo and Coma. We recognize a number of spiral nebulae. This photograph contains among others the three objects M 98, M 99, and M 100. From HANS VEHRENBERG: Mein Messier-Buch. Düsseldorf: Treugesell-Verlag, 1966.

nebula in the constellation Andromeda. In a clear sky it can also be seen with the naked eye and, at a distance of about 2 million light years, is the most distant object we can see without optical aid.

The same remarks as for the observation of galactic nebulae also apply to the extragalactic objects. In most cases they appear in our instrument as diffuse nebular spots and only rarely can we distinguish details of their structure. On the other hand, good photographs of the brighter objects can be taken with a simple astrograph. See Fig. 21–3. A focal length of at least 30 cm, however, is necessary here. The exposure time should be at least one hour, and the photographic material must of course be of the highest sensitivity.

The observation of faint extragalactic objects is a domain of large and powerful instruments. Those who try their luck with lesser means will suffer many disappointments.

The amateur astronomer will find that the extragalactic nebulae listed in Table 21.4 are his best subjects.

Table 21.4. The best Extragalactic Nebulae for Observing

Name	Messier No.	Brightness	Diameter	Constellation
Andromeda Nebula	M 31	4^m	160′	Andromeda
NGC 598	M 33	7	60	Triangle
NGC 5194	M 51	8	12	Canes Venatici
Large Magellanic Cloud	—	1	500	Doradus and Mensa
Small Magellanic Cloud	—	2	200	Tucan

22 / Appendix

General Tables

The Greek Alphabet

A, α	Alpha	H, η	Eta	N, ν	Nu	T, τ	Tau
B, β	Beta	Θ, θ	Theta	Ξ, ξ	Xi	Y, υ	Upsilon
Γ, γ	Gamma	I, i	Iota	O, o	Omicron	Φ, φ	Phi
Δ, δ	Delta	K, κ	Kappa	Π, π	Pi	X, χ	Chi
E, ϵ	Epsilon	Λ, λ	Lambda	P, ρ	Rho	Ψ, ψ	Psi
Z, ζ	Zeta	M, μ	Mu	Σ, σ	Sigma	Ω, ω	Omega

Astronomical and Mathematical-Physical Abbreviations and Symbols

Å	Ångström unit
α	right ascension
A.U.	astronomical unit (distance Sun–Earth)
A	azimuth
a	semimajor axis of orbit, mean distance
b, B, β	latitude
c	velocity of light
Δ	distance from the Earth
δ	declination
E	east
E	eccentric anomaly
PM (μ)	proper motion
e	eccentricity
e	base of natural logarithms
ϵ	obliquity of ecliptic
FI, CI	color index
f	focal length
G (k^2)	constant of gravitation
g	acceleration due to gravity at the Earth's surface
h	altitude
i	inclination
k	Gauss' constant

kpc	kiloparsec
Ly	light year
l, L, λ	longitude
log	Briggs' logarithm (base 10)
ln, (log)	natural logarithm (base e)
λ	wavelength
M	mean anomaly
M	absolute magnitude
Mpc	megaparsec
m	mass
m	apparent magnitude
m_{pv}	apparent magnitude (photo-visual)
m_{pg}	apparent magnitude (photographic)
μ	mean daily motion
μ	micron
1 nm	10^{-6} mm
ω	distance of perihelion from node
P	period of revolution
p	position angle
pc	parsec
π (ω)	longitude of perihelion
π	parallax

π	ratio of circumference of circle to its diameter		T	absolute temperature
			t	hour angle
φ	polar altitude = geographic latitude		t, τ	time
			U	period of revolution
q	distance of perihelion		u	argument of latitude
R, r	distance from the Sun		v	true anomaly
r	radius		X, Y, Z	rectangular coordinates of the Sun in the equatorial system
ρ	Earth's radius			
ρ (d)	distance of a double star (separation of components)		x, y, z	heliocentric rectangular coordinates
ρ	density		ξ, η, ζ	geocentric rectangular coordinates
s	distance			
$s, m, h,$	(as indices) second, minute,		z	zenith distance
d, a	hour, day, year		ZM	central meridian
T	time of perihelion passage			

Signs of the Zodiac and Symbols for the Planets

		Degrees		
			☉	Sun
♈	Aries	0	☾	Moon
♉	Taurus	30	☿	Mercury
♊	Gemini	60	♀	Venus
♋	Cancer	90	♁	Earth
♌	Leo	120	♂	Mars
♍	Virgo	150	♃	Jupiter
♎	Libra	180	♄	Saturn
♏	Scorpius	210	♅	Uranus
♐	Sagittarius	240	♆	Neptune
♑	Capricorn	270		
♒	Aquarius	300		
♓	Pisces	330		

Beginning of the signs in degrees of ecliptic longitude.

Constellations		Phases of the Moon	
☌	Conjunction	●	New Moon
☐	Quadrature	☽	First Quarter
☍	Opposition	○	Full Moon
		☾	Last Quarter
		☊	Ascending Node
		☋	Descending Node

Table 1. Evening and morning elongations valid for a latitude of 54° North

δ	A_3	δ	A_3	δ	A_n	δ	A_n
		−20°	34°1	0°	0°8	+20°	36°0
		19	32.2	+1	2.5	21	38.0
−37°5	90°0	18	30.4	2	4.1	22	40.0
37	86.1	17	28.5	3	5.8	23	42.1
36	76.9	16	26.7	4	7.5		
35	71.9	15	24.9	5	9.2	24	44.2
34	67.8	14	23.1	6	10.9	25	46.4
33	64.4	13	21.4	7	12.6	26	48.6
32	61.3	12	19.6	8	14.3	27	50.9
31	58.4	11	17.9	9	16.0	28	53.4
−30	55.8	−10	16.2	+10	17.8	29	55.9
29	53.2	9	14.4	11	19.5	+30	58.6
28	50.8	8	12.7	12	21.3	31	61.5
27	48.5	11.0	13	13	23.1	32	64.7
26	46.3	6	9.3	14	24.8	33	68.3
25	44.2	5	7.6	15	26.6		
24	42.1	4	5.9	16	28.5	34	72.4
23	40.0	3	4.3	17	30.3	35	77.8
22	38.0	2	2.6	18	32.2	+35°56′6	90.0
21	36.1	−1	0.9	19	34.1		

Table 2. Air mass $M(z)$ and reduction for visual magnitude classes (ΔM_{vis}) as functions of the apparent zenith distance (z)

z	M(z)	Δm_{vis}
0°	1.000	0.00
10	1.015	0.00
20	1.064	0.01
30	1.154	0.03
40	1.304	0.06
50	1.553	0.12
60	1.995	0.23
70	2.904	0.45
75	3.816	0.65
80	5.60	0.99
85	10.40	1.77
87	15.36	2.61

Table 3. Mean refraction according to Bessel's Table (760 mm, 0°)

z obs	R	z obs	R
0°	0′	70°	2′45″
10°	0′11″	75°	3′42″
20°	0′22″	80°	5′31″
30°	0′35″	85°	10′15″
40°	0′51″	88°	19′7″
50°	1′11″	89°	25′36″
60°	1′45″	90°	36′38″

Table 4. Semidiurnal arc

(From Berliner Astronomisches Jahrbuch, 1957, pp. 406–407.)

δ \ φ	+30°	+32°	+34°	+36°	+38°	+40°	+42°	+44°	+46°	+48°	+50°
°	h m	h m	h m	h m	h m	h m	h m	h m	h m	h m	h m
− 30	4 45.4	4 38.8	4 31.8	4 24.4	4 16.5	4 8.1	3 58.9	3 48.9	3 37.9	3 25.7	3 11.8
29	4 48.6	4 42.3	4 35.6	4 28.6	4 21.1	4 13.0	4 4.3	3 54.9	3 44.5	3 33.0	3 20.1
28	4 51.7	4 45.7	4 39.3	4 32.6	4 32.6	4 17.8	4 9.6	4 0.7	3 50.9	3 40.1	3 28.0
27	4 54.7	4 49.0	4 42.9	4 36.5	4 29.8	4 22.5	4 14.7	4 6.2	3 57.0	3 46.9	3 35.5
26	4 57.7	4 52.2	4 46.5	4 40.4	4 33.9	4 27.1	4 19.7	4 11.7	4 3.0	3 53.4	3 42.8
25	5 0.6	4 55.4	4 49.9	4 44.2	4 38.0	4 31.5	4 24.5	4 16.9	4 8.7	3 59.7	3 49.7
24	5 3.5	4 58.5	4 53.3	4 47.8	4 42.0	4 35.8	4 29.2	4 22.0	4 14.3	4 5.8	3 56.5
23	5 6.3	5 1.6	4 56.6	4 51.4	4 45.9	4 40.1	4 33.8	4 27.0	4 19.7	4 11.8	4 3.0
22	5 9.0	5 4.6	4 59.9	4 55.0	4 49.7	4 44.2	4 38.3	4 31.9	4 25.0	4 17.5	4 9.3
21	5 11.7	5 7.5	5 3.1	4 58.4	4 53.5	4 48.3	4 42.7	4 36.7	4 30.2	4 23.2	4 15.4
− 20	5 14.4	5 10.4	5 6.2	5 1.8	4 57.2	4 52.3	4 47.0	4 41.3	4 35.3	4 28.7	4 21.4
19	5 17.0	5 13.3	5 9.3	5 5.2	5 0.8	4 56.2	4 51.2	4 45.9	4 40.2	4 34.0	4 27.3
18	5 19.6	5 16.1	5 12.4	5 8.5	5 4.4	5 0.0	4 55.4	4 50.4	4 45.1	4 39.3	4 33.0
17	5 22.2	5 18.9	5 15.4	5 11.7	5 7.9	5 3.8	4 59.9	4 54.9	4 49.9	4 44.5	4 38.6
16	5 24.7	5 21.6	5 18.4	5 14.9	5 11.4	5 7.5	5 3.5	4 59.2	4 54.6	4 49.5	4 44.1
15	5 27.2	5 24.3	5 21.3	5 18.1	5 14.8	5 11.2	5 7.5	5 3.5	4 59.2	4 54.5	4 49.5
14	5 29.7	5 27.0	5 24.2	5 21.3	5 18.2	5 14.9	5 11.4	5 7.7	5 3.7	4 59.5	4 54.8
13	5 32.1	5 29.7	5 27.1	5 24.4	5 21.5	5 18.5	5 15.3	5 11.9	5 8.2	5 4.3	5 0.0
12	5 34.6	5 32.3	5 29.9	5 27.4	5 24.8	5 22.1	5 19.1	5 16.0	5 12.6	5 9.0	5 5.1
11	5 37.0	5 34.9	5 32.7	5 30.5	5 28.1	5 25.6	5 22.9	5 20.1	5 17.0	5 13.7	5 10.2
− 10	5 39.4	5 37.5	5 35.5	5 33.5	5 31.3	5 29.1	5 26.7	5 24.1	5 21.4	5 18.4	5 15.2
9	5 41.7	5 40.1	5 38.3	5 36.5	5 34.6	5 32.5	5 30.4	5 28.1	5 25.7	5 23.0	5 20.2
8	5 44.1	5 42.6	5 41.1	5 39.5	5 37.8	5 36.0	5 34.1	5 32.1	5 29.9	5 27.6	5 25.1
7	5 46.4	5 45.2	5 43.8	5 42.4	5 41.0	5 39.4	5 37.8	5 36.0	5 34.2	5 32.2	5 30.0
6	5 48.8	5 47.7	5 46.6	5 45.4	5 44.1	5 42.8	5 41.4	5 40.0	5 38.4	5 36.7	5 34.9
5	5 51.1	5 50.2	5 49.3	5 48.3	5 47.3	5 46.2	5 45.1	5 43.9	5 42.6	5 41.2	5 39.7
4	5 53.4	5 52.7	5 52.0	5 51.2	5 50.4	5 49.6	5 48.7	5 47.8	5 46.8	5 45.7	5 44.5
3	5 55.8	5 55.2	5 54.7	5 54.1	5 53.6	5 53.0	5 52.3	5 51.6	5 50.9	5 50.1	5 49.3
2	5 58.1	5 57.7	5 57.4	5 57.1	5 56.7	5 56.3	5 55.9	5 55.5	5 55.1	5 54.6	5 54.1
− 1	6 0.4	6 0.2	6 0.1	6 0.0	5 59.8	5 59.7	5 59.5	5 59.4	5 59.2	5 59.0	5 58.9
0	6 2.7	6 2.7	6 2.8	6 2.9	6 2.9	6 3.0	6 3.1	6 3.2	6 3.4	6 3.5	6 3.6
+ 1	6 5.0	6 5.2	6 5.5	6 5.8	6 6.1	6 6.4	6 6.7	6 7.1	6 7.5	6 7.9	6 8.4
2	6 7.3	6 7.7	6 8.2	6 8.7	6 9.2	6 9.8	6 10.3	6 11.0	6 11.6	6 12.4	6 13.2
3	6 9.6	6 10.3	6 10.9	6 11.6	6 12.3	6 13.1	6 14.0	6 14.8	6 16.8	6 16.8	6 18.0
4	6 11.9	6 12.8	6 13.6	6 14.5	6 15.5	6 16.5	6 17.6	6 18.7	6 20.0	6 21.3	6 22.8
5	6 14.3	6 15.3	6 14.6	6 17.5	6 18.6	6 19.9	6 21.2	6 22.6	6 24.2	6 25.8	6 27.6
6	6 16.6	6 17.8	6 19.1	6 20.4	6 21.8	6 23.3	6 24.9	6 26.6	6 28.4	6 30.4	6 32.5
7	6 19.0	6 20.4	6 21.8	6 23.4	6 25.0	6 26.7	6 28.6	6 30.5	6 32.6	6 34.9	6 37.4
8	6 21.3	6 22.9	6 24.4	6 26.4	6 28.2	6 30.2	6 32.3	6 34.5	6 36.9	6 39.5	6 42.3
9	6 23.7	6 25.5	6 27.4	6 29.4	6 31.4	6 33.7	6 36.0	6 38.5	6 41.2	6 44.1	6 47.3
10	6 26.1	6 28.1	6 30.2	6 32.4	6 34.7	6 37.2	6 39.8	6 42.5	6 45.6	6 48.8	6 52.3
+ 11	6 28.5	6 30.7	6 33.0	6 35.4	6 38.0	6 40.7	6 43.6	6 46.6	6 49.9	6 53.5	6 57.4
12	6 31.0	6 33.4	6 35.9	6 38.5	6 41.3	6 44.3	6 47.4	6 50.8	6 54.4	6 58.3	7 2.5
13	6 33.4	6 36.0	6 38.8	6 41.6	6 44.7	6 47.9	6 51.3	6 54.9	6 58.9	7 3.1	7 7.8
14	6 35.9	6 38.7	6 41.7	6 44.8	6 48.0	6 51.5	6 55.2	6 59.2	7 3.4	7 8.0	7 13.1
15	6 38.4	6 41.4	6 44.6	6 47.9	6 51.5	6 55.2	6 59.2	7 3.5	7 8.1	7 13.0	7 18.5
16	6 41.0	6 44.2	6 47.6	6 51.2	6 54.9	6 58.9	7 3.2	7 7.8	7 12.7	7 18.1	7 23.9
17	6 43.5	6 47.0	6 50.6	6 54.4	6 58.5	7 2.7	7 7.3	7 12.2	7 17.5	7 23.3	7 29.5
18	6 46.1	6 49.8	6 53.7	6 57.7	7 2.0	7 6.6	7 11.5	7 16.7	7 22.4	7 28.5	7 35.3
19	6 48.8	6 52.7	6 56.8	7 1.1	7 5.7	7 10.5	7 15.7	7 21.3	7 27.4	7 33.9	7 41.1
20	6 51.5	6 55.6	6 59.9	7 4.5	7 9.4	7 14.5	7 20.1	7 26.0	7 32.4	7 39.4	7 47.1
+ 21	6 54.2	6 58.6	7 3.1	7 8.0	7 13.1	7 18.6	7 24.5	7 30.8	7 37.6	7 45.1	7 53.3
22	6 56.9	7 1.6	7 6.4	7 11.5	7 17.0	7 22.8	7 29.0	7 35.7	7 42.9	7 50.9	7 59.6
23	6 59.8	7 4.6	7 9.7	7 15.1	7 20.9	7 27.0	7 33.6	7 40.7	7 48.4	7 56.8	8 6.1
24	7 2.6	7 7.7	7 13.1	7 18.8	7 24.9	7 31.3	7 38.3	7 45.8	7 54.0	8 2.9	8 12.9
25	7 5.6	7 10.9	7 16.6	7 22.6	7 29.0	7 35.8	7 43.1	7 51.1	7 59.8	8 9.3	8 19.9
26	7 8.5	7 14.2	7 20.1	7 26.4	7 33.2	7 40.4	7 48.1	7 56.5	8 5.7	8 15.8	8 27.1
27	7 11.6	7 17.5	7 23.8	7 30.4	7 37.5	7 45.0	7 53.2	8 2.1	8 11.8	8 22.6	8 34.7
28	7 14.7	7 20.9	7 27.5	7 34.4	7 41.9	7 49.9	7 58.5	8 7.9	8 18.2	8 29.7	8 42.6
29	7 17.9	7 24.4	7 31.3	7 38.6	7 46.4	7 54.8	8 3.9	8 13.9	8 24.8	8 37.1	8 51.0
+ 30	7 21.2	7 28.0	7 35.2	7 42.9	7 51.1	7 59.9	8 9.5	8 20.1	8 31.7	8 44.8	8 59.7

APPENDIX

Table 4. *Continued*

δ \ φ	+50°	+51°	+52°	+53°	+54°	+55°	+56°	+57°	+58°	+59°	+60°
°	h m	h m	h m	h m	h m	h m	h m	h m	h m	h m	h m
-- 30	3 11.8	3 4.1	2 55.8	2 46.8	2 36.9	2 25.9	2 13.5	1 59.3	1 42.4	1 21.0	0 49.7
29	3 20.1	3 12.9	3 5.3	2 57.0	2 48.0	2 38.1	2 27.1	2 14.7	2 0.4	1 43.4	1 21.9
28	3 28.0	3 21.3	3 14.2	3 6.6	2 58.3	2 49.3	2 39.4	2 28.4	2 15.9	2 1.6	1 44.5
27	3 35.5	3 29.3	3 22.7	3 15.7	3 8.0	2 59.8	2 50.8	2 40.8	2 29.8	2 17.3	2 2.9
26	3 42.8	3 37.0	3 30.8	3 24.2	3 17.2	3 9.6	3 1.4	2 52.4	2 42.4	2 31.3	2 18.8
25	3 49.7	3 44.3	3 38.6	3 32.4	3 25.9	3 18.9	3 11.3	3 3.1	2 54.1	2 44.1	2 33.0
24	3 56.5	3 51.4	3 46.0	3 40.3	3 34.3	3 27.8	3 20.8	3 13.2	3 5.0	2 56.0	2 46.0
23	4 3.0	3 58.2	3 53.2	3 47.9	3 42.3	3 36.2	3 29.8	3 22.8	3 15.3	3 7.1	2 58.0
22	4 9.3	4 4.9	4 0.2	3 55.2	3 50.0	3 44.3	3 38.4	3 31.9	3 25.0	3 17.5	3 9.3
21	4 15.4	4 11.3	4 6.9	4 2.3	3 57.4	3 52.2	3 46.6	3 40.7	3 34.3	3 27.4	3 19.9
− 20	4 21.4	4 17.5	4 13.5	4 9.1	4 4.6	3 59.8	3 54.6	3 49.1	3 43.2	3 36.9	3 30.0
19	4 27.3	4 23.7	4 19.9	4 15.8	4 11.6	4 7.1	4 2.3	3 57.2	3 51.8	3 45.9	3 39.6
18	4 33.0	4 29.6	4 26.1	4 22.3	4 18.4	4 14.2	4 9.8	4 5.1	4 0.1	3 54.7	3 48.9
17	4 38.6	4 35.4	4 32.1	4 28.7	4 25.0	4 21.1	4 17.0	4 12.7	4 8.1	4 3.1	3 57.8
16	4 44.1	4 41.2	4 38.1	4 34.9	4 31.5	4 27.9	4 24.1	4 20.1·	4 15.9	4 11.3	4 6.4
15	4 49.5	4 46.8	4 43.9	4 41.0	4 37.8	4 34.5	4 31.0	4 27.4	4 23.4	4 19.3	4 14.8
14	4 54.8	4 52.3	4 49.7	4 46.9	4 44.1	4 41.0	4 37.8	4 34.4	4 30.8	4 27.0	4 22.9
13	5 0.0	4 57.7	4 55.3	4 52.8	4 50.2	4 47.4	4 44.5	4 41.4	4 38.1	4 34.6	4 30.9
12	5 5.1	5 3.0	5 0.9	4 58.3	4 56.2	4 53.7	4 51.0	4 48.2	4 45.2	4 42.0	4 38.7
11	5 10.2	5 8.3	5 6.4	5 4.3	5 2.1	4 59.8	4 57.4	4 54.9	4 52.2	4 49.3	4 46.3
− 10	5 15.2	5 13.5	5 11.8	5 9.9	5 7.9	5 5.9	5 3.7	5 1.5	4 59.1	4 56.5	4 53.8
9	5 20.2	5 18.7	5 17.1	5 15.2	5 13.7	5 11.9	5 10.0	5 8.0	5 5.8	5 3.6	5 1.2
8	5 25.1	5 23.8	5 22.4	5 21.0	5 19.5	5 17.9	5 16.2	5 14.4	5 12.5	5 10.6	5 8.5
7	5 30.0	5 28.9	5 27.7	5 26.4	5 25.1	5 23.8	5 22.3	5 20.8	5 19.2	5 17.5	5 15.7
6	5 34.9	5 33.9	5 32.9	5 31.8	5 30.7	5 29.6	5 28.4	5 27.1	5 25.7	5 24.3	5 22.8
5	5 39.7	5 38.9	5 38.1	5 37.2	5 36.1	5 35.4	5 34.4	5 33.4	5 32.2	5 31.1	5 29.9
4	5 44.5	5 43.9	5 43.3	5 42.6	5 41.9	5 41.2	5 40.4	5 39.6	5 38.7	5 37.8	5 36.9
3	5 49.3	5 48.9	5 48.4	5 47.9	5 47.4	5 46.9	5 46.3	5 45.8	5 45.2	5 44.5	5 43.8
2	5 54.1	5 53.8	5 53.5	5 53.3	5 52.9	5 52.6	5 52.3	5 52.0	5 51.6	5 51.2	5 50.8
− 1	5 58.9	5 58.8	5 58.7	5 58.6	5 58.4	5 58.3	5 58.2	5 58.1	5 58.0	5 57.9	5 57.7
0	6 3.6	6 3.7	6 3.9	6 4.0	6 4.1	6 4.1	6 4.2	6 4.3	6 4.4	6 4.5	6 4.7
+ 1	6 8.4	6 8.6	6 8.9	6 9.2	6 9.5	6 9.8	6 10.1	6 10.4	6 10.8	6 11.2	6 11.6
2	6 13.2	6 13.6	6 14.0	6 14.5	6 15.0	6 15.5	6 16.0	6 16.6	6 17.2	6 17.8	6 18.5
3	6 18.0	6 18.6	6 19.2	6 19.8	6 20.5	6 21.2	6 22.0	6 22.8	6 23.6	6 24.6	6 25.5
4	6 22.8	6 23.5	6 24.4	6 25.2	6 26.1	6 27.0	6 28.0	6 29.0	6 30.1	6 31.3	6 32.5
5	6 27.6	6 28.6	6 29.6	6 30.6	6 31.7	6 32.8	6 34.0	6 35.3	6 36.6	6 38.1	6 39.6
6	6 32.5	6 33.6	6 34.8	6 36.0	6 37.3	6 38.7	6 40.1	6 41.6	6 43.2	6 44.9	6 46.7
7	6 37.4	6 38.7	6 40.0	6 41.5	6 43.0	6 44.6	6 46.2	6 48.0	6 49.8	6 51.8	6 53.9
8	6 42.3	6 43.8	6 45.3	6 47.0	6 48.7	6 50.5	6 52.4	6 54.4	6 56.5	6 58.8	7 1.2
9	6 47.3	6 48.9	6 50.7	6 52.5	6 54.5	6 56.5	6 58.7	7 0.9	7 3.3	7 5.9	7 8.6
10	6 52.3	6 54.1	6 56.1	6 58.2	7 0.3	7 2.6	7 5.0	7 7.5	7 10.2	7 13.1	7 16.2
+ 11	6 57.4	6 59.4	7 1.6	7 3.9	7 6.3	7 8.8	7 11.4	7 14.2	7 17.2	7 20.4	7 23.8
12	7 2.5	7 4.8	7 7.2	7 9.7	7 12.3	7 15.1	7 18.0	7 21.1	7 24.3	7 27.8	7 31.5
13	7 7.8	7 10.2	7 12.8	7 15.5	7 18.4	7 21.4	7 24.6	7 28.0	7 31.4	7 35.4	7 39.5
14	7 13.1	7 15.7	7 18.6	7 21.5	7 24.6	7 27.9	7 31.4	7 35.1	7 39.0	7 43.2	7 47.7
15	7 18.5	7 21.4	7 24.4	7 27.6	7 31.0	7 34.6	7 38.3	7 42.4	7 46.3	7 51.2	7 56.1
16	7 23.9	7 27.1	7 30.4	7 33.8	7 37.5	7 41.4	7 45.4	7 49.8	7 54.4	7 59.4	8 4.7
17	7 29.5	7 32.9	7 36.5	7 40.2	7 44.1	7 48.3	7 52.7	7 57.4	8 2.5	8 7.9	8 13.7
18	7 35.3	7 38.9	7 42.7	7 46.7	7 50.9	7 55.4	8 0.2	8 5.3	8 10.8	8 16.6	8 23.0
19	7 41.1	7 45.0	7 49.1	7 53.4	7 57.9	8 2.8	8 7.9	8 13.4	8 19.4	8 25.7	8 32.6
20	7 47.1	7 51.3	7 55.6	8 0.3	8 5.2	8 10.4	8 15.9	8 21.9	8 28.3	8 35.2	8 42.8
+ 21	7 53.3	7 57.7	8 2.4	8 7.3	8 12.6	8 18.2	8 24.2	8 30.7	8 37.6	8 45.2	8 53.5
22	7 59.6	8 4.3	8 9.4	8 14.7	8 20.3	8 26.4	8 32.8	8 39.8	8 47.4	8 55.7	9 4.8
23	8 6.1	8 11.2	8 16.6	8 22.3	8 28.3	8 34.9	8 41.9	8 49.5	8 57.7	9 6.8	9 16.9
24	8 12.9	8 18.3	8 24.0	8 30.2	8 36.7	8 43.8	8 51.4	8 59.6	9 8.7	9 18.8	9 30.0
25	8 19.9	8 25.7	8 31.8	8 38.4	8 45.5	8 53.1	9 1.4	9 10.5	9 20.5	9 31.7	9 44.4
26	8 27.1	8 33.4	8 40.0	8 47.0	8 54.7	9 3.0	9 12.1	9 22.1	9 33.2	9 45.9	10 0.6
27	8 34.7	8 41.4	8 48.5	8 56.1	9 4.4	9 13.5	9 23.5	9 34.6	9 47.3	10 1.9	10 19.5
28	8 42.6	8 49.8	8 57.5	9 5.8	9 14.8	9 24.8	9 35.9	9 48.5	10 3.1	10 20.5	10 42.9
29	8 51.0	8 58.7	9 7.0	9 16.1	9 26.0	9 37.1	9 49.6	10 4.1	10 21.5	10 43.7	11 18.1
30	8 59.7	9 8.1	9 17.2	9 27.1	9 38.2	9 50.7	10 5.1	10 22.3	10 44.4	11 18.5	— —

Table 5. Conversion of Mean Time to Sidereal Time

(From Berliner Astronomisches Jahrbuch, 1957, p. 396.)

Red. s	0ᵐ (h m s)	1ᵐ (h m s)	2ᵐ (h m s)	3ᵐ (h m s)	Red. s
0	0 0 0.0	6 5 14.5	12 10 29.1	18 15 43.6	0
1	6 5.2	11 19.8	16 34.3	21 48.8	1
2	12 10.5	17 25.0	22 39.6	27 54.1	2
3	18 15.7	23 30.3	28 44.8	33 59.3	3
4	24 21.0	29 35.5	34 50.0	40 4.6	4
5	30 26.2	35 40.7	40 55.3	46 9.8	5
6	36 31.5	41 46.0	47 0.5	52 15.1	6
7	42 36.7	47 51.2	53 5.8	18 58 20.3	7
8	48 41.9	6 53 56.5	12 59 11.0	19 4 25.5	8
9	0 54 47.2	7 0 1.7	13 5 16.2	10 30.8	9
10	1 0 52.4	6 7.0	11 21.5	16 36.0	10
11	6 57.7	12 12.2	17 26.7	22 41.3	11
12	13 2.9	18 17.4	23 32.0	28 46.5	12
13	19 8.1	24 22.7	29 37.2	34 51.8	13
14	25 13.4	30 27.9	35 42.5	40 57.0	14
15	31 18.6	36 33.2	41 47.7	47 2.2	15
16	37 23.9	42 38.4	47 52.9	53 7.5	16
17	43 29.1	48 43.7	13 53 58.2	19 59 12.7	17
18	49 34.4	7 54 48.9	14 0 3.4	20 5 18.0	18
19	1 55 39.6	8 0 54.1	6 8.7	11 23.2	19
20	2 1 44.8	6 59.4	12 13.9	17 28.4	20
21	7 50.1	13 4.6	18 19.2	23 33.7	21
22	13 55.3	19 9.9	24 24.4	29 38.9	22
23	20 0.6	25 15.1	30 29.6	35 44.2	23
24	26 5.8	31 20.3	36 34.9	41 49.4	24
25	32 11.1	37 25.6	42 40.1	47 54.7	25
26	38 16.3	43 30.8	48 45.4	20 53 59.9	26
27	44 21.5	49 36.1	14 54 50.6	21 0 5.1	27
28	50 26.8	8 55 41.3	15 0 55.9	6 10.4	28
29	2 56 32.0	9 1 46.6	7 1.1	12 15.6	29
30	3 2 37.3	7 51.8	13 6.3	18 20.9	30
31	8 42.5	13 57.0	19 11.6	24 26.1	31
32	14 47.8	20 2.3	25 16.8	30 31.4	32
33	20 53.0	26 7.5	31 22.1	36 36.6	33
34	26 58.2	32 12.8	37 27.3	42 41.8	34
35	33 3.5	38 18.0	43 32.5	48 47.1	35
36	39 8.7	44 23.3	49 37.8	21 54 52.3	36
37	45 14.0	50 28.5	15 55 43.0	22 0 57.6	37
38	51 19.2	9 56 33.7	16 1 48.3	7 2.8	38
39	3 57 24.4	10 2 39.0	7 53.5	13 8.0	39
40	4 3 29.7	8 44.2	13 58.8	19 13.3	40
41	9 34.9	14 49.5	20 4.0	25 18.5	41
42	15 40.2	20 54.7	26 9.2	31 23.8	42
43	21 45.4	27 0.0	32 14.5	37 29.0	43
44	27 50.7	33 5.2	38 19.7	43 34.3	44
45	33 55.9	39 10.4	44 25.0	49 39.5	45
46	40 1.1	45 15.7	50 30.2	22 55 44.7	46
47	46 6.4	51 20.9	16 56 35.5	23 1 50.0	47
48	52 11.6	10 57 26.2	17 2 40.7	7 55.2	48
49	4 58 16.9	11 3 31.4	8 45.9	14 0.5	49
50	5 4 22.1	9 36.6	14 51.2	20 5.7	50
51	10 27.4	15 41.9	20 56.4	26 11.0	51
52	16 32.6	21 47.1	27 1.7	32 16.2	52
53	22 37.8	27 52.4	33 6.9	38 21.4	53
54	28 43.1	33 57.6	39 12.1	44 26.7	54
55	34 48.3	40 2.9	45 17.4	50 31.9	55
56	40 53.6	46 8.1	51 22.6	23 56 37.2	56
57	46 58.8	52 13.3	17 57 27.9	24 2 42.4	57
58	53 4.0	11 58 18.6	18 3 33.1	8 47.7	58
59	5 59 9.3	12 4 23.8	18 9 38.4	24 14 52.9	59

Red. s	m s	Red. s	m s
0.00	0 0.0	0.50	3 2.6
01	3.7	51	6.3
02	7.3	52	9.9
03	11.0	53	13.6
04	14.6	54	17.2
0.05	18.3	0.55	20.9
06	21.9	56	24.5
07	25.6	57	28.2
08	29.2	58	31.8
09	32.9	59	35.5
0.10	36.5	0.60	39.1
11	40.2	61	42.8
12	43.8	62	46.5
13	47.5	63	50.1
14	51.1	64	53.8
0.15	54.8	0.65	3 57.4
16	0 58.4	66	4 1.1
17	1 2.1	67	4.7
18	5.7	68	8.4
19	9.4	69	12.0
0.20	13.0	0.70	15.7
21	16.7	71	19.3
22	20.4	72	23.0
23	24.0	73	26.6
24	27.7	74	30.3
0.25	31.3	0.75	33.9
26	35.0	76	37.6
27	38.6	77	41.2
28	42.3	78	44.9
29	45.9	79	48.5
0.30	49.6	0.80	52.2
31	53.2	81	55.8
32	1 56.9	82	4 59.5
33	2 0.5	83	5 3.2
34	4.2	84	6.8
0.35	7.8	0.85	10.5
36	11.5	86	14.1
37	15.1	87	17.8
38	18.8	88	21.4
39	22.4	89	25.1
0.40	26.1	0.90	28.7
41	29.7	91	32.4
42	33.4	92	36.0
43	37.1	93	39.7
44	40.7	94	43.3
0.45	44.4	0.95	47.0
46	48.0	96	50.6
47	51.7	97	54.3
48	55.3	98	5 57.9
0.49	2 59.0	0.99	6 1.6

The reduction (Red.) must be added to the Mean Time.

Red. s	s	Red. s	s	Red. s	s
0.000		0.003		0.006	
	0.2		1.3		2.4
001		004		007	
	0.5		1.6		2.7
002		005		008	
	0.9		2.0		3.1
003		006		009	
	1.3		2.4		3.5
0.004		0.007		0.010	

Table 6. Conversion of Sidereal Time into Mean Time

(From Berliner Astronomisches Jahrbuch, 1957, p. 397.)

Red.	0ᵐ	1ᵐ	2ᵐ	3ᵐ	Red.
s	h m s	h m s	h m s	h m s	s
0	0 0 0.0	6 6 14.5	12 12 29.1	18 18 43.6	0
1	6 6.2	12 20.8	18 35.3	24 49.9	1
2	12 12.5	18 27.0	24 41.6	30 56.1	2
3	18 18.7	24 33.3	30 47.8	37 2.3	3
4	24 25.0	30 39.5	36 54.0	43 8.6	4
5	30 31.2	36 45.7	43 0.3	49 14.8	5
6	36 37.5	42 52.0	49 6.5	18 55 21.1	6
7	42 43.7	48 58.2	12 55 12.8	19 1 27.3	7
8	48 49.9	6 55 4.5	13 1 19.0	7 33.5	8
9	0 54 56.2	7 1 10.7	7 25.3	13 39.8	9
10	1 1 2.4	7 17.0	13 31.5	19 46.0	10
11	7 8.7	13 23.2	19 37.7	25 52.3	11
12	13 14.9	19 29.4	25 44.0	31 58.5	12
13	19 21.1	25 35.7	31 50.2	38 4.8	13
14	25 27.4	31 41.9	37 56.5	44 11.0	14
15	31 33.6	37 48.2	44 2.7	50 17.2	15
16	37 39.9	43 54.4	50 8.9	19 56 23.5	16
17	43 46.1	50 0.7	13 56 15.2	20 2 29.7	17
18	49 52.4	7 56 6.9	14 2 21.4	8 36.0	18
19	1 55 58.6	8 2 13.1	8 27.7	14 42.2	19
20	2 2 4.8	8 19.4	14 33.9	20 48.5	20
21	8 11.1	14 25.6	20 40.2	26 54.7	21
22	14 17.3	20 31.9	26 46.4	33 0.9	22
23	20 23.6	26 38.1	32 52.6	39 7.2	23
24	26 29.8	32 44.4	38 58.9	45 13.4	24
25	32 36.1	38 50.6	45 5.1	51 19.7	25
26	38 42.3	44 56.8	51 11.4	20 57 25.9	26
27	44 48.5	51 3.1	14 57 17.6	21 3 32.2	27
28	50 54.8	8 57 9.3	15 3 23.9	9 38.4	28
29	2 57 1.0	9 3 15.6	9 30.1	15 44.6	29
30	3 3 7.3	9 21.8	15 36.3	21 50.9	30
31	9 13.5	15 28.0	21 42.6	27 57.1	31
32	15 19.8	21 34.3	27 48.8	34 3.4	32
33	21 26.0	27 40.5	33 55.1	40 9.6	33
34	27 32.2	33 46.8	40 1.3	46 15.8	34
35	33 38.5	39 63.0	46 7.6	52 22.1	35
36	39 44.7	45 59.3	52 13.8	21 58 28.3	36
37	45 51.0	52 5.5	15 58 20.0	22 4 34.6	37
38	51 57.2	9 58 11.7	16 4 26.3	10 40.8	38
39	3 58 3.4	10 4 18.0	10 32.5	16 47.1	39
40	4 4 9.7	10 24.2	16 38.8	22 53.3	40
41	10 15.9	16 30.5	22 45.0	28 59.5	41
42	16 22.2	22 36.7	28 51.2	35 5.8	42
43	22 28.4	28 43.0	34 57.5	41 12.0	43
44	28 34.7	34 49.2	41 3.7	47 18.3	44
45	34 40.9	40 55.4	47 10.0	53 24.5	45
46	40 47.1	47 1.7	53 16.2	22 59 30.8	46
47	46 53.4	53 7.9	16 59 22.5	23 5 37.0	47
48	52 59.6	10 59 14.2	17 5 28.7	11 43.2	48
49	4 59 5.9	11 5 20.4	11 34.9	17 49.5	49
50	5 5 12.1	11 41.2	17 41.2	23 55.7	50
51	11 18.4	17 32.9	23 47.4	30 2.0	51
52	17 24.6	23 39.1	29 53.7	36 8.2	52
53	23 30.8	29 45.4	35 59.9	42 14.5	53
54	29 37.1	35 51.6	42 6.2	48 20.7	54
55	35 43.3	41 58.9	48 12.4	23 54 26.9	55
56	41 49.6	48 4.1	17 54 18.6	24 0 33.2	56
57	47 55.8	11 54 10.3	18 0 24.9	6 39.4	57
58	5 54 2.1	12 0 16.6	6 31.1	12 45.7	58
59	6 0 8.3	12 6 22.8	18 12 37.4	24 18 51.9	59

Red.		Red.	Red.	
m s	s		m s	m s
0.00	0 0.0		0.50	3 3.1
01	3.7		51	6.8
02	7.3		52	10.4
03	11.0		53	14.1
04	14.6		54	17.8
0.05	18.3		0.55	21.4
06	22.0		56	25.1
07	25.6		57	28.8
08	29.3		58	32.4
09	33.0		59	36.1
0.10	36.6		0.60	39.7
11	40.3		61	43.4
12	43.9		62	47.1
13	47.6		63	50.7
14	51.3		64	54.4
0.15	54.9		0.65	3 58.1
16	0 58.6		66	4 1.7
17	1 2.3		67	5.4
18	5.9		68	9.0
19	9.6		69	12.7
0.20	13.2		0.70	16.4
21	16.9		71	20.0
22	20.6		72	23.7
23	24.2		73	27.4
24	27.9		74	31.0
0.25	31.6		0.75	34.7
26	35.2		76	38.3
27	38.9		77	42.0
28	42.5		78	45.7
29	46.2		79	49.3
0.30	49.9		0.80	53.0
31	53.5		81	4 56.7
32	1 57.2		82	5 0.3
33	2 0.9		83	4.0
34	4.5		84	7.6
0.35	8.2		0.85	11.3
36	11.8		86	15.0
37	15.5		87	18.6
38	19.2		88	22.3
39	22.8		89	26.0
0.40	26.5		0.90	29.6
41	30.2		91	33.3
42	33.8		92	36.9
43	37.5		93	40.6
44	41.1		94	44.3
0.45	44.8		0.95	47.9
46	48.5		96	51.6
47	52.1		97	55.3
48	55.8		98	5 58.9
0.49	2 59.5		0.99	6 2.6

Red.		Red.		Red.	
s	s	s	s	s	s
0.000		0.003		0.006	
	0.2		1.3		2.4
001		004		007	
	0.5		1.6		2.7
002		005		008	
0.9		2.0		3.1	
003		006		009	
	1.3		2.4		3.5
0.004		0.007		0.010	

The reduction (Red.) must be subtracted from the Sidereal Time.

Table 7. Julian calendar from 1000 A.D. to 2000 A.D.

(From Berliner Astronomisches Jahrbuch, 1957, p. 401.)

(a) Number of days elapsed since the beginning of the period on January 0, 12^h UT.

Year A.D.	1000	1100	1200	1300	1400	1500	1600	1700	1800	1900
	20	21	21	21	22	22	23	23	23	24
0	86307	22832	59357	95882	32407	68932	05447	41971[1]	78495[1]	15019[1]
4	87768	24293	60818	97343	33868	70393	06908	43432	79956	16480
8	89229	25754	62279	98804	35329	71854	08369	44893	81417	17941
12	90690	27215	63740	00265	36790	73315	09830	46354	82878	19402
16	92151	28676	65201	01726	38251	74776	11291	47815	84339	20863
20	93612	30137	66662	03187	39712	76237	12752	49276	85800	22324
24	95073	31598	68123	04648	41173	77698	14213	50737	87261	23785
28	96534	33059	69584	06109	42634	79159	15764	52198	88722	25246
32	97995	34520	71045	07570	44095	80620	17135	53659	90183	26707
36	99456	35981	72506	09031	45556	82081	18596	55120	91644	28168
40	00917	37442	73967	10492	47017	83542	20057	56105	93105	29629
44	02378	38903	11953	11953	48578	85i03	21518	58042	94566	31090
48	03839	40364	76889	13414	49939	86464	22979	59503	96027	32551
52	05300	41825	78350	14875	51400	87925	24440	60964	97488	34012
56	06761	43286	79811	16336	52861	89386	25901	62425	98949	35473
60	08222	44747	81272	17797	54322	90847	27362	63886	00410	36934
64	09683	46208	82733	19258	55783	92308	28823	65347	01871	38395
68	11144	47669	84194	20719	57244	93769	30284	66808	03332	39856
72	12605	49130	22180	22180	58705	95230	31745	68269	04793	41317
76	14066	50591	87116	23641	60166	96691	33206	69730	06254	42778
80	15527	52052	88577	25102	61627	98152	34667	71191	07715	44239
84	16988	53513	90038	26563	63088	99603	36128	72652	09176	45700
88	18449	54974	91499	64549	64549	01064	37589	74113	10637	47161
92	19910	56435	92960	29485	66010	02525	39050	75574	12098	48622
96	21371	57896	94421	30946	67471	03986	40511	77035	13559	50083
100	22832	59357	95882	32407	68932	05447	41971[1]	78495[1]	15019[1]	51544[1]
	21	21	21	22	22	23	23	23	24	24

[1] The figures denote the number of days elapsed since the beginning of the period on January −1.

(b) Number of days elapsed since the beginning of the leap year cycle on day 0 of each month (equivalent to the last day of the preceding month).[1]

Year	Jan. 0	Feb. 0	March 0	April 0	May 0	June 0	July 0	Aug. 0	Sept. 0	Oct. 0	Nov. 0	Dec. 0
0	0[2]	31[2]	60	91	121	152	182	213	244	274	305	335
1	366	397	425	456	486	517	547	578	609	693	670	700
2	731	762	790	821	851	882	912	943	974	1004	1035	1065
3	1096	1127	1155	1186	1216	1247	1277	1308	1339	1369	1400	1430

[1] From 15 October 1582 to 31 December 1583 the figures must be decreased by 10.
[2] These figures must be increased by 1 in 1700, 1800, and 1900.

Table 8. Conversion of days, hours, and minutes into decimal fractions of the Julian year

$$1 \text{ day} = \frac{1}{365.2422} = 0.0027379 \text{ year}$$

Date Ord. year	Leap year	Jan.	Feb.	March	April	May	June	July	Aug.	Sept.	Oct.	Nov.	Dec.
0	1	0.000	0.085	0.162	0.246	0.329	0.413	0.496	0.580	0.665	0.747	0.832	0.914
1	2	003	088	164	249	331	416	498	583	668	750	835	917
2	3	005	090	167	252	334	419	501	586	671	753	838	920
3	4	008	093	170	255	337	422	504	589	674	756	841	923
4	5	011	096	172	257	340	424	507	591	676	758	843	925
5	6	0.014	0.099	0.175	0.260	0.342	0.427	0.509	0.594	0.679	0.761	0.846	0.928
6	7	016	101	178	263	345	430	512	597	682	764	849	931
7	8	019	104	181	266	348	433	515	600	684	767	851	934
8	9	022	107	183	268	350	435	517	602	687	769	854	936
9	10	025	110	186	271	353	438	520	605	690	772	857	939
10	11	0.027	0.112	0.189	0.274	0.356	0.441	0.523	0.608	0.693	0.775	0.860	0.942
11	12	030	115	192	277	359	444	526	611	695	778	862	945
12	13	033	118	194	279	361	446	528	613	698	780	865	947
13	14	036	120	197	282	364	449	531	616	701	783	868	950
14	15	038	123	200	285	367	452	534	619	704	786	871	953
15	16	0.041	0.126	0.203	0.287	0.370	0.454	0.537	0.622	0.706	0.789	0.873	0.956
16	17	044	129	205	290	372	457	539	624	709	791	876	958
17	18	047	131	208	293	375	460	542	627	712	794	879	961
18	19	049	134	211	296	378	463	545	630	715	797	882	964
19	20	052	137	214	298	381	465	548	632	717	799	884	966
20	21	0.055	0.140	0.216	0.301	0.383	0.468	0.550	0.635	0.720	0.802	0.887	0.969
21	22	057	142	219	304	386	471	553	638	723	805	890	972
22	23	060	145	222	307	389	474	556	641	726	808	893	975
23	24	063	148	225	309	392	476	559	643	728	810	895	977
24	25	066	151	227	312	394	479	561	646	731	813	898	980
25	26	0.068	0.153	0.230	0.315	0.397	0.482	0.564	0.649	0.734	0.816	0.901	0.983
26	27	071	156	233	318	400	485	567	652	736	819	904	986
27	28	074	159	235	320	402	487	569	654	739	821	906	988
28	29	077	162	238	323	405	490	572	657	742	824	909	991
29	30	079		241	326	408	493	575	660	745	827	912	994
30	31	0.082		0.244	0.329	0.411	0.496	0.578	0.663	0.747	0.830	0.914	0.997
31		085		246		413		580	665		832		999

The leap year column is for use in January and February of leap years only.
The fractions of the year may be in error by 0.002, according to the state of the leap year cycle. They apply to 0^h UT.

Table 9. Annual precession P_α in right ascension and P_δ in declination

δ / α	P_α													P_δ
	$+60°$	$+50°$	$+40°$	$+30°$	$+20°$	$+10°$	$0°$	$-10°$	$-20°$	$-30°$	$-40°$	$-50°$	$-60°$	
h	s	s	s	s	s	s	s	s	s	s	s	s	s	''
0	3.07	3.07	3.07	3.07	3.07	3.07	3.07	3.07	3.07	3.07	3.07	3.07	3.07	$+20.0$
1	3.67	3.48	3.36	3.27	3.20	3.13	3.07	3.01	2.95	2.87	2.78	2.66	2.47	$+19.4$
2	4.23	3.87	3.63	3.46	3.32	3.19	3.07	2.95	2.83	2.69	2.51	2.28	1.92	$+17.4$
3	4.71	4.20	3.87	3.62	3.42	3.24	3.07	2.91	2.73	2.53	2.28	1.95	1.44	$+14.2$
4	5.08	4.45	4.04	3.74	3.49	3.28	3.07	2.87	2.65	2.41	2.10	1.69	1.07	$+10.0$
5	5.31	4.61	4.16	3.82	3.54	3.30	3.07	2.84	2.60	2.33	1.99	1.53	0.84	$+\ 5.2$
6	5.39	4.67	4.19	3.84	3.56	3.31	3.07	2.84	2.59	2.30	1.95	1.48	0.76	0.0
7	5.31	4.61	4.16	3.82	3.54	3.30	3.07	2.84	2.60	2.33	1.99	1.53	0.84	$-\ 5.2$
8	5.08	4.45	4.04	3.74	3.49	3.28	3.07	2.87	2.65	2.41	2.10	1.69	1.07	-10.0
9	4.71	4.20	3.87	3.62	3.42	3.24	3.07	2.91	2.73	2.53	2.28	1.95	1.44	-14.2
10	4.23	3.87	3.63	3.46	3.32	3.19	3.07	2.95	2.83	2.69	2.51	2.28	1.92	-17.4
11	3.67	3.48	3.36	3.27	3.20	3.13	3.07	3.01	2.95	2.87	2.78	2.66	2.47	-19.4
12	3.07	3.07	3.07	3.07	3.07	3.07	3.07	3.07	3.07	3.07	3.07	3.07	3.07	-20.0
13	2.47	2.66	2.78	2.87	2.95	3.01	3.07	3.13	3.20	3.37	3.36	3.48	3.67	-19.4
14	1.92	2.28	2.51	2.69	2.83	2.95	3.07	3.19	3.32	3.46	3.63	3.87	4.23	-17.4
15	1.44	1.95	2.28	2.53	2.73	2.91	3.07	3.24	3.42	3.62	3.87	4.20	4.71	-14.2
16	1.07	1.69	2.10	2.41	2.65	2.87	3.07	3.28	3.49	3.74	4.04	4.45	5.08	-10.0
17	0.84	1.53	1.99	2.33	2.60	2.84	3.07	3.30	3.54	3.82	4.16	4.61	5.31	$-\ 5.2$
18	0.76	1.48	1.95	2.30	2.59	2.84	3.07	3.31	3.56	3.84	4.19	4.67	5.39	0.0
19	0.84	1.53	1.99	2.33	2.60	2.84	3.07	3.30	3.54	3.82	4.16	4.61	5.31	$+\ 5.2$
20	1.07	1.69	2.10	2.41	2.65	2.87	3.07	3.28	3.49	3.74	4.04	4.45	5.08	$+10.0$
21	1.44	1.95	2.28	2.53	2.73	2.91	3.07	3.24	3.42	3.62	3.87	4.20	4.71	$+14.2$
22	1.92	2.28	2.51	2.69	2.83	2.95	3.07	3.19	3.32	3.46	3.63	3.97	4.23	$+17.4$
23	2.47	2.66	2.78	2.87	2.95	3.01	3.07	3.13	3.20	3.27	3.36	3.48	3.67	$+19.4$
24	3.07	3.07	3.07	3.07	3.07	3.07	3.07	3.07	3.07	3.07	3.07	3.07	3.07	$+20.0$

Special Tables

Table 10. Position angle P of the Sun's axis and the heliocentric longitude B_0 of the apparent center of the Sun's disc, tabulated at five-day intervals

Date		P	B_0	Date		P	B_0	Date		P	B_0
Jan.	1	+ 2.4°	− 3.0°	May	1	− 24.3°	− 4.2°	Aug.	29	+ 20.1°	+ 7.1°
	6	− 0.1	3.6		6	23.4	3.7	Sep.	3	21.5	7.2
	11	2.5	4.1		11	22.3	3.1		8	22.6	7.3
	16	4.9	4.7		16	21.0	2.6		13	23.6	7.2
	21	7.2	5.1		21	19.5	2.0		18	24.5	7.2
	26	9.4	5.6		26	17.9	1.4		23	25.2	7.0
	31	11.5	6.0		31	16.1	0.8		28	25.8	6.9
Feb.	5	13.6	6.3	June	5	14.2	− 0.2	Oct.	3	26.1	6.6
	10	15.5	6.6		10	12.2	+ 0.4		8	26.3	6.4
	15	17.2	6.8		15	10.3	1.0		13	26.3	6.0
	20	18.9	7.0		20	8.0	1.6		18	26.2	5.7
	25	20.3	7.2		25	5.7	2.2		23	25.8	5.2
Mar.	2	21.7	7.2		30	3.5	2.7		28	25.3	4.8
	7	22.9	7.3	July	5	− 1.2	3.3	Nov.	2	24.5	4.3
	12	23.9	7.2		10	+ 1.1	3.8		7	23.6	3.8
	17	24.6	7.1		15	3.3	4.3		12	22.4	3.2
	22	25.4	7.0		20	5.5	4.8		17	21.1	2.6
	27	25.9	6.8		25	7.7	5.2		22	19.6	2.0
Apr.	1	26.2	6.6		30	9.7	5.6		27	17.9	1.4
	6	26.4	6.3	Aug.	4	11.7	6.0	Dec.	2	16.0	0.8
	11	26.3	5.9		9	13.7	6.3		7	14.0	+ 0.1
	16	26.1	5.5		14	15.5	6.6		12	11.9	− 0.5
	21	25.7	5.1		19	17.1	6.8		17	9.6	1.1
	26	− 25.1	− 4.7		24	+ 18.7	+ 7.0		22	7.3	1.8
									27	+ 4.9	− 2.4

The above values are subject to very small variations due to the 4-yearly cycle of leap years. The changes may amount up to 0.3 for P and up to 0.1 degree for the B_0 values.

Table 11. Principal lunar craters (after H. ROTH)

No.	Name	Selenographic		Rectangular coordinates		
		Longitude l	Latitude b	ξ	η	ζ
1	Mösting A	− 5° 9.8	− 3° 12.9	− 0.08986	− 0.05607	+ 0.99437
2	Bode A	− 1 8.9	+ 8 59.3	− 0.01982	+ 0.15624	+ 0.98752
3	Dionysius	+ 17 19.7	+ 2 46.2	+ 0.29750	+ 0.04832	+ 0.95349
4	Mare Nectaris E	+ 34 58.6	− 17 50.8	+ 0.54567	− 0.30647	+ 0.77995
5	Abulfeda b	+ 13 6.5	− 16 10.7	+ 0.21783	− 0.27864	+ 0.93537
6	Thebit B	− 8 33.0	− 22 20.2	− 0.13752	− 0.38004	+ 0.91469
7	Gassendi A	− 43 33.3	− 18 27.9	− 0.65359	− 0.31673	+ 0.68739
8	Lohrmann A	− 62 37.6	− 0 44.5	− 0.88795	− 0.01288	+ 0.45976
9	Encke b	− 36 44.7	+ 2 20.8	− 0.59775	+ 0.04095	+ 0.80064
10	Euler A	− 36 51.9	+ 20 52.0	− 0.56058	+ 0.35619	+ 0.74758
11	Archimedes A	− 6 24.1	+ 28 1.5	− 0.09843	+ 0.46987	+ 0.87724
12	Cassini C	+ 7 49.3	+ 41 41.6	+ 0.10163	+ 0.66514	+ 0.73977
13	Macrobius B	+ 40 49.6	+ 20 55.9	+ 0.61064	+ 0.35725	+ 0.70676
14	Condaminae a	− 30 3.9	+ 54 21.9	− 0.29188	+ 0.81275	+ 0.50422
15	Hesiod B	− 17 28.5	− 27 5.4	− 0.26735	− 0.45539	+ 0.84920

The mean error amounts to ± 0.00037 Moon radius.

Table 12. Lunar formations in the Berlin System (See page 299)

[IAU numbers and rectangular coordinates according to M. A. Blagg and K. Müller: Named lunar formations I (London, 1935)].

Quadrant	IAU No.	Formation	Rectangular coordinates		Selenographic	
			ξ	η	Longitude l	Latitude b
NW	198	Proclus	+ 0.702	+ 0.277	+ 49°9	+ 16°1
	266	Vitruvius	+ 0.495	+ 0.303	+ 31.3	+ 17.6
	794	Manilius	+ 0.153	+ 0.250	+ 9.1	+ 14.5
NE	1062	Plato	− 0.100	+ 0.782	− 9.2	+ 51.4
	1305	Kap Laplace	− 0.300	+ 0.725	− 25.9	+ 46.5
	1406	Pytheas	− 0.329	+ 0.351	− 20.6	+ 20.6
	1481	Copernicus	− 0.337	+ 0.167	− 20.0	+ 9.6
	1641	Kap Heraklid	− 0.418	+ 0.656	− 33.6	+ 41.0
	1755	Aristarchus	− 0.675	+ 0.402	− 47.7	+ 23.8
SE	2002	Grimaldi	− 0.926	− 0.093	− 68.4	− 5.3
	2127	Billy	− 0.744	− 0.238	− 50.1	− 13.8
	2525	Campanus	− 0.411	− 0.468	− 27.7	− 27.9
	3182	Tycho	− 0.142	− 0.687	− 11.3	− 43.5
SW	4235	Censorinus	+ 0.539	− 0.007	+ 32.7	− 0.4
	4325	Goclenius	+ 0.695	− 0.173	+ 44.9	− 10.0
	4677	Langrenus	+ 0.863	− 0.155	+ 60.9	− 8.9

Table 13. The 128 IAU names of the Mars map of 1958

Lg. and Br. are the aerographic longitude and latitude.

Lg.	Lat.	Name	Lg.	Lat.	Name	Lg.	Lat.	Name
30°	+45°	Acidalium M.	210°	+25°	Elysium	10°	+20°	Oxus
215.	− 5	Aeolis	220.	−45	Eridania	200.	+60	Panchaia
310.	+10	Aeria	40.	−25	Erythraeum M.	340.	−25	Pandorae Fr.
230.	+40	Aetheria	220.	+22	Eunostos	155.	−50	Phaetontis
230.	+10	Aethiopis	335.	+20	Euphrates	320.	+20	Phison
140.	0	Amazonis	0.	+15	Gehon	190.	+30	Phlegra
250.	+ 5	Amenthes	270.	−40	Hadriaticum M.	110.	−12	Phoenicis L.
105.	−45	Aonius S.	290.	−40	Hellas	70.	−40	Phrixi R.
330.	+20	Arabia	340.	− 6	Hellespontica D.	280.	−65	Promethei S.
115.	−25	Araxes	325.	−50	Hellespontus	185.	+45	Propontis
100.	+45	Arcadia	240.	−20	Hesperia	50.	−23	Protei R.
25.	−45	Argyre	345.	+15	Hiddekel	315.	+42	Protonilus
335.	+48	Arnon	60.	+75	Hyperboreus L.	38.	−15	Pyrrhae R.
50.	−15	Auroae S.	295.	−20	Japygia	340.	− 8	Sabaeus S.
250.	−40	Ausonia	130.	−40	Icaria	150.	+60	Scandia
40.	−60	Australe M.	275.	+20	Isidis R.	320.	−30	Serpentis M.
50.	+60	Baltia	330.	+40	Ismenius L.	70.	−20	Sinai
290.	+55	Boreosyrtis	40.	+10	Jamuna	155.	−30	Sirenum M.
90.	+50	Boreum M.	63.	− 5	Juventae F.	245.	+45	Sithonius L.
75.	+ 3	Candor	200.	0	Laestrygon	90.	−28	Solis L.
260.	+40	Casius	200.	+70	Lemuria	200.	+30	Styx
210.	+50	Cebrenia	270.	0	Libya	100.	−20	Syria
320.	+60	Cecropia	65.	+15	Lunae L.	290.	+10	Syrtis Maior
95.	+20	Ceraunius	25.	−10	Margaritifer S.	70.	+50	Tanais
205.	+15	Cerberus	150.	−20	Memnonia	70.	+40	Tempe
0.	−50	Chalce	285.	+35	Meroe	85.	−35	Thaumasia
260.	−50	Chersonesus	0.	− 5	Meridiani S.	255.	+30	Thoth
210.	−58	Chronium M.	350.	+20	Moab	180.	−70	Thyle I
30.	+10	Chryse	270.	+ 8	Moeris L.	230.	−70	Thyle II
110.	−50	Chrysokeras	72.	−28	Nector	10.	+10	Thymiamata
220.	−20	Cimmerium M.	270.	+35	Neith R.	85.	− 5	Tithonius L.
110.	−35	Claritas	260.	+20	Nepenthes	80.	+30	Tractus albus
280.	+55	Copais Pa.	55.	−45	Nereidum Fr.	268.	−25	Trinacria
65.	−15	Coprates	30.	+30	Niliacus L.	198.	+20	Trivium Charontis
230.	− 5	Cyclopia	55.	+30	Nilokeras			
0.	+40	Cydonia	290.	+42	Nilosyrtis	255.	−20	Tyrrhenium M.
305.	− 4	Deltoton S.	260.	+70	Nix Olympica	260.	+70	Uchronia
340.	−15	Deucalionis R.	330.	−45	Noachis	290.	+50	Umbra
0.	+35	Deuteronilus	65.	−45	Ogygis R.	250.	+50	Utopia
180.	+50	Diacria	200.	+80	Olympia	15.	−35	Vulcani Pe.
320.	+50	Dioscuria	65.	−10	Ophir	50.	+10	Xanthe
345.	0	Edom	0.	+60	Ortygia	320.	−40	Yaonis R.
190.	−45	Electris	18.	+ 8	Oxia Pa.	195.	0	Zephyria

Abbreviations:
M = Mare (ocean)
L = Lacus (lake)
R = Regio (landscape)
S = Sinus (bay)
Pa = Palus (marsh)

Pe = Pelagus (sea)
F = Fons (spring)
Fr = Fretum (channel)
D = Depressio (depression)

Table 14. The hourly changes of the central meridian of Mars

h	°	h	°	m	°	m	°	m	°
1	14.6	6	87.7	10	2.4	1	0.2	6	1.5
2	29.2	7	102.3	20	4.9	2	0.5	7	1.7
3	43.9	8	117.0	30	7.3	3	0.7	8	2.0
4	58.5	9	131.6	40	9.7	4	1.0	9	2.2
5	73.1	10	146.2	50	12.2	5	1.2	10	2.4

Table 15. Hourly changes of the central meridian of Jupiter (System I)

h	°	h	°	m	°	m	°	m	°
1	36.6	6	219.5	10	6.1	1	0.6	6	3.7
2	73.2	7	256.1	20	12.2	2	1.2	7	4.3
3	109.7	8	292.7	30	18.3	3	1.8	8	4.9
4	146.3	9	329.2	40	24.4	4	2.4	9	5.5
5	182.9	10	5.8	50	30.5	5	3.0	10	6.1

Table 16. Hourly changes of the central meridian of Jupiter (System II)

h	°	h	°	m	°	m	°	m	°
1	36.3	6	217.6	10	6.0	1	0.6	6	3.6
2	72.5	7	253.8	20	12.1	2	1.2	7	4.2
3	108.8	8	290.1	30	18.1	3	1.8	8	4.8
4	145.1	9	326.4	40	24.2	4	2.4	9	5.4
5	181.3	10	2.6	50	30.2	5	3.0	10	6.0

Table 17. Periodic comets with periods under 200 years

The comets are arranged according to increasing period. The following data are given: name of the comet, time of perihelion passage, period P, perihelion distance q, distance perihelion–node ω, longitude of node Ω, and inclination i of the orbit. The last column gives the magnitude, m_0.

The orbital elements refer as a rule to the last perihelion passage observed up to 1965. Elements can change appreciably, particularly through close approaches to Jupiter. The magnitudes m are referred to $r = \Delta = 1$ and can be transferred to any desired point of the orbit with the formula

$$m = m_0 + 2.5n \log r + 5 \log \Delta.$$

The value of n is as a rule 4, sometimes 6, and occasionally even larger. In certain cases the brightness can deviate even by some magnitudes from the calculated value.

Name	Perihelion passage	P	q	ω	Ω	i	m_0
Wilson-Harrison	1949.78	2.31	1.03	92°	279°	2°	16ᵐ
Encke	1961.10	3.30	0.34	185	335	12	11
Helfenzrieder	1766.32	4.51	0.40	178	76	8	7
Grigg-Skjellerup	1962.00	4.90	0.86	356	215	18	13
Blanpain	1819.89	5.10	0.89	350	79	9	9
Honda-Mrkos-Pajdusakova	1954.10	5.21	0.56	184	233	13	13
Tempel 2	1957.10	5.27	1.37	191	119	12	12
du Toit 2	1945.30	5.28	1.25	202	359	7	12
Barnard 1	1884.62	5.40	1.28	301	6	6	9
Neujmin 2	1927.04	5.43	1.34	194	328	11	11
Schwassman-Wachmann 3	1930.45	5.43	1.01	192	77	17	12
Grischow	1743.02	5.44	0.86	7	89	2	9
Brorsen	1879.24	5.46	0.59	15	102	29	9
Tuttle-Giacobini-Kresak	1962.31	5.48	1.12	38	166	14	12
du Toit-Neujmin-Delporte	1941.55	5.54	1.30	69	230	3	12
Brooks 1	1886.43	5.60	1.33	177	54	13	9
Lexell	1770.62	5.60	0.67	225	134	2	8
Kulin	1939.76	5.64	1.75	293	138	5	12
Tempel-Swift	1908.75	5.68	1.15	114	291	5	13
de Vico-Swift	1894.78	5.86	1.39	297	49	3	10
Pigott	1783.88	5.89	1.46	355	58	45	7
Tempel 1	1879.35	5.98	1.77	160	80	10	10
Pons-Winnecke	1951.69	6.12	1.16	170	94	22	12
Kopff	1958.05	6.32	1.52	162	121	5	10
Taylor	1916.08	6.37	1.56	355	114	16	9
Spitaler	1890.82	6.37	1.82	13	46	13	9
Harrington-Wilson	1951.83	6.38	1.66	343	128	16	12
Forbes	1948.71	6.42	1.54	260	25	5	13
Wolf-Harrington	1958.61	6.51	1.60	187	254	18	11
Schwassmann-Wachmann 2	1961.68	6.54	2.16	358	126	4	11
Giacobini-Zinner	1959.82	6.59	0.94	173	196	31	11
Biela	1852.73	6.62	0.86	223	247	13	8
Barnard 3	1892.94	6.63	1.43	170	207	31	10
Giacobini	1896.82	6.65	1.46	140	194	11	10
Daniel	1950.64	6.66	1.46	7	70	20	13
Wirtanen	1961.29	6.67	1.62	344	86	13	15
d'Arrest	1963.81	6.67	1.37	175	144	18	11
Perrine-Mrkos	1962.12	6.71	1.15	168	243	16	13
Reinmuth 2	1960.90	6.71	1.93	46	296	7	11
Schorr	1918.74	6.71	1.88	279	118	6	11
Brooks 2	1960.46	6.72	1.76	197	177	6	13
Harrington 2	1960.49	6.80	1.58	233	119	9	14

Table 17. *Continued*

Name	Perihelion passage	P	q	ω	Ω	i	m_0
Arend-Rigaux	1965.43	6.82	1.44	329	122	18	11
Holmes	1906.20	6.86	2.12	14	332	21	10
Johnson	1963.44	6.86	2.25	206	118	14	10
Finlay	1960.67	6.90	1.08	322	42	4	11
Borelly	1960.45	7.02	1.45	351	76	31	11
Harrington-Abell	1962.16	7.22	1.78	338	146	17	12
Swift 2	1895.64	7.22	1.30	168	171	3	11
Shajn-Schaldach	1949.90	7.28	2.23	215	167	6	8
Faye	1955.17	7.41	1.65	201	206	10	11
Denning 2	1894.11	7.42	1.15	46	85	6	10
Whipple	1963.33	7.46	2.61	190	188	10	10
Ashbrook-Johnson	1963.75	7.49	2.31	349	2	13	8
Reinmuth 1	1958.23	7.65	2.03	13	124	8	11
Metcalf	1906.77	7.77	1.63	200	195	15	10
Arend	1959.67	7.79	1.83	44	358	22	12
Oterma	1958.44	7.88	3.39	355	155	4	8
Schaumasse	1960.29	8.18	1.20	52	86	12	9
Wolf 1	1959.22	8.43	2.51	161	204	27	13
Jackson-Neujmin	1936.75	8.57	1.46	197	164	13	13
Comas Solá	1961.26	8.59	1.78	40	63	13	9
Denning 1	1881.70	8.69	0.72	313	67	7	9
Swift 1	1889.91	8.92	1.36	70	331	10	12
Kearns-Kwee	1963.93	8.95	2.21	131	315	9	10
Väisälä 1	1960.35	10.46	1.74	44	135	11	13
Neujmin 3	1951.40	10.95	2.03	145	156	4	12
Gale	1938.46	10.99	1.18	209	67	12	11
Slaughter-Burnham	1958.68	11.64	2.54	44	346	8	12
van Biesbroeck	1954.14	12.43	2.42	134	149	7	9
Wild	1960.21	13.19	1.93	167	359	20	13
Peters	1846.42	13.38	1.53	340	262	31	8
Tuttle	1939.86	13.61	1.02	207	270	55	11
du Toit 1	1944.46	14.79	1.28	257	22	19	12
Schwassmann-Wachmann 1	1957.36	16.10	5.54	356	322	10	4
Perrine	1916.45	16.35	0.47	95	224	103	6
Neujmin 1	1948.96	17.97	1.55	347	347	15	8
Crommelin	1956.80	27.87	0.74	196	250	29	11
Tempel-Tuttle	1866.03	33.18	0.98	171	233	163	8
Stephan-Oterma	1942.96	38.96	1.60	358	79	18	7
Westphal	1913.90	61.73	1.25	57	347	41	8
Pons-Gambart	1827.43	63.83	0.81	19	319	136	7
Ross	1883.98	64.63	0.31	138	265	115	7
Dubiago	1921.34	67.01	1.12	97	66	22	11
Brorsen-Metcalf	1919.79	69.06	0.48	130	311	19	9
Olbers	1956.45	69.57	1.18	65	85	45	5
Pons-Brooks	1954.39	70.86	0.77	199	255	74	6
de Vico	1846.18	75.71	0.66	13	79	85	7
Halley	1910.30	76.03	0.59	112	58	162	5
Väisälä 2	1942.13	85.52	1.29	335	172	38	13
Swift-Tuttle	1862.64	119.6	0.96	153	139	114	4
Barnard 2	1889.47	128.3	1.10	60	272	31	9
Mellish	1917.27	145.3	0.19	121	88	33	7
Herschel-Rigollet	1939.60	156.0	0.75	29	355	64	8
Grigg-Mellish	1907.23	164.3	0.92	328	190	110	10

Table 18. The constellations

Latin and English names with their abbreviations. The last column gives the area in square degrees according to internationally accepted boundaries.

Andromeda	And	Andr	Andromeda	721
Antlia	Ant	Antl	The Air Pump	239
Apus	Aps	Apus	The Bird of Paradise	206
Aquarius	Aqr	Aqar	The Water Bearer	980
Aquila	Aql	Aqil	The Eagle	653
Ara	Ara	Arae	The Altar	238
Aries	Ari	Arie	The Ram	441
Auriga	Aur	Auri	The Charioteer	657
Bootes	Boo	Boot	Bootes (The Herdsman)	905
Caelum	Cae	Cael	The Sculptor's Chisel	125
Camelopardus	Cam	Caml	The Giraffe	756
Cancer	Cnc	Canc	The Crab	506
Canes venatici	CVn	CVen	The Hunting Dogs	467
Canis major	CMa	CMaj	The Great Dog	380
Canis minor	CMi	CMin	The Lesser Dog	183
Capricornus	Cap	Capr	The Goat	414
Carina	Car	Cari	The Keel	494
Cassiopeia	Cas	Cass	Cassiopeia	599
Centaurus	Cen	Cent	The Centaur	1060
Cepheus	Cep	Ceph	Cepheus	588
Cetus	Cet	Ceti	The Whale	1231
Chamaeleon	Cha	Cham	The Chameleon	131
Circinus	Cir	Circ	The Compasses	93
Columba	Col	Colm	The Dove	270
Coma (Berenices)	Com	Coma	Berenice's Hair	386
Corona austrina	CrA	CorA	The Southern Crown	128
Corona borealis	CrB	CorB	The Northern Crown	179
Corvus	Crv	Corv	The Crow	184
Crater	Crt	Crat	The Cup	282
Crux	Cru	Cruc	The Cross	68
Cygnus	Cyg	Cygn	The Swan	805
Delphinus	Del	Delf	The Dolphin	189
Dorado	Dor	Dora	The Swordfish	179
Draco	Dra	Drac	The Dragon	1083
Equuleus	Equ	Equl	The Foal (The Little Horse)	72
Eridanus	Eri	Erid	The River Eridanus	1138
Fornax	For	Forn	The Chemical Furnace	397
Gemini	Gem	Gemi	The Twins	514
Grus	Gru	Grus	The Crane	365
Hercules	Her	Herc	Hercules	1225
Horologium	Hor	Horo	The Clock	249
Hydra	Hya	Hyda	The Water Snake	1303
Hydrus	Hyi	Hydi	The Sea Serpent	243
Indus	Ind	Indi	The Indian	294
Lacerta	Lac	Lacr	The Lizard	201
Leo (major)	Leo	Leon	The Lion	947
Leo minor	LMi	LMin	The Lesser Lion	232
Lepus	Lep	Leps	The Hare	290
Libra	Lib	Libr	The Scales (The Balance)	538
Lupus	Lup	Lupi	The Wolf	334
Lynx	Lyn	Lync	The Lynx	545
Lyra	Lyr	Lyra	The Lyre	285
Mensa	Men	Mens	The Table Mountain	153
Microscopium	Mic	Micr	The Microscope	209
Monoceros	Mon	Mono	The Unicorn	481
Musca	Mus	Musc	The Fly	138
Norma	Nor	Norm	The Level (The Square)	165
Octans	Oct	Octn	The Octant	292
Ophiuchus	Oph	Ophi	The Serpent Bearer	948
Orion	Ori	Orio	The Hunter	594
Pavo	Pav	Pavo	The Peacock	377

Table 18. *Continued*

Pegasus	Peg	Pegs	Pegasus	1136
Perseus	Per	Pers	Perseus	615
Phoenix	Phe	Phoe	The Phoenix	469
Pictor	Pic	Pict	The Painter	247
Pisces	Psc	Pisc	The Fishes	890
Piscis austrinus	PsA	PscA	The Southern Fish	245
Puppis	Pup	Pupp	The Poop	673
Pyxis	Pyx	Pyxi	The Mariner's Compass	221
Reticulum	Ret	Reti	The Rhomboidal Net	114
Sagitta	Sge	Sgte	The Arrow	80
Sagittarius	Sgr	Sgtr	The Archer	867
Scorpius	Sco	Scor	The Scorpion	497
Sculptor	Scl	Scul	The Sculptor's Workshop	475
Scutum	Sct	Scut	Sobieski's Shield	109
Serpens	Ser	Serp	The Serpent	637
Sextans	Sex	Sext	The Sextant	313
Taurus	Tau	Taur	The Bull	797
Telescopium	Tel	Tele	The Telescope	251
Triangulum australe	TrA	TrAu	The Triangle	109
Triangulum (boreale)	Tri	Tria	The Southern Triangle	132
Tucana	Tuc	Tucn	The Toucan	294
Ursa major	UMa	UMaj	The Great Bear (The Plough or Dipper)	1279
Ursa minor	UMi	UMin	The Lesser Bear	256
Vela	Vel	Velr	The Sail	500
Virgo	Vir	Virg	The Maiden	1294
Volans	Vol	Voln	The Flying Fish	141
Vulpecula	Vul	Vulp	The Little Fox	268

Table 19. The 170 brightest stars up to magnitude 3.0

In addition to the names and coordinates of the star, the table contains the magnitudes in the 5-color photometry UBVRI of the Arizona-Tonantzintla catalogs of H. L. JOHNSON et al. [Sky and Telescope **30**, No. 1, 24 (1956)], and also the spectral types Sp and the luminosity classes Lc taken from the same catalog (Ia, Ib are supergiants, II bright giants, III normal giants, IV subgiants, V main-sequence stars, p spectral peculiarities). The parallax π is taken from the Catalogue of Bright Stars, 3rd ed., by D. HOFFLEIT. (These are trigonometric or dynamical parallaxes; the former occasionally yield an apparently negative value.)

NOTE: vd and sd denote a visual or a spectroscopic binary system, tr are triple stars, var are variable stars.

Star	α_{1950}	δ_{1950}	V	U − V	B − V	V − R	V − I	Sp Lc	π	Notes
α And	0ʰ05ᵐ8	+ 28°49′	2.06	− 0.58	− 0.11	− 0.02	− 0.12	B8p III	+ 0″024	sd, Sp var
β Cas	06. 9	+ 58 52	2.28	0.45	0.34	0.31	0.52	F2 IV	+ 0. 072	sd
γ Peg	10. 7	+ 14 54	2.86	− 1.08	− 0.21	− 0.08	− 0.31	B2 IV	− 0. 004	
β Hyi	23. 2	− 77 32	2.8					G2 IV	+ 0. 153	
α Phe	23. 8	− 42 35	2.38		1.09	0.81	1.40	K0 III	+ 0. 035	sd
α Cas	37. 7	+ 56 16	2.22	2.30	1.17	0.79	1.38	K0 II−III	+ 0. 009	irr var
β Cet	41. 1	− 18 16	2.00	1.88	1.00	0.72	1.24	K1 III	+ 0. 057	
γ Cas	53. 7	+ 60 27	2.41	− 1.21	− 0.11	0.09	0.00	B0 IV	+ 0. 034	irr var
β And	1 06. 9	+ 35 21	2.04	3.53	1.57	1.24	2.24	M0 III	+ 0. 043	
δ Cas	22. 5	+ 59 59	2.69	0.28	0.13	0.15	0.24	A5 V	+ 0. 029	var
α Eri	35. 9	− 57 29	0.5					B5 IV	+ 0. 023	
α UMi	48. 8	+ 89 02	2.04	0.97	0.59	0.50	0.81	F8 Ib	0. 008	vd u. sd, var
β Ari	51. 9	+ 20 34	2.64	0.23	0.13	0.12	0.18	A5 V	+ 0. 063	sd
α Hyi	57. 2	− 61 49	2.9					F0 V	+ 0. 041	
γ And	2 00. 8	+ 42 05	2.10	2.13	1.21	0.94	1.63	K3 II	+ 0. 005	vd
α· Ari	04. 3	+ 23 14	2.00	2.27	1.15	0.84	1.48	K2 III	+ 0. 043	
β Tri	06. 6	+ 34 45	3.00	0.30	0.16	0.15	0.22	A5 III		
ϑ Eri	56. 4	− 40 30	2.90		0.13	0.15	0.22	A3 V	+ 0. 028	vd u. sd
α Cet	59. 7	+ 03 54	2.52	3.57	1.63	1.35	2.51	M2 III	+ 0. 003	
γ Per	3 01. 2	+ 53 19	2.94	1.16	0.70	0.61	1.06	G8III+A3	+ 0. 011	
β Per	04. 9	+ 40 46	2.15	− 0.44	− 0.06	0.07	0.04	B8 V	+ 0. 031	s tr, var
α Per	20. 7	+ 49 41	1.80	0.88	0.48	0.45	0.78	F5 Ib	+ 0. 029	
δ Per	39. 4	+ 47 38	3.03	− 0.62	− 0.12	0.04	− 0.07	B5 III	+ 0. 007	
η Tau	44. 5	+ 23 57	2.86	− 0.45	− 0.10	0.04	− 0.00	B7 III	+ 0. 005	
ζ Per	51. 0	+ 31 44	2.86	− 0.69	0.10	0.16	0.25	B1 Ib	+ 0. 007	
ε Per	54. 5	+ 39 52	2.89	− 1.18	− 0.18	− 0.06	− 0.24	B0.5V	− 0. 001	sd
γ Eri	55. 7	− 13 39	2.94	3.54	1.60	1.26	2.26	M1 III	+ 0. 003	
α Tau	4 33. 0	+ 16 25	0.86	3.50	1.55	1.22	2.15	K5 III	+ 0. 048	vd
ι Aur	53. 7	+ 33 05	2.67	3.29	1.54	1.06	1.88	K3 II	+ 0. 015	
β Eri	5 05. 4	− 05 09	2.78	0.25	0.12	0.15	0.23	A3 III	+ 0. 042	
β Ori	12. 1	− 08 15	0.15	− 0.66	− 0.03	0.03	− 0.00	B8 Ia	− 0. 003	vd u. sd
α Aur	13. 0	+ 45 57	0.06	1.28	0.81	0.61	1.04	G8III + F	+ 0. 073	sd
γ Ori	22. 4	+ 06 18	1.63	− 1.09	− 0.21	− 0.09	− 0.31	B2 III	+ 0. 026	
β Tau	23. 1	+ 28 34	1.66	− 0.62	− 0.13	− 0.01	− 0.09	B7 III	+ 0. 018	
β Lep	26. 1	− 20 48	2.81	1.31	0.82	0.65	1.09	G5 III	+ 0. 014	vd
ϑ Cen	14 03. 7	− 36 07	2.05		0.96	0.76	1.29	K0 III−IV	+ 0. 059	
α Boo	13. 4	+ 19 27	− 0.06	2.52	1.24	0.98	1.64	K2 IIIp	+ 0. 090	
γ Boo	30. 1	+ 38 32	3.02	0.31	0.20	0.14	0.22	A7 III	+ 0. 016	
η Cen	32. 3	− 41 56	2.30		− 0.21	− 0.09	− 0.27	B1.5V	0. 012	
α Cen	36. 2	− 60 38	− 0.1					G2V + dK1	+ 0. 759	vd
α Lup	38. 6	− 47 10	2.29		− 0.21	− 0.08	− 0.25	B1 V	0. 009	
ε Boo	42. 8	+ 27 17	2.37	1.70	0.97	0.77	1.29	K1 III + A	+ 0. 013	vd
α Lib	48. 1	− 15 50	2.74	0.26	0.15	0.17	0.21	A3p V	+ 0. 049	
β UMi	50. 8	+ 74 22	2.07	3.25	1.47	1.10	1.87	K4 III	+ 0. 031	
β Lup	55. 2	− 42 56	2.67		− 0.21	− 0.09	− 0.26	B2 IV	0. 012	
γ TrA	15 14. 2	− 68 30	3.0					A1 V	+ 0. 005	
β Lib	14. 3	− 09 12	2.61	− 0.48	− 0.11	− 0.04	− 0.13	B8 V	− 0. 012	sd

Table 19. *Continued*

Star	α_{1950}	δ_{1950}	V	U – V	B – V	V – R	V – I	Sp Lc	π	Notes
γ Lup	31. 8	−41 00	2.77		−0.20	−0.15	−0.37	B2 V	0. 008	vd
α CrB	32. 6	+26 53	2.27	−0.06	−0.02	0.09	0.05	A0 V	+0. 043	sd
α Ser	41. 8	+06 35	2.65	2.41	1.17	0.81	1.37	K2 III	+0. 046	
β TrA	50. 7	−63 17	2.9					F2 IV	+0. 078	
π Sco	55. 8	−25 58	2.92	−1.08	−0.19	−0.09	−0.29	B1 V	+0. 005	sd
δ Sco	57. 4	−22 29	2.33	−1.00	−0.10	−0.04	−0.17	B0 V	0. 011	
β Sco	16 02. 5	−19 40	2.55	−0.90	−0.08	−0.00	−0.10	B0.5V +B2 V		v tr u. sd
δ Oph	11. 7	−03 34	2.74	3.55	1.00	1.29	2.32	M0.5 III	+0. 029	
η Dra	23. 3	+61 38	2.74	1.61	0.91	0.61	1.07	G III	+0. 043	vd
α Sco	26. 3	−26 19	0.89	3.17	1.83	1.56	2.79	M2 I	+0. 019	vd
β Her	28. 1	+21 36	2.77	1.63	0.94	0.64	1.11	G8 III	+0. 017	sd
τ Sco	32. 8	−28 07	2.82	−1.27	−0.25	−0.11	−0.36	B0 V	+0. 014	
ζ Oph	34. 4	−10 28	2.57	−0.82	0.02	0.10	0.06	O9.5V	−0. 007	
ζ Her	39. 4	+31 42	2.81	0.86	0.66	0.54	0.85	G0 IV	+0. 110	vd
α TrA	43. 4	−68 56	1.9					K4 III	+0. 024	
ε Sco	46. 9	−34 12	2.28		1.17	0.86	1.46	K2 III−IV	+0. 049	
η Oph	17 07. 5	−15 40	2.42	0.13	0.05	0.06	0.05	A2.5V	+0. 047	vd
β Ara	21. 1	−55 29	2.9					K3 Ib	+0. 026	
υ Sco	27. 4	−37 15	2.70		−0.22	−0.16	−0.40	B3 Ib	0. 010	
α Ara	28. 0	−49 50	2.94		−0.18	−0.10	−0.34	B2.5V	+0. 001	
β Dra	29. 3	+52 50	2.78	1.63	1.00	0.68	1.16	G2 II	+0. 009	
λ Sco	30. 2	−37 04	1.62		−0.18	−0.17	−0.45	B1 V	0. 015	sd
α Oph	32. 6	+12 36	2.07	0.25	0.14	0.14	0.22	A5 III	+0. 056	
ϑ Sco	33. 7	−42 58	1.86		0.41	0.35	0.55	F0 Ib	+0. 020	
κ Sco	39. 0	−39 00	2.41	−1.08	−0.20	−0.08	−0.30	B2 IV	0. 009	
β Oph	41. 0	+04 35	2.77	2.40	1.16	0.81	1.38	K2 III	+0. 023	
ι¹ Sco	44. 1	−40 07	3.0					F2 Ia	+0. 013	
γ Dra	55. 4	+51 30	2.23	3.39	1.51	1.16	2.02	K5 III	+0. 017	
γ Sgr	18 02. 6	−30 26	2.98		1.02	0.73	1.24	K0 III	+0. 018	
δ Sgr	17. 8	−29 51	2.69	2.89	1.38	1.00	1.68	K2 III	+0. 039	
ε Sgr	20. 9	−34 25	1.84		−0.02	−0.00	−0.01	A0 V	+0. 015	
λ Sgr	24. 9	−25 27	2.82	1.94	1.04	0.76	1.32	K2 III	+0. 046	
α Lyr	35. 2	+38 44	0.00	0.03	0.00	−0.04	−0.07	A0 V	+0. 123	
σ Sgr	52. 2	−26 22	2.07	−0.95	−0.22	−0.11	−0.31	B2 V	0. 019	
ζ Sgr	59. 4	−29 57	2.60	0.17	0.08	0.04	0.05	A2 III	+0. 020	vd
ζ Aql	19 03. 1	+13 47	2.98	0.03	0.03	0.04	0.04	A0 V	+0. 036	
π Sgr	06. 8	−21 06	2.87	0.57	0.33	0.34	0.59	F2II−III	+0. 016	vtr
δ Cyg	43. 4	+45 00	2.87	−0.12	−0.02	0.00	−0.02	B9.5III	+0. 021	vd
γ Aql	43. 9	+10 30	2.71	3.21	1.52	1.07	1.82	K3 II	+0. 006	
α Aql	48. 3	+08 44	0.74	0.31	0.23	0.14	0.27	A7 IV, V	+0. 198	
γ Cyg	20 20. 4	+40 06	2.23	1.21	0.67	0.50	0.84	F8 Ib	−0. 006	
α Pav	21. 7	−56 54	2.0					B3 IV	0. 014	sd
α Cyg	39. 7	+45 06	1.25	−0.12	0.09	0.12	0.22	A2 Ia	−0. 013	
ε Cyg	44. 2	+33 47	2.46	1.91	1.03	0.72	1.29	K0 III	+0. 044	
α Cep	21 17. 4	+62 22	2.47	0.33	0.21	0.21	0.32	A7 IV, V	+0. 063	
β Aqr	28. 9	−05 48	2.85	1.42	0.84	0.61	1.02	G0 Ib	+0. 000	
ε Peg	41. 7	+09 39	2.38	3.21	1.53	1.05	1.81	K2 Ib	−0. 005	
δ Cap	44. 3	−16 21	2.81	0.40	0.29	0.24	0.40	A6p	+0. 065	sd
α Gru	22 05. 1	−47 12	1.73		−0.17	−0.08	−0.14	B5 V	+0. 051	
α Tuc	15. 1	−60 31	2.9					K3 III	+0. 019	sd
β Gru	39. 7	−47 09	2.10		1.62	1.91	3.68	M3 II	+0. 003	
η Peg	40. 7	+29 58	2.95	1.46	0.87	0.64	1.11	G8II+F0	−0. 002	sd
α PsA	54. 9	−29 53	1.16	0.14	0.08	0.11	0.13	A3 V	+0. 144	
β Peg	23 01. 3	+27 49	2.42	3.62	1.67	1.51	2.83	M2 II−III	+0. 015	
α Peg	02. 3	+14 56	2.47	−0.06	−0.03	0.01	−0.01	B9.5III	+0. 030	

Table 20. Typical star spectra of the spectral classes O B A F G K M N

The classification according to spectral type Sp and luminosity class Lc is given by comparison with the spectra of standard stars as given, for example, in Table 19. The classification criteria of each observer will depend on the standard star spectra obtained with his particular instrument. The most important criteria of the Yerkes or MK system are given in LANDOLT-BÖRNSTEIN (see Bibliography §23.3); the MKK system can be taken from the "Atlas of Stellar Spectra" by W. W. MORGAN, P. C. KEENAN, and E. KELLMANN, Chicago, 1943.

Characteristic Lines, Bands, and Line Intensities of the Main-Sequence Stars

0 stars: CIII 4650, HeI 4471, HeII 4541, H_γ 4340, SiIV 4089, NIII 4097.

B 0 stars: HeI 4471, HeII 4541, CIII 4540, SiIV 4089, H_δ 4102.

B 5 stars: SiII 4128/4131, HeI 4121.

A 0 stars: Balmer lines H_α 6463, H_β 4861, H_γ 4340, H_δ 4102, H_ε 3970 dominate: the K-line (CaII 3934) is distinct; MgII 4481, next to the Balmer lines, is the most conspicuous line. SiII is at maximum brightness.

A 5 stars: K-line, H-line (CaII 3968), H_δ, FeI 4299/4303, TiII 4303.

F 0 stars: The strength of the Balmer lines is about one half that in A 0 stars: the K- and H-lines, H_ε, H_δ, and the metallic lines are stronger, the G-band 4307 appears.

F 5 stars: CaI 4227, H_γ, G-band.

G 0 stars: Spectrum similar to the spectrum of the Sun (type G2), CaI 4227, H_δ, H_γ, G-band, FeI 4325.

K 0 stars: Spectrum similar to the sunspot spectrum; H- and K-lines, CaI 4227, FeII 4172, FeI 4383, FeI 4325, H_γ.

K 5 stars: CaI 4227, CaII 3934. The G-band is dissolved into lines.

M 0 to M 2 stars: CaI 4227, the band spectrum of TiO dominates.

M 3 to M 10 stars: The TiO bands increase in intensity.

M 0e to M 10e stars: These stars show one or more Balmer lines in emission.

Name	α_{1950}	δ_{1950}	V	U – V	B – V	V – R	V – I	Sp Lc	π	Notes
δ Ori	5 29.5	−00 20	2.21	−1.27	−0.21	−0.07	−0.28	O9.5 II	+0.006	sd
α Lep	30.5	−17 51	2.58	0.47	0.19	0.22	0.43	F0 Ib	+0.002	
ι Ori	33.0	−05 56	2.76	−1.30	−0.23	−0.06	−0.28	O9 III	+0.021	sd
ε Ori	33.7	−01 13	1.70	−1.21	−0.19	−0.01	−0.19	B0 Ia	−0.007	var
ζ Tau	34.7	+21 07	3.03	−0.84	−0.18	−0.00	−0.09	B2 IVp	−0.002	sd
α Col	37.8	−34 06	2.63		−0.15	−0.01	−0.11	B8 Ve	−0.005	
ζ Ori	38.2	−01 58	1.74	−1.26	−0.21	−0.06	−0.27	O9.5 Ib	+0.022	vd u. sd
$\varkappa$ Ori	45.2	−09 41	2.06	−1.18	−0.18	−0.03	−0.21	B0.5 Ia	+0.009	
α Ori	52.5	+07 24	0.69	3.92	1.86	1.60	3.05	M2 Iab	+0.005	var
β Aur	55.9	+44 57	1.90	0.06	0.02	0.09	0.08	A2 V	+0.037	sd
ϑ Aur	56.3	+37 13	2.63	−0.26	−0.08	−0.01	−0.04	B9.5p V	+0.018	vd
β CMa	6 20.5	−17 56	1.98	−1.20	−0.24	−0.11	−0.35	B1 II−III	+0.014	Sp var
α Car	22.8	−52 40	−0.8					F0 Ib	+0.018	
γ Gem	34.8	+16 27	1.91	0.07	−0.00	0.07	0.06	A0 IV	+0.031	sd ?
α CMa	42.9	−16 39	−1.45	−0.03	−0.01	0.00	−0.03	A1 V	+0.373	vd
τ Pup	48.7	−50 33	2.9					K0 III	+0.017	
ε CMa	56.7	−28 54	1.50	−1.13	−0.21	−0.08	−0.29	B2 II	0.001	
δ CMa	7 06.4	−26 19	1.80	1.20	0.67	0.52	0.85	F8 Ia	−0.018	
π Pup	15.4	−37 00	2.70		1.62	1.24	2.15	K5 III	+0.023	
η CMa	22.1	−29 12	2.41	−0.77	−0.07	0.07	0.02	B5 Ia	0.012	
β CMi	24.4	+08 23	2.90	−0.37	−0.09	0.03	−0.04	B7 V	+0.020	
α Gem	31.4	+32 00	1.58	0.04	0.03	0.07	0.06	Am+A3III	+0.072	v tr, je sd

Table 20. *Continued*

Name		α_{1950}	δ_{1950}	V	U – V	B – V	V – R	V – I	Sp Lc	π	Notes
α	CMi	36. 7	+ 05 21	0.35	0.47	0.43	0.41	0.66	F5 IV—V	+ 0.288	vd u. sd
β	Gem	42. 3	+ 28 09	1.13	1.87	1.00	0.75	1.25	K0 III	+ 0.093	
ζ	Pup	8 01. 8	− 39 52	2.25	− 1.34	− 0.29	− 0.06	− 0.27	O5 f	0.004	
ϱ	Pup	05. 4	− 24 10	2.82	0.63	0.44	0.37	0.58	F6 IIp	+ 0.031	var
γ	Vel	08. 0	− 47 11	1.82		− 0.27	− 0.02	− 0.16	WC7+0.7 ?	0.006	vd
ε	Car	21. 5	− 59 21	1.8					K0 II+B	0.010	
δ	Vel	43. 3	− 54 31	1.9					A0 V	+ 0.043	
ι	UMa	55. 8	+ 48 12	3.15	0.27	0.19	0.21	0.29	A7 V	+ 0.066	
λ	Vel	9 06. 2	− 43 15	2.20		1.62	1.24	2.19	K5 Ib	+ 0.015	
β	Car	12. 7	− 69 31	1.7					A1 IV	+ 0.038	
ι	Car	15. 8	− 59 04	2.2					F0 I	+ 0.011	
$\varkappa$	Vel	20. 6	− 54 48	2.5					B2 IV	+ 0.007	
α	Hya	25. 1	− 08 26	1.96	3.20	1.47	1.04	1.81	K3 IIIa	+ 0.017	
ε	Leo	43. 0	+ 24 00	2.98	1.27	0.81	0.66	1.06	G0 II	+ 0.002	
ν	Car	45. 9	− 64 50	3.0					F0	0.020	vd
α	Leo	10 05. 7	+ 12 13	1.35	− 0.47	− 0.12	0.00	− 0.09	B7 V	+ 0.039	
γ	Leo	17. 2	+ 20 06	1.98	2.14	1.15	0.86	1.48	K0 IIIp	+ 0.019	vd
ϑ	Car	41. 2	− 64 08	2.8					O9.5V	0.007	
μ	Vel	44. 6	− 49 09	2.68		0.88	0.69	1.18	G5 III	0.022	vd
β	UMa	58. 8	+ 56 39	2.38	− 0.04	− 0.01	0.08	0.04	A1 V	+ 0.042	
α	UMa	11 00. 7	− 62 01	1.80	2.01	1.07	0.81	1.39	K0 III	+ 0.031	vd
δ	Leo	11. 5	+ 20 48	2.57	0.21	0.11	0.16	0.19	A4 V	+ 0.040	
β	Leo	46. 5	+ 14 51	2.13	0.16	0.08	0.06	0.07	A3 V	+ 0.076	
γ	UMa	51. 2	+ 53 58	2.44	0.04	0.00	0.05	0.02	A0 V	+ 0.020	
δ	Cen	12 05. 8	− 50 27	2.63		− 0.09	0.04	− 0.08	B2 Ve	+ 0.020	var
δ	Cru	12. 5	− 58 28	2.9					B2 IV	− 0.003	
γ	Crv	13. 2	− 17 16	2.57	− 0.45	− 0.10	− 0.02	− 0.11	B8 III	0.024	
α	Cru	23. 8	− 62 49	0.8					B1 IV	0.008	vd, je sd
δ	Crv	27. 3	− 16 14	2.94	− 0.14	− 0.05	− 0.05	− 0.09	B9.5V	+ 0.018	
γ	Cru	28. 4	− 56 50	1.6					M3 II		
β	Crv	31. 8	− 23 07	2.64	1.53	0.88	0.61	1.05	G5 III	+ 0.027	
α	Mus	34. 2	− 68 52	2.8					B3 IV	0.015	
γ	Cen	38. 2	− 48 41	2.16		− 0.01	0.03	0.03	A0 III	+ 0.006	vd
γ	Vir	39. 1	− 01 11	2.73	0.34	0.36	0.29	0.48	F0 V	+ 0.101	vd
β	Cru	44. 8	− 59 25	1.3					B0.5IV	0.007	var
ε	UMa	51. 8	+ 56 14	1.78	− 0.02	− 0.03	− 0.02	− 0.06	A0p	+ 0.008	sd
α	CVe	53. 7	+ 38 35	2.84	− 0.43	− 0.10	− 0.04	− 0.12	F0 V/Ap	0.023	
ε	Vir	59. 7	+ 11 14	2.84	1.68	0.94	0.65	1.09	G8 III	+ 0.036	
ι	Cen	13 17. 8	− 36 27	2.73	0.05	0.02	0.06	0.05	A2 V	+ 0.046	
ζ	UMa	21. 9	+ 55 11	2.27	0.01	0.02	0.06	0.03	A2 V	+ 0.037	vd u. sd
α	Vir	22. 6	− 10 54	0.96	− 1.18	− 0.25	− 0.09	− 0.32	B1 V	+ 0.021	sd
ε	Cen	36. 7	− 53 13	2.99		− 0.24	− 0.15	− 0.40	B1 V	0.011	
η	UMa	45. 6	+ 49 34	1.86	− 0.85	− 0.18	− 0.07	− 0.26	B3 V	+ 0.004	
η	Boo	52. 3	+ 18 39	2.68	0.81	0.58	0.45	0.74	G0 IV	+ 0.102	sd
ζ	Cen	52. 4	− 47 03	2.54		− 0.24	− 0.13	− 0.34	B2 IV	0.013	sd
β	Cen	14 00. 3	− 60 08	0.7					B1 II	+ 0.016	vd, sd ?

Table 21. The classification of variable stars

After B. W. KUKARKIN, P. PARENAGO, J. I. EFREMOV, and P. N. CHOLOPOV:
General Catalog of Variable Stars, 2nd ed., Vol. 1, p. 24.
Moscow: Academy of Sciences of the USSR, 1958.

Main class		Subclass	Brightness variation in m	Period in days	Spectrum max./min.	Peculiarities
Cepheids	C	Classical Cepheids	0.1–2	1–50 or 70	F/G, K	Usually very regular light curve
	Cδ	Classical Cepheids	0.1–2	1–50 or 70	F/G, K	Galactic Cepheids with regular light curve
	CW	Long-period Cepheids	0.1–2	1–50 or 70	F/G, K	1^{m}5 to 2^{m} fainter than C
	I	Irregular variables	—	—	— —	Mainly semiregular variables
	Ia	Irregular variables	—	—	— —	Irregular variables of early spectral type
	Ib	Irregular variables	—	—	— —	Irregular variables of early spectral type
	Ic	Irregular variables	—	—	— —	Slowly and irregularly varying; late spectral type
Long-period variables	M	Mira Ceti stars	2.5–5 and mehr	80–1000	Me, Ce, Se	Well-established period
Red giant variables	SR	Semiregular variables	1–2	30–1000	— —	Period sometimes disturbed
	SRa	Semiregular variables	<2.5	—	M, C, S	Light curve variable
	SRb	Semiregular variables	—	—	M, C, S	Periodicity not clearly defined
	SRc	Semiregular variables	—	—	M, C, S	—
	SRd	Semiregular variables	—	—	F, G, K	—
RR Lyrae variables	RR	Cluster variables	<1–2	0.05–1.2	A, F	Period and light curve usually constant
	RRa	Cluster variables	<1.5	0.5 and 0.7	A, F	Steeply rising asymmetrical light curve
	RRc	Cluster variables	—	0.3	—	Symmetrical light curve
RV Tauri variables	RV	Variable supergiants	3	30–150	G, K, M	Light curve with primary and secondary minima

Table 21. *Continued*

Main class	Subclass		Brightness variation in m	Period in days	Spectrum max./min.	Peculiarities
RV Tauri variables	RVa	Red-giant variables	3	30–150	G, K, M	Constant mean brightness
	RVb	Red-giant variables	3	30–150	G, K, M	Mean brightness varies periodically
	βC	βCephei V. βCanis Maj. V.	0.1	0.1–0.3	B1–B3	Very similar to the Cepheids
	δSc	Scuti variables	<0.25	1.0	F	Variable light curve, similar to RR Lyrae
	α²CV	α²Canis Ven. variables	<0.1	1–25	Ap	Spectrum shows strong lines of Si, Sr, Cr
Eruptive variables	N	Novae	7–16	—	A, F	These stars return to their prenova state
	Na	Novae	7–16	—	A, F	Particularly rapid rise of the light curve
	Nb	Novae	7–16	—	A, F	Slow rise of light curve
	Nc	Novae	7–16	—	A, F	Very slow rise and descent of light curve
	Nd	Recurrent Novae	7–16	—	A, F	Repetition of rise of light
	Ne	Nova-like variables	—	—	—	Very different types of variables
	SN	Supernovae	20	—	—	Emission bands in spectrum
	RCB	R Coronae Borealis variables	1–9	10–100	F, K, R	Slow, unperiodic changes in brightness
	RW	RW Aurigae variables	—	—	—	Irregular variables
	UG	UGeminorum variables (SS Cygni variables	2–6	20–600	—	Light increase in 2–6 days
	UV	UV Ceti variables	1–6	—	dM3e–dM6e	Short light-bursts (ca. 10 min)
	Z	Z Camelopardalis variables	2–5	10–40	—	—
Eclipsing variables	E	Eclipsing variables	—	—	—	—
	EA	Algol variables	—	0.2–10,000	—	Main and usually secondary minima
	EB	βLyrae variables	<2	>1	early Sp. types	No sharp onset of minima
	EW	W Ursae Majoris variables	0.8	1	F–G	The primary and secondary minima are nearly of equal size
	Ell	Ellipsoid variables	—	—	—	Regular light curves of varying character
Unclassifiable variables	—		—	—	—	—

• p = photographic, v = visual.

Table 21. *Continued*

Typical representative		α_{1900}	δ_{1900}	Brightness* in m max./min.		Period in days	Spectrum	Remarks
TW	CMa	7h17m45	−14°7′	9.5	11.0 p	6.99	F5–F8	
δ	Cep	22 25. 45	+57 54	4.1	5.2 p	5.37	F5Ib–G2Ib	See Fig. 19–3
W	Vir	13 20. 83	− 2 52	9.9	11.3 p	17.29	F2c–G6e	
RX	Cep	0 41. 92	+81 25	7.5	7.8 v	—	G5	
V395	Cyg	20 5. 37	+43 46	7.8	8.4 v	—	F8Ib	
CO	Cyg	20 57. 00	+44 22	9.6	10.6 v	—	K5	
TZ	Cas	23 47. 92	+60 27	9.2	10.5 v	—	M2Iab	
o	Cet	2 14. 30	− 3 26	2.0	10.1 v	331.62	M5e–M9e	See Fig. 19–4
VW	UMa	10 52. 18	+70 32	8.4	9.1 p	125	M5	
Z	Aqr	23 47. 08	−16 25	9.5	12.0 p	136.9	M1e–M3e	
AF	Cyg	19 27. 22	+45 56	7.4	9.4 p	94.1	M5e	
μ	Cep	21 40. 45	+58 19	3.6	5.1 v	—	M2eIa	See Fig. 19–5
UU	Her	16 32. 45	+38 10	8.5	10.6 p	—	F2Ib–cF8	
V756	Oph	17 17. 45	+ 1 53	12.3	13.7 p	—	—	
RR	Lyr	19 22. 27	+42 35	6.94	8.03 p	0.567	A2–F1	See Fig. 19–6
SX	UMa	13 22. 33	+56 47	10.6	11.2 p	0.307	A	
EP	Lyr	19 14. 28	+27 40	10.2	11.6 p	83.43	—	

Table 21. *Continued*

Typical representative		α_{1900}	δ_{1900}	Brightness* in m max./min.		Period in days	Spectrum	Remarks
AC	Her	18h26m03	+21°48′	7.4	9.2 p	75.46	F2Ib–K4e	
R	Sge	20 9. 50	+16 25	9.0	11.2 p	70.594	G0Ib–GIb	
β	Cep	21 27. 37	+70 7	3.3	3.35 p	0.190	B2III	
δ	Sct	18 36. 8	− 9 9	4.9	5.19 p	0.194	F2II–IV	
α^2	CVn	12 51. 33	+38 52	3.0	3.1 p	5.47	A0p	
—	—	—	—	—	—	—	—	
V603	Aql	18 43. 80	+ 0 28	−1.1	10.8 p	—	Q	See Fig. 19–7
RR	Pic	6 43. 73	−62 33	1.2	12.8 p	—	Q	
RT	Ser	17 34. 27	−11 53	10.6	16 p	—	Q	
T	GrB	15 55. 32	+26 12	2.0	10.8 v	29,000	gM3ep	
P	Cyg	20 14. 10	+37 43	3	6 v	—	B1eq	
CM (SN 1054 Crab Nebula)	Tau	5 28. 50	+21 57	−6	15.9 p	—	Continuous	
R	CrB	15 44. 45	+28 28	5.8 − 14.8 v		—	cFpep	See Fig. 19–8
RW	Aur	5 1. 43	+30 16	9.6 − 13.6 p		–	dG5ep	
U	Gem	7 49. 17	+22 16	8.9	14.0 v	103	Gep	
UV	Cet	1 34. 00	−18 28	7	12.9 v	—	dM5.5e	
—	—	—	—	—	—	—	—	
QX	Cas	23 53. 45	+60 36	10.2	10.6 p	—	B2	
β	Per	3 1. 67	+40 34	2.2	3.47 v	2.867	B8V+G	See Fig. 19–9
β	Lyr	18 46. 38	+33 15	3.4	4.34	12.908	cB8p	See Fig. 19–13
W	UMa	9 36. 73	+56 24	8.3	9.03 p	0.334	F8p+F8p	
b	Per	4 10. 72	+50 3	4.6	4.66 p	1.527	A2	
V389	Cyg	21 4. 40	+29 48	5.5	5.69 p	—	B8,A0	

Table 22. Double stars

The following catalog of 313 double and multiple systems contains mainly bright objects (with components up to 8^{m}5) with distances of 1″ to 30″. For test purposes some fainter, very wide, and very close pairs, and particularly pairs close to the pole which can be equally well observed throughout the year, have also been included. The southern declination limit is −15°.

The various columns indicate: ADS number, designation by discoverer, coordinates 1950, magnitudes, spectral types (if the spectra of the components differ, only that of the primary component is given), approximate position angles, and distances for 1960 (rapidly increasing or decreasing angles are marked by + or − signs, respectively). The Remarks column gives star names and notes about additional components, for instance, C 8.5 7″ indicates the presence of a third component of magnitude 8.5 and distance 7″; or, AB-C (0″.2) means that the companion is a close system BC of 0″.2 distance. Components fainter than 10^m and spectroscopic binaries are omitted.

F indicates fundamental distances according to P. MULLER (i.e., stars that can be used for the calibration of systematic distance errors). The exact distance for 1960 is given, and, in parentheses, the predicted distance for 1980.

U, for stars with known orbits, denotes the period in years. Where no U values are given, only very small orbital motion has been observed until now, or none at all. [For data about the best known orbits see DIE STERNE **39**, 69 (1963)].

P, Nos. 1–30, are double stars near the pole, according to the map which is Fig. 2–7 [according to Sterne und Weltraum **19**, 118 (1965)].

ADS	Star		α_{1950}	δ_{1950}	m	Sp	θ	ρ	Remarks
32	Σ	3056	0^{h}02^{m}1	+33°59′	7.9−7.9	K	147	0″7	
61	Σ	3062	03. 5	+58 09	6.5−7.3	G	241⁺	1.3	U 107
102	Σ	2	06. 5	+79 26	6.8−7.1	A	34	0.5	U ∼ 400 P 29
191	Σ	12	12. 4	+08 33	6.1−7.7	F	148	12	35 Psc
207	Σ	13	13. 4	+76 40	6.8−7.3	B	63	0.9	U ∼ 1000
558	Σ	46	37. 3	+21 10	5.6−8.5	K	194	6.5	55 Psc
671	Σ	60	46. 1	+57 33	3.5−7.2	G	295	11	η Cas U 480
683	Σ	61	47. 2	+27 26	6.3−6.3	F	296	4.45	(4.45) F 65 Psc
710	Σ	65	49. 6	+68 36	7.9−7.9	A	39	3.1	
824	Σ	79	57. 2	+44 27	6.0−6.8	B	192	7.8	
875	Σ	84	1 01. 2	+01 06	6.2−8.7	F	253	17	26 Cet
899	Σ	88	03. 0	+21 13	5.6−5.8	A	159	30	ψ¹ Psc
903	Σ	90	03. 2	+04 39	6.7−7.6	F	83	33	77 Psc
923	Σ	91	04. 6	−02 00	7.4−8.2	F	317	4.2	
996	Σ	100	11. 1	+07 19	5.6−6.5	A	63	24	ζ Psc
1030	OΣ	28	14. 3	+80 36	8.0−8.0	F	297	0.9	P 26
1081	Σ	113	17. 2	−00 46	6.4−7.4	F	9	1.5	42 Cet
1339	Σ	147	39. 3	−11 34	6.1−7.4	F	90	2.2	χ¹ Cet
1411	OΣ	34	44. 2	+80 38	7.8−8.1	A	259⁺	0.3	U ∼ 400 P 30
1438	Σ	162	46. 2	+47 39	6.5−7.0	A	207	2.0	
1457	Σ	174	47. 4	+22 02	6.2−7.4	F	166	2.8	1 Ari
1477	Σ	93	48. 8	+89 02	2.1−9.1	F	219	18.4	α UMi P 8
1504	Σ	170	50. 6	+75 59	7.4−8.2	A	246	3.3	
1507	Σ	180	50. 8	+19 03	4.7−4.8	A	0	7.8	γ Ari
1538	Σ	186	53. 3	+01 36	6.8−6.8	G	50	1.4	U 158
1563	HV	12	55. 1	+23 20	4.9−7.6	A	46	38	λ Ari
1615	Σ	202	59. 4	+02 31	4.6−5.6	A	295	2.0	α Psc U ∼ 1000
1630	Σ	205	2 00. 8	+42 06	2.3−5.1	K	64	10	γ And A-BC (″4)
1659	OΣ	37	04. 1	+81 15	6.9−9.1	A	210	1.2	P 23
—	—		07. 0	+79 27	6.5−7.1	A	275	55	P 2
1683	Σ	222	07. 8	+38 48	6.0−6.7	A	35	17	59 And
1697	Σ	227	09. 5	+30 04	5.5−6.9	G	72	3.9	ι Tri
1703	Σ	231	10. 2	−02 38	5.9−7.7	G	232	16	66 Cet
1860	Σ	262	24. 9	+67 11	4.7−7.6	A	238	2.4	ι Cas AB U ∼ 1000 (C 8.5 7″)
1878	Σ	268	25. 9	+55 19	6.9−8.2	A	129	2.8	

Table 22. *Continued*

ADS	Star		α_{1950}	δ_{1950}	m	Sp	θ	ρ	Remarks
2046	Σ	295	38. 6	− 00 54	5.8−9.0	F	310	4.1	84 Cet
2080	Σ	299	40. 7	+ 03 02	3.6−7.4	A	295	2.8	γ Cet
2122	Σ	305	44. 6	+ 19 10	7.3−8.2	G	311	3.5	U ∼ 720
2151	Σ	311	46. 5	+ 17 15	5.3−8.7	B	120	3.2	π Ari
2157	Σ	307	47. 0	+ 55 41	3.9−8.5	K	300	28	η Per
2185	Σ	314	49. 3	+ 52 48	7.1−7.3	B	309	1.6	AB-C (″2)
2257	Σ	333	56. 4	+ 21 08	5.2−5.5	A	206	1.5	ε Ari
2270	Σ	331	57. 2	+ 52 09	5.4−6.8	B	85	12	
2294	Σ	320	59. 4	+ 79 13	5.8−9.0	M	229	4.6	47 Cep P 14
2348	Σ	327	3 03. 8	+ 81 17	6.0−10	A	283	24	P 5
2390	Σ	360	09. 0	+ 37 02	8.1−8.3	G	129	2.5	
2443	Σ	369	13. 9	+ 40 18	6.7−8.0	A	28	3.5	
6623	Σ	1187	8ʰ06ᵐ4	+ 32°22′	7.1−8.0	F	30	2.6	
6650	Σ	1196	09. 3	+ 17 48	5.6−6.0	G	4⁻	1.2	ζ Cnc AB U 60 (C 6.3 6″)
6811	Σ	1224	23. 7	+ 24 42	7.1−7.6	A	47	5.8	24 Cnc A-BC (″2)
6815	Σ	1223	23. 8	+ 27 06	6.3−6.3	A	217	5.0	φ² Cnc
6886	Σ	1245	33. 2	+ 06 48	6.0−7.1	F	25	10	
6977	Σ	1270	42. 8	− 02 25	6.5−7.5	F	262	4.7	
6988	Σ	1268	43. 7	+ 28 57	4.2−6.6	G	307	30	ι Cnc
6993	Σ	1273	44. 2	+ 06 36	3.5−6.9	F	270	2.9	ε Hya AB-C (″2 U 15) U ∼ 900
−	Cou	10	45. 9	+ 88 46	7.1−10	A	65	1.7	4 UMi P 21
7034	Σ	1282	47. 6	+ 35 15	7.5−7.5	F	278	3.64	(3.64) F
7071	Σ	1291	51. 2	+ 30 46	6.2−6.5	K	317	1.4	σ² Cnc
7093	Σ	1295	53. 0	− 07 47	6.7−6.9	A	0	4.3	17 Hya
7137	Σ	1298	58. 4	+ 32 27	6.0−8.1	A	137	4.5	66 Cnc
7187	Σ	1311	9 04. 6	+ 23 11	6.9−7.3	F	200	7.7	
7203	Σ	1306	06. 0	+ 67 21	4.9−8.2	F	19	2.5	σ² UMa U ∼ 1000
7251	Σ	1321	11. 4	+ 52 55	8.1−8.1	K	80	18	U ∼ 1000
7286	Σ	1333	15. 4	+ 35 35	6.4−6.7	A	47	1.84	(1.86) F
7292	Σ	1334	15. 8	+ 37 01	3.9−6.5	A	229	2.8	38 Lyn
7307	Σ	1338	17. 9	+ 38 24	6.5−6.7	F	224	1.2	U 220
7324	Σ	1340	19. 2	+ 49 46	7.0−8.7	B	319	6.2	39 Lyn
7348	OΣ	200	21. 5	+ 51 47	6.6−8.3	G	335	1.5	
7352	Σ	1348	21. 8	+ 06 34	7.5−7.6	F	317	1.9	
7380	Σ	1355	24. 7	+ 06 27	7.5−7.5	F	343	2.2	
7425	Σ	1350	30. 2	+ 67 01	8.1−8.2	F	247	10	
7446	Σ	1362	33. 2	+ 73 18	7.2−7.2	F	129	4.9	
7704	OΣ	215	10 13. 6	+ 17 59	7.2−7.4	F	186	1.4	U ∼ 500
7705	Σ	1415	13. 9	+ 71 19	6.7−7.3	A	167	17	
7724	Σ	1424	17. 2	+ 20 06	2.6−3.8	K	121	4.2	γ Leo U ∼ 620
7837	Σ	1450	32. 4	+ 08 55	5.8−8.5	A	157	2.2	49 Leo
7902	Σ	1466	40. 8	+ 05 01	6.3−7.4	K	240	6.5	35 Sex
7936	Σ	1476	46. 8	− 03 46	6.9−7.7	A	10	2″3	40 Sex
7979	Σ	1487	52. 9	+ 25 01	4.5−6.3	A	110	6.5	54 Leo
8119	Σ	1523	11 15. 6	+ 31 49	4.4−4.9	G	152⁻	2.2	ξ UMa U 60
8131	Σ	1529	16. 8	− 01 23	6.9−7.9	F	252	9.4	
8175	Σ	1543	26. 4	+ 39 37	5.3−8.3	A	359	5.5	57 UMa
8220	Σ	1552	32. 1	+ 17 04	6.0−7.3	B	209	3.4	90 Leo
8249	Σ	1559	36. 0	+ 64 37	6.8−7.8	A	322	2.0	
8250	Σ	1561	36. 2	+ 45 23	6.5−8.5	G	252	10	
8347	Σ	1579	52. 5	+ 46 45	6.8−8.0	A	38	3.8	65 UMa AB-C (″3) D 6.8 63″
8406	Σ	1596	12 01. 6	+ 21 44	6.0−7.5	F	237	3.7	2 Com
−	Sh	136	08. 8	+ 81 59	6.5−8.5	K	76	67	P 1
8477	Σ	1619	12. 6	− 06 58	8.0−8.3	G	272	7.2	
8489	Σ	1622	13. 6	+ 40 56	5.9−8.2	K	260	12	2 CVn
8494	Σ	1625	14. 1	+ 80 24	7.3−7.8	F	219	14.4	P 9
8505	Σ	1627	15. 6	− 03 40	6.7−7.0	F	196	20	
8519	Σ	1633	18. 2	+ 27 30	7.0−7.1	F	245	9.0	
8539	Σ	1639	21. 9	+ 25 52	6.7−7.8	A	330	1.3	U ∼ 750
8561	Σ	1645	25. 7	+ 45 04	7.4−8.0	F	158	10	
8568	OΣI	21	26. 4	+ 26 11	5.4−6.7	A	251	145	17 Com
8600	Σ	1657	32. 6	+ 18 39	5.2−6.7	K	271	20	24 Com
8627	Σ	1669	38. 7	− 12 44	6.0−6.1	F	309	5.4	

Table 22. *Continued*

ADS	Star	α_{1950}	δ_{1950}	m	Sp	θ	ρ	Remarks
8630	Σ 1670	39. 1	− 01 11	3.7−3.7	F	308	5.0	γ Vir U 172
8682	Σ 1694	48. 6	+ 83 41	5.3−5.8	A	326	21.6	32 Cam P 6
8375	OΣ 258	53. 6	+ 82 47	7.0−11	K	70	10.5	P 12
8706	Σ 1692	53. 7	+ 38 35	2.9−5.4	A	228	20	α CVn
8708	OΣ 256	53. 9	− 00 41	7.2−7.5	F	90	0.9	
8710	Σ 1695	54. 1	+ 54 22	6.0−7.9	A	283	3.5	
8801	Σ 1724	13 07. 4	− 05 16	4.4−9.0	A	343	7.1	θ Vir
8814	OΣ 261	09. 7	+ 32 21	7.2−7.7	F	340	2.2	
—	Kpr	11. 9	+ 80 44	6.3−10	G	178	1.0	P 25
8891	Σ 1744	21. 9	+ 55 11	2.4−4.0	A	151	14	ζ UMa
8972	Σ 1763	35. 0	− 07 37	7.9−7.9	K	41	2.8	81 Vir
8974	Σ 1768	35. 2	+ 36 33	5.1−7.0	F	106	1.7	25 CVn U 220
8991	Σ 1772	38. 3	+ 20 12	5.7−8.8	A	136	4.7	1 Boo
9000	Σ 1777	40. 6	+ 03 47	5.7−8.0	K	229	3.0	84 Vir
9031	Σ 1785	46. 8	+ 27 14	7.6−8.0	K	144	3.0	U 156
9053	Σ 1788	52. 4	− 07 49	6.5−7.7	F	90	3.3	
9173	Σ 1821	14 11. 7	+ 52 01	4.6−6.6	A	236	13	ϰ Boo
2563	Σ 389	3ʰ26ᵐ1	+ 59°12′	6.4−7.4	A	68	2.7	
2582	Σ 401	28. 3	+ 27 24	6.5−7.0	A	270	11	
2628	β 533	32. 5	+ 31 31	7.6−7.6	F	43	1.1	
2668	Σ 425	37. 0	+ 33 57	7.6−7.6	F	79	2.0	
2850	Σ 470	51. 8	− 03 06	4.9−6.3	A	347	6.8	32 Eri
2867	OΣ 67	52. 9	+ 60 58	5.3−8.5	K	47	1.8	
2888	Σ 471	54. 5	+ 39 52	3.0−8.2	B	10	8.8	ε Per
2926	Σ 479	58. 0	+ 23 04	6.9−7.8	B	128	7.3	
2963	Σ 460	4 01. 7	+ 80 34	5.7−6.4	F	94	0.9	U 415 P 27
2999	Σ 495	04. 8	+ 15 02	6.0−8.8	F	221	3.8	
3082	OΣ 77	12. 8	+ 31 34	8.1−8.1	F	260	0.8	U 200
3085	H VI 98	12. 8	+ 06 04	6.5−7.2	G	315	65	
3188	OΣ 81	21. 4	+ 33 51	5.9−8.7	F	27	4.4	56 Per
3264	Σ 554	27. 3	+ 15 32	5.8−8.1	F	22	1.7	80 Tau U 170
3273	Σ 552	28. 0	+ 39 54	6.9−7.1	B	114	9.0	
3274	Σ 550	28. 1	+ 53 48	5.9−7.0	B	308	10	1 Cam
3297	Σ 559	30. 6	+ 17 55	7.0−7.1	B	277	3.12	(3.12) F
3353	Σ 572	35. 4	+ 26 51	7.2−7.2	F	194	4.04	(4.14) F
3409	Σ 590	41. 2	− 08 53	6.7−6.8	F	317	9.2	55 Eri
3572	Σ 616	55. 8	+ 37 49	5.1−8.0	A	359	5.3	ω Aur
3597	Σ 627	57. 8	+ 03 33	6.6−6.9	A	260	21	
3734	Σ 644	5 06. 9	+ 37 14	6.9−7.0	B	221	1.61	(1.61) F
3764	Σ 652	09. 2	+ 00 59	6.3−7.8	F	182	1.7	
3797	Σ 654	10. 7	+ 02 48	4.7−8.5	K	64	7.0	ϱ Ori
3800	Σ 661	10. 9	− 13 00	4.5−7.4	B	358	2.5	ϰ Lep
3823	Σ 668	12. 1	− 08 15	0.3−7.0	B	202	9.4	β Ori A-BC (″2)
3824	Σ 653	12. 2	+ 32 38	5.3−7.5	A	226	15	14 Aur
3962	Σ 696	20. 2	+ 03 30	5.0−7.1	B	28	32	23 Ori
3991	Wnc 2	21. 3	− 00 55	6.5−7.3	F	162	2.5	A-BC (″3)
4002	Da 5	22. 0	− 02 26	3.8−4.8	B	77	1.5	η Ori
4068	Σ 716	26. 2	+ 25 07	5.9−6.7	A	205	4.7	118 Tau
4123	Σ 729	28. 6	+ 03 15	5.8−7.1	B	27	2.0	33 Ori
4131	Σ 730	29. 3	+ 17 01	6.0−6.5	B	141	10	
4179	Σ 738	32. 4	+ 09 54	3.7−5.6	O	43	4.4	λ Ori
4186	Σ 748	32. 8	− 05 25	6.8−7.9	O	32	9/13	} θ¹ Ori AB/AC/CD
				5.4−6.8		61	/13	
4188	Σ I 16	32. 9	− 05 27	5.2−6.5	B	92	52	θ² Ori
4193	Σ 752	33. 0	− 05 56	2.9−7.1	O	141	11	ι Ori
4200	Σ 742	33. 4	+ 21 58	7.2−7.8	F	268	3.7	
4208	Σ 749	34. 0	+ 26 54	6.4−6.5	B	333	1.1	
4241	Σ 762	36. 2	− 02 38	3.8−7.5	B	84	13	σ Ori AB-D (″3) E 6.5 42″
4263	Σ 774	38. 2	− 01 58	2.0−4.2	B	163	2.5	ζ Ori
4376	Σ 3115	44. 4	+ 62 48	6.5−7.6	A	2	1.0	
4390	Σ 795	45. 3	+ 06 26	6.0−6.0	A	211	1.2	52 Ori
4566	OΣ 545	56. 3	+ 37 13	2.7−7.2	A	319	3.3	θ Aur
4773	Σ 845	6 07. 8	+ 48 43	6.1−6.8	A	356	7.8	41 Aur

Table 22. *Continued*

ADS	Star	α_{1950}	δ_{1950}	m	Sp	θ	ρ	Remarks
4991	Σ 899	19. 9	+ 17 36	7.1 − 8.1	A	19	2.3	
5012	Σ 900	21. 1	+ 04 37	4.5 − 6.5	A	27	13	ε Mon
5107	Σ 919	26. 4	− 07 00	4.7 − 5.2	B	132	7.3	β Mon AB (BC 5.6 3″)
5166	Σ 924	29. 4	+ 17 49	7.1 − 8.0	F	210	20	20 Gem
5322	Σ 950	38. 2	+ 09 57	4.8 − 7.6	O	213	2.8	15 Mon
5400	Σ 948	41. 8	+ 59 30	5.3 − 6.2	A	88	1.7	12 Lyn AB U 700 (C 7.4 9″)
5436	Σ 958	44. 0	+ 55 46	6.3 − 6.3	F	257	4.72	(4.59) F
5559	Σ 982	51. 8	+ 13 15	4.8 − 7.7	F	150	6.9	38 Gem
5605	Σ 997	53. 8	− 13 59	5.2 − 8.5	G	340	3.0	μ CMa
5746	Σ 1009	7 01. 7	+ 52 50	6.9 − 7.0	A	150	3.8	
5983	Σ 1066	17. 1	+ 22 05	3.5 − 8.5	F	215	6.5	δ Gem U ∼ 1200
6004	Σ 1065	18. 4	+ 50 15	7.3 − 7.4	F	254	15	20 Lyn
6012	Σ 1062	18. 8	+ 55 23	5.6 − 6.5	B	315	15	19 Lyn
6060	Σ 1083	22. 6	+ 20 36	7.2 − 8.3	A	44	6.6	
6126	Σ 1104	27. 1	− 14 53	6.2 − 7.8	F	2	2.2	
6175	Σ 1110	31. 4	+ 32 00	2.0 − 2.9	A	167⁻	2.2	αGem AB U ∼ 420 (C 9.2 72″)
6319	Σ 1122	41. 2	+ 65 17	7.7 − 7.7	F	5	15	
6321	OΣ 179	41. 4	+ 24 31	3.7 − 8.4	G	238	7.0	ϰ Gem
6348	Σ 1138	43. 2	− 14 34	6.1 − 6.8	A	339	17	2 Pup
6381	Σ 1146	45. 6	− 12 04	5.7 − 7.7	F	4	2.4	
6454	Σ 1157	52. 0	− 02 40	7.9 − 7.9	F	219	1.0	
6569	Σ 1177	8 02. 6	+ 27 40	6.5 − 7.4	B	351	3.5	
9174	Σ 1816	14ʰ11ᵐ7	+ 29°20′	7.5 − 7.6	F	86	1.0	
9182	Σ 1819	12. 8	+ 03 22	7.9 − 8.0	F	283⁻	1.0	U 360
9192	Σ 1825	14. 2	+ 20 21	6.6 − 8.3	F	162	4.3	
9237	Σ 1833	20. 0	− 07 33	7.5 − 7.5	G	172	5.7	
9247	Σ 1835	20. 9	+ 08 40	5.1 − 6.6	A	192	6.4	A-BC (″2)
9273	Σ 1846	25. 6	− 02 00	5.0 − 9.5	K	110	4.8	φ Vir
9277	Σ 1850	26. 3	+ 28 31	7.0 − 7.4	A	262	25	
9338	Σ 1864	38. 4	+ 16 38	4.9 − 5.8	A	108	5.63	(5.49) F π Boo
9343	Σ 1865	38. 8	+ 13 57	4.6 − 4.7	A	308	1.2	ζ Boo U 123
9358	Σ 1915	40. 9	+ 86 09	7.0 − 10	K	321	2.4	P 18
9372	Σ 1877	42. 8	+ 27 17	2.7 − 5.1	K	338	3.0	ε Boo
9396	β 106	46. 6	− 13 57	5.8 − 6.7	A	355	1.8	μ Lib
9406	Σ 1890	48. 0	+ 48 55	6.1 − 6.8	F	45	3.0	39 Boo
9413	Σ 1888	49. 1	+ 19 19	4.8 − 6.9	G	347	6.9	ξ Boo U 150
9425	OΣ 288	51. 0	+ 15 54	6.9 − 7.6	G	178	1.6	U 215
9445	Hu 908	54. 2	+ 78 23	6.5 − 10	K	254	1.2	P 24
9507	Σ 1910	15 05. 2	+ 09 25	7.4 − 7.4	G	211	4.3	
9626	Σ 1938	22. 6	+ 37 31	7.2 − 7.8	K	22	2.0	μ Boo BC U 260 (A 4.5 108″)
9696	Σ 1972	32. 0	+ 80 37	6.9 − 7.7	G	80	31	π¹ UMi P 3
9701	Σ 1954	32. 4	+ 10 42	4.2 − 5.2	F	177	3.8	δ Ser
9716	OΣ 298	34. 3	+ 39 58	7.4 − 7.6	K	185	1.2	U 56
9728	Σ 1962	36. 0	− 08 38	6.5 − 6.6	F	188	12	
9737	Σ 1965	37. 5	+ 36 48	5.1 − 6.0	B	304	6.24	(6.19) F ζ CrB
9769	Σ 1989	42. 3	+ 80 08	7.4 − 8.2	F	34	0.7	π² UMi P 28 U 160
9778	Σ 1970	43. 9	+ 15 35	3.7 − 9.5	A	265	30	β Ser
9853	Σ 2034	54. 7	+ 83 46	7.6 − 8.1	A	115	1.4	P 22
9909	Σ 1998	16 01. 6	− 11 14	4.2 − 7.2	F	55	8	ξ Sco AB-C
								(AB 4.8 − 5.1 ″8 U 46)
9933	Σ 2010	05. 8	+ 17 11	5.3 − 6.5	G	12	30	ϰ Her
9969	Σ 2021	11. 0	+ 13 40	7.5 − 7.7	K	346	4.17	(4.17) F 49 Ser
9979	Σ 2032	12. 8	+ 33 59	5.7 − 6.7	G	229	6.3	σ CrB U ∼ 1000
10052	Σ 2054	23. 1	+ 61 48	5.9 − 7.1	G	355	1.1	
10058	OΣ 312	23. 3	+ 61 38	2.9 − 8.3	K	141	5.2	η Dra
10129	Σ 2078	35. 0	+ 53 01	5.6 − 6.6	A	108	3.3	17 Dra
10149	Σ I 31	38. 2	+ 04 19	5.7 − 6.9	A	230	70	37/36 Her
10157	Σ 2084	39. 4	+ 31 41	2.9 − 5.9	G	60⁻	1.5	ζ Her U 34
10214	Hu 917	45. 3	+ 77 36	6.1 − 9.4	F	188	2.9	P 17
10312	Σ 2114	59. 6	+ 08 31	6.5 − 7.7	A	181	1.2	
10345	Σ 2130	17 04. 3	+ 54 32	5.8 − 5.8	F	73⁻	2.0	μ Dra U ∼ 500
10418	Σ 2140	12. 4	+ 14 27	3.5 − 5.4	M	107	4.5	α Her
10526	Σ 2161	22. 0	+ 37 11	4.5 − 5.5	A	316	4.02	(4.02) F ϱ Her

Table 22. *Continued*

ADS	Star	α_{1950}	δ_{1950}	m	Sp	θ	ρ	Remarks
10597	Σ 2180	27. 8	+ 50 55	7.6 − 7.8	F	261	3.15	(3.15) F
10628	Σ I 35	31. 3	+ 55 12	5.0 − 5.0	A	312	62	ν^1/ν^2 Dra
10728	Σ 2218	40. 0	+ 63 42	7.1 − 8.3	F	328	1.7	
10750	Σ 2202	42. 0	+ 02 36	6.2 − 6.6	A	93	20	61 Oph
10759	Σ 2241	42. 8	+ 72 11	4.9 − 6.1	F	15	30	ψ Dra
10786	Σ 2220	44. 5	+ 27 46	3.5 − 9.8	G	247	34	μ Her A-BC (″6 U 43)
10849	Σ 2242	49. 7	+ 44 55	8.0 − 8.0	F	327	3.5	
10850	OΣ 338	49. 7	+ 15 20	7.2 − 7.3	K	357	0.8	
10875	β 130	51. 7	+ 40 01	5.2 − 8.5	K	115	1.7	90 Her
10905	Σ 2245	54. 2	+ 18 20	7.3 − 7.3	A	293	2.63	(2.63) F
10993	Σ 2264	59. 4	+ 21 36	5.1 − 5.2	A	258	6.2	95 Her
11005	Σ 2262	18 00. 4	− 08 11	5.4 − 6.0	F	271	2.0	τ Oph U 280
11046	Σ 2272	02. 9	+ 02 32	4.2 − 5.9	K	91⁻	4.4	70 Oph U 88
11061	Σ 2308	03. 8	+ 80 00	5.8 − 6.2	F	232	19.2	40/41 Dra P 7
11089	Σ 2280	05. 8	+ 26 05	5.9 − 6.0	A	183	14	100 Her
11123	Σ 2289	07. 9	+ 16 28	6.5 − 7.5	F	224	1.2	
11324	AC 11	22. 4	− 01 36	6.8 − 7.1	F	357	0.8	U 240
11336	Σ 2323	23. 2	+ 58 46	4.9 − 7.9	A	353	3.8	39 Dra (C 7.4 90″)
11353	Σ 2316	24. 6	+ 00 10	5.4 − 7.7	A	318	3.8	59 Ser
11483	OΣ 358	33. 6	+ 16 56	6.7 − 7.1	G	171	1.8	U 300
11500	Σ 2351	34. 6	+ 41 14	7.6 − 7.6	A	340	5.2	
11558	Σ 2368	37. 7	+ 52 18	7.5 − 7.7	A	322	1.9	
11635	Σ 2382	42. 7	+ 39 37	5.1 − 6.0	A	0	2.8	ε Lyr AB (AC 208″)
11635	Σ 2383	42. 7	+ 39 34	5.1 − 5.4	A	100	2.3	CD U ~ 600
11639	Σ I 38	43. 0	+ 37 33	4.3 − 5.9	A	150	44	ζ Lyr
11640	Σ 2375	43. 0	+ 05 27	6.3 − 6.7	A	117	2.49	(2.54) F AB-CD (je ″1)
11667	Σ 2379	43. 9	− 01 01	5.8 − 7.5	A	121	13	5 Aql
11745	Σ I 39	18ʰ48ᵐ2	+ 33°18′	3 v − 6.7	B	149	46	β Lyr
11853	Σ 2417	53. 7	+ 04 08	4.5 − 5.4	A	104	22	θ Ser
11870	Σ 2452	55. 2	+ 75 43	6.6 − 7.4	A	218	5.7	
12061	Σ 2461	19 05. 5	+ 32 25	5.1 − 9.2	F	299	3.5	17 Lyr
12169	Σ 2486	10. 8	+ 49 46	6.6 − 6.8	G	213	8.6	
12197	Σ 2487	12. 0′	+ 39 03	4.5 − 9.0	B	82	28	η Lyr
12540	Σ I 43	28. 7	+ 27 51	3.2 − 5.4	K	54	35	β Cyg
12594	Σ 2540	31. 1	+ 20 18	7.2 − 8.7	A	147	5.1	
12608	Σ 2571	31. 8	+ 78 09	7.6 − 8.3	F	20	11.3	P 11
12815	Σ I 46	40. 7	+ 50 25	6.3 − 6.4	G	135	38	16 Cyg
12880	Σ 2579	43. 4	+ 45 00	3.0 − 6.5	A	246	2.1	δ Cyg U ~ 540
12913	Σ 2580	44. 5	+ 33 37	5.0 − 9.0	F	70	26	17 Cyg
12962	Σ 2583	46. 4	+ 11 41	6.1 − 6.9	F	110	1.46	(1.46) F π Aql
13007	Σ 2603	48. 3	+ 70 08	4.0 − 7.6	K	13	3.1	ε Dra
13019	Σ 2587	49. 0	+ 03 58	6.7 − 9.3	K	100	4.1	
13087	Σ 2594	51. 9	− 08 21	5.8 − 6.5	B	170	36	57 Aql
13148	Σ 2605	54. 4	+ 52 18	4.9 − 7.4	A	178	3.2	ψ Cyg
13256	Σ 2613	59. 0	+ 10 36	7.5 − 7.7	F	352	4.0	
13277	OΣ 395	20 00. 0	+ 24 48	5.9 − 6.3	F	116	0.9	16 Vul
13312	Σ 2624	01. 6	+ 35 53	7.2 − 8.0	O	174	2.0	C 9.1 42″
13442	Σ 2637	07. 7	+ 20 46	6.4 − 8.7	F	325	12	θ Sge
13506	Σ 2644	10. 0	+ 00 43	6.9 − 7.2	A	208	3.0	
13524	Σ 2675	10. 7	+ 77 34	4.4 − 8.4	B	122	7.4	$\varkappa$ Cep P 13
13672	Σ 2666	16. 4	+ 40 35	6.0 − 8.2	B	245	2.8	
13692	Σ 2671	17. 2	+ 55 14	6.0 − 7.4	A	338	3.5	
13708	Σ 2694	17. 4	+ 80 23	6.8 − 11	A	344	4.0	P 15
14158	Σ 2716	39. 0	+ 32 08	5.9 − 8.0	K	47	2.7	49 Cyg
14259	Σ 2726	43. 6	+ 30 32	4.3 − 9.2	K	67	6.0	52 Cyg
14270	Σ 2725	43. 9	+ 15 43	7.5 − 8.2	K	6	5.5	
14279	Σ 2727	44. 4	+ 15 57	4.5 − 5.5	G	268	10	γ Del
14336	Σ 2732	47. 2	+ 51 43	6.5 − 8.5	B	74	4.1	
14421	OΣ 418	52. 8	+ 32 31	8.1 − 8.2	G	287	1.1	
14504	Σ 2741	56. 9	+ 50 16	5.8 − 7.1	B	27	1.9	
14556	Σ 2742	59. 8	+ 06 59	7.4 − 7.4	F	216	2.8	2 Eql
14573	Σ 2744	21 00. 5	+ 01 20	7.0 − 7.5	F	137	1.4	

Table 22. *Continued*

ADS	Star	α_{1950}	δ_{1950}	m	Sp	θ	ρ	Remarks
14575	Σ 2751	00. 7	+56 28	6.1—7.1	B	354	1.6	
14592	Σ 2745	01. 4	−06 01	5.9—7.3	F	192	2.3	12 Aqr
14636	Σ 2758	04. 4	+38 28	5.4—6.2	K	142	28	61 Cyg U 700
14682	Σ 2762	06. 5	+30 00	5.7—7.7	A	306	3.5	
14749	Σ 2780	10. 5	+59 47	6.0—7.0	B	217	1.0	
14845	Σ 2796	16. 8	+78 24	7.2—10	A	43	26	P 4
14889	OΣ 437	18. 7	+32 14	6.9—7.6	G	28	2.1	
14916	Σ 2801	20. 1	+80 08	7.8—8.5	F	271	2.0	P 20
15007	Σ 2799	26. 4	+10 52	7.4—7.4	F	95	1.6	
15032	Σ 2806	28. 0	+70 20	3 v—8.0	B	250	13	β Cep
15184	Σ 2816	37. 4	+57 16	5.6—8.0	O	121	12	C 8.0 20″
15270	Σ 2822	41. 9	+28 31	4.7—6.1	F	283	1.6	μ Cyg U ∼ 500
15407	Σ 2843	50. 4	+65 31	7.1—7.3	A	143	1.7	
15571	Σ 2873	22 00. 1	+82 38	7.0—7.5	F	69	13.7	P 10
15600	Σ 2863	02. 2	+64 23	4.6—6.5	A	277	7.5	ξ Cep
15767	Σ 2878	12. 0	+07 44	6.8—8.3	A	122	1.5	
15971	Σ 2909	26. 2	−00 17	4.4—4.6	F	261⁻	1.8	ζ Aqr U ∼ 600
15987	Σ I 58	27. 3	+58 10	4 v—7.5	G	191	41	δ Cep
16095	Σ 2922	33. 6	+39 23	5.8—6.5	B	186	22	8 Lac
16228	Σ 2942	41. 8	+39 12	6.3—8.5	K	278	2.9	
16243	OΣ 481	43. 1	+78 15	7.5—9.3	A	272	2.3	P 19
16291	Σ 2947	47. 3	+68 18	7.1—7.1	F	59	4.2	
16294	OΣ 482	47. 7	+82 53	5.0—9.7	K	33	3.5	P 16
16317	Σ 2950	49. 4	+61 25	6.1—7.4	G	295	1.8	
16666	Σ 3001	23 16. 4	+67 50	5.0—7.3	K	211	3.0	o Cep U ∼ 800
16775	Σ 3017	25. 7	+73 49	7.5—8.6	F	25	1.7	
17022	OΣ 508	46. 4	+61 57	5.7—8.2	A	195	1.8	6 Cas
17140	Σ 3049	56. 4	+55 29	5.1—7.2	B	326	3.0	σ Cas
17149	Σ 3050	56. 9	+33 27	6.5—6.7	F	279⁺	1.5	U 320

Table 23. Star clusters and nebulae: Messier's catalog of 1784

The table contains the Messier number, the NGC number (or the IC number marked *), the coordinates for 1950, type of object, visual total brightness, and any other name which may have been given to the object.

The types involved are:

$$O = \text{open clusters}$$
$$G = \text{galactic clusters}$$
$$P = \text{planetary nebulae}$$
$$K = \text{globular clusters}$$
$$E = \text{extragalactic nebulae}$$

The objects 40 and 91 do not exist; the identifications of M 47 and M 48 are doubtful. M 102 is identical with M 101. An object given by Messier without number was numbered M 104 by later authors.

M	NGC	Coordinates		Type	Brightness	Remarks
1	1952	5^h31^m	$+22°0$	G	8^m	Crab Nebula
2	7089	21 31	$-$ 1.1	K	6	
3	5272	13 40	$+28.6$	K	6	
4	6121	16 21	-26.4	K	6	
5	5904	15 16	$+$ 2.3	K	6	
6	6405	17 37	-32.2	O	5	
7	6475	17 51	-34.8	O	4	
8	6523	18 0	-24.4	G	6	
9	6333	17 16	$-18°5$	K	7	Lagoon Nebula
10	6254	16 55	$-$ 4.0	K	7	
11	6705	18 48	$-$ 6.3	O	6	
12	6218	16 45	$-$ 1.9	K	7	
13	6205	16 40	$+36.6$	K	6	
14	6402	17 35	$-$ 3.2	K	8	Globular Cluster in Hercules
15	7078	21 28	$+11.9$	K	6	
16	6611	18 16	-13.8	O	6	
17	6618	18 18	-16.2	G	7	
18	6613	18 17	-17.2	O	8	Omega Nebula
19	6273	17 0	-26.2	K	7	
20	6514	17 59	-23.0	G	9	
21	6531	18 2	-22.5	O	7	Trifid Nebula
22	6656	18 33	-23.9	K	6	
23	6494	17 54	-19.0	O	7	
24	6603	18 16	-18.4	O	5	
25	*4725	18 29	-19.3	O	7	
26	6694	18 43	$-$ 9.4	O	9	
27	6853	19 58	$+22.6$	P	8	
28	6626	18 22	-24.9	K	7	Dumbbell Nebula
29	6913	20 22	$+38.4$	O	7	
30	7099	21 38	-23.4	K	8	
31	224	0 40	$+41.0$	E	4	
32	221	0 40	$+40.6$	E	9	Andromeda Nebula
33	598	1 31	$+30.4$	E	6	Companion of And. Neb.
34	1039	2 39	$+42.6$	O	6	Triangulum Nebula
35	2168	6 6	$+24.3$	O	5	
36	1960	5 33	$+34.1$	O	6	
37	2099	5 49	$+32.5$	O	6	
38	1912	5 25	$+35.8$	O	7	
39	7092	21 30	$+48.2$	O	5	
41	2287	6 45	-20.7	O	5	
42	1976	5 33	$-$ 5.4	G	3	
43	1982	5 33	$-$ 5.3	G	9	Orion Nebula
44	2632	8 37	$+20.0$	O	4	
45	—	3 44	$+24.0$	O	2	Praesepe
46	2437	7 40	-14.7	O	6	Pleiades
47	2422 ?	7 35	-14.5	O	5	

Table 23. *Continued*

M	NGC	Coordinates			Type	Brightness	Remarks
48	?	?		?	?	?	
49	4472	12	27	+ 8.3	E	9	
50	2323	7	1	− 8.3	O	6	
51	5194	13	28	+ 47.4	E	8	
52	7654	23	22	+ 61.3	O	7	
53	5024	13	11	+ 18.4	K	8	
54	6715	18^{h}52^m		− 30°6	K	7	
55	6809	19	37	− 31.1	K	8	
56	6779	19	15	+ 30.1	K	8	
57	6720	18	52	+ 33.0	P	9	
58	4579	12	35	+ 12.1	E	8	
59	4621	12	39	+ 11.9	E	9	
60	4649	12	41	+ 11.8	E	9	
61	4303	12	19	+ 4.8	E	10	
62	6266	16	58	− 30.1	K	9	
63	5055	13	14	+ 42.3	E	10	
64	4826	12	54	+ 21.9	E	7	Ring nebula in Lyra
65	3623	11	16	+ 13.4	E	10	
66	3627	11	18	+ 13.3	E	9	
67	2682	8	48	+ 12.0	O	6	
68	4590	12	37	− 26.5	K	9	
69	6637	18	28	− 32.4	K	9	
70	6681	18	40	− 32.3	K	10	
71	6838	19	52	+ 18.6	K	9	
72	6981	20	51	− 12.7	K	10	
73	6994	20	56	− 12.8	O		
74	628	1	34	+ 15.5	E	10	
75	6864	20	3	− 22.1	K	8	
76	650	1	39	+ 51.3	P	12	
77	1068	2	40	− 0.2	E	9	
78	2068	5	44	+ 0.1	G	8	
79	1904	5	22	− 24.6	K	8	
80	6093	16	14	− 22.9	K	8	Consists of 4 stars only
81	3031	9	52	+ 69.3	E	8	
82	3034	9	52	+ 69.9	E	9	
83	5236	13	34	− 29.6	E	10	
84	4374	12	23	+ 13.2	E	9	
85	4382	12	23	+ 18.5	E	9	
86	4406	12	24	+ 13.2	E	10	
87	4486	12	28	+ 12.7	E	9	
88	4501	12	29	+ 14.7	E	10	
89	4552	12	33	+ 12.8	E	10	
90	4569	12	34	+ 13.4	E	10	
92	6341	17	16	+ 43.2	K	6	
93	2447	7	43	− 23.8	O	6	
94	4736	12	49	+ 41.4	E	8	
95	3351	10	41	+ 12.0	E	10	
96	3368	10	44	+ 12.1	E	9	
97	3587	11	12	+ 55.3	P	12	
98	4192	12	11	+ 15.2	E	11	
99	4254	12	16	+ 14.7	E	10	
100	4321	12	20	+ 16.1	E	11	
101	5457	14	1	+ 54.6	E	10	
103	581	1	30	+ 60.5	O	7	
104	4594	12	37	− 11.4	E	9	

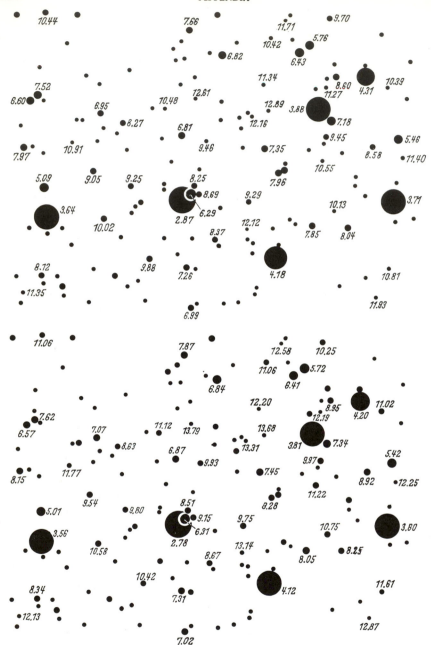

Magnitudes of selected objects at the center of the Pleiades. On top are the visual, at the bottom photographic magnitudes. The data are taken from H. L. JOHNSON and R. J. MITCHELL: Ap. J. **128**, 31 (1958), and are in the UBV system. This map will suffice for most amateur astronomers. For greater detail, reference should be made to the "Atlas der Kapteynschen Areas" (Selected Areas, after Harvard-Groningen by ANTOINE BRUN and HANS VEHRENBERG (see Bibliography, §23.15, page 547).

23 / Bibliography

The following list of books and journals does not lay claim to any degree of completeness. Numerous popular books that have appeared in recent years have had to be omitted. This bibliography is restricted to some selected publications, which in turn contain numerous other references to articles, books, and journals. Most of the publications listed refer to the period after 1945.

The leading bibliography covering all fields in astronomy is "Astronomy and Astrophysics Abstracts," published by Springer-Verlag, Heidelberg–New York. This publication, begun in 1968, succeeded the "Astronomischer Jahresbericht," which had been published since 1899, and which had been preceded by HOUZEAU and LANCASTER, "Bibliographie Générale de l'Astronomie," which carried information up to 1881. (See also § 23.3). Anyone needing a complete survey will find this biannual "Astronomy and Astrophysics Abstracts" indispensable.

23.1. History

ABETTI, G.: The History of Astronomy. London: Sidgwick & Jackson, 1954.
BALL, R. S.: Great Astronomers. London: Isbister & Co., 1895.
BEER, A.: Astronomical Dating of Works of Art. Vistas of Astronomy, Vol. 9. Oxford–New York: Pergamon, 1968.
BEER, A. and P. (Eds.): Kepler: Four Hundred Years. (Symposia Volume.) Oxford–New York: Pergamon, 1974.
BEER, A. and STRAND, K. A. (Eds.): Copernicus—Yesterday and Today (The Washington Symposium.) Oxford–New York: Pergamon, 1974.
BERRY, A.: A Short History of Astronomy. New York: Dover, 1961.
BRODETZKY, S.: Sir Isaac Newton. London: Methuen, 1927.
CLERKE, A. M.: A Popular History of Astronomy during the Nineteenth Century. 4th ed. London: Black, 1908.
CLERKE, A. M.: The Herschels and Modern Astronomy. London: Cassell, 1895.
DRAKE, S.: Galileo. New York: Doubleday, 1957.
DREYER, J. L. E.: Astronomy from Thales to Kepler. New York: Dover, 1953.
GERLACH, W., and LIST, M.: Johannes Kepler; Leben und Werk. Munich: Piper, 1966.
HAWKINS, G. S.: Stonehenge Decoded. London: Fontana-Collins, 1970.
KESTEN, H.: Copernicus and his World. London: Secker and Warburg, 1946.
KING, H. C.: The Background of Astronomy. London: Watts, 1957.
KOESTLER, A.: The Sleepwalkers. London–New York: Hutchinson, 1959.
MACPHERSON, H.: Makers of Astronomy. Oxford: Clarendon Press, 1933.
MEADOWS, A. J.: The High Firmament. (A Survey of Astronomy in English Literature.) Leicester: Leicester Univ. Press, 1969.
MUNITZ, M. K.: Theories of the Universe. Glencoe, Ill.: The Free Press, 1957.

NEUGEBAUER, O.: The Exact Sciences in Antiquity. New York: Dover, 1969.

PANNEKOEK, A.: A History of Astronomy. New York: Interscience, 1961.

SCHMEIDLER, F.: Nikolaus Kopernikus. Stuttgart: Wissenschaftliche Verlags-gesellschaft, 1970.

SHAPLEY, H., and HOWARTH, H. E.: A Source Book in Astronomy. New York: McGraw-Hill, 1929.

STRUVE, O., and ZEBERGS, V.: Astronomy of the Twentieth Century. New York–London: Macmillan, 1962.

TOULMIN, S., and GOODFIELD, J.: The Fabric of the Heavens. (The Ancestry of Science.) London: Hutchinson, 1961.

VAN DER WAERDEN, B. L.: Science Awakening. Groningen: Noordhoff, 1954.

ZINNER, E.: Deutsche und niederländische astronomische Instrumente des 11.–18. Jahrhunderts, 2nd Ed. München: C. H. Beck, 1967.

23.2. Popular Surveys

ABELL, G.: Exploration of the Universe. New York: Holt, Rinehart & Winston, 1964.

CALDER, N.: Violent Universe (An Eye-Witness Account of the Commotion in Astronomy.) London: British Broadcasting Corporation, 1969.

DE VAUCOULEURS, G.: L'Esprit de l'Homme à la Conquête de l'Univers. Paris: Editions Spes, 1951.

DE VAUCOULEURS, G.: Discovery of the Universe. New York: Macmillan, 1957.

EASTWOOD, W.: A Book of Scientific Verse. London: Macmillan, 1961.

ERNST, B., and DE VRIES, T. E.: Atlas of the Universe. London: Nelson, 1961.

GAMOW, G.: The Creation of the Universe. New York: New American Library, 1957.

GAPOSCHKIN, C. PAYNE: Introduction to Astronomy. New York: Prentice-Hall, 1954.

HOERNER, S. VON, and SCHAIFERS, K.: Meyers Handbuch über das Weltall, 4th ed. Mannheim: Bibliographisches Institut, 1964.

HONEGGER, G., and VAN DE KAMP, P.: Space. The Architecture of the Universe. New York: Dell, 1962.

HOYLE, F.: The Nature of the Universe. Oxford: Oxford Univ. Press, 1960.

HOYLE, F.: Astronomy. London: Macdonald, 1962.

HOYLE, F.: Frontiers of Astronomy. London: Heinemann, 1955; and New York: Mercury Books, 1961.

JEANS, J. H.: Physics and Philosophy. Cambridge: Cambridge Univ. Press, 1942.

KING, H. C.: Exploration of the Universe. London: Secker & Warburg, 1964.

MARTIN, M. E.: The Friendly Stars. D. H. Menzel, Ed. New York: Dover, 1964.

MINNAERT, M. G. J.: Dichters over Sterren. Arnhem: van Loghum Slaterus, 1949.

MOORE, P.: Atlas of the Universe. Introduction by A. C. B. Lovell; Epilogue by T. O. Paine, NASA. London: G. Philip and Son, and Mitchell Beazley Ltd., 1969.

PECKER, J. C.: Orion Book of the Sky. New York: Orion Press, 1960.

RICHARDSON, R. S.: The Fascinating World of Astronomy. London: Faber and Faber, 1962.

ROHR, H.: The Radiant Universe. Translated and revised from the German version by A. Beer, London: F. Warne, 1972.

SMART, W. M.: The Riddle of the Universe. New York: Wiley, 1968.

SMART, W. M.: Some Famous Stars. New York: Longmans Green, 1950.

THOMAS, O.: Astronomie—Tatsachen und Probleme, 7th Ed. Salzburg: Bergland Verlag, 1956.

VAN DE KAMP, P.: Basic Astronomy. New York: Random House, 1952.

WEIGERT, A., and ZIMERMANN, H.: ABC of Astronomy, J. Home Dickson, Trans. London: Hilger, 1967.

WOOD, H.: Unveiling the Universe. New York: American Elsevier, 1968.

23.3. Textbooks and Encyclopedias

BEER, A., Ed.: Vistas in Astronomy, Vols. 1–20, pp. 177–223. Oxford–New York: Pergamon, 1955–1975.

DANJON, A.: Astronomie Générale (Astronomie sphérique et éléments de mécanique céleste). Paris: Sennac, 1959.

DRAKE, F. D.: Intelligent Life in Space. New York: Macmillan, 1962.

EDDINGTON, A. S.: The Internal Constitution of the Stars. Cambridge: Cambridge University Press, 1926; and New York: Dover, 1959.

EVANS, D. S.: Observation in Modern Astronomy. London: English Universities Press, 1968.

FLAMMARION, G. C., and DANJON, A.: The Flammarion Book of Astronomy. A. and B. Pagel, Trans. London: Allen and Unwin, 1964.

FLÜGGE, S., Ed.: Encyclopaedia of Physics, Vols. 50–54 (Astrophysics I–V). Heidelberg: Springer, 1958–1962.

FOWLER, W. A., and HOYLE, F.: Nucleosynthesis in Massive Stars and Supernovae. Chicago: University of Chicago Press, 1965.

GOLDBERG, L., Ed.: Annual Review of Astronomy and Astrophysics. Palo Alto, Calif.

HOUZEAU, J. C., and LANCASTER, A.: Bibliographie Générale de l'Astronomie. 3 Vols. D. W. Dewhirst, Ed., Introduction by A. Beer. London: Holland Press, 1964.

HUFFER, C. M., TRINKLEIN, F. E., and BUNGE, M.: An Introduction to Astronomy. New York: Holt, Rinehart & Winston, 1967.

KLECZEK, J.: Astronomical Dictionary in Six Languages (English, Russian, German, French, Italian, and Czech). Prague: Academy of Sciences, 1961.

KROGDAHL, W. S.: The Astronomical Universe (An Introductory Text in College Astronomy). New York: Macmillan, 1961.

KUIPER, G. P., and MIDDLEHURST, B. M., Eds.: Stars and Stellar Systems. 9 Vols. University of Chicago Press, 1960.

LANDOLT-BÖRNSTEIN: Zahlenwerte und Funktionen aus Naturwissenschaft und Technik, Neue Serie: Gruppe VI; Astronomie, Astrophysik und Weltraumforschung. Heidelberg–New York: Springer, 1965.

McCREA, W. H.: Physics of the Sun and Stars. London: Hutchinson, 1950.

McLAUGHLIN, D. B.: Introduction to Astronomy. Boston: Houghton Mifflin, 1961.

MIDDLEHURST, B. M., and KUIPER, G. P., Eds.: The Solar System. 5 Vols. University of Chicago Press, 1953–1966.

MINNAERT, M. G. J.: Practical Work in Elementary Astronomy. Dordrecht: Reidel, 1969.

PECKER, J. C.: Le Ciel. Encyclopédie Essentielle. Paris: Delpire, 1960.

PECKER, J. C., and SCHATZMAN, E.: Astrophysique Générale. Paris: Masson, 1959.

RUDAUX, L., and DE VAUCOULEURS, G.: Larousse Encyclopedia of Astronomy. Introduction by F. L. Whipple. London: Hamlyn, 1962.

RUSSELL, H. N., DUGAN, R. S., and STEWART, J. Q.: Astronomy. 2 vols. New York: Ginn, 1962.

SCHATZMAN, E., Ed.: Astronomie. Encyclopédie de la Pléiade. Paris: Gallimard, 1962.

SCHWARZSCHILD, M.: Structure and Evolution of the Stars. Princeton, N.J.: Princeton University Press; and New York: Dover, 1958.

SMART, W. M.: Textbook on Spherical Astronomy, 5th ed. Cambridge University Press, 1962.

STRUVE, O.: Elementary Astronomy. (B. Lynds, Ed.) Oxford: Oxford Univ. Press, 1968.

STUMPF, K.: Astronomy A to Z. English version by A. Beer. New York: Grosset & Dunlap, 1964.

SWIHART, T. L.: Astrophysics and Stellar Astronomy. New York: Wiley, 1968.

TAYLER, R. J.: The Stars: Their Structure and Evolution. London: Wykeham Publications, 1970.

TRUMPLER, R. J., and WEAVER, H. F.: Statistical Astronomy. Berkeley: University of California Press, 1953.

UNSÖLD, A.: The New Cosmos. W. H. McCrea, Trans. London: Longmans, Green & Co.; and New York: Springer-Verlag, 1969.

VON WEIZSÄCKER, C. F.: The History of Nature. London: Routledge & Kegan Paul, 1951.

VORONTSOV-VELYAMINOV, B. A.: Astronomical Problems: An Introductory Course in Astronomy. English version by A. Beer, in collaboration with J. B. Hutchings and P. M. Rabbit. Oxford–New York: Pergamon, 1969.

WOOLARD, E. W., and CLEMENCE, G. M.: Spherical Astronomy. New York–London: Academic Press, 1966.

WYATT, S. P.: Principles of Astronomy. Boston: Allyn & Bacon, 1964.

23.4. Astronomical Tools

BEER, A.: The Medium-type Planetarium. Jena: Carl Zeiss, 1962.

BORN, M., and WOLF, E.: Principles of Optics. London–New York: Pergamon Press, 1959.

BROWN, S.: All About Telescopes. Barrington, N.Y.: Edmund Scientific Co., 1968.

CRAWFORD, D. L., Ed.: The Construction of Large Telescopes. London: Academic Press, 1966.

DE VAUCOULEURS, G.: Astronomical Photography. London: Faber and Faber, 1961.

DIMITROFF, G. F. and BAKER, J. G.: Telescopes and Accessories. Cambridge, Mass.: Harvard University Press, 1954.

EMERSON, M. N.: Amateur Telescope Mirror Making. New York: Carlton Press, 1969.

HORWARD, N. E.: Handbook for Telescope Making. London: Faber and Faber, 1969.

INGALLS, A. G.: Amateur Telescope Making: I, II, III. New York: Scientific American, 1953.

KING, H. C.: The History of the Telescope. London: Griffin, 1955.

KUIPER, G. P., and MIDDLEHURST, B. M., Eds.: Telescopes. University of Chicago Press, 1960.

MERTON, G.: Photography and the Amateur Astronomer. London: British Astronomical Association, 1953.

MICZAIKA, G. R., and SINTON, W. M.: Tools of the Astronomer. Cambridge, Mass.: Harvard University Press, 1961.

MINNAERT, M. G. J.: The Nature of Light and Colour in the Open Air. New York: Dover, 1954.

PAGE, T. and L. W.: Telescopes: How to Make Them and Use Them. London: Macmillan, 1969.

PAUL, H. E.: Telescopes for Skygazing. New York: Amphoto, 1970.

RACKHAM, T.: Astronomical Photography and the Telescope. London: Faber and Faber, 1961.

ROHR, H.: Das Fernrohr für Jedermann: Building a Reflector, 4th ed. Zurich: Rascher, 1964.

ROTH, G. D., Ed.: Refraktor-Selbstbau. München: Verlag Uni-Druck, 1965.

SELWYN, E. W. H.: Photography in Astronomy. Rochester: Eastman Kodak, 1950.

SIDGWICK, J. B.: Amateur Astronomer's Handbook, 3rd ed. London: Faber & Faber, 1958.

STAUS, A.: Fernrohrmontierungen und ihre Schutzbauten für Sternfreunde, 3rd ed. München: Verlag Uni-Druck, 1971.

STRONG, J., et al.: Modern Physical Laboratory Practice. London; Blackie & Son, 1939.

TEXEREAU, J.: How to Make a Telescope. New York: Interscience, 1957.

THACKERAY, A. D.: Astronomical Spectroscopy. London: Eyre & Spottiswoode, 1961.

WEIGERT, A., and ZIMERMANN, H.: ABC of Astronomy. London: Hilger & Watts, 1967.

WERNER, H.: From the Aratus Globe to the Zeiss Planetarium. Stuttgart: Fischer, 1957.

WOOD, F. B.: Photoelectric Astronomy for Amateurs. New York: Macmillan, 1963.

WOODBURY, D. O.: The Glass Giant of Palomar. New York: Dodd-Mead, 1963.

23.5. Radio Astronomy

AARONS, J.: Solar System Radio Astronomy. New York: Plenum Press, 1965.

BROWN, HANBURY R., and LOVELL, A. C. B.: The Exploration of Space by Radio. London: Chapman & Hall, 1957.

DAVIES, R. D., and PALMER, H. P.: Radio Studies of the Universe. Princeton, N.J.: Van Nostrand, 1959.

EVANS, J. V., and HAGFORS, T., Eds.: Radar Astronomy. New York: McGraw-Hill, 1968.

GINZBURG, W. L., and SYROVATSKII, S. I.: The Origin of Cosmic Rays. London–New York: Pergamon, 1964.

HYDE, F. H.: Radio Astronomy for Amateurs. London: Lutterworth Press, 1962.

KRAUS, J. D.: Radio Astronomy. New York: McGraw-Hill, 1966.

Large Radio-Telescopes (Symposium). Organization for Economic Cooperation and Development, Paris, 1961.

PALMER, H. P., DAVIES, R. D., and LARGE, M. I., Eds.: Radio Astronomy Today. Manchester: University Press, 1963.

PAWSEY, J. L., and BRACEWELL, R. N.: Radio Astronomy. Oxford: Clarendon Press, 1955.

PIDDINGTON, J. H.: Radio Astronomy. London: Hutchinson, 1961.

SHKLOVSKY, I. S.: Cosmic Radio Waves. Cambridge, Mass.: Harvard University Press, 1960.

SMITH, F. GRAHAM: Radio Astronomy, 3rd ed. London: Penguin, 1966.

VERSCHUUR, G. L., and KELLERMANN, K. I.: Galactic and Extra-Galactic Radio Astronomy. New York: Springer-Verlag, 1974.

WOLFENDALE, A. W.: Cosmic Rays. London: Lewnes, 1963.

23.6. Guides for the Observer

BRANDT, R.: The Sky through Binoculars. (Himmelswunder im Feldstecher.) Leipzig: Barth, 1968.

Burnham's Celestial Handbook (8 parts).

HOWARD, N. E.: The Telescope Handbook and Star Atlas. New York: Crowell, 1967.

MAYALL, R. N., and M. W.: Skyshooting—Hunting the Stars with your Camera. New York: Ronald Press, 1949.

MENZEL, D. H.: A Field Guide to the Stars and Planets. Boston: Houghton Mifflin, 1964.

MINNAERT, M. G. J.: Practical Work in Elementary Astronomy. Dordrecht: Reidel, 1969.

MOORE, P.: The Sky at Night (Television Talks). London: Eyre & Spottiswoode, 1964.

MOORE, P.: The Observer's Book of Astronomy, 4th ed. London: Frederick Warne & Co., 1971.

MOORE, P., Ed.: Practical Amateur Astronomy. London: Lutterworth Press, 1963.

MUIRDEN, J.: Astronomy with Binoculars. London: Faber and Faber, 1963.

MUIRDEN, J.: The Amateur Astronomer's Handbook. Cassell Publishers, London, 1969.

NORTON, A. P.: Norton's Star Atlas and Reference Handbook. Edinburgh: Gall & Inglis, 1964.

OLCOTT, W. T.: Olcott's Field Book of the Skies. R. N. and M. W. Mayall, Eds. New York: Putnam, 1954.

ROTH, G. D.: BLV-Himmelsführer Sterne und Planeten. München-Bern-Wien: BLV-Verlagsgesellschaft, 1972.

ROTH, G. D.: The Amateur Astronomer and his Telescope, 2nd ed. London: Faber and Faber, 1973.

ROTH, G. D.: Handbook for Planet Observers. London: Faber and Faber, 1970.

SIDGWICK, J. B.: Observational Astronomy for Amateurs, 2nd ed. London: Faber and Faber, 1957.

VEHRENBERG, H.: Mein Messier-Buch. Düsseldorf: Treugesell Verlag, 1966.

VEHRENBERG, H.: Atlas of Deep-Sky Splendors. Cambridge, Mass.: Sky Publishing Corp., 1967.

WEBB, T. W.: Celestial Objects for Common Telescopes. 2 Vols. M. W. Mayall, Ed. New York: Dover, 1962.

WIDMANN, W., and SCHÜTTE, K.: The Young Specialist Looks at Stars and Planets. A. Beer, Trans. and Ed. University of Cambridge. London: Burke, 1969.

WOOD, H.: The Southern Sky: Astronomy Through Your Own Eyes. Sydney: Angus & Robertson, 1965.

23.7. Numerical Work

AHNERT, P.: Astronomisch–Chronologische Tafeln für Sonne, Mond und Planeten. Leipzig: J. A. Barth, 1965.

ALLEN, C. W.: Astrophysical Quantities, 3rd ed. London: Athlone Press, 1973.

BAUSCHINGER, J.: Die Bahnbestimmung der Himmelskörper. Leipzig: Engelmann, 1906.

BROUWER, D., and CLEMENCE, G. M.: Methods of Celestial Mechanics. New York: Academic Press, 1961.

DUBYAGO, A.: Determination of Orbits. New York: Macmillan, 1962.

HERGET, P.: Computation of Orbits. Ann Arbor: University of Michigan Press, 1948.

Interpolation and Allied Tables. (Prepared by H.M. Nautical Almanac Office.) H.M. Stationery Office, London, 1956.

KULIKOV, K. A.: Fundamental Constants of Astronomy. London: Oldbourne Press, 1964.

PODOBED, V. V.: Fundamental Astrometry. University of Chicago Press, 1965.

RYABOV, Y.: An Elementary Survey of Celestial Mechanics. New York: Dover, 1961.

SMART, W. M.: Celestial Mechanics. London: Longmans, 1960.

STRACKE, G.: Bahnbestimmung der Planeten und Kometen. Berlin, Springer, 1929:

23.8. Earth and Atmosphere

BATES, D. R.: The Planet Earth, 2nd ed. Oxford–New York: Pergamon, 1964.

GAMOW, G.: Biography of the Earth. New York: Viking Press, 1943.

GAMOW, G.: A Planet called Earth. New York: Viking Press, 1963.

JORDAN, P.: The Expanding Earth. A. Beer, Trans. and Ed. Oxford–New York: Pergamon Press, 1971.

KUIPER, G. P., Ed.: The Atmospheres of the Earth and Planets, 2nd ed. University of Chicago Press, 1952.

KUIPER, G. P., Ed.: The Earth as a Planet. University of Chicago Press, 1954.

ORTNER, J., and MASELAND, H.: Introduction to Solar-Terrestrial Relations. Dordrecht: Reidel, 1965.

23.9. Sun

ABETTI, G.: The Sun. J. B. Sidgwick, Trans. London: Faber and Faber, 1957.

BAXTER, W. M.: The Sun and the Amateur Astronomer. David & Charles, Newton Abbot, Devon, Great Britain, 1973.

BEYNON, W. J. G., and BROWN, G. M.: Solar Eclipses and the Ionosphere. ICSU Symposium 1955. London: Pergamon Press, 1956.

BRANDT, J. C., and HODGE, P. W.: Solar System Astrophysics. New York: McGraw-Hill, 1964.

BRAY, R., and LOUGHHEAD, R. E.: Sunspots. London: Chapman & Hall, 1964.

BURGESS, R. D., and HULTS, M. E.: A Shadow-band Experiment. Sky and Telescope 38, 95 (1969).

COUSINS, F. W.: Sundials. New York: Universe Books Inc., 1970.

DE JAGER, C.: The Solar Spectrum. Dordrecht: Reidel, 1965.

DUNCOMBE, J. S.: Solar Eclipses. (a) 1971–1975, (b) 1976–1980). U.S. Naval Obs. Circ. Nos. 101 and 113, 1964 and 1966.

DYSON, F. W., and WOOLLEY, R. v. d. R.: Eclipses of the Sun and Moon. Oxford University Press, 1937.

ELLISON, M. A.: The Sun and its Influence, 2nd ed. London: Routledge & Kegan Paul, 1959.

GAMOW, G.: A Star called the Sun. New York: Viking Press, 1964.

GAMOW, G.: Birth and Death of the Sun. New York: Macmillan, 1941.

JASTROW, R., and CAMERON, A. G. W.: Origin of the Solar System. New York: Academic Press, 1963.

KUIPER, G. P., Ed.: The Sun. University of Chicago Press, 1961.

LASKE, L. M.: Die Sonnenuhren. Berlin: Springer-Verlag, 1959.

LYTTLETON, R. A.: Mysteries of the Solar System. Oxford: Clarendon Press, 1968.

MEADOWS, A. J.: Early Solar Physics: Selected Readings in Physics. Oxford–New York: Pergamon Press, 1970.

MEEUS, J., GROSJEAN, C. C., and VANDERLEEN, W.: Canon of Solar Eclipses: Solar Eclipses Between 1898 and 2510. Oxford–New York: Pergamon Press, 1966.

MENZEL, D. H.: Our Sun. Cambridge, Mass.: Harvard University Press, 1959.

MOORE, C. E., et al.: The Solar Spectrum (2935–8770 A). U.S. Government Printing Office, Washington, D.C., 1966.

MÜLLER, R.: Die Sonne. Orion Books, Vol. 118. Murnau: Publishers Lux, 1958.

SHKLOVSKII, I. S.: Physics of the Solar Corona. A. Beer, Ed. Oxford–New York: Pergamon Press, 1965.

SMITH, H. J., and SMITH, E.: Solar Flares. New York: Macmillan, 1963.

VON KLÜBER, H.: On the Prediction of Sunspot Maxima. In Henseling's Himmelskalender. Berlin: Condor Verlag, 1947.

VON OPPOLZER, T.: Kanon der Finsternisse. New York: Dover, 1963.

WALDMEIER, M.: Tafeln zur heliographischen Ortsbestimmung. Basel: Birkhäuser, 1950.

WALDMEIER, M.: Ergebnisse und Probleme der Sonnenforschung, 2nd ed. Leipzig: Akademische Verlagsgesellschaft, 1955.

WALDMEIER, M.: The Sunspot Activity in the Years 1610–1960. (English and German text.) Zürich: Schultess & Co., 1961.

WHIPPLE, F. L.: Earth, Moon and Planets. Cambridge, Mass.: Harvard University Press, 1963.

ZIRIN, H.: The Solar Atmosphere. Waltham–Toronto–London: Blaisdell, 1966.

23.10. The Moon

ALTER, D.: Lunar Atlas. New York: Space Science Division, NASA, 1968.

ARTHUR, D. W. G., and AGNIERAY, A. P.: Lunar Designation and Positions. Tucson: University of Arizona Press, 1964.

BALDWIN, R. B.: The Measure of the Moon. University of Chicago Press, 1963.

BALDWIN, R. B.: A Fundamental Survey of the Moon. New York: McGraw-Hill, 1965.

CHERRINGTON, E. H.: Exploring the Moon Through Binoculars. New York: McGraw-Hill, 1969.

Communications of the Lunar and Planetary Laboratory, 3, No. 50 (1965), etc. Tucson: University of Arizona.

COMRIE, L. J.: A Short Semi-graphical Method of Predicting Occultation. London: H.M. Stationery Office, 1926.

DAVIDSON, M.: The Reduction of Occultations for Stars Fainter than Magnitude 7.5. J. Brit. Astron. Assoc. 48, 120, 1938.

FIELDER, G.: Structure of the Moon's Surface. Oxford–New York: Pergamon Press, 1961.

Geological Maps of the Moon. (Scales 1:1,000,000 and 1:3,800,000.) Distribution Section, U.S. Geological Survey, 1200 South Eads St., Arlington, Va. 22202.

KOPAL, Z., Ed.: Physics and Astronomy of the Moon. New York–London: Academic Press, 1962.

KOPAL, Z.: The Moon. Dordrecht: Reidel, 1969.

KOPAL, Z.: A New Photographic Atlas of the Moon. Introduction by H. C. Urey. London: R. Hall & Co., 1971.

KOPAL, Z., CARDER, R. W.: Mapping the Moon—Past and Present. Dordrecht: D. Reidel, 1974.

KOPAL, Z., KLEPESTA, J., and RACKHAM, T. W.: Photographic Atlas of the moon. London: Academic Press, 1965.

KUIPER, G. P., Ed.: Photographic Lunar Atlas. Based on photographs taken at the Mount Wilson, Lick, Pic du Midi, McDonald, and Yerkes Observatories. University of Chicago Press, 1960.

KUIPER, G. P., Ed.: The Moon, Meteorites and Comets. University of Chicago Press, 1962.

Lunar Planning Chart Series, 1st ed. USAF Aeronautical Chart and Information Center, 1969.

MARSDEN, B. G., and CAMERON, A. G. W.: The Earth–Moon System. New York: Plenum Press, 1966.

MIDDLEHURST, B. M., and KUIPER, G. P.: The Moon, Meteorites and Comets. The Solar System, Vol. IV. University of Chicago Press, 1963.

MIYAMOTO, S., and HATTORI, A.: Photographic Atlas of the Moon, 2nd ed. University of Kyoto, 1964.

The Moon. Circulars of the Lunar Section of the British Astronomical Association, London.

MOORE, P.: Moon Flight Atlas. With Moon Maps: Hallwag Ltd., Berne. London: G. Philip & Son Ltd., 1970.

MOORE, P.: A Survey of the Moon. Norton, New York, 1963.

MOORE, P., and CATTERMOLE, P.: The Craters of the Moon. Lutterworth Press, London, 1967.

The National Geographic Society. Map of the Moon. Washington, D.C. 20036.

RACKHAM, T.: The Moon in Focus. Oxford and New York, Pergamon Press, 1968.

SALISBURY, J. W., and GLAESER, P. E.: The Lunar Surface Layer. New York: Academic Press, 1964.

WEIL, N. A.: Lunar and Planetary Surface Conditions. New York: Academic Press, 1965.

WHITAKER, E. A., et al.: Rectified Lunar Atlas. Tucson: University of Arizona Press, 1963.

WILKINS, H. P., and MOORE, P.: The Moon. London: Faber and Faber, 1955.

23.11. Planets

ALEXANDER, A. F. O'D.: The Planet Saturn. London: Faber and Faber, 1962.

ALEXANDER, A. F. O'D.: The Planet Uranus: A History of Observation, Theory and Discovery. London: Faber and Faber, 1965.

DE VAUCOULEURS, G.: The Planet Mars. New York: Macmillan, 1954.

KUIPER, G., and MIDDLEHURST, B. M., Eds.: Planets and Satellites. University of Chicago Press, 1961.

Lowell Observatory, Flagstaff, Ariz. 86001: Map of Mars.

MICHAUX, C. M., FISH, F. F., JR., MURRAY, F. W., SANTINA, R. E., and STEFFEY, P. C.: Handbook of the Physical Properties of the Planet Jupiter. NASA: Washington, D.C. (NASA SP-3031), 1967.

MOORE, P.: The Planet Venus, 3rd ed. London: Faber and Faber, 1961.

MOORE, P.: Guide to the Planets. London: Eyre & Spottiswoode, 1955.

PEEK, B. M.: The Planet Jupiter, 2nd ed. London: Faber and Faber, 1958.

ROTH, G. D.: The System of Minor Planets. London: Faber and Faber, 1963.

ROTH, G. D.: Taschenbuch für Planetenbeobachter. Mannheim: Bibliographisches Institut, 1966.

SANDNER, W.: The Planet Mercury. London: Faber and Faber, 1963.

SANDNER, W.: Planeten—Geschwister der Erde. Weinheim: Verlag Chemie, 1971.

SANDNER, W.: Satellites of the Solar System. London: Faber and Faber, 1965.

SHARONOV, V. V.: The Nature of the Planets. Jerusalem: Israel Program for Scientific Translations, 1964.

SLIPHER, E. C.: The Photographic Story of Mars. Cambridge, Mass.: Sky Publishing Corporation, 1962.

SLIPHER, E. C.: A Photographic Study of the Brighter Planets. Lowell Observatory, Flagstaff, Ariz.; The National Geographic Society, Washington, D.C., 1964.

UREY, H. C.: The Planets, their Origin and Development. Oxford University Press, 1952.

WATSON, F. G.: Between the Planets. Cambridge, Mass.: Harvard University Press, 1956.

WHIPPLE, F. L.: Earth, Moon and Planets. Cambridge, Mass.: Harvard University Press, 1963.

23.12. Comets and Meteors

BEER, A., LYTTLETON, R. A., and RICHTER, N. B.: The Nature of Comets. (English version of N. B. RICHTER's "Statistik und Physik der Kometen". Leipzig: Barth, 1954). London: Methuen, 1963.

HAWKINS, G. S.: Meteors, Comets, and Meteorites. New York: McGraw-Hill, 1964.

KRINOV, E. L.: Principles of Meteoritics. Oxford–New York: Pergamon, 1960.

KRINOV, E. L.: Giant Meteorites. Oxford–New York: Pergamon, 1966.

KUIPER, G. P., and MIDDLEHURST, B. M.: The Solar System, Vol. V. University of Chicago Press, 1963.

LOVELL, A. C. B.: Meteor Astronomy. Oxford: Clarendon Press, 1954.

LYTTLETON, R. A.: The Comets and Their Origin. Cambridge University Press, 1953.

MASON, B.: Meteorites. New York: Wiley, 1962.

MOORE, C. B., Ed.: Researches on Meteorites. New York: Wiley, 1962.

ÖPIK, E. J.: Physics of Meteor Flight in the Atmosphere. New York: Wiley, 1958.

PORTER, J. G.: Comets and Meteor Streams. London: Chapman & Hall; and New York: Wiley, 1952.

WATSON, F. G.: Between the Planets, revised ed. Cambridge, Mass.: Harvard University Press, 1956.

WHIPPLE, F. L.: The Harvard Photographic Meteor Program. Sky and Telescope **8**, 90 (1949).

23.13. Astronautics

CARTWRIGHT, E. M., Ed.: Exploring Space with a Camera. NASA, Washington, D.C., 1968.

HARGREAVES, J. K.: The coverage of satellite passages from a ground station. Planet Space Sci. **14**, 617–622 (1966).

HAVILAND, R. P., and HOUSE, C. M.: Handbook of Satellites and Space Vehicles. Princeton: Van Nostrand, 1965.

INCE, A. N., and TROLLOPE, L. T.: A method for the determination of visibility, range and Doppler shift in studies of satellite systems. Brit. Interplanet. Soc. **19**, 433–446 (1964).

KING-HELE, D.: Observing Earth Satellites. London: Macmillan, 1966.

LIEMOHN, H. B.: Optical Observations of Apollo 8. Sky and Telescope **37**, 156–160 (1969).

MASSEVITCH, A. G., and LOSINKSY, A. M.: Photographic tracking of artificial satellites. Space Sci. Rev. **11**, 308–340 (1970).

OVENDEN, M. W.: Artificial Satellites. London: Penguin, 1960.

OVENDEN, M. W.: Life in the Universe. London: Heinemann, 1963.

PILKINGTON, J. A.: The visual appearance of artificial earth satellites. Planet. Space Sci. **12**, 597–606 (1964); **13**, 541–550 (1965); **14**, 1281–1289 (1966); **15**, 1895–1911 (1967).

Space Research. Amsterdam: North Holland, 1960, etc.

TAYLOR, G. E.: A brief review of visual satellite tracking. Brit. Astron. Assoc. **77**, 196–198 (1967).

VAN ALLEN, J. A.: Scientific Uses of Earth Satellites. London: Chapman & Hall, 1956.

VEIS, G.: Optical tracking of artificial satellites. Space Sci. Rev. **2**, No. 2 (1963).

VERTREGT, M.: Principles of Astronautics, 2nd ed. Amsterdam: Elsevier, 1963.

WILLIAMS, J. G., and McCUE, G. A.: An analysis of satellite optical characteristics data. Planetary Space Sci. **14**, 839–847 (1966).

YOUNG, H., SILCOCK, B., and DUNN, P.: Journey to Tranquillity: The History of Man's Assault on the Moon. London: Jonathan Cape, 1969.

23.14. Stars—Galaxies—Cosmology

ABETTI, G., and HACK, M.: Nebulae and Galaxies. V. Barocas, Trans. London: Faber and Faber, 1964.

ALLER, L. H.: Gaseous Nebulae. London: Chapman & Hall, 1956.

ARP, H.: Atlas of Peculiar Galaxies. Pasadena: California Institute of Technology, 1966.

BAADE, W.: Evolution of Stars and Galaxies. C. Payne-Gaposchkin, Ed. Cambridge, Mass.: Harvard University Press, 1963.

BEYER, M.: Variable Stars. In Landolt-Börnstein, Group VI, Vol. 1; Astronomy and Astrophysics, pp. 517–563. Berlin–Heidelberg–New York: Springer-Verlag, 1965.

BLAAUW, A., and SCHMIDT, M., Eds. Galactic Structure. University of Chicago Press, 1965.

BOK, B. J., and BOK, P. F. The Milky Way, 3rd ed. Cambridge, Mass.: Harvard University Press, 1957.

BONDI, H.: Rival Theories of Cosmology. New York: Oxford University Press, 1960.

BONDI, H.: Cosmology, 2nd ed. Cambridge University Press, 1961.

CAMERON, A. G. W., and STEIN, R. F.: Stellar Evolution. New York: Plenum Press, 1966.

CAMPBELL, L., and JACCHIA, L.: The Story of Variable Stars. Philadelphia: Blakiston Co., 1945.

DUFAY, J.: Galactic Nebulae and Interstellar Matter. London: Hutchinson, 1957.

FANNING, A. E.: Planets, Stars and Galaxies. New York: Dover, 1966.

GAMOW, G.: The Creation of the Universe. New York: Viking Press, 1952.

GINZBURG, V. L., and SYROVATSKII, S. I.: The Origin of Cosmic Rays. Oxford–New York: Pergamon, 1964.

GLASBY, I. S.: Variable Stars. Cambridge, Mass., Harvard University Press, 1968.

HAYASHI, C., et al.: Evolution of the Stars. University of Kyoto, 1962.

HEINTZ, W. D.: Doppelsterne. München: Goldmann, 1971.

HEINTZ, W. D., Ed.: Proceedings of the 5th IAU Colloquium (Nice, 1969). Astrophys. Space Sci. 11, No. 1 (1971).

HEINTZ, W. D.: Doppelsterne. München: Goldmann, 1971.

HODGE, P. W.: Galaxies and Cosmology. New York: McGraw-Hill, 1966.

HOYLE, F.: Of Man and Galaxies. London: Heinemann, 1965.

HOYLE, F.: Galaxies, Nuclei and Quasars. London: Heinemann, 1966.

KUIPER, G. P., and MIDDLEHURST, B. M., Eds.: Stars and Stellar Systems, 9 Vols. University of Chicago Press, 1960.

KUIPER, G. P., and MIDDLEHURST, B. M., Eds.: Stars and Stellar Systems, Vol. 5, Galactic Structure. A. BLAAUW and M. SCHMIDT, Eds. University of Chicago Press, 1965.

MCCREA, W. H.: Physics of the Sun and Stars. London: Hutchinson, 1950.

MCVITTIE, G. C.: Fact and Theory in Cosmology. London: Eyre & Spottiswoode, 1961.

MCVITTIE, G. C.: General Relativity and Cosmology, 2nd ed. London: Chapman & Hall, 1965.

MEADOWS, A. J.: Stellar Evolution. Oxford–New York: Pergamon Press, 1967.

NORTH, J. D.: The Measure of the Universe: A History of Modern Cosmology. Oxford: Clarendon Press, 1965.

OGORODNIKOV, K. F.: Dynamics of the Stellar System. A. Beer, Ed. Oxford–New York: Pergamon, 1965.
PAGE, T., Ed.: Stars and Galaxies: Birth, Ageing and Death in the Universe. Englewood Cliffs, N.J.: Prentice Hall, 1962.
PAGE, T., and PAGE, L. W., Eds.: Starlight: What It Tells about the Stars. Macmillan Library of Astronomy No. 5. New York: Macmillan, 1967.
PAGE, T., and PAGE, L. W.: The Evolution of Stars. Macmillan Library of Astronomy No. 6. New York: Macmillan, 1968.
PAYNE-GAPOSCHKIN, C.: Stars in the Making. Cambridge, Mass.: Harvard University Press, 1952.
PAYNE-GAPOSCHKIN, C.: Variable Stars and Galactic Structure. London: Athlone Press, 1954.
PAYNE-GAPOSCHKIN, C.: The Galactic Novae. Amsterdam: North-Holland, 1957.
REDDISH, V. C.: Evolution of the Galaxies. Edinburgh: Oliver & Boyd, 1967.
ROBINSON, I., et al., Eds.: Quasi-stellar Sources and Gravitational Collapse. University of Chicago Press, 1965.
ROWE, A. P.: Astronomy and Cosmology. London: Thames and Hudson, 1968.
SANDSTRÖM, A. E.: Cosmic Ray Physics. Amsterdam: North-Holland, 1965.
SCHWARZSCHILD, M.: Structure and Evolution of the Stars. Princeton, N.J.: Princeton University Press, 1958.
SHAPLEY, H.: The Inner Metagalaxy. New Haven: Yale University Press, 1957.
STROHMEIER, W.: Variable Stars. A. J. Meadows, Ed., Pergamon Press, Oxford, 1972.
STRUVE, O.: Stellar Evolution. Princeton, N.J.: Princeton University Press, 1958.
VON DER PAHLEN, E.: Stellar Statistik. Leipzig: Barth, 1937.
WHITROW, G. T.: Structure and Evolution of the Universe. London: Hutchinson, 1959.
WOLFENDALE, A. W.: Cosmic Rays. London: Lewnes, 1963.
WOLTJER, J., Ed.: Galaxies and the Universe. New York: Columbia University Press, 1968.

23.15. Maps

ALTER, G., RUPRECHT, J., and VANÝSEK, V.: Catalogue of Star Clusters and Associations (in 9 parts), Card Index. Prague: Czechoslovak Academy of Sciences, 1958–1964.
ARGELANDER, F. W.: Atlas des nördlichen gestirnten Himmels für den Anfang des Jahres 1855. (Charts of the Bonner Durchmusterung, B.D.), 3rd ed. Bonn: Dümmler, 1954.
BEČVÁŘ, A.: Atlas Coeli 1950.0. Prague: Czechoslovak Academy of Sciences, 1956.
BEČVÁŘ, A.: Atlas Eclipticalis 1950.0. Prague: Czechoslovak Academy of Sciences, 1958.
BEYER, M., and GRAFF, K.: Sternatlas of all stars up to 9^m, clusters, and nebulae, between the North Pole and $23°$ declination, 3rd ed. Bonn: Dümmler, 1950.
BRUN, R., and VEHRENBERG, H. Atlas of Selected Areas—163 charts for the Northern and 43 charts for the Southern sky. Düsseldorf: Treugesell-Verlag, 1971.
DE CALLATAY, V.: Goldmanns Himmelsatlas. W. Jahn, Ed. Munich: Goldmann, 1959.
FELSMANN, G., and KOHL, O.: Atlas des gestirnten Himmels für das Aequinoktium 1950. Berlin:Akademie-Verlag, 1956.
KLEPESTA, J.: Himmelskarten. Stuttgart: Franckh.
KOHL, O., and FELSMAN, G.: Atlas des gestirnten Himmels. Berlin: Akademie-Verlag, 1956.

NORTON, A. P.: A Star Atlas and Reference Handbook for Students and Amateurs, 15th ed. Edinburgh: Gall & Inglis, 1964.

SANDAGE, A.: The Hubble Atlas of Galaxies. Washington, D.C.: Carnegie Institution, 1961.

SCHAIFERS, K.: Himmelsatlas (Tabulae Caelestes), 8th ed. Revision of the Schurig-Götz Atlas, stars up to 6^m. Mannheim: Bibliographisches Institut, 1960.

SCHÖNFELD, E.: Atlas der Himmelszone zwischen $-1°$ und $-23°$ Deklination für den Anfang des Jahres 1855 (Charts of the Southern Bonner Durchmusterung, B.D.), 2nd ed. Bonn: Dümmler, 1951.

SCHURIG, R., and GÖTZ, P.: Tabulae Caelestes, 8th ed. K. Schaifers, Ed. Mannheim: Bibliographisches Institut, 1960.

Smithsonian Astrophysical Observatory. Star Atlas. Cambridge, Mass.: MIT Press, 1970.

SUTER, H.: Drehbare Sternkarte "Sirius". Bern: Schweizerische Astronomische Gesellschaft, 1952.

VEHRENBERG, H.: Photographischer Stern-Atlas. 303 Star maps of the Northern Sky between $+90°$ and $-26°$ declination. Düsseldorf: Treugesell, 1962.

VEHRENBERG, H.: Photographischer Stern-Atlas (Falkauer-Atlas). 161 star maps of the Southern Sky between $-14°$ and $-90°$ declination. Düsseldorf: Treugesell, 1962.

VEHRENBERG, H.: Atlas of Kapteyn's Selected Areas. Northern and Southern Parts. Düsseldorf: Treugesell, 1965.

VEHRENBERG, H.: Mein Messier-Buch. Photographischer Atlas samtlicher Messier-Objekte und vieler anderer heller Objecte. Düsseldorf: Treugesell, 1966.

VEHRENBERG, H., and BLANK, D.: Handbuch der Sternbilder. Düsseldorf: Treugesell, 1970.

VEHRENBERG, H.: Milchstrassen-Mosaiks. 2 Series. Düsseldorf: Treugesell, 1971.

VEHRENBERG, H.: Atlas Stellarum 1950.0. One volume for the North, one for the South. Düsseldorf: Treugesell, 1971.

WIDMANN, W.: Kosmos rotatable star map. Stuttgart: Frankh, 1967.

23.16. Catalogs

AITKEN, R. G.: New General Catalogue of Double Stars Within 129° of the North Pole. Carnegie Institution: Washington, D.C., 1934.

ALTER, G., and RUPRECHT, J.: Atlas of the Open Star Clusters. Prague: Czechoslovak Academy of Sciences, 1963.

BOSS, B., et al.: General Catalogue of 33,342 Stars for the Epoch 1950. Carnegie Institution: Washington, D.C., 1937.

CANNON, A. J., and PICKERING, E. C.: The Henry Draper Catalogue (HD). Cambridge, Mass: Harvard Observatory, 1918, etc.

Catalogue of Bright Stars, 3rd ed. D. Hoffleit, Ed. New Haven: Yale University Press, 1965.

DE VAUCOULEURS, G., and DE VAUCOULEURS, A.: Reference Catalogue of Bright Galaxies. Austin: The University of Texas Press, 1964.

DREYER, J. L. E.: New General Catalogue of Nebulae and Clusters of Stars. Memoirs of the Royal Astronomical Society, London, 1888.

FRICKE, W., and KOPFF, A.: Fourth Fundamental Catalogue (FK4). Karlsruhe: Braun, 1963.

General Catalogue of Stellar Radial Velocities (by WILSON, R. E.). Washington, D.C., 1953.

General Catalogue of Trigonometric Parallaxes (by JENKINS, L.). New Haven: Yale Observatory, 1952–1963.

General Catalog of Variable Stars. Obshii Katalog peremennych zvezd—OKPZ (by B. V. Kukarkin et al. 3rd Ed. 3 volumes. With English versions of the remarks. Akademija Nauk SSSR, Moscow, 1969–1971.

HAGEN, G. L.: An Atlas of Open Cluster Colour-Magnitude Diagrams, Publications of the David Dunlap Observatory, Vol. 4. Toronto, 1970.

Henry-Draper-Catalogue (HD). Harvard Annals Vols. 91–99.

HOFFLEIT, D., Ed.: Catalogue of Bright Stars. New Haven: Yale University Press, 1965.

IRIARTE, B., JOHNSON, H. L., MITCHELL, R. I., and WISNIEWSKI, W. K.: On the general technique of photoelectric photometry; extensive catalogue of UBVRI magnitudes for stars brighter than 5^m. Sky and Telescope 30, 21 (1965).

JENKINS, L. F.: General Catalogue of Trigonometric Stellar Parallaxes. New Haven: Yale University Press, 1952.

KUKARKIN, et al.: Catalogue of Variable Stars. 3 Vols. Moscow: Astronomical Council of the Academy of Sciences of the USSR, 1971.

New General Catalogue of Nebulae and Clusters (NGC) and Index Catalogue (IC). Memoirs Roy. Astron. Soc. 49, 51, and 59, London, 1890, 1895, and 1910.

PEREK, L., and KOHOUTEK, L.: Catalogue of Galactic Planetary Nebulae. Academia Prague, 1967.

PRAGER, R., and SCHNELLER, H.: Geschichte und Literatur des Lichtwechsels veränderlicher Sterne. Volumes I–V. 2nd ed. Berlin: Akademie-Verlag, 1952–1957.

Publications of the U.S. Naval Observatory. Compilation of photoelectric UBV magnitudes for more than 2,000 stars down to about 15^m. Second Series, Vol. XXI, Washington, D.C., 1968.

ROBERTSON, J.: Catalog (NZC) of 3539 Zodiacal Stars. Astronomical Papers of the American Ephemeris, Vol. 10, II.

SANDAGE, A. The Hubble Atlas of Galaxies. Carnegie Institution: Washington, D.C., 1961.

Scheinbare Oerter der 1535 Sterne des Vierten Fundamental-Katalogs (FK 4). Heidelberg: Astronomisches Rechen-Institut, 1963.

Stellar Data Center Strasbourg: Compilations of photometric and other data. Observatoire, 11 rue de l'Université, Strasbourg 67.

Visual Double Stars. Index Catalogue, IDS, in 2 volumes. Lick Observatory, Mt. Hamilton, Calif., 1963.

WILSON, R.: General Catalogue of Stellar Radial Velocities. Carnegie Institution, Washington, D.C., 1953.

ZWICKY, F.: Catalogue of Galaxies and Clusters of Galaxies. Zürich: Speich, 1963.

23.16. Almanacs

Astronomer's Handbook (IAU). Reidel, Dordrecht.

The Astronomical Ephemeris. H.M. Nautical Almanac Office; Nautical Almanac Office of the United States Naval Observatory. London and Washington.

Explanatory Supplement to the Astronomical Ephemeris and the American Ephemeris and Nautical Almanac. London, 1961.

Handbook of the Astronomical Society of Southern Africa.

Handbook of the British Astronomical Association, c/o Royal Astron. Soc., Burlington House, London, W.1.

Kalender für Sternfreunde. P. AHNERT, Ed. Leipzig: J. A. Barth.

The Nautical Almanac. London.

Observer's Handbook of the Royal Astronomical Society of Canada.

Der Sternhimmel. R. NAEF, Ed. Sauerländer, Aarau, Switzerland.

23.18. Journals

Annales d'Astrophysique. Revue Internationale. Centre National de la Recherche Scientifique. Paris.

Association Française d'Astronomie. Observatoire d'Astrophysique, 61 Saint-Aubin-de-Courteraie, Orne.

The Astronomical Journal. American Astronomical Society, American Institute of Physics, New York.

L'Astronomie et Bulletin de la Société Astronomique de France. Paris.

Astronomische Nachrichten. Akademie-Verlag, Berlin.

Astronomy and Astrophysics. A European Journal. Springer-Verlag, Heidelberg.

Astronomy and Space. P. MOORE, Ed. David & Charles, Newton Abbot.

The Astrophysical Journal. University of Chicago Press.

Ciel et Terre. Bulletin de la Société Belge d'Astronomie, de Météorologie et de Physique du Globe. Uccle-Bruxelles.

Coelum. Bologna.

Hemel en Dampkring. Den Haag.

Hermes. Publication of the "Junior Astronomical Society." J. G. Porter, Patron; I. Ridpath, Ed. Ilford, Essex, England.

Icarus. International Journal of the Solar System. Academic Press, New York.

Journal des Observateurs. Centre National de la Recherche Scientifique. University of Marseille.

The Journal of the British Astronomical Association. c/o Royal Astronomical Society, Burlington House, London, W.1.

Journal of the Royal Astronomical Society of Canada. Toronto.

Mitteilungen der Astronomischen Gesellschaft. Karlsruhe.

Modern Astronomy. Buffalo, N.Y.

Monthly Notices of the Royal Astronomical Society. Burlington House, London, W.1.

Nachrichtenblatt der Vereinigung der Sternfreunde e.V. (See Sterne und Weltraum.)

The Observatory. Royal Greenwich Observatory, Herstmonceux Castle, Sussex.

Orion. Journal of the "Schweizerische Astronomische Gesellschaft."

Photo-Bulletin of the American Astronomical Society (technical details of astronomical photography). Smithsonian Astrophysical Observatory, 60 Garden Street, Cambridge, Mass. 02138.

Planetary and Space Science. Pergamon Press, New York.

Publications of the Astronomical Society of the Pacific. San Francisco.

Sky and Telescope. C. A. Federer, Jr., Ed. Sky Publishing Corporation, Cambridge, Mass.

Soviet Astronomy. (English translation of the Russian "Astronomical Journal.") American Institute of Physics, New York.

Spaceflight. London.

Die Sterne. H. Lambrecht, Ed. Barth, Leipzig.

Sterne und Weltraum. H. Elsässer, G. D. Roth, K. Schaifers, and H. Vehrenberg, Eds. Verlag Sterne und Weltraum, Düsseldorf.

The Strolling Astronomer. Box AZ, University Park, New Mexico.

Transactions of the International Astronomical Union (IAU). Reidel, Dordrecht.

Weltraumfahrt. Zeitschrift für Astronautik and Raketentechnik. Umschau-Verlag, Frankfurt a.M.

23.19. Astronomical Societies

American Astronomical Society, Princeton, New Jersey 08540.

Astronomical Society of Japan, c/o Astronomical Observatory, Mitaka, Tokyo, Japan.

Astronomical Society of the Pacific, c/o California Academy of Sciences, Golden Gate Park, San Francisco, Calif. 94118.
British Astronomical Association, c/o Colin Ronan, R.A.S. Burlington House, London, W.1.
International Union of Amateur Astronomers (IUAA). Dr. Luigi Baldinelli, Piazza Martiri 1, I-40121 Bologna, Italy.
Junior Astronomical Society, I. Ridpath, Secretary, 35 Oakwood Gardens, Seven Kings, Ilford, Essex, England.
Royal Astronomical Society, Burlington House, Piccadilly, London, W.1.
Royal Astronomical Society of Canada, 252 College Street, Toronto 130, Ontario, Canada.
Schweizerische Astronomische Gesellschaft. Gen. Secr., Fichtenweg 6, CH–3400 Burdorf, Switzerland.
Société Astronomique de France, 28 rue Saint-Dominique, Paris 7.
Vereinigung der Sternfreunde, H. Oberndorfer, Volkssternwarte, Anzingerstrasse 1, D-8 München 80, Germany.

23.20. Abbreviations

AAS   = American Astronomical Society, USA
AAVSO   = American Association of Variable Star Observers, USA
AG   = Astronomische Gesellschaft, Germany
AJ   = Astronomical Journal, New Haven, USA
AJB   = Astronomischer Jahresbericht, Heidelberg
ALPO   = Association of Lunar and Planetary Observers, USA
AN   = Astronomische Nachrichten, Berlin
ApJ   = Astrophysical Journal, USA
ASP   = Astronomical Society of the Pacific
BA   = Bulletin Astronomique, France
BAA   = British Astronomical Association
BAC   = Bulletin of the Astronomical Institutes of Czechoslovakia
BAN   = Bulletin of the Astronomical Institutes of the Netherlands
BAV   = Berliner Arbeitsgemeinschaft für veränderliche Sterne, Berlin
BJ   = Berliner Astronomisches Jahrbuch, Berlin
BSAF   = L'Astronomie, Bulletin de la Société Astronomique de France, Paris
CdT   = Connaissance des Temps, France
DOB   = Documentation des Observateurs, Paris
Gaz. astr. = Gazette astronomique, Antwerp
HA   = Annals of the Harvard College Observatory, USA
HC   = Harvard College Observatory Circular, USA
IAU   = International Astronomical Union
JBAA   = Journal of the British Astronomical Association
JO   = Journal des Observateurs, France
LOB   = Lick Observatory Bulletin, USA
MfP   = Mitteilungen für Planetenbeobachter, Mannheim
MN   = Monthly Notices of the Royal Astronomical Society, England
NA   = Nautical Almanac and Astronomical Ephemeris, London
Obs   = The Observatory, London
PASJ   = Publications of the Astronomical Society of Japan
PASP   = Publications of the Astronomical Society of the Pacific, USA
PAT   = Populär Astronomisk Tidskrift, Stockholm
RAS   = Royal Astronomical Society, England
RAJ   = Astronomisches Journal der UdSSR, Moscow

SAG = Schweizerische Astronomische Gesellschaft
SuW = Sterne und Weltraum, Mannheim
VdS = Vereinigung der Sternfreunde, Germany
ZfA = Zeitschrift für Astrophysik, Berlin-Göttingen-Heidelberg
ZfI = Zeitschrift für Instrumentenkunde

23.21. Teaching Aids

Photographs

Black and white photographs from the Hale Observatories from the Caltech Bookstore, Pasadena, California, for the United States; in all other countries from the Secretary of the Royal Astronomical Society, Burlington House, London, W.1.

Color photographs (Palomar) only from the Caltech Bookstore. U.S. Naval Observatory, Flagstaff, Arizona: from the General Secretary of the Schweizerische Astronomische Gesellschaft, Fichtenweg 6, CH–3400 Burgdorf, Switzerland.

S. A. Boulter. Films and Film Strips on Astronomy. J. Brit. Astron. Assoc. **70**, 219 (1960).

Special Catalogue of the Sky Publishing Corp., 49–50–51 Bay State Road, Cambridge, Mass. 02138.

Demonstration Equipment.

Celestial land lunar globes are obtainable from several firms.
In recent years small planetaria have become available, and many are now in use in schools and colleges to give a better understanding of astronomical phenomena. Such instruments are made by Baader Planetarium KG, Hartelstrasse 30, D–8000, München 21, Germany.

23.22. Planetaria

As well as major planetaria, many smaller ones exist throughout the world, often attached to schools and public observatories. Some 150 so-called medium-sized planetaria have been made by the firm VEB Zeiss in Jena, East Germany, up to the present time. Among them we may mention those in Egypt (Alexandria and Cairo), Bulgaria (Varna), Belgium (Antwerp), Burma (Rangoon Naval College), West Germany (Bremerhaven, Berlin, Hanover, Recklinghausen, and Cologne), Cuba (Havana, Santiago, Mariel, Santa Clara, and Las Villas), East Germany (Stralsund, Berlin, Schwerin, Rostock, Dresden, Strausberg, Halle, Senftenberg, Herzberg, Eilenburg, Wustrow, Bautzen, Leipzig, Dessau, Hoyerswerda, Radebeul, Suhl, Magdeburg, and Potsdam), England (London), India (New Delhi and Porpandar), Japan (Gifu), Mexico (Mexico D.F.), Mongolia (Ulan Bator), New Zealand (Auckland), Spain (Barcelona, Cadiz, Bilbao, Vigo, Teneriffe, and Lanzarote), Russia (Moscow, Minsk, Novosibirsk, Kharkov, Kiev, and 80 other places).

This number is growing continually. It is of great interest for the amateur astronomer to get in touch with one of the planetaria; he will gain valuable information and advice on the situation of amateur astronomy is his particular region, and thus be able to make contact with other enthusiasts, private observatories, and so on.

The principal manufacturers of planetaria publish lists that indicate the present distribution of their instruments. Here we mention Carl Zeiss, Oberkochen, West Germany; VEB Optische Werke, Jena, East Germany; Goto Optical Manufacturing Co., Tokyo, Japan; Spitz Laboratories, Yorklyn/Chadds Ford, Pennsylvania.

All these data are based on figures kindly provided by Donald D. Davis, Director of the Planetarium, Manitoba Museum of Man and Nature, 190 Rupert Avenue, Winnipeg 2, Manitoba, Canada; and Hermann Mucke, Wissenschaftlicher Leiter, Planetarium, Prater Hauptallee, Vienna. Both these places maintain detailed statistical records of modern planetarium work all over the world.

Current Planetarium Activities

[From D. H. GALLAGHER, "Planetaria of the World," Manitoba Museum of Man and Nature (Planetarium), Winnipeg, Canada, 1969.]

Country	City	Planetarium name	Instrument	Opening date	Seating capacity	Languages of programs	Public telescopes available?
Canada	Calgary, Alberta	Centennial	Zeiss-Jena 67	July 67	258	English	Yes, 6
	Vancouver, B.C.	H. R. MacMillan	Zeiss-Jena 68	Oct. 68	265	English	Not yet
	Winnipeg, Manitoba	Manitoba Museum of Man and Nature	Zeiss V-S W.G.	May 68	287	English, French	No
	Toronto, Ont.	McLauchlin	Zeiss-Jena 67	Oct. 68	350	English	6" Mak., 8" Cass.
	Montreal, Que.	Dow	Zeiss V W.G.	April 66	400	French, English	No
Mexico	Mexico, D.F.	Luis Enrique Erro	Zeiss IV W.G.	Oct. 67	446	Spanish	No
U.S.A.	Montgomery, Ala.	W. A. Gayle	Spitz STP	Oct. 68	236	English	Yes
	De Anya, Calif.	Tashima	Minolta MS-15	Jan. 70	300	English	16", 6", 4", Coronograph
	Los Angeles, Calif.	Griffith Observatory	Zeiss IV W.G.	May 35	663	English	Yes
	San Francisco, Calif.	Morrison	Academy	Nov. 52	375	English	No
	Colorado Springs, Colo.	U.S.A.F. Academy	Spitz B	Feb. 59	350	English	No
	Denver, Colo.	Chas. C. Gates	Spitz STP	July 68	236	English	22" Cass.
	Bradenton, Fla.	Bishop	Spitz STP	Oct. 66	200	English	Yes
	Atlanta, Ga.	Fernbank Science Center	Zeiss V W.G.	Dec. 67	500	English	36" Cass.
	Chicago, Ill.	Adler	Zeiss VI W.G.	May 30	430	English	Yes
	Baton Rouge, La.	La. Arts & Science Center	Zeiss IV W.G.	June 67	248	English	No
	Boston, Mass.	Charles Hayden	Korkosz	Oct. 58	276	English	No
	E. Lansing, Mich.	Abrams	Spitz STP	Feb. 64	261	English	Yes
	Flint, Mich.	Robt. T. Longway	Spitz B	June 58	292	English	Yes
	St. Louis, Mo.	McDonnell	Goto Sat. L-1	April 63	400	English	Yes
	New York, N.Y.	American Hayden	Zeiss VI W.G.	Oct. 35	800	English	No

Country	City	Planetarium name	Instrument	Opening Date	Seating capacity	Languages of programs	Public telescopes available?
	Rochester, N.Y.	Strasenburgh	Zeiss VI W.G.	Sept. 68	240	English	Yes
	Chapel Hill, N.C.	Morehead	Zeiss VI W.G.	May 49	450	English	
	Philadelphia, Pa.	Fels	Zeiss IV W.G.	Nov. 33	503	English	Yes
	Pittsburgh, Pa.	Buhl	Zeiss II W.G.	Oct. 39	490	Eng. (Fr., Sp., Ger., Rus.)	Yes
	Houston, Tex.	Burke Baker	Spitz STP	July 64	232	English	No
	Salt Lake City, Utah	Hansen	Spitz STP	Nov. 65	235	English	Yes, 2
Austria	Vienna	Urania	Zeiss IV W.G.	June 64	240	German	
Belgium	Brussels	Alberteum Scientiae	Zeiss II	1953	400	German, French	Occasionally
Czechoslovakia	Bratislava	Bratislava	Zeiss-Jena 55	Summer 71		Czechoslovakian	
	Prague	Prag	Zeiss-Jena 54	Nov. 60	460	Czechoslovakian, 5 others	Yes
England	London	London	Zeiss IV W.G.	March 58	550	English	No
France	Paris	Palais de la Découverte	Zeiss III W.G.	May 52	372	French	No
Germany-E	Jena	Zeiss-Jena	Zeiss-Jena 67	June 26	402	German, occasionally others	No
Germany-W	Berlin	Wilhelm Foerster Sternwarte	Zeiss V W.G.	June 65	300	German, English, French	Yes
	Bochum	Bochum Sternwarte	Zeiss V W.G.	Nov. 64	248	German, occasionally others	Yes
	Hamburg	Hamburg	Zeiss V W.G.	April 30	356	German (English)	Yes
	Munich	Deutsches	Zeiss IV W.G.	May 60	160	Ger., Eng., Fren., Ital., Span.	Yes
	Nuremburg	Nürnberg	Zeiss II Mod.	Dec. 61	286	German (English)	Assoc. with obs.
Greece	Athens	Eugenides	Zeiss IV W.G.	June 66	250	Grk., Ger., Eng., Fren.	Yes
Italy	Milan	Hoepli	Zeiss IV	May 30	307	Italian	No
	Rome	Sala	Zeiss II	Oct. 28	387	Italian	No
Netherlands	The Hague	Haagsche Courant	Zeiss I	Feb. 34	150	Dutch, English, German	Not yet
Poland	Chorzow	Mikolaja Kopernika	Zeiss-Jena 55	Dec. 55	400	Polish, 16 others	Yes
Portugal	Lisbon	Lisbon	Zeiss-Jena 64	July 65	340	Portuguese	Not yet

Country	City	Institution	Instrument	Date	Seats	Languages	Refractor/Notes
Switzerland	Lucerne	Longines	Zeiss V-S W.G.	July 69	300	German, French, Italian, Spanish, English (simult.)	No
USSR	Kiev	Kiev	Zeiss-Jena	1962		Russian	
	Leningrad	Leningrad	Zeiss-Jena 58	Oct. 59		Russian	
	Moscow	Moscow	Zeiss-Jena 64	Nov. 29	600	Russian	
	Riga	Riga	Zeiss-Jena 63	July 64		Russian	
	Wolgograd	Wolgograd	Zeiss-Jena 53	Sept. 54		Russian	
Ceylon	Colombo	Colombo	Zeiss-Jena 65	March 65	570	Sinhalese, Tamil, English	No
China	Peking	Peking	Zeiss-Jena 56	Sept. 57		Chinese	
India	Calcutta	Birla	Zeiss-Jena 61	Sept. 62	500	Eng., Bengali, Hindi, Oriya, Tamil	Yes
Indonesia	Djakarta	Djakarta	Zeiss-Jena 65	Oct. 65			
Iraq	Baghdad		Zeiss-Jena 69	1970			
Japan	Akashi	Akashi	Zeiss-Jena	June 60		Japanese	
	Nagoya	Nagoya Science Mus.	Zeiss V W.G.	Nov. 62	445	Japanese	Yes
	Osaka	Zeiss	Zeiss II	March 37	330	Japanese	
	Tokyo	Gotoh	Zeiss IV W.G.	March 57	453	Japanese	
Thailand	Bangkok	Bangkok	Zeiss IV W.G.	Aug. 64	500	Thai, English	Yes
Argentina	Buenos Aires	De La Ciudad de Buenos Aires	Zeiss IV W.G.	April 68	360	Spanish, English	No
	Rosario	Rosario	Zeiss IV W.G.	Under const.	650	Spanish	Yes, 150 mm +108 mm Refr
Columbia	Bogota	De Bogota	Zeiss-Jena 69	Nov. 69	500	Spanish	No
Brazil	Sao Paulo	Municipal do Ibirapvera	Zeiss IV W.G.	Jan. 57	374	Portuguese	Yes
Uruguay	Montivideo	Municipal	Spitz B	June 54		Spanish	
Venezuela	Caracas	Humboldt	Zeiss IV W.G.	July 61	317	Spanish	Yes
Egypt	Cairo	Cairo	Zeiss-Jena 64	Jan. 67	550		
S. Africa	Johannesburg	University of Witwatersrand	Zeiss II	Oct. 60	422	English, African	No

Index